Handbook of Research on Environmental Education Strategies for Addressing Climate Change and Sustainability

Ahmed Karmaoui
Bioactives, Health and Environmental Lab, Faculty of Sciences and Techniques, University of Moulay Ismail, Morocco & Southern Center of Culture and Science, Morocco

Abdelkrim Ben Salem
Mohammed V University, Morocco

Mohd Talha Anees
University of Malaya, Malaysia

IGI Global
PUBLISHER of TIMELY KNOWLEDGE

A volume in the Advances in Educational Technologies and Instructional Design (AETID) Book Series

Published in the United States of America by
IGI Global
Engineering Science Reference (an imprint of IGI Global)
701 E. Chocolate Avenue
Hershey PA, USA 17033
Tel: 717-533-8845
Fax: 717-533-8661
E-mail: cust@igi-global.com
Web site: http://www.igi-global.com

Library of Congress Cataloging-in-Publication Data

Names: Karmaoui, Ahmed, 1982- editor. | Ben Salem, Abdelkrim, 1981- editor. | Anees, Mohd Talha, 1988- editor.
Title: Handbook of research on environmental education strategies for addressing climate change and sustainability / Ahmed Karmaoui, Abdelkrim Ben Salem, and Mohd Talha Anees, editors.
Description: Hershey, PA : Engineering Science Reference, [2021] | Includes bibliographical references and index. | Summary: "This book provides innovative approaches, methods, and ideas in education for climate change, environment strategies, and sustainability along with the development of curriculum and strategies for sustainable development goals"-- Provided by publisher.
Identifiers: LCCN 2020046728 (print) | LCCN 2020046729 (ebook) | ISBN 9781799875123 (hardcover) | ISBN 9781799875192 (ebook)
Subjects: LCSH: Environmental education--Study and teaching. | Environmental education--Research.
Classification: LCC GE70 .H365 2021 (print) | LCC GE70 (ebook) | DDC 333.7071--dc23
LC record available at https://lccn.loc.gov/2020046728
LC ebook record available at https://lccn.loc.gov/2020046729

This book is published in the IGI Global book series Advances in Educational Technologies and Instructional Design (AETID) (ISSN: 2326-8905; eISSN: 2326-8913)

British Cataloguing in Publication Data
A Cataloguing in Publication record for this book is available from the British Library.

Advances in Educational Technologies and Instructional Design (AETID) Book Series

Lawrence A. Tomei
Robert Morris University, USA

ISSN:2326-8905
EISSN:2326-8913

Mission

Education has undergone, and continues to undergo, immense changes in the way it is enacted and distributed to both child and adult learners. In modern education, the traditional classroom learning experience has evolved to include technological resources and to provide online classroom opportunities to students of all ages regardless of their geographical locations. From distance education, Massive-Open-Online-Courses (MOOCs), and electronic tablets in the classroom, technology is now an integral part of learning and is also affecting the way educators communicate information to students.

The **Advances in Educational Technologies & Instructional Design (AETID) Book Series** explores new research and theories for facilitating learning and improving educational performance utilizing technological processes and resources. The series examines technologies that can be integrated into K-12 classrooms to improve skills and learning abilities in all subjects including STEM education and language learning. Additionally, it studies the emergence of fully online classrooms for young and adult learners alike, and the communication and accountability challenges that can arise. Trending topics that are covered include adaptive learning, game-based learning, virtual school environments, and social media effects. School administrators, educators, academicians, researchers, and students will find this series to be an excellent resource for the effective design and implementation of learning technologies in their classes.

Coverage

- Online Media in Classrooms
- E-Learning
- Web 2.0 and Education
- Bring-Your-Own-Device
- Virtual School Environments
- Classroom Response Systems
- Instructional Design
- Digital Divide in Education
- Curriculum Development
- Instructional Design Models

IGI Global is currently accepting manuscripts for publication within this series. To submit a proposal for a volume in this series, please contact our Acquisition Editors at Acquisitions@igi-global.com or visit: http://www.igi-global.com/publish/.

The Advances in Educational Technologies and Instructional Design (AETID) Book Series (ISSN 2326-8905) is published by IGI Global, 701 E. Chocolate Avenue, Hershey, PA 17033-1240, USA, www.igi-global.com. This series is composed of titles available for purchase individually; each title is edited to be contextually exclusive from any other title within the series. For pricing and ordering information please visit http://www.igi-global.com/book-series/advances-educational-technologies-instructional-design/73678. Postmaster: Send all address changes to above address.

Titles in this Series

For a list of additional titles in this series, please visit:
http://www.igi-global.com/book-series/advances-educational-technologies-instructional-design/73678

Advancing Online Course Design and Pedagogy for the 21st Century Learning Environment
Daniel Chatham (Middlebury Institute of International Studies at Monterey, USA)
Information Science Reference • © 2021 • 382pp • H/C (ISBN: 9781799855989) • US $195.00

Affordances and Constraints of Mobile Phone Use in English Language Arts Classrooms
Clarice M. Moran (Appalachian State University, USA)
Information Science Reference • © 2021 • 229pp • H/C (ISBN: 9781799858058) • US $175.00

Examining an Operational Approach to Teaching Probability
Alessio Drivet (Geogebra Institute of Turin, Italy)
Information Science Reference • © 2021 • 369pp • H/C (ISBN: 9781799838715) • US $195.00

4C-ID Model and Cognitive Approaches to Instructional Design and Technology Emerging Research and Opportunities
Guilhermina Maria Lobato Ferreira de Miranda (Instituto de Educação, Universidade de Lisboa, Portugal) Manuel Joaquim Henriques Rafael (Universidade de Lisboa, Portugal) Mário Marcelino Luis de Melo (UI, DEF, Instituto de Educação, Universidade de Lisboa, Portugal) Joana Martinho de Almeida Costa Pardal (ISTAR, Instituto Universitário de Lisboa (ISCTE-IUL), Lisboa, Portugal) and Thiago Bessa Pontes (Universidade Federal do Cariri, Brazil)
Information Science Reference • © 2021 • 243pp • H/C (ISBN: 9781799840961) • US $175.00

Implementing Augmented Reality Into Immersive Virtual Learning Environments
Donna Russell (Walden University, USA)
Information Science Reference • © 2021 • 315pp • H/C (ISBN: 9781799842224) • US $195.00

Handbook of Research on Innovations in Non-Traditional Educational Practices
Jared Keengwe (University of North Dakota, USA)
Information Science Reference • © 2021 • 462pp • H/C (ISBN: 9781799843603) • US $245.00

Handbook of Research on Using Educational Robotics to Facilitate Student Learning
Stamatios Papadakis (University of Crete, Greece) and Michail Kalogiannakis (University of Crete, Greece)
Information Science Reference • © 2021 • 651pp • H/C (ISBN: 9781799867173) • US $275.00

IGI Global
PUBLISHER of TIMELY KNOWLEDGE

701 East Chocolate Avenue, Hershey, PA 17033, USA
Tel: 717-533-8845 x100 • Fax: 717-533-8661
E-Mail: cust@igi-global.com • www.igi-global.com

Editorial Advisory Board

List of Contributors

Table of Contents

Detailed Table of Contents

The chapter highlights the importance of the concept of "ecological sanitation" (Ecosan) in environmental education. It also describes the implementation and management of the "Ecosan" project in several primary and middle schools and explores the lessons learned from reusing organic waste to promote sustainability in arid communities in the southeastern region of Morocco. The project planned and built Ecosan systems that recycle human wastes in five schools from five oasis villages of the Draa-Tafilalt region. The results suggest that the Ecosan system may be a sustainable option to support water, soil, and biodiversity in the vulnerable arid areas, but the scale-up of the project and adoption by the local communities will require more work and coordination with the government and local stakeholders.

The Indian Government's Smart Cities Mission intends to develop citizen-friendly, sustainable cities. Despite its laudatory objective, such a top-down scheme, doesn't reflect any grassroots level work to address water and waste issues or promote energy efficiency/saving at city-scale in which citizens also become participatory change agents. Out of overall food produced in the world, one-third is wasted. It's a matter of global concern with economic, social and environmental implications. Every tonne of food waste diverted from landfills saves 1.9tonnes CO2e (Carbon dioxide equivalent). This action research pilot project is a climate change mitigation initiative to treat wet waste at source through demonstration and techno-economic feasibility involving Rajkot citizens as participatory change agents. While currently awareness about such environmental issues and its environment-friendly solutions among Rajkot citizens is comparatively low. There is a potential of mass adoption of such initiatives, supplemented with government incentives making cities smart while also reducing Greenhouse Gas emissions. This paper describes successes and lessons from this pilot.

study identifies which methods are perceived as being most effective. The findings show that integrating sustainability into universities' strategic plans and implementing sustainability initiatives that actively involve the student through the development of strategic alliances with international universities and institutions, combined with the inclusion of field trips to learn about sustainability are perceived as the most relevant learning approaches.

Environmental education (EE) for sustainable development remains a valuable subject of contemporary society, which is characterized with environmental issues such as climate change, pollution, loss of biodiversity, and resource degradation. The delivery of EE is based on the North American Association for environmental education values of knowledge, dispositions, competencies, and responsible behavior towards the environment. EE is a transformative tool to learners since it prepares learners with skills, attitudes, knowledge, and values to resolve environmental problems. It promotes environmental activism and action-oriented resolution of environmental issues. The full benefits of EE are challenged by limited human capacity, questionable professionalism, limited resources, and poor transformation of knowledge to practice. These challenges however can be alleviated through community engagement in formulating EE programs, multidisciplinary engagements, and research on EE delivery and quality.

The current agricultural and food system needs to become more sustainable. Scientific evidence shows that organic agriculture is particularly environmental beneficial and can contribute to the achievement of the SDGs. Yet, it is a knowledge intensive practice and all actors along the value chain need to be accordingly trained. Focusing on farmers, this chapter assesses available training initiatives in the sector. In order to become more effective, public bodies should engage more in capacity building and training on organics. Moreover, additional training initiatives should focus on specific crops or agricultural systems, which also addresses the training need of farmers in conversion. The use of ICT has the potential to facilitate the development of more efficient training interventions. The further development of training initiatives is a vehicle to promote organic agriculture and ultimately to contribute to achieving the SDGs and climate mitigation and adaptation in particular.

Foreword

Handbook of Research on Environmental Education Strategies for Addressing Climate Change and Sustainability is a joint production between Moulay Ismail University in Meknes (& FST Errachidia), Mohammed V University in Rabat, the Southern Center for Culture and Science in Zagora, and the University of Malaya in Malaysia. It is in this capacity, an outstanding example of the synergy between university and civil society in environmental preservation field.

Environmental education is a process of recognizing values and clarifying concepts that develop various necessary skills and attitudes to understanding and appreciating the relationships between people, their culture, and their biological and physical contexts. It aims to enable people to understand their environment, to behave in it in a positive way and to correctly assess the impact on the environment of their own lifestyle. It also aims to promote active participation in public life with the intention of promoting a quality environment, based on sustainable development.

This book is designed to respond to the growing trend of international interest in climate change education and environmental education. It has a didactic design on the best methods and tools for integrating environmental and climate change education and sustainability concept into school and university curricula. Environmental education can contribute to the relevance and quality of science education. Good results in science and awareness of global environmental problems often go together and combine with a sense of responsibility for managing a sustainable environment.

Prof. Karmaoui, Prof. Ben Salem, and Prof. Talha Anees lead us on a journey covering a wide range of topics through chapters hand-selected by distinguished authors. These subjects prepare us, educate us, inspire us, and empower us to create the future together. The material ranges from philosophy to practice, and from general to specific case studies. Chapters cover what the future will look like, individual role, society position, education and sustainable development, climate change adaptation strategies, as well as species conservation and carbon sequestration and environmental sustainability governance.

Handbook of Research on Environmental Education Strategies for Addressing Climate Change and Sustainability is a recommended read for students, serving teachers and beginners, administrators, trainers, practitioners, professionals, researchers, and anyone interested in educational strategies and programs for climate change and sustainability.

Mohammed Regragui
Faculty of Science, Mohammed V University in Rabat, Morocco

Preface

The present handbook of research deals with the issue of education generally and environmental education (EE) specifically. Through this book, we seek to create a serious scientific discussion among researchers from six continents and to encourage and support environmental awareness among learners and educators in educational and academic institutions, and universities. This collective work is an opportunity to shed light on the importance of the environment and sustainable development and alert to the threat of climate change.

This book includes various interesting topics about environment, education and the threat of climate change such as environmental sustainability governance, ecological sanitation, wet waste composting, pro-environmental practices, ecotourism, smart cities, agriculture, organic agriculture, garden-based learning, water, energy, pollution, and climate change adaptation. Considering climatic and socio-economic global changes, the necessity for academic research, stakeholders and public participation, and decision-makers' support to address the environmental issues have become an urgent need. The global aim of this book is to produce an important reference providing innovative approaches, methods, and ideas in environmental education that contribute to the compilation of learning resources for climate change, environment, and sustainability.

This handbook involves educational, technical, and scientific approaches to develop an integrated awareness of the problem of climate change and the role of environmental education in finding solutions to achieve sustainable development. It elaborates on a set of studies on the positive role that ecosystems play in preserving local biodiversity, storing carbon, preserving local woody species, and mitigating the effects of climate change. The present book highlights the negative impacts of urbanization on agricultural activities and soil biodiversity due to mineral pollution (e.g. Heavy metals) and pollution from anthropological activities. Smart cities are a model for developing sustainable cities that are friendly to citizens and the environment by adopting initiatives to mitigate the effects of climate change.

Demonstrating the joint technical and economic feasibility of a group of projects such as ecological sanitation and treating wet waste at source and converting it into organic fertilizer by making the citizen an agent of participatory change are concrete solutions. Additionally, the decision of individuals and families to conduct recycling and adopt green technology remains one of the determinants of friendly environmental home practices. Among the aspects that the book elaborates are the role of sectoral partnership and community involvement in promoting environmental sustainability and the governance and the adoption of approved environmental methods and strategies. Associated agriculture and organic agriculture as a key to organic gardening that carefully selects compatible plants to grow together for their common benefits requires training strategies. Intense practice of all actors with a focus on farmers as a pathway to achieving the Sustainable Development Goals (SDGs) is another aspect that was developed in this collective work.

Environmental education remains a crucial solution for every change and sustainable development. It allows stimulating the involvement of the human element in every constructive initiative that preserves the environment. Environmental education should start at an early age, from kindergartens and primary schools to universities. In addition to the environmental education concept, the current book investigates the notion of climate education and climate change. In fact, it also draws attention to the impact of climate change on life and nature, and the extent to which climate change adaptation and sustainability are included in school curricula.

Chapter 1 is about the promoting environmental sustainability governance: a critical exposure from cross-sector partnership. In the last decade, the emerging needs for social responsibility on environmental concern has been considerably transmitted into the initiative of firm alliance. This paper attempts to examine the essential points of university, industry and community-based strategic partnership for further collaboration alliance. This paper focuses on the key role of strategic partnership with cooperating into the concern in driving the procedural stage on sustainable development. The finding reveals that outstanding value of strategic partnership would give insights into empowering sustainable-based institutional arrangements assigned with the wide attempts to contribute the prudent public policy formulation to implement the good service to solve environmental and related issues. The value is that strategic partnership trend incorporated in this context would be the point of view for sustainable development agenda rooted into the way of living processes paradigm together with demonstrating the wide range of sustainable governance.

Chapter 2 explored the implementation of culturally relevant science-based projects in preschools and primary schools: from roots to wings. This chapter develops a conceptual and operational framework on education for sustainability based on a teacher training program located in the south of Chile. Nurtured by the experience collected by the programs' participation in different international networks, it put emphasis on the idea of the importance of combining local knowledge and an integrated science approach to foster action for climate and sustainability from an early stage. After presenting the conceptual and empirical framework, the chapter describes the implementation of culturally-relevant research projects based on the experience of eight schools. It concludes by proposing an operational model and declaring the main constraints and opportunities of project-based-learning for sustainability and climate action. The authors present some recommendations to combine local and scientific knowledge for teacher practice and educational policy, considering the importance of having a critical look at our daily practices as educators given the global socio-ecological crisis.

Chapter 3 investigates Turkish pre-service primary teachers' environmental moral reasoning patterns in environmental dilemmas: pre-service primary teachers' environmental moral reasoning patterns. This chapter explored preservice primary teachers' moral reasoning patterns in local and non-local environmental dilemmas. 47 preservice primary teachers enrolling in an environmental education course voluntarily participated in the study. The data were collected via preservice teachers' written reports about local and non-local environmental dilemmas and analyzed by means of qualitative and quantitative methods. The results revealed that preservice teachers mostly focused on ecocentric and anthropocentric moral reasoning on sea pollution case while using ecocentric and non-environmental reasoning on the deforestation case. The t-test results also revealed that preservice teachers used more anthropocentric reasoning patterns in the sea pollution case when compared to the deforestation case. The inclusion of local and non-local dilemmas into the environmental education courses in undergraduate teacher education programs can facilitate preservice teachers' moral reasoning.

Chapter 4 focuses on the ecological sanitation (ECOSAN) in oasis communities. This current chapter highlights the importance of the concept of "ecological sanitation" (Ecosan) in environmental education. It also describes the implementation and management of the "Ecosan" project in several primary and middle schools and explores the lessons learned from reusing organic waste to promote sustainability in arid communities in the southeastern region of Morocco. The project planned and built Ecosan systems that recycle human wastes in five schools from five oasis villages of the Draa-Tafilalt region. This chapter reports the successes and limitations from this project including the design and installation of the Ecosan system as well as the sensitization workshops conducted before and during the project with the participants. In each school, an Ecosan system was designed and installed. Liquid waste produced was stored until its disinfection and used in the irrigation of a designed school garden with a drip irrigation system and the composted excreta was used as a solid fertilizer. The results suggest that the Ecosan system may be a sustainable option to support water, soil, and biodiversity in the vulnerable arid areas, but the scale-up of the project and adoption by the local communities will require more work and coordination with the government and local stakeholders.

Chapter 5 demonstrates the wet waste composting as a greening society initiatives (a step towards climate resilient societies). In fact, Indian Government's Smart Cities Mission intends to develop citizen-friendly, sustainable cities. In-spite of such top-down schemes, no grassroots level work is done to address water and waste woes or promote energy efficiency/saving at city-scale where citizens also become participatory change agents. Awareness about these environmental issues and adoption for environment friendly initiatives in Rajkot is comparatively low. One-third of all the food produced in the world is wasted. It's a matter of global concern having economic, social and environmental implications. For every ton of food waste diverted from landfill a saving of 1.9ton CO2e is made. This action research project, a climate change mitigation initiative through demonstration shared techno-economic feasibility for treating wet waste at source by making Rajkot citizens participatory change agents. There is a potential of mass adoption through such initiatives along-with government incentives making cities smart while also reducing Greenhouse Gas emissions. This paper describes successes and lessons from this pilot.

Chapter 6 highlights the ecotourism as a sustainable and active citizenship approach. In this chapter the authors reports that tourism offers both advantages and disadvantages to the involved countries. Ecotourism emerges as a tourism segment based on the principles of sustainable tourism, oriented to the preservation of nature, respect for local traditions and customs, interest in the history and culture of the destination, encouragement of intercultural communication and aid to the local economy. This study aims to understand how students of Tourism from the Porto Polytechnic Institute, in Portugal, know and define it. Additionally, it aims to understand how many of these students have participated in typology activities and the reasons that led them to do so. A quantitative research with 123 students (44% of the universe) was conducted and the results showed that 23% of students know the four basic principles of Ecotourism and about half have experience in the practice of it. The main motivation is related to "Contact with nature". Finally, it is discussed how Ecotourism can be seen as a way of empowerment of students, regarding sustainability and the well-being of our environment and its inhabitants.

Chapter 7 explores the impact of agriculture and garden based learning on students of secondary school. Agriculture and Garden-Based learning is the need of the hour as it helps in the cognitive and psychomotor development of the student. It provides skill-based education which not only ensures learning by doing but also helps in educating the child about growing various types of plants. The study for this is conducted by taking 500 students from the Secondary Level by providing them a Self-prepared Questionnaire by the research scholar in two schools of Panipat. Students are taken from Science with Maths, Botany and Agriculture fields. Based on the questionnaire for plant grower statistical Analysis is done to find if there is any significant difference between learning agriculture in school and not learning it in school. On Statistical Analysis using the Anova test it was found that learning Agriculture was beneficial for children in holistic development. It also helps the schools in improving the nutritive value of mid-day meals, involving the students in some skill-based education which can help in solving the problems of malnutrition and providing jobs to millions of Indian Youth.

Chapter 8 elucidates effective education methods for sustainability teaching. This chapter, through an exploratory snapshot, aims to provide the first empirical work in the European context about the perceived effectiveness of sustainability education methods. Thus, based on a sample of 432 students from the Economics and Business Faculty of an Andalusian university (southern Spain), this study identifies which methods are perceived as being most effective. The findings show that integrating sustainability into universities' strategic plans and implementing sustainability initiatives that actively involve the student through the development of strategic alliances with international universities and institutions, combined with the inclusion of field trips to learn about sustainability are perceived as the most relevant learning approaches.

Chapter 9 investigates the environmental education and its effects on environmental sustainability. Environmental education (EE) for sustainable development remains a valuable subject of contemporary society, which is characterized with environmental issues such as climate change, pollution, loss of biodiversity and resource degradation. The delivery of EE is based on the North American association for environmental education values of knowledge, dispositions, competencies and responsible behavior towards the environment. EE is a transformative tool to learners since it prepares learners with skills, attitudes, knowledge and values to resolve environmental problems. It promotes environmental activism and action-oriented resolution of environmental issues. The full benefits of EE are challenged by limited human capacity, questionable professionalism, limited resources and poor transformation of knowledge to practice. These challenges however can be alleviated through community engagement in formulating EE programs, multidisciplinary engagements and research on EE delivery and quality.

Chapter 10 reports training strategies for organic agriculture as a pathway to achieve the SDGs. In this book chapter, the authors advance that the current agricultural and food system needs to become more sustainable. Scientific evidence shows that organic agriculture is particularly environmental beneficial and can contribute to the achievement of the SDGs. Yet, it is a knowledge intensive practice and all actors along the value chain need to be accordingly trained. Focusing on farmers, this chapter assesses available training initiatives in the sector. In order to become more effective, public bodies should engage more in capacity building and training on organics. Moreover, additional training initiatives should focus on specific crops or agricultural systems, which also addresses the training need of farmers in conversion. The use of ICT has the potential to facilitate the development of more efficient training interventions. The further development of training initiatives is a vehicle to promote organic agriculture and ultimately to contribute to achieving the SDGs and climate mitigation and adaptation in particular.

Chapter 11 explores the students' understanding of climate change in prone areas exposed to the impact of climate change through a case study of Lembata and North Timor Tengah Districts, Indonesia. Information about climate change is useful for children to anticipate the potential impacts of climate change. However, until 2016 in Lembata and North Timor Tengah Districts, Indonesia, which is the vulnerable area exposed to the effects of climate change, was found that children's understanding is still low, even though 98% claims that they can feel some changes in their environment. The result of this research were obtained by collecting data from student from both regions and were proceeded to find Climate Change Awareness Index (CAI). It showed that most students (89%) have more knowledge about climate change and the rest (11%) do not understand. This condition is influenced by facilities and infrastructure factors, access to books and internet, teacher's quality, and gender. It is also known that information about climate change in Indonesian textbooks have not reached 50% according to UNICEF recommendations to provide students with good understanding of climate change.

Chapter 12 is entitled "Teaching Climate Change at Primary and Secondary Levels for Sustainable Development". The authors reports that the young minds are provided with climate change and sustainable development education through various means. Environmental Education is being taught at the Primary Level in Central Board of Secondary Education Schools while it is being taught at Secondary Level in the Social Science and Science Subject in C.B.S.E. Schools of India with various tools and methods. In the present study, we have incorporated some methods which help in teaching about Climate change at Primary and Secondary levels. The study shows how the adopted methods are effective in disseminating knowledge about climate change. The investigator decided to adopt the survey method of research to know the method of teaching being adopted for teaching of Climate Change in environmental science at primary level and in Science or Social Science, Eco club at Secondary Level. Findings were that-Students should be taught in the school curriculum itself about Climate Change and its effects. The teaching strategies should be adopted based on the level of the child, his level of cognitive ability, assessability, and ease of use.

Chapter 13 reviews the climate change adaptation and sustainability in the Bangladeshi school curriculum. Bangladesh is one of the most affected countries by climate change and extremely vulnerable to environmental degradation. Young people, as the new generation, need to understand the consequences of climate change and its impact on their lives and the lives of their fellow citizens. This chapter presents the findings from a systematic evaluation of Bangladeshi school curriculum, and how it prepares the future generation for climate change adaptation and ensures a sustainable future for them. This chapter highlights the recommendations from the findings and suggests how the National Curriculum could address climate change mitigation and adaptation for sustainability by providing contents to provide grounding for the young people as global citizens to act in the local context. Although there is some content included in the school textbooks related to climate change and sustainable development issues, there is a need for inclusion of more relevant contents in the national school curriculum so that school textbooks provide enough contents on climate change, its mitigation and adaptation.

Chapter 14 studies determinants of household pro-environmental practices through an application of Australian survey data. The decision by individuals and households to undertake recycling and to adopt green technology is driven by a large array of factors. Understanding the role of these factors is in turn important for developing appropriate green policies and in enacting sustainable and efficient education and information campaigns. The main aim of this chapter is to outline the factors impacting on pro-environmental behavior of a sample of Australian households. Using data drawn from a survey undertaken across the State of Queensland, the chapter will consider how a factors such as socio-economic background, beliefs around climate change, previous exposure to environmental hazards and beliefs about individuals and governments ability to determine outcomes help explain patterns of pro-environmental behavior in various settings.

Chapter 15 analyses the role of cities in native woody species conservation and carbon storage Insight from Niger. Loss of native tree species constitutes a major socio-economic and ecological problem.

Nevertheless, there is a scanty data on the role of Sahelian cities in native woody species conservation and climate change mitigation. There this study tried to determine the native woody diversity, structure and carbon stocks in Niamey, and Maradi, Niger. The study used the plot method to inventory dendrometric variables based on a stratified sampling approach from peri-urban areas to the built environment. Biodiversity indices and allometric models were used for the data analysis. A total of 63 native woody species representing 47 genera and 24 families were recorded in the overall urban forests in two cities. In Niamey 46 species belonging to 35 genera and 19 families were recorded. While in Maradi 50 species, 41 genera, and 19 families were inventoried. Fabaceae was the dominant families in both cities with equal richness (13 species).

Chapter 16 studies the effects of urbanization on the degradation of the Palm grove around the city of Marrakech. The city of Marrakech is experiencing a very significant development in terms of urbanization which leads to a significant reduction of the agricultural activity and increase of the rate of metallic pollution in soils. The impact of this metallic pollution on the microbial activity in these soils, can negatively affect the growth and the development of the date palm (*Phoenix dactylifera L*). In addition, the urbanization has direct effects on the degradation of palm groves in the Marrakech region, following a decrease in agricultural and forestry productivity after anthropogenic activities which are at the origin of high metallic trace element (MTE) contents in soils .In order to assess the impacts of this pollution, the authors carried out a study in the rhizospheric soils of the date palm (*Phoenix dactylifera L*) monitoring the levels of heavy metals (Pb, Zn,) Eight sites were chosen, five urban sites (three gardens, two boulevards) and two peri-urban sites (palm groves).

Chapter 17 studies the Heavy Metals Pollution in Coimbatore district, Tamil Nadu. This chapter is a review of published research papers, books, and book chapters, unpublished scientific reports prepared by research organisations, and expert talk about heavy metal pollution. This book chapter also summarises the research studies conducted in Coimbatore and a summary of the studies says that very tiny information is available on heavy metals in soil, water, and air components. Till now heavy metals in plants, plant products, animals, animal products, bioaccumulation in humans, and its impacts are not accounted in this region. Further, these chapters identify the research gaps, research interventions are required to assess the heavy metals contaminations in different ecosystems. It also highlights the need for long-term monitoring of heavy metals in the ground, surface water, food materials, agricultural products produced in the contaminated regions. The chapter concludes with a need for study on human health impacts, restoration, and remediation measures required in Coimbatore regions.

Acknowledgment

There is an increasing international interest in education on climate change and the environment in recent years. This handbook offers suggestions on how to integrate climate change education and sustainability in school programs. It explores empirical and conceptual models to assist those who are interested in learning and teaching environmental knowledge and climate change education with hands-on projects and participatory methods.

The handbook includes scientific and technical methods, advanced technologies related to climate change, sustainable development, and examples of participatory environmental activities for students. The findings of this book are suitable for a large audience including government ministries, NGOs, education organizations, students, educators, professors, teachers, and researchers.

Special thanks go to my co-editors, my colleagues, Prof. Ben Salem A. from Mohammed V University in Rabat, and Prof. Mohd Talha Anees from the University of Malaya in Malaysia for their hard work and patience to edit with me this handbook of research. Big thanks go also to all the contributors who wrote the chapters from Australia, Spain, Portugal, Belgium, Germany, Morocco, and India, and to the reviewers to support us in the review process.

We appreciate the support of the IGI Global team, the encouragements and support of Prof. Moulay Brahim Sedra, the Dean of the Faculty of Science and techniques Errachidia (Moulay Ismail University), Prof. Mohammed Regragui, the Dean of the Faculty of Science (Mohammed V University in Rabat), and the staff of the Southern Center for Culture and Science of Zagora, Morocco. Many thanks go also to the whole Editorial Advisory Board, Prof. Siham Zerouali, Prof. Lhoussain Hajji, Prof. Samir El Jaafari, Prof. Ali Aït Boughrous, Prof. Moustapha Arahou, and Prof. Rachida Hassikou.

Lastly, I would like to thank the members of my family, my wife Siham, my two children Yassin and Amine, and my dad for their patience and continuous support.

Ahmed Karmaoui
Bioactives, Health and Environmental Lab, Faculty of Sciences and Techniques, University of Moulay Ismail, Morocco & Southern Center of Culture and Science, Morocco

Section 1

Sustainable Experiences and the School

Chapter 1
Promoting Environmental Sustainability Governance:
A Critical Exposure From Cross-Sector Partnership

Miftachul Huda
https://orcid.org/0000-0001-6712-0056
Universiti Pendidikan Sultan Idris, Malaysia

Azmil Hashim
Universiti Pendidikan Sultan Idris, Malaysia

Mohd Aderi Che Noh
Universiti Pendidikan Sultan Idris, Malaysia

Mohd Hairy Ibrahim
Universiti Pendidikan Sultan Idris, Malaysia

Budi Rismayadi
Universitas Buana Perjuangan, Karawang, Indonesia

Sari Laelatul Qodriah
Universitas Muhammadiyah Cirebon, Indonesia

Abdul Rozak
Universitas Swadaya Gunung Jati, Cirebon, Indonesia

Tassim Abu Bakar
Universiti Brunei Darussalam, Brunei

Yabit Alas
Universiti Brunei Darussalam, Brunei

Saidai Hitam
Universiti Brunei Darussalam, Brunei

Exzayrani Sulaiman
Universiti Brunei Darussalam, Brunei

Fauziah Che Leh
Universiti Pendidikan Sultan Idris, Malaysia

ABSTRACT

In the last decade, the emerging needs for social responsibility on environmental concern has been considerably transmitted into the initiative of firm alliance. This chapter attempts to examine the essential points of university-, industry-, and community-based strategic partnership for further collaboration alliance. This chapter focuses on the key role of strategic partnership with cooperating into the concern in driving the procedural stage on sustainable development. The findings reveal that outstanding value

DOI: 10.4018/978-1-7998-7512-3.ch001

of strategic partnership would give insights into empowering sustainable-based institutional arrangements assigned with the wide attempts to contribute the prudent public policy formulation to implement the good service to solve environmental and related issues. The value is that strategic partnership trend incorporated in this context would be the point of view for sustainable development agenda rooted into the way of living processes paradigm together with demonstrating the wide range of sustainable governance.

INTRODUCTION

In the last decade, the serious concern on environmental sustainability has been widely boarded due to the massive global transformation of firms' condition and also workplace environment (Aminudin et al., 2018a; Anggraeni et al., 2018a; Kapp, 2018; Rajagopalan, 2015). This is needed of the strategically responsible awareness on the way to solve the emerging issues on environmental sector (Aminin et al., 2018; Ansell & Gash, 2017; Fauzi et al., 2018). The most influential basis on sustainable development distribution could be viewed in detail into the new platform for daily human interaction and communication through digital setting (Abadi et al., 2019; Linkov et al., 2018). With this digitalization media, this provides an easy way to deliver the information through finger touched pathway, such as smartphone (Anshari et al., 2017; Razzaq et al., 2018). Along with such progress to give an insightful value in the attempts on transmitting the potentials of sustainability value with responsibility, the key point to achieve this attainment should be pointed out sustaining both corporate values assigned into the academic environment (Anggraeni et al., 2018b; Datta, Gopalakrishna-Remani & Bozan, 2015). This is an urgent need to respond in both technical and theoretical foundation plan to achieve the solving attainment, since some considerable issues shown into the number of challenges such as environmental concern, lack of moral enhancement in the organization sphere, and also other problematic issues. In this view, this should be considered into giving the cooperation on the way to expand sustainable-based arrangement in the strategic partnership context. The inextricable involvement should do with considering the emerging issues to give the solution appropriately in the context of academic enhancement.

This becomes a comprehensive essence with a significant point to go through dealing with the attempts as the initiative of strategic partnership to give an insightful value in disseminating the responsibility of environmental engagement. The initiative with such identification by scholarly enhancement refers to assist the sustainable commitment (Abadi et al., 2018a; Baker & Mehmood, 2015; Puspitasari et al., 2020). Distributed along with such approaches, the wide exploration in supplementing the sustainability engagement could be potentially supplemented in an effort to help the strategic partnership role on running the governance initiative (Brandi, 2017). Among the numbers are cooperating with some certain programs, such as community-firm engagement for environmental pollution plan (Shi & Xu, 2018), service learning enhancement initiative (Huda et al., 2018a; Molderez & Fonseca, 2018), academia-industry work plan project for sustainability empowerment (Adams et al., 2016; Evans et al., 2015), and also academia-community service engagement program (Erzurumlu, & Erzurumlu, 2015). All such initiatives may become good supplement in enabling the positive feedback to considerably transmit on the way of applying for the policy made by the stakeholder in transferring the universal values of corporate citizenship.

In addition, attempts on exploring initially fundamental pattern shift of sustainable governance on both public sector and secondary, and tertiary institution need to have a sufficiently mutual line in particular through paying a particular attention the responsible awareness. Such this concern should bring along with identifying the essential point of sustainable governance to be more engaged into the problem solution towards the emerging environmental issues. The strategic initiation on partnership context, in particular, into the way to control the management setting on embedding the environmental governance to commit with sustainable development needs to determine the initial plan to distribute the key values of corporate-academic engagement plan. Such this initiative could be started simply through the classroom setting, faculty scale and also entire campus with the responsible commitment on maintaining the environmental society in strategic partnership. Among the number of scholarly approaches with the creative basis on implementation stage and also theoretical exploration (Abadi et al., 2018b; Dögl & Behnam, 2015; Kaur& Singh, 2018; Leal Filho & Rayman-Bacchus, 2019; Mensah, Ayarkwa & Nani, 2018; Ortiz-de-Mandojana, Aguilera-Caracuel & Morales-Raya, 2016; Yadlapalli, Rahman & Gunasekaran, 2018), less attention was given on examining the way on expanding the sustainable governance on environmental concern through strategic partnership. As such, this chapter attempts to examine in a particular way on sustainable governance with strategic partnership initiative to have an insightful value on managing the effective procedure to resolve the problem of environmental concern. This chapter is aimed at critical analysis about the strategic integration of governance enhancement in strategic partnership context in referring to concern in driving the procedural stage on sustainable development.

Towards Sustainable Governance in Environmental Issues

The number of global challenges to impact on the natural sources, in recent decades, has to do with the expansion shift widely brought up into the environmental issues. Several cases could be viewed along with the serious concern about condition of current society in facing such global challenges. Among the significant numbers, there are few such as in the attempts to deal with the crises of economic sector (Dögl & Behnam, 2015), degradation of environmental sphere with climate challenges (Abadi et al., 2018c; Kapp, 2018), inequality access of water supply due to geographical deal (Galaitsi et al., 2016), desertification issues together with deforestation condition (Geist, 2017), and also facing the wide range of eradication in the poverty basis (Wu & Si, 2018). Such number showed the significant potentials on the way to gesture valuable points of implementing the sustainable commitment in developing the responsible awareness to handle with an integrative practice. In this view, the integrative way to conduct includes the points of sustainability in the context of the global challenges through maximising the number of initiatives to elaborate together with the sustainable development.

In the way to have a sufficient commitment through observing the potential issues of environmental concern, it is necessary to consider the key solution with a fundamental notion of sustainability in developing to manage the overall system of governance, mainly through strategic partnership context. The concisely particular identification towards the environmental concern should bring along with committed awareness on sustainable governance initiative to solve the emerging issues on environmental pollution (Abadi et al., 2018d; Irviani et al., 2019; Mulawarman et al., 2018). In directing the sustainability-responsible awareness governed with overall guideline, the systematic basis on achieving the responsible awareness to deal with the sustainable challenges needs to have a mutual work consistency within the organization context (Ali et al., 2020; Din et al., 2020; Sacchetti & Tortia, 2016). The inclusive structure to gather all stakeholders between university and industry partnership would lead to synergy the control management with steering appliances distributed into the societal context (Susilowati et al., 2018b; Yadlapalli et al., 2018). This is because of the way of such attainment which sustained the more systematic integration to run the process well with the capacity and capability to handle such issues. The governmental organization should take point of consolidating the entire work together with partnership alliance with the private sector or community service in ensuring the responsibility and integrity towards the program released under the project. The responsible commitment to come up with sustainable issues should be embedded early and consistently through the program initiative such as community service (Ali et al., 2020; Huda et al., 2018; Erzurumlu & Erzurumlu, 2015). The sufficient exploration of distribution enhancement of marketing supply and society demand may become an initial assessment in the attempts to share the mutual responsibility through policy making arena to come up complementing each other in achieving the sustainable governance for environmental concern.

Environmentally sustainable governance for strategic partnership initiative refers to the intentional corporation as the initiative on expanding the environmentally ecological control. It should bring along with the cooperative management through multi-sector partnership in addressing the green problems. With the significance of moderately innovative platforms on empowering the strategic partnership, taking consideration in underpinning the multiple forms of widely spread on worldwide achievement linkages on strategic partnership is entirely taking the variety of experiential forms of governance sustainability with its pivotal role in both public and private sector (Aminudin et al., 2018b; Bolton & Hannon, 2016; Warner & Sullivan, 2017). The briefly determined partnership could be reflected through the various models on individual and social responsibility in governing the networks engagement and monitoring systems.

Expanding Strategic Partnership Commitment for Environmentally Sustainable Responsibility

The initial value of expanding the strategic partnership alliance to enlarge the initiative of sustainable responsibility may begin with continued learning service through taking place in reflecting on the present social consciousness. In this view, the uniqueness on getting involved with the various activities into their systems both teamwork and organization should place on the practically expansive learning in expanding the mutual understanding towards the certain object with the entire processes to achieve the professional and ethical responsibility for situation management (Abadi et al., 2018e; Fitrian et al., 2019; Susilowati et al., 2018a; Trevino & Nelson, 2016; Wittmer, 2016). In order to sustain the initiative of sustainable responsibility engagement, the teamwork commitment should bring the activity awareness to concern the environmental ethos together with motivating to contribute into solving the real complexity situations. In particular, considering such attainment in the way to enlarge the entire commitment to work together in focusing on what to conduct properly within the activity engagement refers to point out reflecting a wide substance towards action arrangement to develop the rational thinking basis and critical consciousness skills in enabling the individual and social quality performance to solve the emerging issues on environmental concern (Amin et al., 2018; Assahubulkahfi et al., 2018; Rathod et al., 2020; Schaltegger & Burritt, 2017). Through focusing on surviving the reflections on managing the way of activity system in ensuring the flow of insights development to enhance the learning inquiry, the assignment procedure to adopt the innovative engagements on reflecting the cultural diversity conditions for instance refers to expand the initiative to gain the strategic partnership should begin with analysing the emphasis on the way to pursue the conceptual development in understanding the dialogue together with the multiple partnership to achieve the goal-oriented procedure guideline.

In line with obtaining the strategic partnership across multi-sector both private and public, rebuilding the integrated systems to interact with activity systems could be started into positioning the teamwork commitment indicated in the organisational culture. When getting involved in the entire procedure gained through professional and ethical responsibility, the need to emphasise the wider context of conceptualising an integrative theoretical approach in addressing the areas of the combination on organisational learning should be extensively recognized in detail into proposing the procedure guideline to consider in particular conceived as the awareness to achieve the expansive learning initiative (Abadi et al., 2018f; Bassano et al., 2018; Berkes, 2017; Huda et al., 2017). As such, this attainment may lead to the environmental concern by proposing the individual quality in learning continuation with disciplines maintenance on innovative and creative ideas to the organisational culture assigned into the teamwork commitment. Moreover, the transformative initiation on achieving the learning principle to work together within the team could be expanded through effective leadership for governance sustainability commitment to allow strategic partnership to develop an integrated process of understanding, reflecting and conducting what necessary deed with an appropriate basis (Bentley, Pugalis & Shutt, 2017; Colbert, Nicholson & Kurucz, 2018; Huda et al., 2019a). Since the performance on proceeding the realisation of jointly supporting the sustainability awareness to commit with the innovative work assigned into the effective management control on environmental stability, the advantage of key ideas on underlying the way to expand better understanding together with transforming appropriately useful platform on governance sustainability (Aminudin et al., 2018c; Bringezu et al., 2016; Cizel, Ajanovic & Cakar, 2016; Kholili et al., 2020). With this regard, the design on directing to the combination towards the valuable agenda of sustainability continued into the active direction to involve with addressing the key ideas on the contextual basis at the institution for instance may arrange for the essential foundations on how to behave a mutually consistent management of individual and social level on governance sustainability (Atmotiyoso and Huda, 2018; Ansell & Gash, 2017). The sustainability committed appropriately within picturing the mechanism of obtaining the quality of enhanced understanding mainly among the individuals to transmit the rational ability would lead to sustain the shared vision as a strategic role model in improving the existing practices in the context of organisation teamwork.

In the attempts to enhance the initiative of strategic partnership to expand the involvement of organisational learning commitment, the wider capability concerned into the organisation stability should be embedded through a holistically consistent sustainability in showing the strategic indicator for governance sustainability. Through this rational overview with theoretical ideas, better understanding may be achieved to foster the organisational learning enhancement within the teamwork. As such, this attainment derives from the initiative point of view in empowering the committed awareness initiated to envisioning an integrative approach to achieve the environmental sustainability with engaging the digital based construction (Fauzi et al., 2019; Huda et al., 2016; Ortiz-de-Mandojana et al., 2016). Manifesting the glance from the lessons learned could sustain through such various projects committed as the initiative of strategic partnership in order to drive the organisational culture with better knowledge understanding to assist problem solving creativity (Cucuzzella, 2016; Mitchell & Walinga, 2017). With this regard, the theoretical perspective of maintaining the structure overview on sustainable responsibility to help understand the particular way through the wisdom-based action culture with the developments of inculcating the systemic transformation of institutional and organizational culture (Baumgartner & Rauter, 2017;

Huda et al., 2018; Marzuki et al., 2020). In order to expand wide association with sustainability in the core strategies, bringing the holistic structure in transmitting the valuable insight on taking progression on the way to empower the strategic partnership should be taken into consideration with continuing the sustainable governance. With an essential value of organisational culture to transmit the environmental responsibility, quality performance assigned with the learning theory could have an insight into providing a useful platform basis to reach the better comprehension in the process of expanding the sustainability across multisector partnership.

In addition, the initiative of managerial culture on dealing with the sustainability commitment refers to the essential point of bridging the valuable insights between theoretical and practical stage on knowledge understanding to advance the efforts in handling the complex situation by achieving the solution wisely within the structural context. In assisting the governance sustainability through organisational culture, the complex issues on environmental concern with pointing out the strategic partnership alliance should be achieved into focusing on the professionality and morality aspect (Bosselmann, 2016; Dacholfany et al., 2020; Huda et al., 2017a; Maseleno et al., 2018a; Roos, 2017). Since there were new fields on the development to involve into the governance sustainability, the organisational culture has to bring the spirit of teamwork commitment as a significant element to create the organisation culture. As such, such challenges on environmental concern may be handled by focusing on learning organisation procedure together with individual and social skills in the attempts to strengthen the sustainability. With growing the essential target to achieve, the necessity to adopt the skilful quality in underlying what to do wisely should go through the process of transformation in order to create the organisational culture enhancement (Huda et al., 2018; Moksin et al., 2018; Oktafianto et al., 2018). The point is that the organisational culture principle on strategic partnership would provide the sequences of clear picture in the way to understand the processes in embedding sustainable awareness as the key responsibility towards environmental concern amidst the society at large and institution in particular (Evans et al., 2015; Maseleno et al., 2018b; Muslihudin et al., 2019; Rajagopalan, 2015). The number of various institutions' activities could be transformed into the initiative of service learning for instance used to enhance the particular notion into governance sustainability through strategic partnership initiation.

Empowering Environmentally Sustainable Governance through Industry, University and Community-based Strategic Partnership

In this occasion, strategic partnership of university and industry alliance could be improved in the attempts on building the collaboration initiative for sustainable governance. This is to ensure the environmental concern issues may be handled together with achieving the wide range of corporate values in facing the sustainable problems (Ansell & Gash, 2017; Antadze & McGowan, 2017; Huda et al., 2017b; Maseleno et al., 2018c). With considering the emerging issues on complex circumstance of environmental problems, university and industry-strategic alliance partnership should be addressed into demanding the active engagement for the society basis in resolving the issues appropriately within the wise basis. Sustainable awareness in governing the systems in running the program project is entirely rooted with the mutual achievement of fulfilling the need for sustainability-based moral quality. In this view, the encouragement process of governing the sustainable expansion should bring along with stressing the responsibility awareness in attaining the environmental conduciveness (Maseleno et al., 2019; Orts & Smith, 2017; Othman et al., 2016). The way of getting improved with a real implementation of corporate values attainment assigned into the social responsibility should be committed to enhance the initiative of responding the environmental concern. Critical exposure on committing with the responsibility awareness has to begin with gaining the societal community through joining partnership alliance between private and public sector in ensuring the integrity development for sustainable development.

In addition, the significant approach in empowering the sustainable development through strategic partnership engagement in determining an important instrument to adopt the positive feeling involved with optimistic mechanism progression to achieve the societal balance. Through the critical mechanism on considering the initiative of sustainability enhancement, the commitment on expanding the rise of partnerships in the setting of societal steering mechanisms could be initially incorporated with self-control management systems in enabling the accessibility of decision making policy to open recommendation for new progress initiative reform (Djuric & Filipovic, 2015; Huda et al., 2017c; Lueg & Radlach, 2016; Putra et al., 2018). The variety of joining the cooperation along with the environmentally sustainable concern should be provided with the new arrangement assigned into the distinctive engagement process for strategic mechanisms in achieving the goal achievement. In this context, joining strategic partnerships incorporated with the emerging trends and approaches on the environmental governance arrangements should be employed with framing the appropriate instrument to deal with problem solving skills together with rational thinking scale (Antadze & McGowan, 2017; Erzurumlu & Erzurumlu, 2015; Malik et al., 2019; Ristiani et al., 2018). As such, this emerging trend combined with the nature point of determining the partnership to accelerate the goal decision refers to attain the necessity of knowledge understanding on the way to maximise on the functioning of partnerships. With exploring the perspective from focusing on the framing initiative as the strategic instruments to support the goal achievement, the individual actors in advancing the skills of problem solving should be transmitted into the knowledge understanding with the partnership engagement.

In line with incorporating the partnership engagement through academia and industry alliance, the goal achievement of sustainable responsibility could expand the entire assessment in organising the mutual commitment to advance the initiative of knowledge base on the current academic trend within the contextual comprehension (Datta et al., 2015; Rosa et al., 2019; Sacchetti & Tortia, 2016; Yadlapalli et al., 2018). In the effort to advance the strategic partnerships to enlarge the commitment initiated to enhance the responsible awareness in environmental concern, the way of organizing the procedural context should be addressed along with reviewing the current basis in addressing the knowledge type gained through identifying the principal understanding on environmental issues. Through fulfilling the components of this knowledge gap, the critical exposure in the attempts of reviewing process should bring along with beginning into the basic issues reflected through contextual approach in order to determine the necessary goal achievement dealt with the strategic plan to be applied among the emerging issues wisely and appropriately within the principal guideline of sustainable development responsibility (Antadze & McGowan, 2017; Huda et al., 2017d; Orts & Smith, 2017; Rosli et al., 2018). In this view, the role of strategic planning management is important to give an insightful value of what to do and how to achieve in the context of sustainability enhancement for environmental context through approaching the smart-based instruction (Huda and Hashim, 2021). This is to ensure in enabling the balance between content and context within a mutual line into the practical and theoretical engagement assimilated with enlarging the characterization of the significant way to explore the knowledge understanding basis. In the attempts on building the valuable points of strategic partnership, the comprehensive assessment should be undertaken in a mutual line of examining the necessary contribution to be achieved through elaborating the significant essence in sustainability commitment.

Expanding Clear Vision Achievement-based Environmentally Sustainable Governance

On the basis of dealing with sustainability commitment, the goal achievement on social aspect, effective incorporation, procedural guideline and declarative enhancement could be created to the knowledge set out through feeling acquisition together with experiential learning (Polk, 2015; Salamah et al., 2019; Sari et al., 2018; Septiropa et al., 2018). In this view, the basic principles in comprehending such competencies should bring along with creating the sustainable governance for environmental sector. The strategic mutual line on making the stability through cognitive, psychomotor and affective aspect would lead to enhance the learning skills in realizing the accurately sustainable training expertise (Ajmain et al., 2019; Dubey & Gunasekaran, 2015; Huda, 2021; Wu, 2017). The competency to transmit the community-based experiential learning has to go through widespread appreciation in the way to enlarge the environmental consciousness. The sufficient contribution in incorporating the community-industry-and academia to the widespread of enhancing the applied value for sustainable governance refers to the systematic arrangement in building the environmental sustainability (Berkes, 2017; Dögl & Behnam, 2015; Huda et al., 2017; Ortiz-de-Mandojana et al., 2016; Susilowati et al., 2018c). With this regard, the capability to acquire the contextual basis on determining the way to do through advancing the experiential skills and also knowledge expertise could be applied together with the strategic plan management. In the attempts to gain extensive appreciation in underlying the technique to conduct basically in ensuring the sustainability to advance the quality basis for sustainable responsibility, the environmental care should be committed into the balance of conceptualising the knowledge understanding together with practicing the further act to enhance the extent of sustainability awareness (Issa, 2017; Mahat et al., 2016; Sivaram et al., 2020; Sugiyarti et al., 2018). In this context, the essentials to empower the extensive responsiveness in enhancing the capacity of strategic approach planned through the achievement point of view should deal with bringing the community initiative as the preventive actions to combat the environmental pollution.

In particular approach, the agenda to apply for the sustainable awareness as an essential element to solve the global challenges, namely environmental pollution from the firm waste has to begin with sustaining the ability to manage individual quality to acquire the discipline commitment (Bebbington & Unerman, 2018; Howlett, Ferreira & Blomfield, 2016). The outstanding value is that responsible awareness to gain the sustainability-experiential skills could be deployed through academic emphasis towards the variety of fields of studies including digital ethics, economics, sciences and policy studies in underlying the valuable insights of such connection to supply the theoretical basis and scientifically practical inquiry to generate the advance on sustainable governance (Too & Bajracharya, 2015). With expanding the essential points of various subject areas, the outstanding contribution might be achieved to advocate the challenging issues on environmental space. Moreover, contributing to such transmission into the advance systematic knowledge together with adopting other knowledge forms could lead to enhance the advocates of sustainability in solving the appropriate solution for environmental, social, economic and communication challenges (Bringezu et al., 2016). The comprehensive way to enlarge the sustainability could be formed into the education setting through which the human resource quality would have a sufficient point of view in transmitting the knowledge value into the field worldwide. This is because the way of incorporating the initial attempts refers to improve the capability with skilful experts on facing the environmental challenges from the firm's waste pollution (Kapp, 2018; Kaur & Singh, 2018). Transforming the technical points in reducing the poverty together with continuing to apply for the major agent in incorporating the current situation in the society should be cooperated into empowering the quality of sustainable responsibility commitment.

With this point of view, the rightful points to embed the way of how to achieve such attainment need to go through representing the process of transforming the responsible awareness together with the strategic plan in managing the sustainable governance. The significant attribution of critical analysis for solving the problematic issues should build the initiative for the way of teaching skills assigned into the pedagogical together with technological competence as the outstanding value to generate the wellbeing achievement (Howlett et al., 2016; Susilowati et al., 2019; Wahyuni et al., 2020). The ability to adopt the technology for instance has to do a wise performance including the way to transmit the information into others shared with the strategic principles of professional and ethical balance (Huda et al., 2018a; Pardimin et al, 2018). With this regard, integrating the sustainable governance assigned into representing the necessary act to respond the challenging issues has to begin with the responsible feeling committed to the quality performance to underlie the sustainable self in order to lead to the social consciousness through the appropriate information gained from the human interaction base. As the entire processes from the strategic plan management on the principles of sustainable-based education setting to the development levels on transmitting the values of responsibility commitment, the need to enhance the social construction of having the integrity incorporated into the quality skills of tolerance and empathy may lead to achieve the basis of mutual respect across the multi-cultural society (Altinay, Dagli & Altınay, 2017; Huda et al., 2017c; Pane & Patriana, 2016; Thoresen, 2019; Wulandari et al., 2018). With committing the efficiency of gaining the corporate values distributed into the accountable consciousness, getting involved with a sufficient knowledge sharing to have a well responsible enhancement is entirely determined to play a basic performance in leading to the strategic partnership responsibility.

CONCLUSION

The emerging needs for social responsibility on environmental concern considerably transmitted into the initiative of firm alliance such as environmental concern, lack of moral enhancement in the organization sphere, and also other problematic issues should be considered into giving the cooperation on the way to expand sustainable-based arrangement. This chapter did elaborate with critical analysis on dealing with the essential points of university, industry and community-based strategic partnership for further collaboration alliance. In focusing on the key role of strategic partnership with cooperating into the concern in driving the procedural stage on sustainable development, there are five core stages in expanding strategic partnership commitment for environmentally sustainable responsibility. Those are empowering environmentally sustainable governance through industry, university and community-based strategic partnership; expanding clear vision achievement-based environmentally sustainable governance; enhancing clearly responsible commitment for environmentally sustainable governance; transmitting strategic cooperation management for environmentally sustainable governance; fostering professional and moral-based strategic partnership in environmentally sustainable governance. As such, the outstanding value of strategic partnership would give insights into empowering sustainable-based institutional arrangements assigned with the wide attempts to contribute the prudent public policy formulation to implement the good service to solve environmental and related issues. The valuable insight refers to the ultimate point that strategic partnership trend incorporated in this context would be the point of view for sustainable development agenda rooted into the way of living processes paradigm together with demonstrating the wide range of sustainable governance with strategic integration in driving the procedural stage on sustainable development. The outstanding value of strategic governance in the way to empower sustainable-based program initiative is supposed to give a detailed point in ensuring the primary contribution to shape the commitment to build the partnership into some programs to be incorporated within the institutional arrangements. In particular, this work will give insights into the wide attempts to contribute into prudent public policy formulation in the way to implement the good service along with meeting the needs to solve environmental and related issues. Moreover, strategic partnership trend incorporated with mutual cooperation interconnected to work with creating the sustainable engagement should consequently bring the solution in generating the innovative approach in the attempts in encouraging the creative participation in working together to solve problems. Here the inner process could be transmitted into giving insights in providing the number of great potentials for planning together with implementation and monitoring the strategies as key responsibilities to ensure the sustainable development could be achieved.

REFERENCES

Abadi, S., Hawi, A., Dacholfany, I., Huda, M., Teh, K. S. M., Walidi, J., ... Maseleno, A. (2019). Identification of Sundep, Leafhopper and Fungus of Paddy by Using Fuzzy SAW Method. *International Journal of Pharmaceutical Research*, *11*(1), 695–699.

Abadi, S., Huda, M., Basiron, B., Ihwani, S.S., Jasmi, K.A., Hehsan, A., Safar, J., Mohamed, A.K., Embong, W.H.W., Mohamad, A.M., Noor, S.S.M., Novita, D., Maseleno, A., Irviani, R., Idris, M., & Muslihudin, M. (2018a). Implementation of fuzzy analytical hierarchy process on notebook selection. *International Journal of Engineering and Technology*, *7*(2.27), 238-243.

Abadi, S., Huda, M., Hehsan, A., Mohamad, A.M., Basiron, B., Ihwani, S.S., Jasmi, K.A., Safar, J., Mohamed, A.K., Embong, W.H.W., Noor, S.S.M., Brahmono, B., Maseleno, A., Fauzi, A.N., Aminudin, N., & Gumanti, M. (2018b). Design of online transaction model on traditional industry in order to increase turnover and benefits. *International Journal of Engineering and Technology*, *7*(2.27), 231-237.

Abadi, S., Huda, M., Jasmi, K.A., Noor, S.S.M., Safar, J., Mohamed, A.K., Embong, W.H.W., Mohamad, A.M., Hehsan, A., Basiron, B., Ihwani, S.S., Maseleno, A., Muslihudin, M., Satria, F., Irawan, D., & Hartati, S. (2018c). Determination of the best quail eggs using simple additive weighting. *International Journal of Engineering and Technology*, *7*(2.27), 225-230.

Abadi, S., Huda, M., Teh, K. S. M., Haron, Z., Ripin, M. N., Hehsan, A., Sarip, S., Hehsan, M. R., Amrullah, M., & Maseleno, A. (2018d). Hazard Level of Vehicle Smoke by Fuzzy Multiple Attribute Decision Making with Simple Additive Weighting Method. *International Journal of Pharmaceutical Research*, *10*(4), 58–71.

Abadi, S., Teh, K.S.M., Huda, M., Hehsan, A., Ripin, M.N., Haron, Z., Muhamad, N.H.N., Rianto, R., Maseleno, A., Renaldo, R., & Syarifudin, A. (2018e). Design of student score application for assessing the most outstanding student at vocational high school. *International Journal of Engineering and Technology*, *7*(2.27), 172-177.

Abadi, S., Teh, K.S.M., Nasir, B.M., Huda, M., Ivanova, N.L., Sari, T.I., Maseleno, A., Satria, F., & Muslihudin, M. (2018f). Application model of k-means clustering: insights into promotion strategy of vocational high school. *International Journal of Engineering and Technology*, *7*(2.27), 182-187.

Adams, R., Jeanrenaud, S., Bessant, J., Denyer, D., & Overy, P. (2016). Sustainability-oriented innovation: A systematic review. *International Journal of Management Reviews*, *18*(2), 180–205. doi:10.1111/ijmr.12068

Adela, A., Jasmi, K. A., Basiron, B., Huda, M., & Maseleno, A. (2018). Selection of dancer member using simple additive weighting. *IACSIT International Journal of Engineering and Technology*, *7*(3), 1096–1107. doi:10.14419/ijet.v7i3.11983

Ali, A. B., Tajuddin, S. N. A. A., Marzuki, M., & Huda, M. (2020). (in press). Digital Approach of Malay Literature through Interactive Computer Method (ICM): A Proposed Framework Model. *Solid State Technology*.

Ali, A. H., Tajuddin, S. N. A. A., Marzuki, M., & Huda, M. (2020). Digital Approach of Malay Literature through Interactive Computer Method (ICM): A Proposed Framework Model. *Solid State Technology*, *63*(6), 12651–12666.

Altinay, F., Dagli, G., & Altınay, Z. (2017). Role of technology and management in tolerance and reconciliation education. *Quality & Quantity*, *51*(6), 2725–2736. doi:10.100711135-016-0419-x

Amin, M. M., Nugratama, M. A. A., Maseleno, A., Huda, M., & Jasmi, K. A. (2018). Design of cigarette disposal blower and automatic freshner using mq-5 sensor based on atmega 8535 microcontroller. *IACSIT International Journal of Engineering and Technology*, *7*(3), 1108–1113. doi:10.14419/ijet.v7i3.11917

Aminin, S., Huda, M., Ninsiana, W., & Dacholfany, M. I. (2018). Sustaining civic-based moral values: Insights from language learning and literature. *International Journal of Civil Engineering and Technology*, *9*(4), 157–174.

Aminudin, N., Fauzi, Huda, M., Hehsan, A., Ripin, M.N., Haron, Z., Junaidi, J., Irviani, R., Muslihudin, M., Hidayat, S., Maseleno, A., Gumanti, M., & Fauzi, A.M. (2018a). Application program learning based on android for students experiences. *International Journal of Engineering and Technology*, *7*(2.27), 194-198.

Aminudin, N., Huda, M., Ihwani, S.S., Noor, S.S.M., Basiron, B., Jasmi, K.A., Safar, J., Mohamed, A.K., Embong, W.H.W., Mohamad, A.M., Maseleno, A., Masrur, M., Trisnawati, & Rohmadi, D. (2018b). The family hope program using AHP method. *International Journal of Engineering and Technology*, *7*(2.27), 188-193.

Aminudin, N., Huda, M., Mohamed, A.K., Embong, W.H.W., Mohamad, A.M., Basiron, B., Ihwani, S.S., Noor, S.S.M., Jasmi, K.A., Safar, J., Natalie, L., Ivanova, Maseleno, A., Triono, A., & Nungsiati. (2018c). Higher education selection using simple additive weighting. *International Journal of Engineering and Technology*, *7*(2.27), 211-217.

Anggraeni, E.Y., Pardimin, Dacholfany, M.I., Akla, Huda, M., Teh, K.S.M., Hehsan, A., Junaidi, J., Yusof, F.M., Abas, H., Husin, M.F.A., Apriani, D., & Maseleno, A. (2018b). Modelling effectivenes of IS learning methodology with AHP method. *IACSIT International Journal of Engineering and Technology*, *7*(4), 4708–4714.

Anggraeni, E.Y., Huda, M., Maseleno, A., Safar, J., Jasmi, K.A., Mohamed, A.K., Hehsan, A., Basiron, B., Ihwani, S.S., Embong, W.H.W., Mohamad, A.M., Noor, S..S.M., Fauzi, A.N., Wijaya, D.A., & Masrur, M. (2018a). Poverty level grouping using SAW method. *International Journal of Engineering and Technology*, *7*(2.27), 218-224.

Ansell, C., & Gash, A. (2017). Collaborative platforms as a governance strategy. *Journal of Public Administration: Research and Theory*, *28*(1), 16–32. doi:10.1093/jopart/mux030

Anshari, M., Almunawar, M. N., Shahrill, M., Wicaksono, D. K., & Huda, M. (2017). Smartphones usage in the classrooms: Learning aid or interference? *Education and Information Technologies*, *22*(6), 3063–3079. doi:10.100710639-017-9572-7

Antadze, N., & McGowan, K. A. (2017). Moral entrepreneurship: Thinking and acting at the landscape level to foster sustainability transitions. *Environmental Innovation and Societal Transitions*, *25*, 1–13. doi:10.1016/j.eist.2016.11.001

Assahubulkahfi, M., Sam, Y. M., Maseleno, A., & Huda, M. (2018). LQR Tuning by Particle Swarm Optimization of Full Car Suspension System. *International Journal of Engineering & Technology*, *7*(2.13), 328-331.

Atmotiyoso, P., & Huda, M. (2018). Investigating Factors Influencing Work Performance on Mathematics Teaching: A Case Study. *International Journal of Instruction*, *11*(3), 391–402. doi:10.12973/iji.2018.11327a

Baker, S., & Mehmood, A. (2015). Social innovation and the governance of sustainable places. *Local Environment*, *20*(3), 321–334. doi:10.1080/13549839.2013.842964

Bassano, C., Carotenuto, A., Ferretti, M., Pietronudo, M. C., & Coskun, H. E. (2018, July). Smart University for sustainable governance in smart local service systems. In *International Conference on Applied Human Factors and Ergonomics* (pp. 206-216). Springer.

Baugher, J. E., Osika, W., & Robert, K. H. (2016). Ecological Consciousness, Moral Imagination, and the Framework for Strategic Sustainable Development. In *Creative Social Change: Leadership for a Healthy World* (pp. 119–142). Emerald Group Publishing Limited. doi:10.1108/S2058-88012016037

Baumgartner, R. J., & Rauter, R. (2017). Strategic perspectives of corporate sustainability management to develop a sustainable organization. *Journal of Cleaner Production*, *140*, 81–92. doi:10.1016/j.jclepro.2016.04.146

Bebbington, J., & Unerman, J. (2018). Achieving the United Nations Sustainable Development Goals: An enabling role for accounting research. *Accounting, Auditing & Accountability Journal*, *31*(1), 2–24. doi:10.1108/AAAJ-05-2017-2929

Bentley, G., Pugalis, L., & Shutt, J. (2017). Leadership and systems of governance: The constraints on the scope for leadership of place-based development in sub-national territories. *Regional Studies*, *51*(2), 194–209. doi:10.1080/00343404.2016.1181261

Berkes, F. (2017). Environmental governance for the anthropocene? Social-ecological systems, resilience, and collaborative learning. *Sustainability*, *9*(7), 1232. doi:10.3390u9071232

Bolton, R., & Hannon, M. (2016). Governing sustainability transitions through business model innovation: Towards a systems understanding. *Research Policy*, *45*(9), 1731–1742. doi:10.1016/j.respol.2016.05.003

Bosselmann, K. (2016). *The principle of sustainability: transforming law and governance*. Routledge.

Brandi, C. A. (2017). Sustainability standards and sustainable development–synergies and trade-offs of transnational governance. *Sustainable Development*, *25*(1), 25–34. doi:10.1002d.1639

Bringezu, S., Potočnik, J., Schandl, H., Lu, Y., Ramaswami, A., Swilling, M., & Suh, S. (2016). Multi-scale governance of sustainable natural resource use—Challenges and opportunities for monitoring and institutional development at the national and global level. *Sustainability*, *8*(8), 778. doi:10.3390u8080778

Budiyanto, G., Ipnuwati, S., Al Gifari, S. A., Huda, M., Jalal, B., Maseleno, A., & Hananto, A. L. (2018). Web based expert system for diagnosing disease pest on banana plant. *IACSIT International Journal of Engineering and Technology*, *7*(4), 4715–4721.

Cizel, B., Ajanovic, E., & Cakar, K. (2016). Prerequisites for effective and sustainable destination governance. *Anatolia*, *27*(2), 155–166. doi:10.1080/13032917.2015.1058831

Colbert, B. A., Nicholson, J., & Kurucz, E. C. (2018). Humanistic Leadership for Sustainable Transformation. In Evolving Leadership for Collective Wellbeing: Lessons for Implementing the United Nations Sustainable Development Goals (pp. 33-47). Emerald Publishing Limited. doi:10.1108/S2058-880120180000007004

Cucuzzella, C. (2016). Creativity, sustainable design and risk management. *Journal of Cleaner Production*, *135*, 1548–1558. doi:10.1016/j.jclepro.2015.12.076

Dacholfany, M.I., & Islahudin, A.N., Kuliyatun, Ilyas, U., Sudrajat, A., Siregar, M., Cholis, N., Jandra, M., Nurmartiani, E., Huda, M., Maseleno, A., Hashim, A., Leh, F.C., Rosman, A.S., Ab Rahim, N.M.Z., Razzak, M.A., Rashid, M.H., Darawi, A.B.S., & Nasir, B.M. (2020). Investigating Islamic-based Moral and Professional Principles of Stem Cell-Cloning Technolog. *International Journal of Pharmaceutical Research*, *12*(SI2), 4478–4488.

Datta, P., Gopalakrishna-Remani, V., & Bozan, K. (2015). The impact of sustainable governance and practices on business performance: An empirical investigation of global firms. *International Journal of Sustainable Society*, *7*(2), 97–120. doi:10.1504/IJSSOC.2015.069912

Din, G., Omar, A., Said, C. S., Hanif, M. H. M., Ali, A. B., Huda, M., & Maseleno, A. (2020). (in press). Developing Software Application of Sasteraya: An Empirical Research. *International Journal of Pharmaceutical Research*.

Djuric, M., & Filipovic, J. (2015). Human and social capital management based on complexity paradigm: Implications for various stakeholders and sustainable development. *Sustainable Development, 23*(6), 343–354. doi:10.1002d.1595

Dögl, C., & Behnam, M. (2015). Environmentally sustainable development through stakeholder engagement in developed and emerging countries. *Business Strategy and the Environment, 24*(6), 583–600. doi:10.1002/bse.1839

Dubey, R., & Gunasekaran, A. (2015). Shortage of sustainable supply chain talent: An industrial training framework. *Industrial and Commercial Training, 47*(2), 86–94. doi:10.1108/ICT-08-2014-0052

Erzurumlu, S. S., & Erzurumlu, Y. O. (2015). Sustainable mining development with community using design thinking and multi-criteria decision analysis. *Resources Policy, 46*, 6–14. doi:10.1016/j.resourpol.2014.10.001

Evans, J., Jones, R., Karvonen, A., Millard, L., & Wendler, J. (2015). Living labs and co-production: University campuses as platforms for sustainability science. *Current Opinion in Environmental Sustainability, 16*, 1–6. doi:10.1016/j.cosust.2015.06.005

Fauzi, Huda, M., Teh, K.S.M., Haron, Z., Ripin, M.N., Hehsan, A., Abas, H., Hehsan, M.R., Irawan, J., & Maseleno, A. (2018). The Design of Fuzzy Expert System Implementation for Analyzing Transmissible Disease of Human. *International Journal of Pharmaceutical Research, 10*(4), 68–78.

Fauzi, I., Irviani, R., Muslihudin, M., Satria, F., Huda, M., Kamenez, N. V., & Maseleno, A. (2019). Revolutionizing Education through Artificial Intelligence: Fuzzy Multiple Attribute Decision Making Approach for Determining the Best Vocational High School. *Applied Mechanics and Materials, 892*, 234–239. doi:10.4028/www.scientific.net/AMM.892.234

Fitrian, Y., Huda, M., Muhtar, A., Arifin, A. Y., Musa, N., Teh, M., ... Maseleno, A. (2019). Application Design for Determining Suitable Cosmetics with the Facial Skin Type Using Fuzzy Logic Approach. *Journal of Computational and Theoretical Nanoscience, 16*(5-6), 2153–2158. doi:10.1166/jctn.2019.7866

Florini, A., & Pauli, M. (2018). Collaborative governance for the sustainable development goals. *Asia & the Pacific Policy Studies, 5*(3), 583–598. doi:10.1002/app5.252

Galaitsi, S., Russell, R., Bishara, A., Durant, J., Bogle, J., & Huber-Lee, A. (2016). Intermittent domestic water supply: A critical review and analysis of causal-consequential pathways. *Water (Basel), 8*(7), 274. doi:10.3390/w8070274

Geist, H. (2017). *The causes and progression of desertification*. Routledge. doi:10.4324/9781315240855

Hamid, A., Sudrajat, A., Kawangit, R. M., Don, A. G., Huda, M., Jalal, B., ... Maseleno, A. (2018). Determining basic food quality using SAW. *IACSIT International Journal of Engineering and Technology, 7*(4), 3548–3555.

Howlett, C., Ferreira, J. A., & Blomfield, J. (2016). Teaching sustainable development in higher education: Building critical, reflective thinkers through an interdisciplinary approach. *International Journal of Sustainability in Higher Education, 17*(3), 305–321. doi:10.1108/IJSHE-07-2014-0102

Huda, M., & Hashim, A. (2021). (in press). Smart Learning Environment (SLE) in Fourth Industrial Revolution (IR 4.0). *International Journal of Asian Business and Information Management.*

Huda, M. (2021). (in press). Towards Professional and Ethical Balance: Insights into Application Strategy on Media Literacy Education. *Kybernetes.*

Huda, M. (2019). Empowering application strategy in the technology adoption: Insights from professional and ethical engagement. *Journal of Science and Technology Policy Management, 10*(1), 172–192. doi:10.1108/JSTPM-09-2017-0044

Huda, M., Ulfatmi, Luthfi, M.J., Jasmi, K.A., Basiron, B., Mustari, M.I., Safar, A., Embong, H.W.H., Mohamad, A.M., & Mohamed, A.K. (2019d). Adaptive online learning technology: Trends in big data era. In Diverse Learning Opportunities Through Technology-Based Curriculum Design. (pp.163-195), Hershey, PA: IGI Global. doi:10.4018/978-1-5225-5519-3.ch008

Huda, M., Almunawar, M. N., Hananto, A. L., Rismayadi, B., Jasmi, K. A., Basiron, B., & Mustari, M. I. (2018a). Strengthening Quality Initiative for Organization Stability: Insights From Trust in Cyberspace-Based Information Quality. In *Cases on Quality Initiatives for Organizational Longevity* (pp. 140–169). IGI Global. doi:10.4018/978-1-5225-5288-8.ch006

Huda, M., Anshari, M., Almunawar, M. N., Shahrill, M., Tan, A., Jaidin, J. H., ... Masri, M. (2016). Innovative Teaching in Higher Education: The Big Data Approach. *The Turkish Online Journal of Educational Technology, 15*(Special issue), 1210–1216.

Huda, M., Haron, Z., Ripin, M. N., Hehsan, A., & Yaacob, A. B. C. (2017a). Exploring Innovative Learning Environment (ILE): Big Data Era. *International Journal of Applied Engineering Research: IJAER, 12*(17), 6678–6685.

Huda, M., Hehsan, A., Basuki, S., Rismayadi, B., Jasmi, K. A., Basiron, B., & Mustari, M. I. (2019a). Empowering Technology Use to Promote Virtual Violence Prevention in Higher Education Context. In *Intimacy and Developing Personal Relationships in the Virtual World* (pp. 272–291). IGI Global. doi:10.4018/978-1-5225-4047-2.ch015

Huda, M., Jasmi, K. A., Alas, Y., Qodriah, S. L., Dacholfany, M. I., & Jamsari, E. A. (2017b). Empowering Civic Responsibility: Insights From Service Learning. In S. Burton (Ed.), *Engaged Scholarship and Civic Responsibility in Higher Education* (pp. 144–165). IGI Global. doi:10.4018/978-1-5225-3649-9.ch007

Huda, M., Jasmi, K. A., Basiran, B., Mustari, M. I. B., & Sabani, A. N. (2017c). Traditional Wisdom on Sustainable Learning: An Insightful View From Al-Zarnuji's Ta 'lim al-Muta 'allim. *SAGE Open, 7*(1), 1–8. doi:10.1177/2158244017697160

Huda, M., Jasmi, K. A., Embong, W. H., Safar, J., Mohamad, A. M., Mohamed, A. K., Muhamad, N. H., Alas, Y., & Rahman, S. K. (2017d). Nurturing Compassion-Based Empathy: Innovative Approach in Higher Education. In M. Badea & M. Suditu (Eds.), *Violence Prevention and Safety Promotion in Higher Education Settings* (pp. 154–173). IGI Global. doi:10.4018/978-1-5225-2960-6.ch009

Huda, M., Jasmi, K. A., Hehsan, A., Shahrill, M., Mustari, M. I., Basiron, B., & Gassama, S. K. (2017e). Empowering Children with Adaptive Technology Skills: Careful Engagement in the Digital Information Age. *International Electronic Journal of Elementary Education, 9*(3), 693–708.

Huda, M., Jasmi, K. A., Mustari, M. I., Basiron, B., Mohamed, A. K., Embong, W., ... Safar, J. (2017f). Innovative E-Therapy Service in Higher Education: Mobile Application Design. *International Journal of Interactive Mobile Technologies*, *11*(4), 83–94. doi:10.3991/ijim.v11i4.6734

Huda, M., Maseleno, A., Jasmi, K. A., Mustari, I., & Basiron, B. (2017g). Strengthening Interaction from Direct to Virtual Basis: Insights from Ethical and Professional Empowerment. *International Journal of Applied Engineering Research: IJAER*, *12*(17), 6901–6909.

Huda, M., Maseleno, A., Muhamad, N. H. N., Jasmi, K. A., Ahmad, A., Mustari, M. I., & Basiron, B. (2018b). Big Data Emerging Technology: Insights into Innovative Environment for Online Learning Resources. *International Journal of Emerging Technologies in Learning*, *13*(1), 23–36. doi:10.3991/ijet.v13i01.6990

Huda, M., Maseleno, A., Teh, K. S. M., Don, A. G., Basiron, B., Jasmi, K. A., Mustari, M. I., Nasir, B. M., & Ahmad, R. (2018c). Understanding Modern Learning Environment (MLE) in Big Data Era. *International Journal of Emerging Technologies in Learning*, *13*(5), 71–85. doi:10.3991/ijet.v13i05.8042

Huda, M., Muhamad, N. H. N., Isyanto, P., Kawangit, R. M., Marni, N., Mohamed, A. K., & Safar, A. J. (2020). *Building Harmony in Diverse Society: Insights from Practical Wisdom*. International Journal of Ethics and Systems. doi:10.1108/IJOES-11-2017-0208

Huda, M., Mulyadi, D., Hananto, A. L., Nor Muhamad, N. H., Mat Teh, K. S., & Don, A. G. (2018d). Empowering corporate social responsibility (CSR): Insights from service learning. *Social Responsibility Journal*, *14*(4), 875–894. doi:10.1108/SRJ-04-2017-0078

Huda, M., Qodriah, S. L., Rismayadi, B., Hananto, A., Kardiyati, E. N., Ruskam, A., & Nasir, B. M. (2019c). Towards Cooperative with Competitive Alliance: Insights into Performance Value in Social Entrepreneurship. In *Creating Business Value and Competitive Advantage with Social Entrepreneurship* (p. 294). IGI Global. doi:10.4018/978-1-5225-5687-9.ch014

Huda, M., Sabani, N., Shahrill, M., Jasmi, K. A., Basiron, B., & Mustari, M. I. (2017h). Empowering Learning Culture as Student Identity Construction in Higher Education. In A. Shahriar & G. Syed (Eds.), *Student Culture and Identity in Higher Education* (pp. 160–179). IGI Global. doi:10.4018/978-1-5225-2551-6.ch010

Huda, M., Shahrill, M., Maseleno, A., Jasmi, K. A., Mustari, I., & Basiron, B. (2017i). Exploring Adaptive Teaching Competencies in Big Data Era. *International Journal of Emerging Technologies in Learning*, *12*(3), 68–83. doi:10.3991/ijet.v12i03.6434

Huda, M., Sudrajat, S., Kawangit, R. M., Teh, K. S. M., & Jalal, B. (2019b). Strengthening divine values for self-regulation in religiosity: Insights from Tawakkul (trust in God). *International Journal of Ethics and Systems*, *35*(3), 323–344. doi:10.1108/IJOES-02-2018-0025

Huda, M., & Teh, K. S. M. (2018). Empowering Professional and Ethical Competence on Reflective Teaching Practice in Digital Era. In K. Dikilitas, E. Mede, & D. Atay (Eds.), *Mentorship Strategies in Teacher Education* (pp. 136–152). IGI Global. doi:10.4018/978-1-5225-4050-2.ch007

Huda, M., Teh, K. S. M., Nor, N. H. M., & Nor, M. B. M. (2018e). Transmitting Leadership Based Civic Responsibility: Insights from Service Learning. *International Journal of Ethics and Systems*, *34*(1), 20–31. doi:10.1108/IJOES-05-2017-0079

Irviani, R., Muslihudin, M., Satria, F., Huda, M., Kamenez, N. V., & Maseleno, A. (2019). Revolutionizing Education through Artificial Intelligence: Fuzzy Multiple Attribute Decision Making Approach for Determining the Best Vocational High School. *Applied Mechanics and Materials*, *892*, 234–239. doi:10.4028/www.scientific.net/AMM.892.234

Issa, T. (2017). Sustainability Perspective and Awareness Amongst Higher Education in Australia. *Sustainability, Green IT and Education Strategies in the Twenty-first Century*, 17-33.

Issa, T., Isaias, P., & Issa, T. (Eds.). (2017). *Sustainability, Green IT and Education Strategies in the Twenty-first Century*. Springer. doi:10.1007/978-3-319-57070-9

Kapp, K. W. (2018). Environmental disruption: General issues and methodological problems. In *Green Accounting* (pp. 31–48). Routledge. doi:10.4324/9781315197715-2

Kaur, H., & Singh, S. P. (2018). Environmentally sustainable stochastic procurement model. *Management of Environmental Quality*, *29*(3), 472–498. doi:10.1108/MEQ-04-2017-0039

Kholili, H. (in press). Adaptive Learning Technology Through Social Network Site (SNS)- Strategy Enhancement. *Walailak Journal of Science and Technology*.

Kovbasyuk, O., & Blessinger, P. (Eds.). (2013). *Meaning-centered education: International perspectives and explorations in higher education*. Routledge. doi:10.4324/9780203115084

Laurie, R., Nonoyama-Tarumi, Y., Mckeown, R., & Hopkins, C. (2016). Contributions of education for sustainable development (ESD) to quality education: A synthesis of research. *Journal of Education for Sustainable Development*, *10*(2), 226–242. doi:10.1177/0973408216661442

Leal Filho, W., & Rayman-Bacchus, L. (2019). Sustainability Governance in Traditional Crafts Communities: A Project Proposition. In *Social Responsibility and Sustainability* (pp. 277–287). Springer. doi:10.1007/978-3-030-03562-4_14

Linkov, I., Trump, B., Poinsatte-Jones, K., & Florin, M. V. (2018). Governance strategies for a sustainable digital world. *Sustainability*, *10*(2), 440. doi:10.3390u10020440

Lueg, R., & Radlach, R. (2016). Managing sustainable development with management control systems: A literature review. *European Management Journal*, *34*(2), 158–171. doi:10.1016/j.emj.2015.11.005

Mahat, H., Saleh, Y., Hashim, M., & Nayan, N. (2016). Model development on awareness of education for sustainable schools development in Malaysia. *The Indonesian Journal of Geography*, *48*(1), 37–46. doi:10.22146/ijg.12446

Malik, S., Fatima, F., Imran, A., Chuah, L. F., Klemeš, J. J., Khaliq, I. H., ... Akbar, M. M. (2019). Improved project control for sustainable development of construction sector to reduce environment risks. *Journal of Cleaner Production*, *240*, 118214. doi:10.1016/j.jclepro.2019.118214

Marzuki, M., Wan Hussin, W. N., Amin, N. M., & Huda, M. (2020). The Sustainability of Passenger Boat Services in Malaysia. *Solid State Technology*, *63*(5), 4361–4372.

Maseleno, A., Pardimin, Huda, M., Ramlan, Hehsan, A., Yusof, Y.M., Haron, Z., Ripin, M.N., Nor, N.H.M., & Junaidi, J. (2018b). Mathematical Theory of Evidence to Subject Expertise Diagnostic. *ICIC Express Letters*, *12*(4), 369. doi:10.24507/icicel.12.04.369

Maseleno, A., Huda, M., Jasmi, K. A., Basiron, B., Mustari, I., Don, A. G., & Ahmad, R. (2018a). *Hau-Kashyap approach for student's level of expertise*. Egyptian Informatics Journal., doi:10.1016/j.eij.2018.04.001

Maseleno, A., Sabani, N., Huda, M., Ahmad, R., Jasmi, K. A., & Basiron, B. (2018c). Demystifying Learning Analytics in Personalised Learning. *IACSIT International Journal of Engineering and Technology*, *7*(3), 1124–1129. doi:10.14419/ijet.v7i3.9789

Maseleno, A., Shankar, K., Huda, M., Othman, M., Khoir, P., & Muslihudin, M. (2019). *Citizen Economic Level (CEL) using SAW. In Expert Systems in Finance: Smart Financial Applications in Big Data Environments, 97*. Routledge.

Mensah, S., Ayarkwa, J., & Nani, G. (2018). A theoretical framework for conceptualizing contractors' adaptation to environmentally sustainable construction. *International Journal of Construction Management*, 1-11.

Mitchell, I. K., & Walinga, J. (2017). The creative imperative: The role of creativity, creative problem solving and insight as key drivers for sustainability. *Journal of Cleaner Production*, *140*, 1872–1884. doi:10.1016/j.jclepro.2016.09.162

Moksin, A. I., Shahrill, M., Anshari, M., Huda, M., & Tengah, K. A. (2018). The Learning of Integration in Calculus Using the Autograph Technology. *Advanced Science Letters*, *24*(1), 550–552. doi:10.1166/asl.2018.12067

Molderez, I., & Fonseca, E. (2018). The efficacy of real-world experiences and service learning for fostering competences for sustainable development in higher education. *Journal of Cleaner Production*, *172*, 4397–4410. doi:10.1016/j.jclepro.2017.04.062

Mulawarman, A., Sudrajat, A., Hendri, N., Kamar, K., Mulyadi, D., Budiyanto, G., & Huda, M., and Maseleno. (2018). FMADM for determining superior commodity at agroindustry area. *IACSIT International Journal of Engineering and Technology*, *7*(4), 4667–4673.

Muslihudin, M., Ilayaraja, M., Kumar, K. S., Shankar, K., Jamilah, J., Novitasari, D., & ... & Maseleno. (2019). A. Decision Support System in Kindergarten Selection using TOPSIS Method. *International Journal of Recent Technology and Engineering*, *8*(1), 3291–3298.

Oktafianto, Sudrajat, A., Kawangit, R.M., Don, A.G., Huda, M., Saputri, A.D., & Maseleno, A. (2018). Determining housing location using weighted product. *IACSIT International Journal of Engineering and Technology*, *7*(4), 3563–3568.

Ortiz-de-Mandojana, N., Aguilera-Caracuel, J., & Morales-Raya, M. (2016). Corporate governance and environmental sustainability: The moderating role of the national institutional context. *Corporate Social Responsibility and Environmental Management*, *23*(3), 150–164. doi:10.1002/csr.1367

Orts, E. W., & Smith, N. C. (Eds.). (2017). *The moral responsibility of firms*. Oxford University Press. doi:10.1093/oso/9780198738534.001.0001

Othman, R., Shahrill, M., Mundia, L., Tan, A., & Huda, M. (2016). Investigating the Relationship Between the Student's Ability and Learning Preferences: Evidence from Year 7 Mathematics Students. *The New Educational Review*, *44*(2), 125–138. doi:10.15804/tner.2016.44.2.10

Pane, M. M., & Patriana, R. (2016). The significance of environmental contents in character education for quality of life. *Procedia: Social and Behavioral Sciences*, *222*, 244–252. doi:10.1016/j.sbspro.2016.05.153

Pardimin, A., Ninsiana, W., Dacholfany, M.I., Kamar, K., Teh, K.S.M., Huda, M., Hananto, A.L., Muslihudin, M., Shankar, K., & Maseleno, A. (2018). Developing Multimedia Application Model for Basic Mathematics Learning. *Journal of Advanced Research in Dynamical and Control Systems*, *10*(14), 1347–1356.

Polk, M. (Ed.). (2015). *Co-producing knowledge for sustainable cities: Joining forces for change*. Routledge. doi:10.4324/9781315748030

Puspitasari, R., Riadi, D., Huda, M., & Maseleno, A. (2020). Self-Management of Youth Transmigrant in the Use of Information Technology. *European Journal of Molecular & Clinical Medicine*, *7*(11), 247–249.

Putra, D. A. D., Jasmi, K. A., Basiron, B., Huda, M., Maseleno, A., Shankar, K., & Aminudin, N. (2018). Tactical Steps for E-Government Development. *International Journal of Pure and Applied Mathematics*, *119*(15), 2251–2258.

Rajagopalan, R. (2015). Environmental studies: from crisis to cure (No. 3). Oxford University Press.

Rathod, H. S., Chauhan, R., Huda, M., & Maseleno, A. (2020). A Conceptual Research Study on Identification of Biases with towards Assets Allocation and Portfolio Management of Investors. *European Journal of Molecular & Clinical Medicine*, *7*(10), 985–994.

Razzaq, A., Samiha, Y., & Anshari, M. (2018). Smartphone Habits and Behaviors in Supporting Students Self-Efficacy. *International Journal of Emerging Technologies in Learning*, *13*(2), 94–109. doi:10.3991/ijet.v13i02.7685

Ristiani, P., Teh, K.S.M., Fauzi, A., Hananto, A.L., Huda, M., Muslihudin, M., Shankar, K., & Maseleno, A. (2018). Decision Support System Model for Selection of Best Formula Milk for Toddlers Using Fuzzy Multiple Attribute Decision Making. *Journal of Advanced Research in Dynamical and Control Systems*, *2*(Special issue), 2075–2088.

Roos, J. (2017). Practical wisdom: Making and teaching the governance case for sustainability. *Journal of Cleaner Production*, *140*, 117–124. doi:10.1016/j.jclepro.2015.10.135

Rosa, A. T. R., Pustokhina, I. V., Lydia, E. L., Shankar, K., & Huda, M. (2019). Concept of Electronic Document Management System (EDMS) as an Efficient Tool for Storing Document. *Journal of Critical Reviews*, *6*(5), 85–90.

Rosli, M. R. B., Salamon, H. B., & Huda, M. (2018). Distribution Management of Zakat Fund: Recommended Proposal for Asnaf Riqab in Malaysia. *International Journal of Civil Engineering and Technology*, *9*(3), 56–64.

Sacchetti, S., & Tortia, E. (2016). The extended governance of cooperative firms: Inter-firm coordination and consistency of values. *Annals of Public and Cooperative Economics*, *87*(1), 93–116. doi:10.1111/apce.12058

Salamah, P., Jandra, M., Sentono, T., Huda, M., & Maseleno, A. (2019). The effects of Emotional Intelligence, Family Environment and Learning Styles on Social-Science Learning Outcomes: An Empirical Analysis. *TEST Engineering and Management*, *81*, 4374 - 4386.

Sari, N.Y., Huda, M., & Teh, K.S.M., Ristiani, & Maseleno, A. (2018). Decision support system model for selection of best formula milk for toddlers using fuzzy multiple attribute decision making. *IACSIT International Journal of Engineering and Technology*, *7*(4), 3556–3562.

Schaltegger, S., & Burritt, R. (2017). *Contemporary environmental accounting: issues, concepts and practice*. Routledge. doi:10.4324/9781351282529

Septiropa, Z., Osman, M.H., Abd. Rahman, A.B., Ariffin, M.A.M., Huda, M., & Maseleno, A. (2018). Profile of cold-formed steel for compression member design a basic combination performance. *International Journal of Engineering and Technology*, *7*(2.27), 284-290.

Shi, X., & Xu, Z. (2018). Environmental regulation and firm exports: Evidence from the eleventh Five-Year Plan in China. *Journal of Environmental Economics and Management*, *89*, 187–200. doi:10.1016/j.jeem.2018.03.003

Sivaram, M., Shanmugapriya, S., Yuvaraj, D., Porkodi, V., Akbari, A., Hashim, W., & Huda, M. (2020). Decision Support System for Determining Academic Advisor Using Simple Additive Weighting. In *Cognitive Informatics and Soft Computing* (pp. 149–156). Springer. doi:10.1007/978-981-15-1451-7_16

Sugiyarti, E., Jasmi, K. A., Basiron, B., Huda, M., Shankar, K., & Maseleno, A. (2018). Decision support system of scholarship grantee selection using data mining. *International Journal of Pure and Applied Mathematics*, *119*(15), 2239–2249.

Susilowati, T., Dacholfany, M.I., Aminin, S., Ikhwan, A., Nasir, B.M., Huda, M., Prasetyo, A., Maseleno, A., Satria, F., Hartati, S., & Wulandari. (2018a). Getting parents involved in child's school: using attendance application system based on SMS gateway. *International Journal of Engineering and Technology*, *7*(2.27), 167-174.

Susilowati, T., Jasmi, K.A., Basiron, B., Huda, M., Shankar, K., Maseleno, A., & Julia, A., & Sucipto. (2018b). Determination of Scholarship Recipients Using Simple Additive Weighting Method. *International Journal of Pure and Applied Mathematics*, *119*(15), 2231–2238.

Susilowati, T., Manickam, P., Devika, G., Shankar, K., Muslihudin, M., Hashim, W., ... Maseleno, A. (2019). Decision support system for determining lecturer scholarships for doctoral study using CBR (Case-based reasoning) method. *International Journal of Recent Technology and Engineering*, *8*(1), 3281–3290.

Susilowati, T., Teh, K.S.M., Nasir, B.M., Don, A.G., Huda, M., Hensafitri, T., Maseleno, A., Oktafianto, & Irawan, D. (2018c). Learning application of Lampung language based on multimedia software. *International Journal of Engineering and Technology, 7*(2.27), 175-181.

Thoresen, V. W. (2019). Empathy, Adaptability, Moderation and Sharing. *Teacher Education for Sustainable Development and Global Citizenship: Critical Perspectives on Values, Curriculum and Assessment*, 217.

Too, L., & Bajracharya, B. (2015). Sustainable campus: Engaging the community in sustainability. *International Journal of Sustainability in Higher Education*, *16*(1), 57–71. doi:10.1108/IJSHE-07-2013-0080

Trevino, L. K., & Nelson, K. A. (2016). *Managing business ethics: Straight talk about how to do it right*. John Wiley & Sons.

Wahyuni, N., Kulik, A. A., Lydia, E. L., Shankar, K., & Huda, M. (2020). Developing Region to Reduce Economic Gap and to Support Large Environment Activities. *Journal of Environmental Treatment Techniques*, *8*(1), 540–545.

Wals, A. E. (Ed.). (2009). *Review of contexts and structures for education for sustainable development: 2009*. Unesco.

Warner, M., & Sullivan, R. (Eds.). (2017). *Putting partnerships to work: Strategic alliances for development between government, the private sector and civil society*. Routledge. doi:10.4324/9781351281249

Wittmer, D. P. (2016). Developing a behavioral model for ethical decision making in organizations: Conceptual and empirical research. In *Ethics in public management* (pp. 57–77). Routledge.

Wu, G. C. (2017). Effects of socially responsible supplier development and sustainability-oriented innovation on sustainable development: Empirical evidence from SMEs. *Corporate Social Responsibility and Environmental Management*, *24*(6), 661–675. doi:10.1002/csr.1435

Wu, J., & Si, S. (2018). Poverty reduction through entrepreneurship: Incentives, social networks, and sustainability. *Asian Business & Management*, *17*(4), 243–259. doi:10.105741291-018-0039-5

Wulandari, Aminin, S., Dacholfany, M.I., Mujib, A., Huda, M., Nasir, B.M., Maseleno, A., Sundari, E., Fauzi, & Masrur, M. (2018). Design of library application system. *International Journal of Engineering and Technology*, *7*(2.27), 199-204.

Yadlapalli, A., Rahman, S., & Gunasekaran, A. (2018). Socially responsible governance mechanisms for manufacturing firms in apparel supply chains. *International Journal of Production Economics*, *196*, 135–149. doi:10.1016/j.ijpe.2017.11.016

Chapter 2
Implementation of Culturally Relevant Science-Based Projects in Preschools and Primary Schools:
From Roots to Wings

Martin Bascope
Pontificia Universidad Catolica de Chile, Chile

Kristina Reiss
School of Education, Technische Universität München, Germany

Josefina Cortés
Pontificia Universidad Católica de Chile, Villarrica, Chile

Pablo Gutierrez
Pontificia Universidad Católica de Chile, Chile

ABSTRACT

This chapter develops a conceptual and operational framework on education for sustainability based on a teacher training program located in the south of Chile. Nurtured by the experience collected by the programs' participation in different international networks, it put emphasis on the idea of the importance of combining local knowledge and an integrated science approach to foster action for climate and sustainability from an early stage. After presenting the conceptual and empirical framework, the chapter describes the implementation of culturally-relevant research projects based on the experience of eight schools. It concludes by proposing an operational model and declaring the main constraints and opportunities of project-based-learning for sustainability and climate action. The authors present some recommendations to combine local and scientific knowledge for teacher practice and educational policy, considering the importance of having a critical look at our daily practices as educators given the global socio-ecological crisis.

DOI: 10.4018/978-1-7998-7512-3.ch002

INTRODUCTION

In the context of the current climate crisis and socio-ecological dilemmas, humans face future scenarios that put into question our lifestyles, economical activities, development models, and, therefore, our educational settings and how we prepare younger generations to confront highly uncertain futures. Intricately connected global problems and devastating climate science reports and forecasts show that we must stop procrastinating and act now. Change must start from early childhood and children must be active agents for change, alongside their teachers, parents, and communities. In addition, cultural backgrounds, worldviews, and traditional knowledge perspectives provide an opportunity to enrich the curriculum and to create place-based meaningful core educational activities, through which children can learn about socio-ecological issues and become agents of change in their own localities.

Project-based learning (PBL) offers a good opportunity to move toward connecting schools with local and global socio-ecological challenges related to climate change and sustainability. In this chapter, the authors review PBL literature and international experiences, resulting in a list of conditions for the implementation of culturally relevant and scientifically coherent projects. Within this framework, the qualitative results from the implementation of eight school research projects involving a collaboration between these schools and a university in the south of Chile are presented.

The research was conducted in Villarrica, a city in the Araucanía Region, homeland of the Mapuche people, Chile's largest indigenous population. The Villarrica Campus of the Pontificia Universidad Católica de Chile provides both initial and in-service teacher training to preschool and primary school teachers. Over the last seven years, it has implemented courses, diplomas, and training opportunities to foster inquiry-based science learning with a focus on sustainable development and intercultural and inclusive practices. This chapter describes the methodology used to create pedagogical materials and methodologies to support education for sustainable development opportunities and in particular, how place-based and student-centered PBL can be a useful tool for coping with complex climate change and sustainability challenges.

This chapter contributes to the field by presenting an operational framework, which, on the one hand, will be useful at both the policy and the school level and, on the other hand, will enrich academic discussion of PBL in early childhood education. It also offers guidance for implementation at the school level, and shares what has been learned from a year of monitoring the opportunities and constraints of implementing these projects in different schools. In addition, accounts of the collaborative practices that led to the creation of pedagogical materials and methodologies can be applied to other contexts and provide inspiration for the fostering of further projects in the future.

RESEARCH FRAMEWORK AND RELATED INTERNATIONAL EXPERIENCES

Sustainable thinking requires proper scaffolding, starting with providing education on attitudes, knowledge, and skills for sustainability, beginning in preschool. According to a systematic review of education for sustainable development (ESD) for children aged four to ten years (Bascopé et al., 2019), ESD must be transversal and not discipline-bound, exploring the fundamentals of cosmopolitan citizenship and the ways we interact in the context of our everyday lives. Therefore, it is important to focus on the problem of how to promote local actions that will train learners to think globally from an early age, in order to develop resilience. The authors highlight the importance of transdisciplinary, science-integrated opportunities, value-oriented learning and PBL, and argue that PBL and outdoor education are two suitable methodologies for achieving this objective. In this sense, there is a need to develop educational opportunities that combine indigenous knowledge, intergenerational exchange, scientific knowledge, traditional lifestyles, and local practices. These integrative inter- and transdisciplinary methods also provide opportunities to foster agency and collaborative skills among children.

Unfortunately, there is limited empirical literature that tests and demonstrates the dynamic, underlying processes of project work in the context of the early childhood classroom (Clark, 2006). Therefore, research on project-based teaching needs to be expanded.

Experiential Outdoor Education (Free Exploration) or STEM Early Childhood Education?

Waite (2020) has reviewed a database of more than 3200 K-12 schools in the United States that have implemented PBL, revealing the growing importance of this approach. Her analysis of the dataset found differences in implementation and in the relationship between educational approaches, suggesting a close relationship between PBL and "experiential learning" and "maker learning." The latter is especially prominent in schools with students from higher socioeconomic backgrounds. In many cases, PBL is related with 'explore and do', going outside, leaving aside, and allowing children to take a closer look at the process of scientific inquiry.

Campbell et al. (2018) argue for the importance of outdoor and playful settings to conduct STEM (science, technology, engineering, and mathematics) in early childhood, to nurture learning interest and self-learning. Using a validated observation protocol for STEM education activities (Katz, 2010), they observed the use of meaningful early childhood education opportunities to develop a variety of basic scientific skills key to scaffolding future scientific literacy and enhancing children's dispositions to lifelong learning . With regard to scientific thinking and skills, Sodian et al. (1991) found out that even first grade students could distinguish between a "hypothetical belief" and "evidence," and could draw simple tests to prove conflicting hypotheses. PBL is a means to developing this way of thinking based on projects designed and data collected by children themselves. It has been proven to be a powerful tool to foster scientific thinking and skills (Koslowski, 1996; Kuhn et al., 1988; Kuhn & Ho, 1980).

Besides important basic scientific skills, there is also evidence that the use of PBL from an early age promotes better communication and language skills and effective engagement with deeper learning. Murtaza and Mahmood (2018) conducted a case study with fifty-four head teachers and school teachers in Pakistan that involved the development of PBL approaches to promote active learning among Class I English language students. In this dynamic classroom approach, students actively explored real-world problems and challenges and acquire deeper knowledge, resulting in a marked improvement in effective teaching and meaningful learning. These authors also found some evidence that student engagement increased throughout the entire project when the driving question was relevant to their geographical context.

There is also evidence for the importance of outdoor educational experiences in generating both the motivation to live sustainability and to take action for the natural world (Gray & Birrell, 2015). Early immersive experiences in natural environments have been found to affect future environmental appreciation and action (Liefländer et al., 2013).

PBL has been linked to an enhanced sense of community and making connections between the school curriculum and local practices. Keskitalo (2019) explored how place-based education projects help to bridge the gap between the Sámi indigenous worldview and the school curriculum in Finland. Research in Chile has concluded that the science curriculum can benefit from engaging children in learning about local culture (Bascopé & Caniguan, 2016), and such projects enhance community and family involvement and make learning more attractive and engaging for both students and parents.

Involvement with the community, with the natural environment, and with indigenous cultures are possible by means of a well-defined PBL process. It is both very important and a challenge for implementation that research projects balance outdoor exploration, scientific inquiry, and community involvement, in order to achieve integrative projects that focus on the development of sustainable mindsets and the fostering of lifelong learning.

Keys in PBL Approaches for Sustainability: Operational Framework

Bascopé et al.'s (2019) systematic review and this chapter's literature review highlight some key elements of PBL for sustainability. This operational framework is presented as a research-based framework to help practitioners to plan research projects. The main objectives are to produce improved engagement of students, to bring projects closer to the school curriculum, to connect schools with their social and natural environments, to produce longstanding learning and attitudes, and to generate action and agency with regard to climate and sustainability issues.

The following perspectives were considered as guidance for the implemented projects that will be presented in the results section, where an analysis of the constraints and opportunities of putting these perspectives into practice will be presented.

- **Context-relevant and Feasible Projects**: The connection with local challenges, problems, and dilemmas is very important. In early years learning, in particular, making learning visible in everyday routines is essential. In particular, children should be aware of contributing to and making a difference to their communities. This also involves learning 'from and about local communities', based on what Duhn (2012) refers to as a 'pedagogy of place', generating conditions to establish fruitful dialogue and co-learning with community actors in a bidirectional exchange (Bascopé et al., 2019; Duhn, 2012; Zachariou & Symeou, 2009).
- **Utilize Student Interest and Encourage Students to Be Decision-makers**: It is important to involve students in project ideation, both to provide capacities to generate relevant questions and to define relevant problems, and to enable students to be active participants in the project from the start (Mitchell et al., 2009). This means generating activities with children in order to define the problem, generate research questions, build research objectives, work on planning fieldwork and analysis activities, and thinking about possible solutions and contributions.
- **Provide Hands-on and Outdoor Experiences:** Encounters with nature and the local community are widely supported in the research literature for their impact on future attitudes toward the environment and local culture, and for their power to engage students in active learning (Nigh & Bertely, 2018; Sykes, 2008; Tilbury & Wortman, 2008; Zachariou & Symeou, 2009). Defining a project based on local problems in the environs of the educational institution, includes providing direct contact with the place where the project is conducted, developing sensitivity for the specific problem or challenge, gathering information, and conducting proper empirical research (Green, 2015).
- **Inquiry-based Science Learning**: It is important to keep in mind that in addition to the direct impact of outdoor education, free exploration and nature experiences by themselves are not enough

to start building the capacity to confront complex future scenarios. Bridging STEM competence from an early stage is essential to fostering interest in future learning and to building scientific capacities and the systematic thinking required to address complex problems (Campbell et al., 2018). Engaging in real scientific-based activities fosters the formation of future agents of change, and fostering agency is widely recognized as a key element of early childhood education (Caiman & Lundegard, 2014; Davis, 2015; Sawitri, 2017). A problem-based approach helps to develop scientific competence in a constructivist way, using an interdisciplinary focus to address the specific problem defined. This means in practice, to adapt curricular objectives from diverse domains, to contribute to the different stages of the project (Bascopé et al., 2019).

- **Value-oriented and Politically-aware Projects:** It is important that children gain an understanding that these interventions are situated within a defined value system, a specific historical context, and real-world social circumstances. When defining objectives, it is important to understand which sustainability values are being formed and to promote reflection on actual circumstances, and future consequences, if no action is taken. Bascopé et al. (2019) emphasize the importance of forming both "ethic" and "aesthetic" values. Ethics include values for democratic citizenship that combine individual and collective rights, protection of the commons, interdependence and connection with nature, and respect for diverse worldviews (Elliott & Young, 2016; Hägglund & Samuelsson, 2009; Wals, 2017). Aesthetic values, meanwhile, create a lasting connection with natural and cultural contexts and positive attitudes toward life. Aesthetics activates the learning process through intuition and creativity, and works as a meta-structure to develop connections with distant objects. Therefore, developing aesthetic values has an impact on the cognitive development necessary to forge a sense of belonging to the local environment (Eernstman & Wals, 2013; Gray & Birrell, 2015; Vecchi, 2010).

International STEM Education for Sustainable Development (STEM4SD) Experiences

The following is a review of three international networks working on ESD and education for climate change.

The Experimento, a Latin American STEM network, is a collaborative network funded by the Siemens Foundation (Germany), involving Latin American universities and non-governmental organizations with science education expertise. The objective of the network is to enhance educational linkages in the region through a strategy to promote multi-sectorial alliances called STEAM (science, technology, engineering, art, mathematics) Territories.

In STEAM Territories, civil society, science, economics, and state actors unite to form local alliances to foster and strengthen STEAM education throughout the educational chain within a geographically defined space. This systemic approach focuses on specific local problems, and responds to global challenges such as the promotion of science and technology education, the stimulation of 21st Century skills, and integrated social and environmental sustainable development. STEAM Territories host and are incubators for innovative education programs. The format is based on the experience of the German network model MINT Regionen, which is run by the Koerber Foundation and the National STEM Forum of Germany. The Siemens Foundation adapted this model internationally and initiated the STEAM Territory project in Latin America. At the time of writing, there are seven STEAM Territories in operation: STEAM Territory Valparaiso (Chile), STEAM Territory Macro Zona Sur (Chile), STEAM Territory

Tacna Region (Peru), STEAM Territory Miraflores, Lima (Peru), STEAM Territory Arequipa Region (Peru), STEM+H Territory Medellín (Colombia) and STEM Territory Mexico State (Mexico)[1].

The second education for sustainability program is the International Dialogue on STEM Education (IDoS), an international platform that focuses on early childhood education and compiles experiences of STEM education for sustainability from five continents. IDoS includes early childhood practitioners, educators, and decision-makers from around the world, and is organized by the German Haus der Kleinen Forscher (Little Scientists' House) foundation, which operates in Germany to foster inquiry-based science learning and exploration opportunities in the early years' context.

In the 2019 version of IDoS, innovative ideas to promote the use of science for social good were shared and communicated. As a result of the organization and implementation of the dialogue, a position paper on STEM ESD (STEM4SD) was written and endorsed by more than one hundred practitioners and decision-makers from five continents (Pahnke et al., 2019). The paper presents seven guiding principles to orient STEM4SD programs and activities, together with a summary of initiatives implemented in different contexts to inspire new programs and promote international exchange to address the global sustainability challenges.[2]

Finally, the third international network is the Office for Climate Education (OCE), a global collaborative initiative to promote climate change education that provides a highly scientific-based platform to develop and transfer educational resources and methodologies for climate education[3]. Even though this network focuses on children aged between 9 and 15 years, it highlights the high quality of this scientific-based climate education platform. Most OCE resources are produced through the collaborative efforts of Intergovernmental Panel on Climate Change scientists and science education pedagogical experts. It contains a multiplatform set of materials, combining, among others, computer based, game-based, and outdoor education.

Despite these networks, methodologies, principles, and educational resources produced to address climate change and sustainability issues, it is important to note the lack of direction and systematization of ESD inquiry-based learning programs and initiatives, especially in the context of Latin America, and worldwide in the case of early childhood education (Bascopé, et al., 2020). The following research findings, which systematize and present an orientation for the implementation of STEM4SD learning opportunities, learning through empirical data, and practical experiences, are an effort to address this shortcoming.

SCHOOL EXPERIENCES: A CULTURALLY RELEVANT PROJECT-BASED ESD PROGRAM

The Program Context

In contemporary Latin America, historical challenges in education, including inequality, poverty, social justice, and interculturality, are now compounded by global challenges, such as climate change and a growing crisis of democracy. In the Araucanía Region, located in Chile, the environmental challenges (e.g. water scarcity, deforestation, loss of biodiversity), are closely related to historical land use practices and ethnic conflict. Environmental and intercultural issues, therefore, are interlinked and at the heart of the social, political, and economic agenda. The region is characterized by its national parks and natural landscapes. It has also gained recognition as the homeland of the Mapuche people, the largest indigenous population in Chile, who are engaged in ongoing political struggle.

The University Campus of Pontificia Universidad Católica de Chile, located in Villarrica, offers undergraduate degrees in early childhood education and primary education. Campus Villarrica plays a key role in pre- and in-service teacher training in the central-south Chilean region, which is characterized by a relatively high proportion of indigenous and non-indigenous rural populations and high rates of poverty relative to the rest of the country.

The Campus runs two research centers, the Center for Indigenous and Intercultural Research (CIIR) and the Center for Local Development (CEDEL). Because of the work of these two centers, ESD has become a central topic. With the support of the Siemens Foundation, the Campus has developed professional development programs for teachers on inquiry-based STEM subjects. These programs focus on the importance of the development of 21st Century skills key to understanding and evaluating phenomena such as climate change and the exploitation of natural resources. The objective of these programs is not to teach the transmission of declarative knowledge, but provide inquiry-based methods that support teachers in the use and understanding of constructivist problem-oriented approaches, such as PBL. Along with the methodological focus, the programs address environmental and intercultural education topics that can be addressed from a STEM perspective.

Having already trained more than 400 teachers, CEDEL is conducting research, such as that presented in this chapter, to measure the impact of the educational innovations on students' skills, abilities, and values in the last five years. A further aim of this research is to understand the key attitudinal and organizational perspectives that support the success of such programs in educational institutions.

Epu Trokin Kimun: A Culturally Relevant Inquiry-based Learning Program

This program arose from a realization among rural educators that they lacked the skills and support to teach scientific inquiry using inclusive and interdisciplinary methods consistent with the local reality of the children in their care. In addition, an erosion and lack of interest in indigenous Mapuche knowledge was detected, with primary teachers and preschool educators feeling unable to teach local knowledge, and intercultural educators, tasked with teaching indigenous language and culture, reporting feeling isolated from the rest of the school team (Bascopé & Canihuan, 2016). Through the collaborative work of the educational team and the use of game-based resources, the aim of the project was to support children in the creation and execution of interdisciplinary scientific research projects based on local knowledge and traditions.

The specific objectives of the program were:

1. To support learning processes based on local knowledge that engage with local history and promote a culture of care for natural and cultural heritage.
2. To develop culturally relevant instruction addressing academic success, cultural competencies, and sociopolitical awareness through an inquiry-based learning approach (Ladson-Billings, 1995).
3. To bring together the fields of education and science in a reciprocal collaboration to transform the traditional educational setting.

Through this project, children would learn to value and recognize the local and scientific knowledge of their territories, while also developing creativity and critical thinking skills, and connecting their schools to local contexts and problems.

Even though the historical context of the Mapuche is powerful and attractive, an indigenous context is not a necessary condition for the implementation of approaches of this kind. Richness of community and environment can also be found in urban and non-indigenous settings.

Program Methodology

After four years of collaboration between the Pontificia Universidad Católica de Chile, education practitioners and local communities, the Epu Trokin Kimun (ETK) ("exchange between two knowledges" in Mapudungun, the Mapuche language) kit, in both Spanish and Mapudungun, was designed by an interdisciplinary team to exchange knowledge between different environments. The process started with in-depth field research in six schools from different parts of the region. Group interviews were conducted with teachers, intercultural educators, parents, and members of indigenous communities. As a result of this process, five possible dimensions were found to establish a dialogue between local and scientific knowledge: 1) health and the human body, 2) food and culinary processes, 3) crafts, art, and tool-making, 4) flora, fauna, and ecosystems, and 5) the universe and worldviews (Bascopé & Canihuan, 2016).

After systematizing the results, 12 sets of teacher and student working guides adapted to the Chilean national science grade 1–6 primary curriculum, were designed and implemented for validation in 13 schools in the area. All of these guides were designed for students to conduct structured inquiry incorporating both scientific and indigenous knowledge on different topics within this five-dimension framework. The implementation was piloted at schools and activities were modified into the final design.[4] Despite a lot of effort, and high levels of teacher satisfaction with the working guides, the expected impact on children was low, as activities lasted for only 90 minutes and took place only once or twice in the school year. Moreover, activities were isolated and, only connected to the curriculum in a few science lessons. The challenge, therefore, was finding ways to increase impact, extend the connection between local context and curricular objectives, and make the activities more meaningful for the children.

In this way, the idea for a more open kit to promote PBL emerged. Working with an ecologist, Mapudungun specialists, teachers, and designers, a resource was created containing 326 coins, each featuring the Spanish, Mapudungun, and scientific name of a different species, along with six boards representing natural cycles (e.g., phases of the day, the seasons, and moon phases), geographical representations (e.g., cardinal coordinates and Mapuche territories), and a board to build a "network of relationships" among species (Figure 1). The intention of these boards and coins was to create a space for symbolic representation of hypothesis, pre-existing knowledge, evidence sorting and communication of results, to encourage teachers and students to go out and conduct research in and about their territories, and to consider complex sustainability challenges involving environmental, cultural, and economic issues.

Next, eight rural schools in the Araucanía Region began a training program for teachers to implement PBL as a vehicle for dialogue between knowledges. In the first stage, in order to develop this pedagogic tool, teachers were trained in scientific inquiry methodologies, and learned how to generate research questions about problems in the area or community surrounding their school, use mapping strategies, and the ETK kit. Next, monitoring processes were initiated in each school. This was a crucial phase, as it provided information on how the adaptations were working. Finally, in order to showcase the results,

Figure 1. Sample boards from the Epu Trokin Kimun Kit. Source: Authors

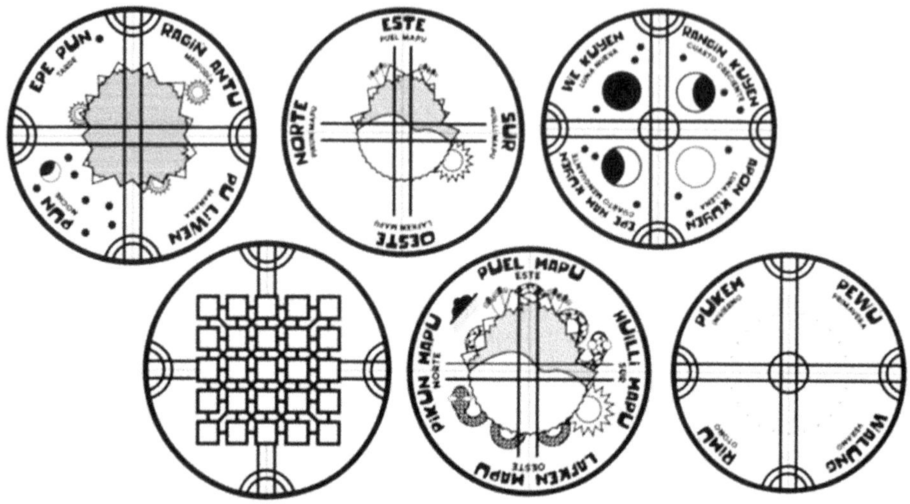

three school fairs were organized. These were open to the entire community with the objective of empowering students and teachers as researchers and custodians of their environment.

Each project had to state the relevance of the problem, the impact of the project, expected outcomes or solutions, activities, actions, and what the students were expected to learn, in accordance with the school curriculum. Examples of local projects included bird-tree relationship networks, the revalorization of local medicinal herbs, the recovery of local wild fruits and related traditional recipes, the elaboration of a Mapuche medicine kit, the protection and regeneration of seeds, revalorization of local wetlands, revalorization of native species, and reforestation projects.

During the implementation process, structured class observations and field work activities were observed, and registered using a standardized protocol. Afterwards, participating teachers were interviewed to systematize the results and to provide feedback. The following section presents the resulting PBL model, emphasizing the critical nodes that emerged during the research.

RESULTS

There are two parts to this section. The first summarizes the results in terms of the operational model, with a structure based on the sequential steps taken to conduct a research project. The second summarizes findings about the interactions between schools, local knowledge, and scientists in what was defined as 'culturally relevant project-based learning', and presents insights that focus on constraints and opportunities in the implementation of these projects.

Project/Place-Based Operational Model: Fostering Local and Scientific Knowledge

Following an extensive literature review of the international experience and knowledge gained during the implementation of school projects, the authors proposed a model for the implementation of relevant place-based research projects, as shown in Figure 2. The model is divided into four operational steps. The spiral path represents the divergence from the first stages of a project, tackling big sustainability challenges at the beginning, converging towards specific curriculum-related activities, to finally achieve one practical local solution. Starting from global problems towards local specific tasks.

Figure 2. Project-based learning operational model. Source: Authors

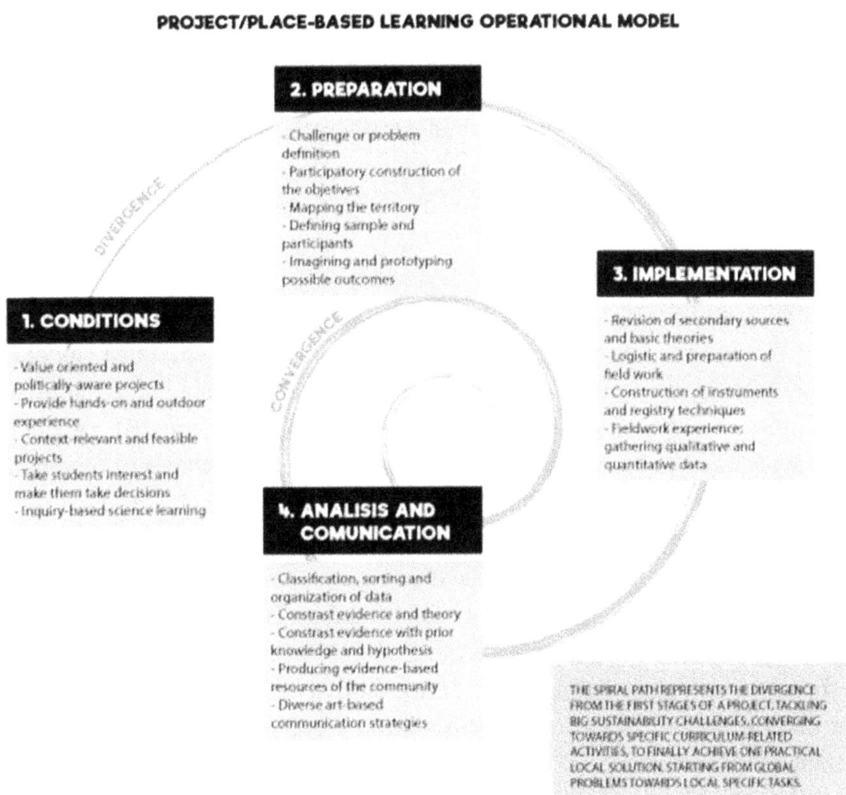

The four steps for the model were defined following project implementation. Each step is briefly defined and, from the teachers' perspective, areas where more support was needed and where problems were encountered during implementation, are described. This will inform further research and provide support mechanisms for other teachers interested in PBL

Conditions for Implementation

The conditions for project implementation were defined at the beginning of the chapter, based on an extensive literature review. These conditions, essential to starting and developing PBL, also clarify the context in which projects were to be developed. None of the schools had optimal conditions for the implementation of a PBL methodology. This, therefore, was the primary challenge. It was important to work with teachers to develop an awareness of different local challenges, and to recognize the strengths and weaknesses of their territories.

In the construction of context-relevant and feasible projects, students and teachers had to compromise and participate. Teachers acknowledged local knowledge, thus connecting themselves with local problems. Children were encouraged to be aware of and to contribute to their communities, and to develop value-oriented and politically-aware projects in collaboration with their teachers.

According to the teachers, arousing the students' interest and encouraging them to make decisions was challenging. It proved difficult to involve all children in project ideation, in defining the problem, and in thinking about possible solutions. In addition, it was important that each schools' Institutional Educational Project was connected to inquiry-based science learning, or at least recognize and support those school actors (e.g., teachers, principal, administrative team) interested and motivated in developing this learning approach.

Preparation

This step is essential to building a consistent and contextualized project. Students and teachers began to familiarize themselves with the project, and plan its stages. Teachers stated that students had difficulty raising their voice, as many were experiencing research for the first time and unfamiliar with the work performed by scientists or universities. At the same time, teachers found the PBL methodology challenging and declared a need for support in the preparation process. Even though problem definition arose from activities with children based on what they wanted to understand or research in their territories, teachers found the participatory construction of objectives difficult, resulting in few changes from the initial research objectives constructed during the training program. Teachers stated that the articulation of work between peers would help in the preparation of materials to support the definition of problems and objectives.

At first, defining the sample and participants proved difficult, and the importance of mapping techniques of prior knowledge and as sources of information was highlighted. Teachers and students were familiar with their territories and local knowledge, but found it difficult to convert this into a research project with a defined problem and research objectives.

Determining possible outcomes was facilitated through the collaboration of scientists, teachers, and students, with each contributing from their own understanding and perspective. Although innovative outcomes can result from such interaction, the focus was placed on developing scientific inquiry skills, encouraging thinking about outcomes communication proved difficult.

Implementation

Implementation was one of the more successful parts of the projects. Teachers discovered learning opportunities outside the classroom, in the territory, while children expressed a strong interest in outdoor experiences, especially those in which they interacted with nature and the community at the same time. This success was aided by part of the training program involving collaborative planning of fieldwork activities by teachers and scientists. In addition, the close relationship that some schools had with communities made the preparation of fieldwork and outdoor experiences easier.

The weakest part of the process was the registry and instruments used to gather information. Due to a lack of scientific inquiry skills, teachers and children faced difficulties not only in the choice of method, but also in the construction of the method, which had to fit with an analysis technique. In the end, some very interesting abstract and artistic registry and creative post-field work writing processes emerged that were later presented to a wider audience. An additional difficulty was the children's different levels of involvement in and contribution to the research.

Analysis and Communication

One of the weaknesses of this step was the lack of time to adequately prepare the communication product. As already mentioned, this may have been because planning and outcomes can be difficult and demanding stages when the process is not fully understood. Teachers stated that they put a great deal of effort into preparing the presentation and showcase, but there was no collaboration between different teachers, parents, students, or institutions.

More work was needed to contrast theory and evidence, or to contrast pre-field work knowledge and post-field work evidence. To achieve this, deeper means of verification were required, and more and better data were needed for analysis. In particular, systematic methods of analysis and registration were often absent, thus complicating the analysis. Moreover, greater clarity was required on the part of the teachers regarding analysis and registration methods, and how to conduct these, in order to design effective analysis and registration learning strategies with children. This is something that develops with research experience and, for majority of both teacher and student participants, this was their first experience of implementing PBL with scientists.

Various communication initiatives emerged from the schools. Children were highly innovative in the strategies they chose to represent their results. For example, there were musical and artistic performances, a mobile nursery suitcase, leaf-shaped keychains, and books recipes, demonstrating a wide variety of ways to disseminate results. These examples show that there is a desire to innovate integrative and comprehensive ways to present research results involving artistic expression.

Constraints and Opportunities for the Implementation of "Culturally Relevant Project-based Learning"

A Balance Between the Curricular Anxious Teacher and the Irreverent 'Green Sheep'.

A primary constraint at the beginning of the teacher training workshops was the attempt by a group of educators to understand how they could combine this project with national curriculum content. For more traditionally-oriented teachers, starting from the problem rather than from the curriculum was problematic. Once teachers and students had defined the problem and the guiding questions, the next step was planning the research fieldwork and analysis. It was surprising to see how impressed teachers and educators were by the wide variety of curricular objectives that could be covered in the implementation of different projects.

Another group of teachers, the so-called 'green sheep', were very keen on environmental education and taking the children outside of the school. However, these teachers lacked science-based research experience and sometimes had little interest in the official curriculum. With these teachers, the objectives of the research projects became fuzzy or were changed along the way, and no clear connection was made between the project and other school activities. This behavior was risky with respect to the children's post-project understanding, what they perceived to be important or serious, or how curricular objectives could be achieved.

These finding are in line with other research. Early childhood educators report a number of challenges. For example, Borko and Putnam (1996) found that teachers were opposed to project-based teaching methods, leading to inertia regarding the replacement of traditional teaching approaches. Even those teachers who express an interest in adopting project-based approaches have reported concerns about the lack of specific guidelines, such as a teacher's manual or an objectives-driven lesson plan (Katz & Chard, 2000; Clark, 2006). The lack of clarity regarding, how to guide children's research, how to plan a research project, and how to evaluate it, have proven to be entrance restrictions in adopting the project-based approach, particularly among novice teachers (Clark, 2006).

The Development of Guided Agency for Sustainability and Action for Climate Among Both Students and Teachers

Developing agency to meet the socio-ecological crisis at the same time as children acquire diverse scientific skills is critical in the current climate crisis. This provides children with the skills not only to tackle environmental problems, but also to learn how to deal with the anxiety generated by an uncertain future and a socio-environmental tragedy that can only be overcome if we work together. From a teacher's perspective, taking action on local issues helps to overcome feelings of inadequacy in the face of much bigger global problems. Children understand that their small actions can contribute to tackling bigger problems, helping them to overcome passive and sometimes desolate attitudes toward complex sustainability issues.

However, the development of this sense of agency was not limited to children. Teachers revealed that the project went beyond their expectations regarding the impact they could have on their communities, both inside and outside of schools. Observing how teams were strengthened, how parents became involved, and key community members became interested in school activities, empowered teachers in their capacity to lead meaningful learning processes and take positive action in the local community.

Successful Bidirectional Exchange Between Schools and Scientists

Teachers valued their own and their students' acquisition of empirical social and natural science skills, and the opportunities they were given to better understand the work of scientists, causing them to reassess their perception of science as a complex black box. The scientists involved reflected on this as a meaningful opportunity to transmit knowledge and transfer skills, and to learn how to communicate science in simpler and more accessible ways. They developed a deeper understanding of how to address and interact with people in diverse contexts, and made the effort to land ideas that were comprehensible to children. This exercise is not as simple as it sounds, as it can take scientists out of their comfort zone, giving them also the opportunity to critically evaluate their work and its social impact.

The transition from naïve perceptions of what a research project entails, to more complex understanding was a positive result recognized by the participating teachers. Most of those interviewed reported a change in their understanding of scientific research from three perspectives: 1) the range of possible scientific projects; 2) the importance of the complete step-by-step process of research; and 3) the role of the students' participation in the research. Teachers' initial views of PBL were centered on the field experience, as they had neither a variety of techniques and research methods at their disposal, nor any previous experience of interdisciplinary research. The working sessions with researchers helped the teachers to develop new ideas, building on the complexity of research and gaining familiarity with the scientific process.

CONCLUSION

In conclusion, this chapter offers teachers, educators, academics, and organizations interested in environmentally and culturally relevant education, an opportunity to understand the concept of ESD both from the experience of these Chilean program and by means of the acquisition of a conceptual ESD framework and through the international experience. We hope this frame will be useful in validating everyday action against climate change in educational institutions, when addressing educational communities, families, and authorities.

This is a collective and multi-sector responsibility. Teachers and educational institutions must take responsibility. But a collective effort from public and private sectors is also called for to build a society that is better prepared to confront a very complex future scenario. The efforts of schools and kindergartens alone will not be sufficient; additional external support is essential to develop more up-to-date educational opportunities. This chapter provided an example of an interesting way for academia and schools to connect and collaborate. This example was based on the experience of connecting schools with an interdisciplinary research center; however, many other alliances are possible.

The importance of starting in early years education must be emphasized, creating early connections with sustainability and climate challenges, having discussions about moral dilemmas, conversations about restoration, conservation, and care for the commons, all accompanied by research-based action to foster curiosity and agency. However, certain institutional and professional capacities are needed to ensure better research projects. These include changing teachers' perspectives of PBL before and after the projects, reinforcing the importance of an inquiry-based approach to understanding scientific reasoning, and building better-oriented projects.

Some questions remain that will guide the direction of further research. Are students with place-based research experience more capable of recognizing and acting in response to complex local socio-ecological issues? How can we foster scientific-school collaboration on a greater scale? Can transdisciplinary collaboration be fostered digitally? How do we make space for a dialogue about epistemological beliefs when working in transdisciplinary and intercultural scenarios? Given that these projects were conducted in rural areas and places with rich natural environments, we must consider how these kinds of experiences can be transferred to urban scenarios. The present research sought to present an operational framework and deepen attitudinal-sustainability-related dynamics for both teachers and students. But what happen to scientific skills and other cognitive competencies throughout this process?

The integrated fashion of the frame and experiences presented in this chapter aim to contribute, from the very beginning of formal education, to the promotion of actions to enhance the resilience of socio-ecological systems to climate change and toward sustainability. Multi-actor and transdisciplinary learning opportunities give children the opportunity to both know and experience reality, and give teachers and schools the opportunity to be agents for change in the face of this important and irrevocable challenge.

REFERENCES

Bascopé, M., & Caniguan, N. I. (2016). Propuesta pedagógica para la incorporación de conocimientos tradicionales de Ciencias Naturales en Primaria. *Revista Electrónica de Investigación Educativa*, *18*(3), 162–175. https://redie.uabc.mx/redie/article/view/1143

Bascopé, M., Perasso, P., & Reiss, K. (2019). Systematic review of education for sustainable development at an early stage: Cornerstones and pedagogical approaches for teacher professional development. *Sustainability*, *11*(3), 719. doi:10.3390u11030719

Caiman, C., & Lundegård, I. (2014). Pre-school children's agency in learning for sustainable development. *Environmental Education Research*, *20*(4), 437–459. doi:10.1080/13504622.2013.812722

Campbell, C., Speldewinde, C., Howitt, C., & MacDonald, A. (2018). STEM practice in the early years. *Creative Education*, *09*(1), 11–25. doi:10.4236/ce.2018.91002

Clark, A.-M. (2006). Changing classroom practice to include the project approach. *Early Childhood Research & Practice*, *8*(2), n2.

Davis, J. M. (2015). *What is early childhood education for sustainability and why does it matter?* (J. M. Davis, Ed.). Cambridge University Press.

Duhn, I. (2012). Making "place" for ecological sustainability in early childhood education. *Environmental Education Research*, *18*(1), 19–29. doi:10.1080/13504622.2011.572162

Eernstman, N., & Wals, A. E. (2013). Locative meaning-making: An arts-based approach to learning for sustainable development. *Sustainability*, *5*(4), 1645–1660. doi:10.3390u5041645

Elliott, S., & Young, T. (2016). Nature by default in early childhood education for sustainability. *Australian Journal of Environmental Education*, *32*(1), 57–64. doi:10.1017/aee.2015.44

Gray, T., & Birrell, C. (2015). 'Touched by the Earth': A place-based outdoor learning program incorporating the Arts. *Journal of Adventure Education and Outdoor Learning*, *15*(4), 330–349. doi:10.1080/14729679.2015.1035293

Green, C. J. (2015). Toward Young children as active researchers: A critical review of the methodologies and methods in early childhood environmental education. *The Journal of Environmental Education*, *46*(4), 207–229. doi:10.1080/00958964.2015.1050345

Hägglund, S., & Samuelsson, I. P. (2009). Early childhood education and learning for sustainable development and citizenship. *International Journal of Early Childhood*, *41*(2), 49–63. doi:10.1007/BF03168878

Katz, L. G. (2010). STEM in the early years. *Early Childhood Research & Practice*, *12*(2), 11–19.

Keskitalo, P. (2019). Place and space in Sámi education. *Policy Futures in Education*, *17*(4), 560–574. doi:10.1177/1478210319848530

Koslowski, B. (1996). *Theory and evidence: The development of scientific reasoning*. MIT Press.

Kuhn, D., Amsel, E., O'Loughlin, M., Schauble, L., Leadbeater, B., & Yotive, W. (1988). *The development of scientific thinking skills*. Academic Press.

Kuhn, D., & Ho, V. (1980). Self-directed activity and cognitive development. *Journal of Applied Developmental Psychology*, *1*(2), 119–133. doi:10.1016/0193-3973(80)90003-9

Liefländer, A. K., Fröhlich, G., Bogner, F. X., & Schultz, P. W. (2013). Promoting connectedness with nature through environmental education. *Environmental Education Research*, *19*(3), 370–384. doi:10.1080/13504622.2012.697545

Mitchell, S., Foulger, T. S., Wetzel, K., & Rathkey, C. (2009). The negotiated project approach: Project-based learning without leaving the standards behind. *Early Childhood Education Journal*, *36*(4), 339–346. doi:10.100710643-008-0295-7

Murtaza, T., & Mahmood, M. (2018). Active learning through project based learning approach in English language lessons for early age groups. *Journal of Early Childhood Care and Education. Islamabad, 2*. Retrieved from https://search.proquest.com/docview/2362277270/abstract/4F2C79042A944842PQ/1

Nigh, R., & Bertely, M. (2018). Conocimiento y educación indígena en Chiapas, México: Un método intercultural. *Diálogos Sobre Educación*, *9*(16). Retrieved from http://www.scielo.org.mx/scielo.php?script=sci_abstract&pid=S2007-21712018000100003&lng=es&nrm=iso&tlng=es

O'Donnell, C., & Bascopé, M. (2019). *Using science to do social good: STEM education for sustainable development.* Position paper developed in preparation for the second "International Dialogue on STEM Education" (IDoS) in Berlin. Retrieved from www.haus-der-kleinen-forscher.de

Sawitri, D. R. (2017). Education for sustainable development: How early is too early? *Advanced Science Letters*, *23*(3), 2559–2560. doi:10.1166/asl.2017.8699

Sodian, B., Zaitchik, D., & Carey, S. (1991). Young children's differentiation of hypothetical beliefs from evidence. *Child Development*, *62*(4), 753–766. doi:10.2307/1131175

Sykes, H. (2008). Editorial: Narratives in aboriginal, history and place-based education. *Curriculum Inquiry*, *38*(5), 541–544. http://www.jstor.org/stable/25475937

Tilbury, D., & Wortman, D. (2008). How is community education contributing to sustainability in practice? *Applied Environmental Education and Communication*, *7*(3), 83–93. doi:10.1080/15330150802502171

Vecchi, V. (2010). *Art and creativity in Reggio Emilia: Exploring the role and potential of ateliers in early childhood education*. Routledge.

Waite, C. (2020). Spotlight on Project-Based Learning: Seeing the forest and the trees. *Childhood Education*, *96*(2), 30–41. doi:10.1080/00094056.2020.1733860

Wals, A. E. J. (2017). Sustainability by default: Co-creating care and relationality through early childhood education. *International Journal of Early Childhood*, *49*(2), 155–164. doi:10.100713158-017-0193-5

Zachariou, A., & Symeou, L. (2009). The local community as a means for promoting education for sustainable development. *Applied Environmental Education and Communication, 7*(4), 129–143. doi:10.1080/15330150902744152

KEY TERMS AND DEFINITIONS

Culturally Relevant Learning: Based on the framework proposed by Ladson-Billing (1995), it refers to learning opportunities addressing academic success, cultural competencies, and sociopolitical awareness through an inquiry-based learning approach. In the context of education for sustainability, it puts an emphasis in making environmental and STEM education approaches culturally aware.

Mapuche People: Group of indigenous inhabitants of present-day south-central Chile and southwestern Argentina. The collective term refers to a wide-ranging ethnicity composed of various groups who shared a common linguistic heritage as Mapudungun speakers. They are the largest ethnic group in Chile.

Place-Based Education: The notion of "place" refers to not only the natural but also the social surroundings of the school or educational institution. Place-based education, put the schools' context in the center of the educational process, giving voice and importance to the excluded, revindicating silent actors, narratives and knowledges. It intends to foster students' sense of agency and to give a concrete propose to the learning experience.

Project-Based Learning: In the context of this chapter it refers to research projects implemented by kids and guided by teachers and community partners, to tackle local challenges in the school community or surroundings. It has a direct link with Education for Sustainability and climate change education, since it is focus on children's action towards complex socioecological issues. It has an interdisciplinary focus, considering diverse knowledge, tools and skills to affront local problems.

STEAM Education: In accordance with the classical STEM approach (Science, Technology, Engineering and Mathematics), this acronym incorporates the A for "Arts". It corresponds to educational approaches with a holistic approach, considering the applied science STEM approach, (problem or project oriented), and incorporating artistic expressions to give meaningful educational experiences.

ENDNOTES

1 More information about educational strategies and innovations can be found in the web site of the Latin American STEM network https://experimento.lat.siemens-stiftung.org/

2 The paper is available in https://www.haus-der-kleinen-forscher.de/en/international-dialogue-on-stem-education/idos2019/position-paper

3 For more information look at www.oce.global.

4 All the guides can be found in www.eputrokinkimun.org

Chapter 3
Investigating Turkish Pre-Service Primary Teachers' Environmental Moral Reasoning Patterns on Environmental Dilemmas

Umran Betul Cebesoy
https://orcid.org/0000-0001-7753-1203
Usak University, Turkey

ABSTRACT

This chapter explored preservice primary teachers' moral reasoning patterns on local and non-local environmental dilemmas. Forty-seven preservice primary teachers enrolling in an environmental education course voluntarily participated in the study. The data were collected via preservice teachers' written reports about local and non-local environmental dilemmas and analyzed by means of qualitative and quantitative methods. The results revealed that preservice teachers mostly focused on ecocentric and anthropocentric moral reasoning on sea pollution case while using ecocentric and non-environmental reasoning in the deforestation case. The t-test results also revealed that preservice teachers used more anthropocentric reasoning patterns in the sea pollution case when compared to the deforestation case. The inclusion of local and non-local dilemmas into the environmental education courses in undergraduate teacher education programs can facilitate preservice teachers' moral reasoning.

INTRODUCTION

Today we are living in a world where we are facing serious environmental problems. Human pressure on nature has accelerated the impact of climate change and its associated factors including desertification, deforestation, droughts, and water salinity (Karmaoui, 2019). Moreover, the demand for energy, food, water, and agricultural land has increased (Karmaoui, 2019; Minucci & Karmaoui, 2016). This high

DOI: 10.4018/978-1-7998-7512-3.ch003

demand also threatens climate change adaptation strategies and sustainability efforts (Chelleri, Minucci, Ruiz & Karmaoui, 2014). The concept of 'sustainability' has become part of the environmental education agenda since the 1990s (Tilbury, 1995). The need of providing a balance between economic development and environmental conservation created a necessity of including sustainability in the environmental education agenda (World Commission on Environment and Development, 1987). Another report 'the Rio Earth Summit's Agenda 21 calls for promoting sustainable development through education (UN Conference on Environment and Development [UNCED], 1992). Consequently, sustainable development has become a guiding concept in environmental education (Bonnet, 1999).

While there is no commonly accepted definition of environmental education, it can be defined as providing means to the individuals to gain knowledge about the environment, to get essential skills and attitudes for protecting and improving the environment (North American Association for Environmental Education [NAAEE], 2004). Thus, environmental education can be assumed as a response to the effort of achieving environmental protection and improvement (Loughland, Reid, Walker & Petocz, 2003). Moreover, there is an ongoing effort to raise environmentally literate citizens who are informed about environmental problems and ready to make informed decisions (NAAEE, 2004). If so, this assumes teachers are already well equipped to raise future citizens and today's students as environmentally literate by having the necessary understanding, skills, and attitudes towards the environment and environmental problems (Desjean-Perrotta, 2013). With this respect, higher education institutions that teachers are raised have gained importance. Higher education institutions are recognized as places where possible links between environmental awareness and learning opportunities for students are established (Tilbury, 2004). UNESCO also explicitly recognized the significant role of formal and higher education for providing means to establish sustainable development efforts. In this report, raising active and knowledgeable citizens who are informed in making decisions about complex, economic, social, and environmental issues is emphasized (UNESCO, 2002). On the other hand, making informed decisions require morality aspect (Sadler, 2004). Kortenkamp and Moore (2001) argue that environmental education exploring the relationship between humans and nature should include the morality aspect. Values and morality are considered to be important in environmental education (Tilbury, 1995). This is also consistent with Zeidler and Schafer's (1984) findings which reported that that college students provided higher levels of moral reasoning on environmental issues. With this respect, moral reasoning comes to the fore.

BACKGROUND

Environmental Education

The foundations of environmental education have been established in the Tbilisi Declaration (UNESCO-UNEP, 1977). In this report, the ultimate goal of environmental education has been defined as:

The ultimate aim of environmental education is to enable people to understand the complexities of the environment and the need for nations to adapt their activities and pursue their development in ways which are harmonious with the environment. (p.12)

Along with this goal, the term 'environmental literacy' has gained significant importance in environmental education. Disinger and Roth (1992) defined environmental literacy as the capacity to understand the structure of environmental systems while being able to take suitable actions to preserve, restore, and improve these systems. As a result, environmental literacy requires individuals' sufficient understanding of environmental issues, the processes and systems shaping the environment as well as taking personal responsibility and make informed decisions about environmental issues (NAAEE, 2004). The same guidelines also define the environmentally literate individuals as 'can live with nature and act equitably toward each other' (p. 2). While environmental literacy ensures to raise environmentally literate individuals who can deal with the environmental challenges, environmental education is a mean to raise environmentally literate individuals (Stevenson, Carrier & Peterson, 2014; Stevenson, Peterson, Bondell, Mertig & Moore, 2013; Roth, 1992; Hollweg et al. 2011). To active this, knowledge, skills, affect, and behavior which are assumed to be key elements of environmental literacy needs to be addressed and enhanced through education (Disinger & Roth, 1992). Supporting this, Cutter and Smith (2001) indicated that environmental education has a crucial role at primary, secondary, and tertiary levels to raise environmentally literate students. This is, in fact, in line with the Tbilisi Declaration call for the integration of environmental education to the whole stages of formal education (UNESCO-UNEP, 1977). This call includes pre-service teacher education programs.

Pre-service Teachers

Existing studies reported that teachers' perceptions and their knowledge about the environment affect their teaching practices (Alvarez-Garcia et al. 2015; Chang & So, 2015; Rickinson, 2001; Stevenson et al. 2014; Sund & Wickman, 2008). However, studies also indicated that preservice teachers had limited understanding of many environmental issues (i.e. Amirshokoohi, 2010; Arslan et al. 2012; Desjean Perrotta, 2013; Teksoz, Sahin, & Ertepinar, 2010; Tuncer, Boone, Tuzun & Oztekin, 2014; Tuncer et al. 2009). For instance, Arslan et al. (2012) investigated preservice teachers' misconceptions about global warming, the greenhouse effect, ozone layer depletion, and acid rain by using a two-tier diagnostic test. The results revealed that preservice teachers had a limited understanding of the atmosphere related to environmental problems. In another study, exploring preservice teachers' drawings about the environment, Desjean-Perrotta (2013) revealed that most of the preservice teachers showed limited understanding of the components (living, human, biotic, and built) in the environment. They did not include humans in their drawings indicating that they did not acknowledge the effect of humans on the environment. Lack of fundamental knowledge in environmental issues as stated above can be resulted in teachers feeling incompetent to teach environmental issues in their classes (Arslan et al. 2012). In line with this finding, teachers are found to be not well prepared to teach and lead the discussion about environmental dilemmas in their classes as Morin and colleagues (2013) indicated. Moreover, many preservice teacher education programs provide insufficient training for environmental education during undergraduate education (Crim et al. 2017; Desjean-Perrotta, 2013). Supporting this, examining teachers' and school administrators' opinions, Yates and colleagues (2019) revealed similar results. Yates et al. (2019) found out that less than half of the participants had taken environmental education courses during their undergraduate education while 40% indicated that undergraduate programs did not provide sufficient preparation for teaching environmental education courses. They indicated that environmental education should be part of preservice teacher education programs. Supporting this finding, research studies revealed that pre-service teachers who had training about environmental education during the undergraduate education showed a willingness to teach environmental issues in their classes (Derman, Sahin & Hacieminoglu, 2016; Groves & Pugh, 1999; Haight, 2015; Morin, Tytler, Barraza, Simonneaux & Simonneaux, 2013;

Teksoz et al. 2010). Similarly, this kind of training has positive effects on preservice teachers' perceptions of environmental uses and self-efficacy beliefs (Saribas, Doganca Kucuk & Ertepinar, 2017). Teachers' teaching of environmental issues also has positive effects on their students (Shepardson, Wee, Priddy & Harbor, 2005). These positive effects can be listed: Environmental education courses can enhance students' reasoning (Morin et al. 2013), argumentation skills (Dawson & Carson, 2017), environmental literacy (Stevenson et al. 2013) and socioscientific reasoning and environmental literacy (Kinslow, Sadler & Nyugen, 2019). These environmental education courses should be carefully designed (Kinslow et al. 2019). Guidelines for the Preparation and Professional Development of Environmental Educators (2010) also calls for the inclusion of environmental education into the preservice teacher education programs. Specifically, this study focused on preservice primary teachers for two reasons: First, classroom/primary teachers are usually the first environmental educators for primary students (Crim et al. 2017). Second, primary schools are often assumed to be the places where environmentally literate and active citizens are raised (Cutter & Smith, 2001).

Moral Reasoning

Kortemkamp and Moore (2001) stressed the importance of including morality in environmental issues. If an individual perceives a situation as morally relevant, then, she/he can hold a position showing moral sensitivity towards the situation and can discuss the moral concerns (Sadler, 2004). Thus, recognition of moral principles is important in environmental dilemmas.

While investigating the morality of environmental dilemmas, distinct levels of environmental attitudes were classified as egocentric, anthropocentric, and ecocentric (Merchant, 1992). A similar classification was used biocentric (in the exchange of ecocentric), social-altruistic (in the exchange for anthropocentric), and egocentric (Schultz & Zelezny, 1999; Stern & Dietz, 1994). While egoistic environmental attitudes were based on perceived environmental concerns of individuals that directly affect them individually, social-altruistic environmental attitudes mainly focus on the protection of the environment for the sake of human beings. In other words, humans are in the center of everything in social-altruistic environmental attitudes. Lastly, biocentric environmental attitudes focus on the environment itself for the sake of the environment and the species living in the environment (Kopnina, Washington, Taylor, & Piccolo, 2018; Stern & Dietz, 1994). Based on the tripartite classification of environmental attitudes, Kortemkamp and Moore (2001) prosed another tripartite classification for moral reasoning as ecocentric, anthropocentric, and non-environmental. While individuals who have ecocentric moral reasoning patterns value the environment intrinsically, the individuals with anthropocentric values are concerned with the environment as the destruction of the environment mainly influences human beings. Individuals with non-environmental reasoning patterns refer the social contracts and norms.

This tripartite classification is used in numerous studies (i.e. Karpiak & Baril, 2008; Kortemkamp & Morre, 2011; Tuncay et al. 2012). For instance, Tuncay et al. (2012) investigated Turkish preservice science teachers' environmental moral reasoning patterns in local and non-local environmental dilemmas. The findings revealed that the participants mostly used ecocentric and anthropocentric moral reasoning patterns and their reasoning patterns were not related to the issues being local or non-local. Exploring college students' moral reasoning and environmental concern, Karpiak and Baril (2008) found that moral reasoning was positively correlated with ecocentrism while being negatively correlated with environmental apathy and uncorrelated with anthropocentrism. In another study, Kortenkamp and Moore

(2001) revealed that pro-environmental attitudes were more related to ecocentric and anthropocentric moral reasoning but were less related to non-environmental reasoning.

The context of the moral problem can influence individuals' moral reasoning. That is the individuals can display a different level of moral reasoning (Zeidler & Schafer, 1984). Thus, the context is considered as important for framing and enhancing the reasoning quality (Morin et al. 2013). Here, we chose to explore the locality of environmental issues as a context. Environmental problems can be labeled as local which happens in individuals' own country and as non-local which happens in any county outside that the individual lives in (Tuncay et al. 2012). Goksen, Adaman, and Zenginoguz (2002) found out that participants' environmental concerns can be differentiated based on geographical proximity. They defined two types of environmental concerns as local and global. Their study revealed that participants' local environmental concern is significantly higher than their global environmental concern. In another study conducted by Unal (2008) revealed that preservice teachers were found to be more concerned about the global environmental problem when compared to local ones. Based on the context, individuals' moral concerns can change. Thus, Sadler (2004) proposed to investigate what kind of moral concerns that the individuals have. Based on this, this study adopted two local and non-local environmental dilemmas on sea pollution and deforestation to investigate preservice primary teachers' environmental moral reasoning patterns.

The Purpose and the Research Questions

Teacher preparation for effective environmental education has been identified as one of the overarching goals by NAAEE (2004). Supporting this, existing studies reported that teachers' perceptions and their knowledge about the environment influence their teaching practices (Alvarez-Garcia, Sureda-Negre, & Comas-Forgas, 2015; Rickinson, 2001; Sund & Wickman, 2008). However, relevant studies also suggest that pre-service teachers lack sufficient training during their undergraduate education (Arslan, Cigdemoglu & Moseley, 2012; Crim, Moseley & Desjean-Perrotta, 2017; Desjean-Perrotta, 2013; Yates et al. 2019). In line with these findings, Alvarez-Garcia and colleagues (2015) revealed the same pattern in their meta-analysis study which included 22 journal articles and two doctoral theses. Those results support the limited space for including environmental education courses in preservice teacher education programs.

Morality, on the other hand, is assumed to be part of the relationship between humans and nature (Kortenkamp & Moore, 2011). As a result, many studies explored moral considerations (Sternäng & Lundholm, 2011), moral reasoning (Karpiak & Baril, 2008; Tuncay, Yılmaz Tuzun & Tuncer Teksoz, 2012; Zeidler & Schafer, 1984), and moral sensitivity (Sadler, 2004) in environmental issues. Exploring preservice teachers' moral reasoning patterns are important for two reasons: first of all, if it is aimed to raise environmentally literate individuals, then, teachers themselves need to be environmentally literate and need to be aware of their concerns and actions while making decisions in environmental problems. Second, if teacher education programs realize the pre-service teachers' moral reasoning patterns, then, more effective undergraduate courses could be designed and implemented. Thus, first, there is an urgent need to explore future teachers' moral reasoning patterns when dealing with environmental issues. This study responds to this urgent need by exploring the moral reasoning patterns of pre-service primary teachers' moral reasoning patterns and the factors which influence their reasoning.

In this study, pre-service primary teachers' environmental moral reasoning patterns were investigated throughout using two local and non-local environmental dilemmas presented in the context of the environmental education course. The research questions guided this study are:

1. How preservice primary teachers' decisions are influenced by local and non-local environmental dilemmas?
2. What factors influence preservice primary teachers' decisions in local and non-local environmental dilemmas?
3. What are the moral reasoning patterns of preservice primary teachers toward local and non-local environmental dilemmas?
4. Is there any significant difference between preservice primary teachers' use of ecocentric, anthropocentric, and non-environmental moral reasoning patterns in different cases (sea pollution and deforestation)?

METHOD

In this study, a mixed-method approach, specifically, exploratory sequential design, was used (Figure 1). This design is characterized by an initial phase of qualitative data collection and followed by quantitative data analysis (Creswell & Plano-Clark, 2011).

Figure 1. The exploratory sequential design

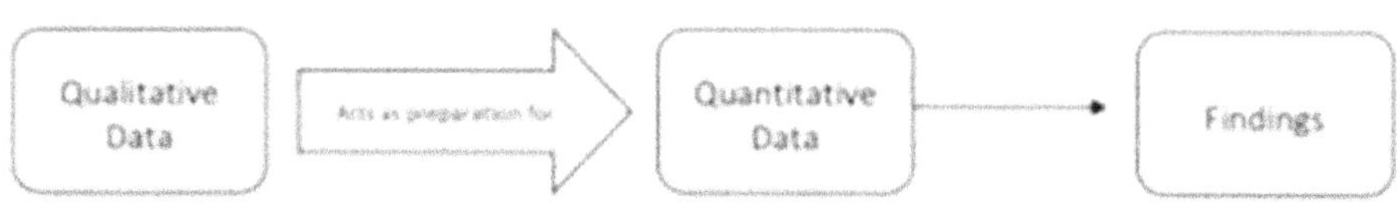

In the present study, qualitative data based on pre-service primary teachers' written responses were collected, and then, the qualitative data transformed into quantitative form. We adopted a quantizing approach which includes transforming qualitative data into numerical codes (Miles & Huberman, 1994; Teddlie & Tashakkori, 2006; Tashakkori & Teddlie, 1998). This helps the researcher to quantify the patterns in pre-service primary teachers' moral reasoning patterns for making comparisons between different cases. In this study, first, qualitative approach to code pre-service primary teachers' written reports in two cases based on the tripartite classification of Kortemkamp and Moore (2001) was used as an analytical framework for qualitative analysis. Then, qualitative methods were used to further elaborate and exemplify the patterns in preservice primary teachers' moral reasoning. The analytical framework used in to code preservice primary teachers' moral reasoning patterns in their written responses were presented along with appropriate examples in Table 1:

Participants

47 second-grade (sophomore) students on a four-year degree primary teacher education program of a public university located in the Western part of Turkey participated in the study. The study was conducted within an Environmental Education course. This course is a compulsory course offered in the 4th semester in the primary teacher education program.

Table 1. The analytical framework used in determining participants' moral reasoning in environmental dilemmas (based on Kortemkamp & Moore's study (2001))

Moral reasoning pattern	Definition	Example
Ecocentric	PTs referred to the rights or intrinsic value of nature or proposed protecting nature for nature's sake.	'The sea accident that happened in Turkish territorial waters affected me more. Because the health of living creatures is in danger and this accident could lead to the extinction of these animals.' (PT-41, sea pollution case)
Anthropocentric	PTs referred to preserving nature for the sake of human beings as humans could not survive without nature in their responses.	'In the short term, only forests and animals appear to be affected by this event [referred to the deforestation of rain forests as a non-local case], but in the long run, people will be affected more. As a result of forest destruction, carbon dioxide emissions will increase and the greenhouse effect will be felt more. Consequently, new diseases that are fatal to humans will spawn.' (PT-15, deforestation case)
Non-environmental	PTs referred to social contracts, truthfulness, legal regulations, or laws in their responses while dealing with the dilemmas.	'Such accidents [referred to the local and non-local accidents in the seas] have occurred as a result of insufficient supervision, not applying the applicable regulations properly, and the people who are in charge of these issues not performing their duties well.' (PT-33, sea pollution case)

The Context of the Study and Data Collection

This study was conducted in the Environmental Education course, which was offered in the primary teacher education program. The first part of the course was an introductory environmental education course including the core concepts of the environment: The ecosystems, the biotic and abiotic factors forming ecosystems, and the interactions in the ecosystems. Then the environmental problems such as water, air, soil, noise, light, and radioactive pollutions threatening the environment were introduced along with climate change and global warming issues which is the biggest threat to the world that humans live on. Lastly, sustainable development, education for sustainable development, and the importance of environmental awareness and responsibility have been introduced. The second part of the course included introducing environmental dilemmas each week. The scenarios were about the local and non-local real-world problems about oil spills in the United States and Turkey (first case) and deforestation in Indonesia and Turkey (second case) were sent a week before the course. There were several questions along with the scenarios and the pre-service teachers were asked to prepare a written report about the scenarios.

While the scenarios were prepared by reviewing the environmental disasters around the world and Turkey, the content and the questions used in the scenarios were prepared by reviewing previous studies (Sadler & Zeidler, 2004; Tuncay et al. 2012). Each scenario was discussed as a whole group discussion for two weeks (a total of two hours each week). These scenarios were pilot tested in the previous semester (2015 spring semester). The data collection was based on the second part of the course. A total of 94 written reports (for two scenarios) were collected from 47 participants. The cases used in the study were presented below:

Case 1: Sea Pollution

Pollution of waterways, seas, and oceans is one of the most important environmental problems (Cutter & Smith, 2001). It is also subject to international reports of UNESCO-UNEP (1997). Kobayashi (1991) calls for coastal seas as a resource for environmental education. Sea or river pollution is used as a case for investigating students' (Strong, 1998) or adults' environmental concerns (Goksen et al. 2002). For instance, Goksen et al. (2002) used sea pollution in Istanbul (a city in Turkey) as a local issue to investigate participants' environmental concerns. Similarly, we used sea pollution in Istanbul Bosporus by providing different sea accidents that happened in Istanbul Bosporus for the past 50 years. This case constituted a local issue for preservice teachers. In line with this case, a non-local issue (the Deepwater Horizon petroleum platform explosion in Mexican Gulf) was presented to participants. The information in both cases including the casualties, the environmental damage caused by the sea accidents and explosion in the Deepwater Horizon petroleum platform, the effects to the seas and oceans, the living creatures in the seas and oceans as well as economic and social damages to the nearby countries, cities, and districts. After providing this information, a series of questions asking which case was more influential, why the participants thought are asked and the preservice teachers handled written reports about their opinions.

Case 2: Deforestation

Like sea pollution, deforestation and degradation of forests are also considered as one of the important environmental problems by causing additional warming for the world's climate (Cutter & Smith, 2001; Feddema et al. 2005). Along with climate change, deforestation introduce problems like droughts and desertification (Karmaoui, 2019). These important problems are subject to international reports (UNEP, 1983; World Commission on the Environment Development, 1987). Thus, these issues are pointed out in numerous environmental education studies (i.e. Bahk, 2010; Cebesoy, 2019; Cebesoy & Karisan, 2020). For instance, Bahk (2010) used narrative films to investigate college students' views about deforestation. In another study conducted by Cebesoy (2019), a two-day climate change education workshop focusing on the aspects of deforestation, forests strengths, and weakness as well as forest policies was implemented. In a similar manner, the deforestation case was used as a local and non-local issue in this study. While deforestation in different districts of Turkey was used as a local issue, deforestation of rain forests and plantation of palm oil trees in Indonesia was used as a non-local case. Along with cases, additional information about causes and consequences in both local and non-local issues were presented. After providing this information, a series of questions asking which case was more influential, why the participants thought are asked and the preservice teachers handled written reports about their opinions.

Data Analyses

Each written report was analyzed by using the content analysis approach (Krippendorf, 2004) along with descriptive statistics (percentages and frequencies) in this study. Among content analysis approaches, directed content analysis (where existing theory and research are used to further explore the phenomena in detail) was used in the study (Hsieh & Shannon, 2005). The frequency of appearance for each subject area and aspect was calculated (Kortenkamp & Moore, 2001). To answer the first and third research questions of this study, each participants' written response was coded as 'ecocentric', 'anthropocentric', or 'non-environmental' for moral reasoning. The frequencies of each moral reasoning pattern were calculated. For the second research question, the author used a qualitative data analysis procedure to code the data. Specifically, the unit of analysis was used for coding preservice teachers' written explanations (Merriam, 1998). The unit of analysis can be described as the major entity that is being analyzed in the research. It could be individuals, groups, or participants' other artifacts such as books or photos (Trochim, 2020). In this study, participants' each written response represented a coding unit for analyses. The frequencies were calculated. For answering the last research question, a paired sample t-test was used.

While 'PT' indicated the pre-service primary teacher, a number is also given identifying s specific pre-service teacher (i.e. PT-3). The square bracelets ([]) were used to complete the meaning. Different scenarios were coded as C-1 and C-2.

The Trustworthiness of the Study

The validity and reliability issues were addressed by adopting several methods. First, research triangulation was ensured for preventing researcher misinterpretation. Another researcher who had experience in socioscientific issues and environmental dilemmas was invited to code the preservice teachers' written responses to ensure a consensus between researchers. While the author of this chapter coded half of the data, 10% of the random-selected data were coded by the second researcher. The inter-rater reliability (Miles & Huberman, 1994) which shows the consistency of agreement between different coders was calculated by using the below-stated formula:

Inter-rater reliability = number of agreements / the total number of agreements and disagreements

Inter-rater reliability was calculated as 90% showing that the data analysis was compatible and reliable. After that, the author coded the other half of the data to complete the analysis.

RESULTS

First, preservice primary teachers (PTs hereafter) were asked which case (local or non-local environmental dilemmas) was more influential in their decisions. PTs' responses in the Sea Pollution case were presented in Figure 2. While more than half of the participants indicated local dilemma about sea pollution was more influential on their decisions (52.19%), 38.30% of participants indicated the non-local issue presented was more influential. Lastly, 8.51% of participants (n=4) indicated both dilemmas influenced their decisions.

To answer the second research question, we explored how PTs' decisions were influenced by local and non-local environmental dilemmas. We examined how they explained why different cases were more influential and what affected them the most. In the sea pollution case, there were two sub-cases which were mainly related to sea pollution including one local and one non-local case (see Figure 3). The PTs who indicated that the local issue was more influential explained their reasons in two ways: While most of them (64%, n=16) indicated that the locality of the issue was more influential compared to the second scenario which focused on the sea pollution after the Deepwater Horizon accident in 2010. Sample excerpts were provided below:

I am much more affected by the disasters happening in my country. Because we need to start to be sensitive to our close environment. Then, we can start doing something for the world. (PT-2)

Actually, it is sad to read that there were casualties and oil spread to a large area in the first case [referring to the Deepwater Horizon sea accident case-non-local issue]. Still, I am much more affected by

Figure 2. Preservice teachers' decisions in Sea Pollution case

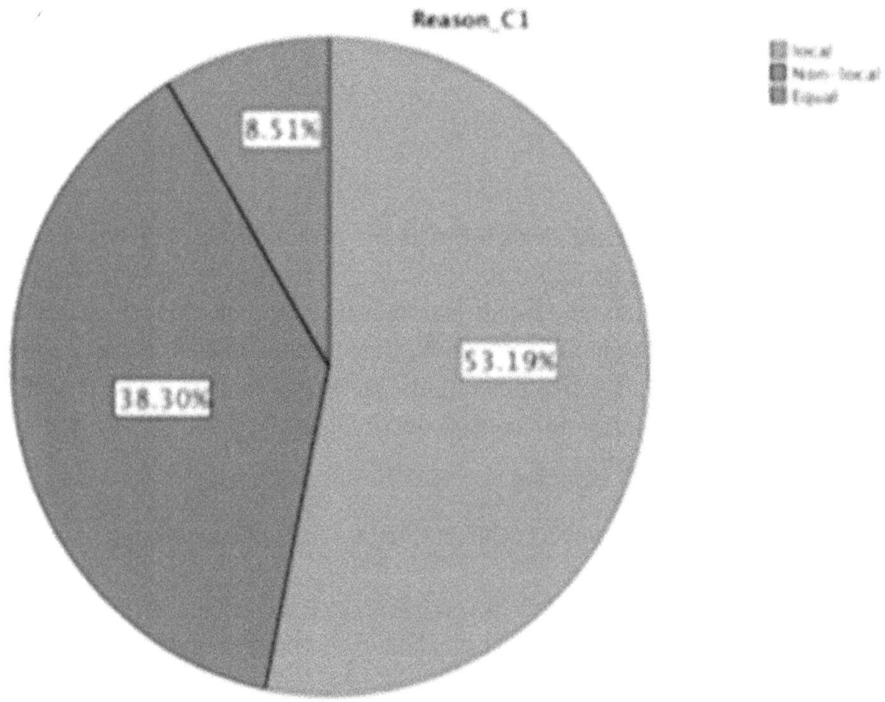

the second one which happened in my country as the consequences and damages were directly affect my country. (PT-19)

Among these two cases, the second case [referring to the sea accidents in Turkey territorial waters case-local issue] affected me the most for two reasons: First, all the accidents happened in my country and the second, the insurance companies which did not cover the damages make our country to be more vulnerable to sea accidents. (PT-37)

The second case happening in our territorial waters affected me the most. (PT-40)

As seen from the excerpts, PTs mainly expressed the locality of the scenario (i.e. happening in Turkish territorial areas). Besides these explanations, some PTs also explained the severity of environmental harm. Sample excerpts are provided below:

Among these cases, the events in the second case were more impressive. For instance, three sea accidents happened in the last six months according to the case. This causes marine pollution in Turkish territorial areas and affects all the living creatures in the sea. It also possesses a great risk for us. (PT-9)

The accidents happening in Turkey affected me more. The insurance companies did not take full responsibility and this caused huge environmental harm besides causing damage to our economy. (PT-25)

Figure 3. Pre-service teachers' explanations about their decisions in the Sea Pollution sub-case

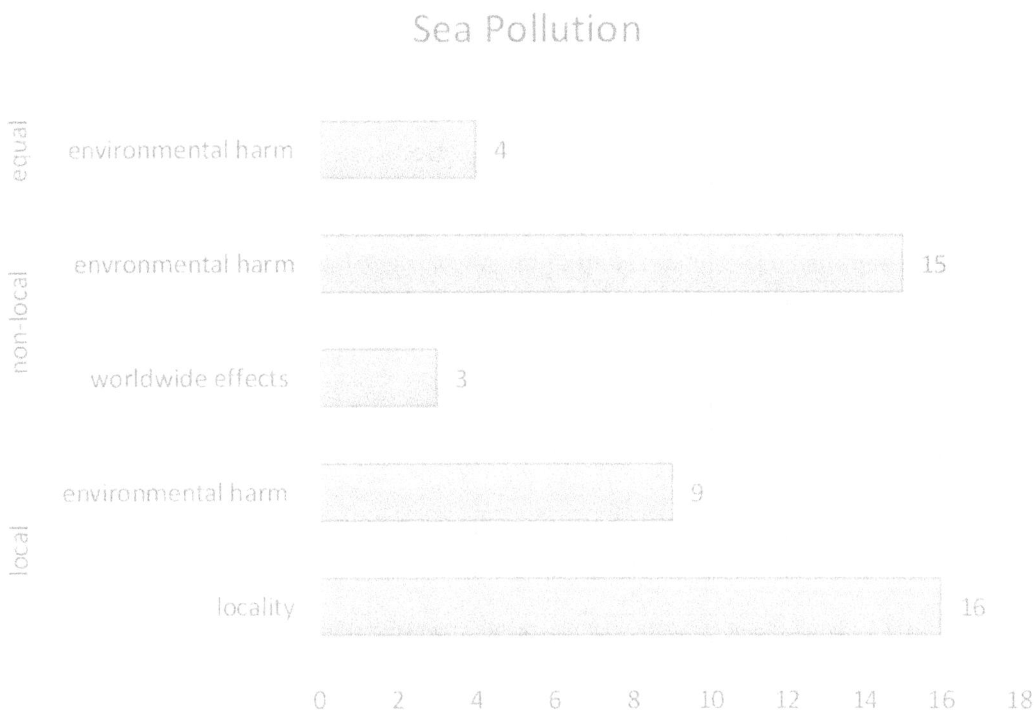

The accidents happening in Istanbul Bosporus affects me more. It just causes huge environmental harm to the living things within our borders. (PT-35)

These excerpts provided evidence that PTs also considered the environmental harms caused by sea accidents happening in Turkish territorial borders. Besides, 38.30% of the PTs indicated that the case focusing on the non-local issue was more influential. Then we explored why they thought so in a similar way. Most of the participants (61.11%, n=11) who indicated that the first case happening in Mexico gulf (non-local issue) emphasized the severity of the Deepwater Horizon accident and how it affected the environment in general. Some excerpts are presented below:

The first case [referring to the Deepwater Horizon accident in the Mexico gulf] affected me the most since there were lots of dead people and huge environmental damage after the accident. (PT-7)

The explosion on the Deepwater Horizon platform affected me the most. Because a fire occurred after the explosion which caused the loss of life. The fire caused air pollution. Moreover, the platform sank after the explosion which endangered the sea creatures. In general, the explosion caused a huge environmental disaster. (PT-15)

The first case was more impressive since the accident was bigger when compared to the accidents that happened in Turkey. It even caused a state of emergency in the United States. The fire lasted for 36 hours which caused the death of a few workers, too. Besides, tons of petroleum spread into the sea which is incredibly bad for the living things in the sea. (PT-44)

These experts showed how PTs feel about the Deepwater Horizon accident. Also, three preservice teachers indicated that the case presented has worldwide effects as:

The explosion on the Deepwater Horizon platform affected me the most. I believe that the second case [referring to the Deepwater Horizon accident in the Mexico gulf] was gloomier for its effects. There were tons of petroleum spread into the sea which will have worldwide effects as we are living in a global world. (PT-43)

The second case affected me more since it has a wider influence area. The consequences will be felt worldwide. (PT-24)

Lastly, four participants indicated that both cases (local and non-local) affected them in an equal manner. Sample excerpts were presented below:

Both cases affected me very much. Both the ideas of sea pollution and the living things in the sea would be harmed made me sad about the issues. (PT-3)

Both cases caused serious damage not only to the environment in which the accidents happened but also to the whole environment. The chemicals which spread to the sea would evaporate and would return to the ground as rain. The seas all around the world are connected and the amount of petroleum that spread into the sea in those accidents is huge. (PT-20)

Similarly, we explored how preservice primary teachers were affected by local and non-local environmental dilemmas regarding the deforestation case (see Figure 4). The majority of participants (76.60%) indicated that the non-local dilemma about the deforestation of rain forests and plantation of palm oil trees in Indonesia were more influential in their decisions. While 14.89% of PTs indicated that the local dilemma about deforestation in Turkey was more influential, 8.51% of PTs indicated both of the dilemmas were influential (n=4).

After determining which case (local or non-local) was more influential, we explored how PTs were influenced by the presented dilemmas. We examined how they explained why different cases were more influential and what affected them the most. In the deforestation case, there were two sub-cases which were mainly related to deforestation including one local and one non-local case (see Figure 5).

Unlike the sea pollution case, preservice primary teachers mainly indicated that the non-local environmental dilemma focusing on deforestation and palm tree plantations in Indonesia affected them more when compared to the local case focusing on the deforestation in Turkey. When we explored their reasons, we revealed that most of the PTs (58.33%, n=21) indicated that the effects on the forests, animals, and the people were severe. Sample excerpts are provided as:

Figure 4. Preservice teachers' decisions in the Deforestation case

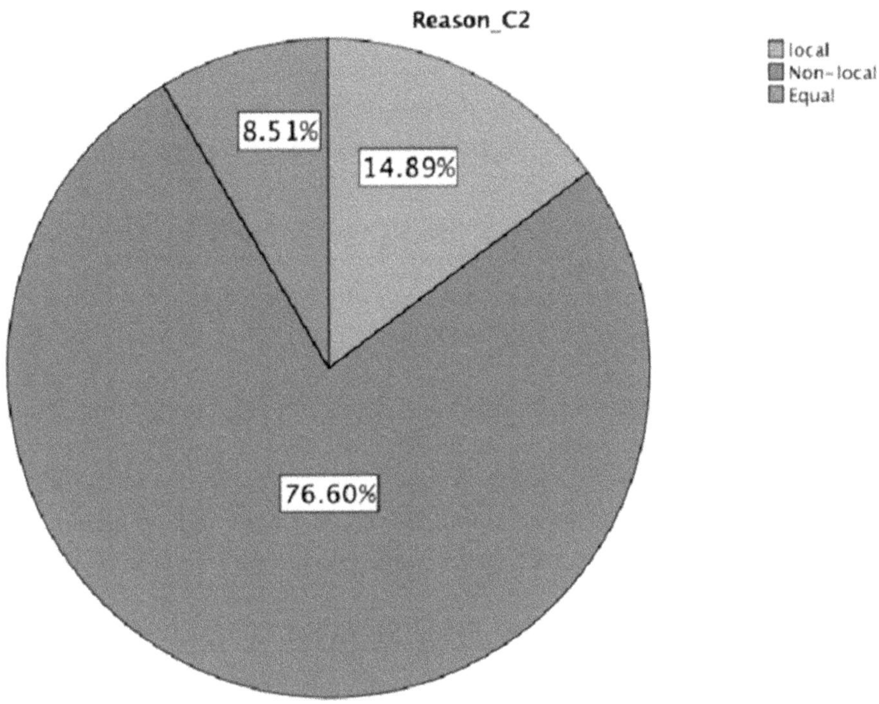

The second case [referring the deforestation in Indonesia] affected me. I think the effect of the second case is greater on our world. Amazon forests in the area are considered as livers of the world and there

Figure 5. Pre-service teachers' explanations about their decisions in the Deforestation sub-case

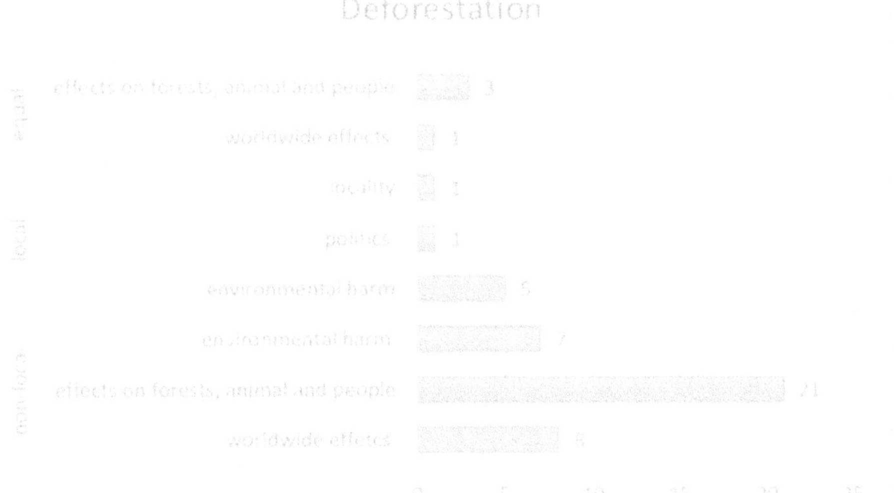

is a great biological diversity including a huge number of animals and plant species. Destructing the forests in Indonesia will endanger the plant and animal species. Eventually, the ecological balance will deteriorate which will also cause global warming. (PT-6)

The second case affected me more. Destroying the forests to plant more palm oil trees endanger the animals and the locals living in that area. A degradation in the ecosystem will affect all of us. (PT-30)

The second case affected me more since we are using products containing palm oil in our lives unconsciously. I also think the consequences are more severe when compared to the first case [referring to the deforestation in Turkey case] as the consequences include climate change, soil erosion, the exploitation of local people, and torturing animals living in the area. (PT-39)

In addition to the aforementioned effects on animals, forests, and local people, some preservice primary teachers (n=8) also emphasized the consequences of the non-local issue would have worldwide effects as:

The second case affected me more since the second case has a greater impact on our world. Since the Amazon forests are considered the livers of the world, the consequences will be felt worldwide. (PT-28)

The second case affected me more. Because I think the second case damaged the world more which has a greater impact that is felt worldwide. (PT-18)

Besides the aforementioned effects on animals, forests, and local people, some preservice primary teachers (n=5) also emphasized the environmental harm in general. Some excerpts are provided as:

The second case affected me the most. The consequences to the environment are more severe. (PT-15)

The second case affected me more since the destruction of forests in the area cause a huge environmental disaster. (PT-45).

Some preservice primary teachers (n=7) indicated that the local deforestation dilemma is more influential for them. Their reasons differed for locality, politics, and environmental harm. Sample excerpts referring to different reasons are presented below:

The first case affects me more. Destruction of forests will damage the balance of nature. (PT-27- environmental harm)

The most impressive case that affected me was the first case [referring to the deforestation case in Turkey] since it is especially tolerated as called public interest. I think it is political ignorance. (PT-19-politics)

Honestly, the first scenario [referring to the deforestation case in Turkey] which happened in my country influenced me the most. This is because we are directly being affected by these disasters. (PT-34-locality)

Lastly, there are few preservice primary teachers (n=4) who indicated that both cases affected them equally. Sample excerpts are provided below:

Both presented cases affected me in the same way since deforestation in both cases affects every inch of our world. (PT-5)

Both [referring to the local and non-local sub-cases] made me sad as the main problem was the same in both cases: Deforestation. (PT-35)

After determining which local or non-local environmental dilemma was more influential and their reasons, we explored PTs' moral reasoning patterns to answer the third research question. The frequencies for the ecocentric, anthropocentric, and non-environmental moral reasoning patterns were calculated. The mean frequencies and the standard deviations for local and non-local environmental dilemmas are presented in Table 2.

Table 2. The mean frequencies and the standard deviations for local and non-local environmental dilemmas (N=47)

	Moral reasoning patterns					
	Ecocentric		Anthropocentric		Non-environmental	
Cases	Mean	SD	Mean	SD	Mean	SD
Sea Pollution	1.89	1.60	1.81	1.30	1.77	1.43
Deforestation	2.49	2.22	1.32	1.43	1.87	1.08

According to the mean frequencies for Sea pollution and deforestation environmental dilemmas (Table 1), the moral reasoning patterns of preservice primary teachers showed similar patterns in the sea pollution dilemma. PTs mostly focused on ecocentric, anthropocentric, and non-environmental reasoning patterns in a similar manner (M_{eco}=1.89, M_{ant}=1.81, and M_{NE}=1.43 respectively) in the sea pollution case. Still, ecocentric moral reasoning was the highest mean score among other reasoning patterns while PTs used non-environmental reasoning patterns to a lesser extent in this dilemma. For the deforestation dilemma, PTs mostly focused on ecocentric moral reasoning (M_{eco}=2.49) when compared to anthropocentric (M_{ant}=1.32) and non-environmental moral reasoning (M_{NE}=1.87). Interestingly, the mean score of non-environmental moral reasoning was higher when compared to the anthropocentric moral reasoning patterns. In both dilemmas, PTs moral reasoning patterns mostly focused on ecocentric moral reasoning. The descriptive statistics showed that PTs were more concerned about the value of the environment aside from its usefulness for humans (ecocentric). Still, PTs also showed some concern that the degradation of the environment possesses a threat to human wellbeing (anthropocentric). Lastly, PTs also showed their reliance on the need for laws and some legal regulations for decreasing the effect of environmental problems (non-environmental). Sample excerpts providing examples for PTs' moral reasoning patterns are shown below:

We used a paired sample t-test to determine if there were significant mean differences between two environmental dilemmas (sea pollution and deforestation cases) to answer the last research question of this study. The findings are presented in Table 4.

The t-test results showed that there was statistically significant difference between PTs' anthropocentric moral reasoning patterns in sea pollution (M=1.81, SD=1.30) and deforestation (M=1.32, SD=

Table 3. Sample excerpts showing PTs' moral reasoning patterns

Cases	Moral reasoning patterns		
	Ecocentric	Anthropocentric	Non-environmental
Sea Pollution	'These environmental problems [referring to the local and non-local cases] will affect all the living things in the ecosystems. The natural environment, plants, animals, and people will be affected eventually.' (PT-9) 'I am very worried that the environment, water, and the sea are so polluted, that both living creatures in the sea and living species around are affected.' (PT-31)	'The seas and oceans are the property of all human beings. Polluting and destroying natural resources are against moral values.' (PT-26) 'The innocent people would be affected by these environmental problems [referring to the local and non-local cases]. Of course, the living creatures in these areas will be affected.' (PT-37)	'I think the governors of countries are primarily responsible for these environmental problems. They should act more responsibly by relating necessary laws.' (PT-24) 'Governments should conduct more research and control. The number of inspections should be increased and the old ships should not let to be in the seas.' (PT-32)
Deforestation	'Forests have never experienced a period of extinction as they have experienced in the last century. Therefore, biodiversity is under a huge threat.' (PT-27) 'Nature is being destroyed unconsciously. Neither the animals, not the local people are being considered. Moreover, nature will be affected by soil and water pollution. This is worrying. (PT-43)	'What worries me the most is the destruction of forests which are the main source of breathing. We are doing the worst to humanity by destroying the forests.' (PT-2) 'The whole world would be affected by destroying the forests for industrial purposes. Forests which are life sources would disappear and the species would perish as a result of climate change and this would affect people negatively.' (PT-4)	'Logging should be done under the government's control. The laws which can protect the forests should be prepared and implemented immediately.' (PT-20) 'To prevent environmental damage, there needs to effective laws determined by the governments.' (PT-40)

1.42) cases; $t_{(46)}$ =2.06, $p = 0.04$. The results suggest that PTs used more anthropocentric reasoning patterns in the sea pollution case when compared to the deforestation case. On the other hand, no statistical difference was found between ecocentric and non-environmental moral reasoning patterns ($p > 0.05$).

SOLUTIONS AND RECOMMENDATIONS

This chapter explored how preservice primary teachers' decisions were influenced by local and non-local environmental dilemmas and their moral reasoning patterns by using a mixed-method approach. To investigate the research questions, the environmental education course focusing on local and non-local environmental dilemmas was designed. The results revealed that different cases evoked different concerns. While local issue about sea pollution was more influential on PTs' decisions (52.19%), the non-local issue about deforestation and forest degradation was found to be more influential (76.6%). This study also explored what evoked these opinions. While most of the PTs (64%) who thought the local issue was more influential indicated that the locality of the issue (i.e. being in their own country) was the main reason for their decisions. Some participants also expressed the severity of environmental harm. The PTs who stated that the non-local issue was important also expressed environmental harm in this case. In the second case (deforestation) participants mostly stated they were influenced by the effects of deforestation on the forests, animals, and the people (58.33%). The descriptive statistics revealed that the PTs mostly focused on ecocentric, anthropocentric, and non-environmental reasoning patterns

Table 4. Paired samples t-test findings for moral reasoning patterns (N = 47).

Reasoning patterns	*t*	*p*
Ecocentric	-1.74	0.08
Anthropocentric	2.06	0.04
Non-environmental	-0.38	0.70

in a similar manner (M_{eco}=1.89, M_{ant}=1.81, and M_{NE}=1.43 respectively) in sea pollution. On the other hand, PTs mostly focused on ecocentric moral reasoning (M_{eco}=2.49) when compared to anthropocentric (M_{ant}=1.32) and non-environmental moral reasoning (M_{NE}=1.87). The paired-sample t-test results revealed that PTs used more anthropocentric reasoning patterns in the sea pollution case when compared to the deforestation case. On the other hand, no statistical difference was found between ecocentric and non-environmental moral reasoning patterns.

This study revealed that the means scores for ecocentric, anthropocentric, and non-environmental reasoning patterns were close to each other in the sea pollution case. While the mean score for ecocentric moral reasoning was the highest, the mean score for non-environmental moral reasoning was the lowest. This finding was in line with Yuksel et al.'s (2012) study which reported that preservice science teachers' moral reasoning was mostly ecocentric. They also found that non-environmental moral reasoning was the least referred reasoning pattern. This contradicts our findings which revealed a higher use of non-environmental moral reasoning in written responses. Actually, Kortemkamp and Moore (2001) indicated that pro-environmental (environment friendly) attitudes were more related to ecocentric and anthropocentric attitudes and less related to non-environmental attitudes. Thus, we expected that if PTs in this study had more environment-friendly attitudes, they would use less non-environmental moral reasoning. However, they used non-environmental moral reasoning patterns as frequently as ecocentric and anthropocentric moral reasoning patterns. In the second case of this study, PTs' moral reasoning was mostly ecocentric as in line with Yuksel et al.'s (2012) study.

We revealed that PTs' decisions about the most influential case changed: While PTs expressed the local environmental dilemma was more influential in the sea pollution case, They indicated that the non-local environmental issue was more influential in the deforestation case. This finding shows that the PTs' reasoning can change based on the context being discussed. Supporting this, Tuncay et al. (2012) found out that preservice science teachers' moral reasoning was not related to the issues being local or non-local. The importance of context for environmental education is emphasized in numerous studies (Littledyke, 2004; Morin et al. 2013; Sadler, 2004). Thus, to enhance participants' interest, different contexts can be used for investigating participants' moral reasoning.

We revealed different patterns in PTs' decisions for local and non-local cases. While most PTs found the local issue presented was more influential in the sea pollution case, most of them thought the non-local issue was influential in the deforestation case. Local and global environmental problems were found to evoke different feelings. However, the available studies reported contradicting findings in the literature (Goksen et al. 2002; Tuncay et al. 2012; Unal, 2008). For instance, Goksen et al. (2002) found that participants' local environmental concerns were higher than global environmental concerns. On the other hand, Unal (2008) revealed that preservice teachers were more concerned about global environmental problems when compared to local environmental problems. Contrary to Unal's (2008) findings, Yuksel et al. (2012) found no difference between preservice mathematics and science teachers' reasoning patterns in local and non-local environmental reasoning. They concluded that preservice teachers were concerned for both local and non-local environmental problems. There might be several reasons for these contradicting findings: the first one is that these studies were conducted with different participants. While Unal's (2008) study was conducted with preservice science mathematics and early childhood teachers and explored their perceptions towards local and non-local environmental dilemmas, Yuksel et al.'s (2012) study included only preservice science teachers. Indeed, supporting this assumption, Unal (2008) revealed departmental differences. On the other hand, Goksen et al. (2002) conducted their study with randomly selected individuals living in Istanbul. Also, this study was conducted with preservice primary teachers- a different sample from previous studies. Thus, the differences in the participants' background (being teacher candidate or regular individuals) might be the reason why differences in the reasoning patterns and environmental concerns were revealed. Another reason might be the design of the study. While both Unal's (2008) and Goksen et al.'s (2012) studies were survey-based, Yuksel et al. (2012) used a mixed-method approach in their study. However, there were not any intervention like in this study. In contrast, this study was part of an environmental education course, and the data were collected during the course. This might be the reason why PTs had differences in different cases in the present study. Another plausible explanation of differences of decisions about local and non-local environmental problems revealed in different cases might be Van Liere and Dunlap's (1980) argument: Different environmental problems can evoke different feelings which in result had dissimilar meanings for the individuals. In line with Van Liere and Dunlap's (1980) argument, Kortenkamp and Moore (2001) also revealed that different environmental dilemmas evoked different moral reasoning patterns. For instance, participants use more ecocentric moral reasoning in an environmental dilemma about land-use. This finding also supports our explanation.

The total mean frequencies of moral reasoning patterns indicated that PTs mostly focused on ecocentric, anthropocentric, and non-environmental reasoning patterns in a similar manner. Finding similar mean scores for non-environmental moral reasoning with anthropocentric and ecocentric moral reasoning patterns were surprising. Even though a correlational analysis was beyond the scope of this chapter, we were expecting less reliance on non-environmental moral reasoning as Yuksel et al.'s (2012) study. Indeed, Kortenkamp and Moree (2001) also indicated that environmentally friendly attitudes were more related to ecocentric and anthropocentric moral reasoning but were less related to non-environmental reasoning. On the other hand, Thompson and Barton (1994) indicated that environmentally friendly attitudes were related to ecocentric attitudes while no relationship was found with anthropocentric attitudes.

Lastly, the statistical analysis in this study revealed that the only difference between cases (sea pollution and deforestation) was found for anthropocentric moral reasoning patterns implying that PTs used more anthropocentric moral reasoning in the deforestation case when compared to the sea pollution case. This finding is in line with Tuncay et al.'s (2012) study which indicated that there was no single pattern for moral reasoning among local and non-local environmental dilemmas.

FUTURE RESEARCH DIRECTIONS

This study has several limitations that can provide direction for further research. First of all, this study included PTs' perceptions and opinions about two local and environmental dilemmas, namely sea pollution and deforestation. As stated above different environmental problems can evoke different perceptions as Van Liere and Dunlap (1980) indicated. Thus, the use of different cases can evoke different moral reasoning patterns. Increasing the number of environmental dilemmas included in environmental education courses can enhance and develop participants' moral reasoning and moral development.

Second, this study used participants' written reports as a data collection tool. Further data collection tools such as individual or focus group interviews can be conducted to provide space for preservice teachers to further elaborate and express their opinions.

Third, the importance of including environmental education courses in teacher education programs is emphasized in the relevant literature (Crim et al. 2017; Desjean- Perrotta, 2013). Moreover, findings of existing literature also show the fruitfulness of these kinds of environmental education courses (Derman et al. 2016; Groves & Pugh, 1999; Haight, 2015; Morin et al. 2013; Teksoz et al. 2010). With this respect, this study was an attempt to include local and non-local environmental dilemmas to enhance preservice primary teachers' moral reasoning patterns. Still, there needs to be an improvement in the curriculum of preservice teacher education programs. Environmental education courses usually focus on standardized topics including biotic, abiotic factors, different types of environmental pollution excluding the morality aspect of environmental education. When the importance of morality in environmental education is considered (Sadler, 2004), the number of studies focusing on the morality aspects needs to be increased.

Lastly, the focus of the present study was to explore whether locality/geographical proximity influence preservice teachers' opinions about environmental issues. Total numbers of moral reasoning patterns in each case (sea pollution and deforestation) were calculated for this purpose. As a further step, the participants' moral reasoning patterns can be classified as local or non-local to compare whether there is any difference among reasoning patterns between local and local environmental dilemmas.

CONCLUSION

This chapter explored preservice primary teachers' opinions about local and non-local environmental dilemmas and their moral reasoning patterns in different cases. The results revealed that different cases evoked different concerns for preservice primary teachers: While local issue about sea pollution was more influential on PTs' decisions in the first case, the non-local issue about deforestation and forest degradation was found to be more influential in the second case. Close examination revealed that the locality of the issue (i.e. being in their own country) was the main reason for the PTs who indicated that the local issue presented in the first case was more influential. On the other hand, PTs expressing that the non-local issue was more influential in their decisions stated that the effects of deforestation on the forests, animals, and people were more severe. This was the main reason for their decisions. These findings support that PTs' environmental moral reasoning can change based on the context being investigated. Another important finding revealed in this study was that even though PTs mostly used ecocentric moral reasoning patterns, they still referred to anthropocentric and non-environmental reasoning patterns in their decisions. While existing studies reported that pre-service teachers had insufficient training during their undergraduate education (i.e. Arslan et al. 2012; Crim et al. 2017), the environmental education course adopted in this study helped to reveal pre-service primary teachers' moral reasoning patterns, their decisions in local and non-local environmental dilemmas, and the factors influencing their decisions. Hence, these findings suggest the need for more emphasis on including environmental education courses in pre-service teacher education programs. When the importance of classroom teachers in raising environmentally literate individuals (Cutter & Smith, 2001) is considered, this chapter is an attempt to fill an important gap by focusing on preservice primary teachers who are usually the first environmental educators of primary children (Crim et al. 2017). As the number of environmental problems that are faced with is enhanced day by day, raising environmentally literate individuals who can deal with these problems by making informed decisions becomes crucial for educators. Teachers as well as future teachers are at the very core of this process. Thus, there needs to be an ongoing effort to enhance teachers' as well as preservice teachers' knowledge and understanding of environmental issues.

REFERENCES

Álvarez-García, O., Sureda-Negre, J., & Comas-Forgas, R. (2015). Environmental education in pre-service teacher training: A literature review of existing evidence. *Journal of Teacher Education for Sustainability*, *17*(1), 72–85. doi:10.1515/jtes-2015-0006

Amirshokoohi, A. (2010). Elementary pre-service teachers' environmental literacy and views toward Science, Technology, and Society (STS) Issues. *Science Educator*, *19*(1), 56–63.

Arslan, H. O., Cigdemoglu, C., & Moseley, C. (2012). A three-tier diagnostic test to assess pre-service teachers' misconceptions about global warming, greenhouse effect, ozone layer depletion, and acid rain. *International Journal of Science Education*, *34*(11), 1667–1686. doi:10.1080/09500693.2012.680618

Bahk, C. M. (2010). Environmental education through narrative films: Impact of medicine man on attitudes toward forest preservation. *The Journal of Environmental Education*, *42*(1), 1–13. doi:10.1080/00958960903479811

Bonnett, M. (1999). Education for sustainable development: A coherent philosophy for environmental education? *Cambridge Journal of Education*, *29*(3), 313–324. doi:10.1080/0305764990290302

Cebesoy, Ü. B. (2019). Pre-service teachers' opinions about a two-day climate change education. *International Research in Geographical and Environmental Education*, *28*(3), 211–227. doi:10.1080/10382046.2019.1579982

Cebesoy, Ü.B. & Karisan, D. (2020). Teaching the role of forests in mitigating the effects of climate change using outdoor educational workshop. *Research in Science and Technological Education*. doi:10.1080/02635143.2020.1799777

Chelleri, L., Minucci, G., Ruiz, A., & Karmaoui, A. (2014). Responses to drought and desertification in the Moroccan Drâa Valley Region: Resilience at the expense of sustainability? *International Journal of Climate Change: Impacts & Responses*, *5*(2), 17–33. doi:10.18848/1835-7156/CGP/v05i02/37199

Creswell, J. W., & Plano Clark, V. L. (2011). *Designing and conducting mixed methods research* (2nd ed.). Sage Publications.

Crim, C., Moseley, C., & Desjean-Perrotta, B. (2017). Strategies toward the inclusion of environmental education in educator preparation programs: Results from a national survey. *School Science and Mathematics*, *117*(3-4), 104–114. doi:10.1111sm.12211

Cutter, A., & Smith, R. (2001). Gauging primary school teachers' environmental literacy: An issue of 'priority'. *Asia Pacific Education Review*, *2*(2), 45–60. doi:10.1007/BF03026290

Dawson, V., & Carson, K. (2017). Using climate change scenarios to assess high school students' argumentation skills. *Research in Science & Technological Education*, *35*(1), 1–16. doi:10.1080/02635143.2016.1174932

Derman, A., Sahin, E., & Hacieminoglu, E. (2016). Does outdoor education make any difference in environmental literacy of pre-service classroom teachers? *International Journal of Environmental and Science Education*, *11*(15), 8491–8506.

Desjean-Perrotta, B. (2013). Raising a generation of environmentally literate children: Are preservice teachers ready? *Childhood Education*, *89*(4), 260–267. doi:10.1080/00094056.2013.815556

Disinger, J. F., & Roth, C. E. (1992). *Environment Literacy*. ERIC Clearinghouse for Science, Mathematics and Environmental Education. Retrieved from https://files.eric.ed.gov/fulltext/ED351201.pdf

Feddema, J. J., Oleson, K. W., Bonan, G. B., Mearns, L. O., Buja, L. E., Meehl, G. A., & Washington, W. M. (2005). The importance of land-cover change in simulating future climates. *Science*, *310*(5754), 1674–1678. doi:10.1126cience.1118160 PMID:16339443

Goksen, F., Adaman, F., & Zenginobuz, E. Ü. (2002). On environmental concern, willingness to pay, and postmaterialist values: Evidence from Istanbul. *Environment and Behavior*, *34*(5), 616–633. doi:10.1177/0013916502034005003

Groves, H., & Pugh, F. (1999). Elementary preservice teacher perception of the greenhouse effect. *Journal of Science and Technology*, *8*(1), 75–81.

Haight, A. (2015). *Exploring teachers' perception of environmental education preparation experiences on their incorporation of environmental education in the profession* (Unpublished Doctoral dissertation). Northern Kentucky University, Highland Heights, KY.

Hollweg, K. S., Taylor, J., Bybee, R. W., Marcinkowski, T. J., McBeth, W. C., & Zoido, P. (2011). *Developing a Framework for Assessing Environmental Literacy*. North American Association for Environmental Education. Retrieved from http://www.naaee.net/framework

Hsieh, H. F., & Shannon, S. E. (2005). Three approaches to qualitative content analysis. *Qualitative Health Research*, *15*(9), 1277–1288. doi:10.1177/1049732305276687 PMID:16204405

Karmaoui, A. (2019). Drought and desertification in Moroccan Pre-Sahara, Draa valleys: Exploring from the perspective of young people. *Geoenvironmental Disasters*, *6*(2), 1–13. doi:10.118640677-019-0118-8

Karpiak, C. P., & Baril, G. L. (2008). Moral reasoning and concern for the environment. *Journal of Environmental Psychology*, *28*(3), 203–208. doi:10.1016/j.jenvp.2007.12.001

Kinslow, A. T., Sadler, T. D., & Nguyen, H. T. (2019). Socio-scientific reasoning and environmental literacy in a field-based ecology class. *Environmental Education Research*, *25*(3), 388–410. doi:10.1080/13504622.2018.1442418

Kobayashi, T. (1991). A suggestion about environmental education using the five senses. *Marine Pollution Bulletin*, *23*, 623–626. doi:10.1016/0025-326X(91)90745-E

Kohlberg, L. (1986). A current statement of some theoretical issues. In S. Modgil & C. Modgil (Eds.), *Lawrence Kohlberg: Consensus and Controversy* (pp. 485–546). Falmer.

Kopnina, H., Washington, H., Taylor, B., & Piccolo, J. J. (2018). Anthropocentrism: More than just a misunderstood problem. *Journal of Agricultural & Environmental Ethics*, *31*(1), 109–127. doi:10.100710806-018-9711-1

Kortenkamp, K. V., & Moore, C. F. (2001). Ecocentrism and anthropocentrism: Moral reasoning about ecological commons dilemmas. *Journal of Environmental Psychology, 21*(3), 261–272. doi:10.1006/jevp.2001.0205

Littledyke, M. (2004). Primary children's views on science and environmental issues: Examples of environmental cognitive and moral development. *Environmental Education Research, 10*(2), 217–235. doi:10.1080/1350462042000198186

Loughland, T., Reid, A., Walker, K., & Petocz, P. (2003). Factors influencing young people's conceptions of environment. *Environmental Education Research, 9*(1), 3–19. doi:10.1080/13504620303471

Merchant, C. (1992). *Radical ecology: The search for a livable world.* Routledge, Chapman & Hall.

Merriam, S. B. (1998). *Qualitative research and case study applications in education.* Jossey-Bass.

Miles, M. B., & Huberman, A. (1994). *Qualitative data analysis: An expanded sourcebook.* Sage.

Minucci, G., & Karmaoui, A. (2016). Exploring the water-food-energy and climate nexus: Insights from the Moroccan Draa Valley. In A. Colucci A., M. Magoni, & S. Menoni (Eds.), *Peri-Urban Areas and Food-Energy-Water Nexus* (pp. 89-97). Springer. http://doi-org-443.webvpn.fjmu.edu.cn/10.1007/978-3-319-41022-7_11

Morin, O., Tytler, R., Barraza, L., Simonneaux, L., & Simonneaux, J. (2013). Cross cultural exchange to support reasoning about socio-scientific sustainability issues. *Teaching Science, 59*(1), 16–22.

North American Association for Environmental Education. (2004). *Guidelines for the preparation and professional development of environmental educators.* Author.

Rickinson, M. (2001). Learners and learning in environmental education: A critical review of the evidence. *Environmental Education Research, 7*(3), 207–320. doi:10.1080/13504620120065230

Roth, C. E. (1992). *Environmental literacy: its roots, evolution and directions in the 1990s.* ERIC Clearinghouse for Science, Mathematics and Environmental Education. Retrieved from https://files.eric.ed.gov/fulltext/ED348235.pdf

Sadler, T. D. (2004). Moral sensitivity and its contribution to the resolution of socio-scientific issues. *Journal of Moral Education, 33*(3), 339–358. doi:10.1080/0305724042000733091 PMID:16317839

Sadler, T. D., & Zeidler, D. L. (2004). The morality of socioscientic issues: Construal and res- olution of genetic engineering dilemmas. *Science Education, 88*(1), 4–27. doi:10.1002ce.10101

Saribas, D., Kucuk, Z. D., & Ertepinar, H. (2017). Implementation of an environmental education course to improve pre-service elementary teachers' environmental literacy and self-efficacy beliefs. *International Research in Geographical and Environmental Education, 26*(4), 311–326. doi:10.1080/10382046.2016.1262512

Schultz, P. W., & Zelezny, L. (1999). Values as predictors of environ- mental attitudes: Evidence for consistency across 14 countries. *Journal of Environmental Psychology, 28*(3), 255–265. doi:10.1006/jevp.1999.0129

Shepardson, D. P., Wee, B., Priddy, M., & Harbor, J. (2007). Students' mental models of the environment. *Journal of Research in Science Teaching*, *44*(2), 327–348. doi:10.1002/tea.20161

Stern, P. C., & Dietz, T. (1994). The value basis of environmental concern. *The Journal of Social Issues*, *50*(3), 65–84. doi:10.1111/j.1540-4560.1994.tb02420.x

Sternäng, L., & Lundholm, C. (2011). Climate change and morality: Students' perspectives on the individual and society. *International Journal of Science Education*, *33*(8), 1131–1148. doi:10.1080/09500693.2010.503765

Stevenson, K. T., Carrier, S. J., & Peterson, M. N. (2014). Evaluating strategies for inclusion of environmental literacy in the elementary school classroom. *The Electronic Journal for Research in Science & Mathematics Education*, *18*(8), 1–17.

Stevenson, K. T., Peterson, M. N., Bondell, H. D., Mertig, A. G., & Moore, S. E. (2013). Environmental, institutional, and demographic predictors of environmental literacy among middle school children. *PLoS One*, *8*(3), e59519. doi:10.1371/journal.pone.0059519 PMID:23533631

Strong, C. (1998). The impact of environmental education on children's knowledge and awareness of environmental concerns. *Marketing Intelligence & Planning*, *16*(6), 349–355. doi:10.1108/02634509810237523

Sund, P., & Wickman, P. (2008). Teachers' objects of responsibility: Something to care about in education for sustainable development? *Environmental Education Research*, *14*(2), 145–163. doi:10.1080/13504620801951681

Tashakkori, A., & Teddlie, C. (1998). *Mixed methodology: Combining qualitative and quantitative approaches*. Sage.

Teddlie, C., & Tashakkori, A. (2006). A general typology of research designs featuring mixed methods. *Research in the Schools*, *13*(1), 12–28.

Teksoz, G., Sahin, E., & Ertepinar, H. (2010). Environmental literacy, pre-service teachers, and a sustainable future. *Hacettepe Journal of Education*, *39*(39), 307–320.

Thompson, S. C. G., & Barton, M. A. (1994). Ecocentric and anthropocentric attitudes toward the environment. *Journal of Environmental Psychology*, *14*(2), 149–157. doi:10.1016/S0272-4944(05)80168-9

Tilbury, D. (1995). Environmental education for sustainability: Defining the new focus of environmental education in the 1990s. *Environmental Education Research*, *1*(2), 195–212. doi:10.1080/1350462950010206

Tilbury, D. (2004). Environmental education for sustainability: A force for change in higher education. In P. B. Corcoran & A. E. J. Wals (Eds.), *Higher education and the challenge of sustainability* (pp. 97–112). Kluwer Academic Publishers. doi:10.1007/0-306-48515-X_9

Trochim, W. (2020). *Research methods data base*. Retrieved from https://conjointly.com/kb/

Tuncay, B., Yılmaz-Tuzun, O., & Tuncer-Teksoz, G. (2012). Moral reasoning patterns and influential factors in the context of environmental problems. *Environmental Education Research*, *18*(4), 485–505. doi:10.1080/13504622.2011.630576

Tuncer, G., Boone, J. W., Tuzun, O. Y., & Oztekin, C. (2014). An evaluation of the environmental literacy of pre-service teachers in Turkey through Rasch analysis. *Environmental Education Research*, *20*(2), 202–227. doi:10.1080/13504622.2013.768604

Tuncer, G., Tekkaya, C., Sungur, S., Cakiroglu, J., Ertepinar, H., & Kaplowitz, M. (2009). Assessing pre-service teachers' environmental literacy in Turkey as a mean to develop teacher education programs. *International Journal of Educational Development*, *29*(4), 426–436. doi:10.1016/j.ijedudev.2008.10.003

UN Conference on Environment and Development (UNCED). (1992). *Agenda 21*. New York: UNCED. Retrieved from https://sustainabledevelopment.un.org/content/documents/Agenda21.pdf

Unal, N. (2008). *Preservice teachers' perceptions toward global versus local environmental issues* (Unpublished Master thesis), Middle East Technical University, Ankara, Turkey.

UNEP. (1983). *The state of the environment*. Nairobi: Author.

UNESCO. (2002). *Education for sustainability, From Rio to Johannesburg: Lessons learnt from a decade of commitment*. Report presented at the Johannesburg World Summit for Sustainable Development, Paris: UNESCO. Retrieved from https://unesdoc.unesco.org/ark:/48223/pf0000127100

UNESCO-UNEP. (1978). The Tbilisi Declaration: Final report of Intergovernmental Conference on Environmental Education. *Connect*, *3*(1), 1–8.

UNESCO-UNEP. (1997). The oceans: Diminishing resources, degraded environment and loss of biodiversity. *Connect*, *22*(3/4), 1–33.

van Liere, K. D., & Dunlap, R. E. (1980). The social bases of environmental concern: A review of hypotheses, explanations and empirical evidence. *Public Opinion Quarterly*, *44*(2), 181–199. doi:10.1086/268583

World Commission on Environment and Development. (1987). *Our common future*. Oxford University Press. Retrieved from https://sustainabledevelopment.un.org/content/documents/5987our-common-future.pdf

Yates, K., Reefer, A., Robertson, D., Hubbard-Sanchez, J., Huss, J., & Wilder, M. (2019). Educators' perceptions of environmental education and professional development in teacher preparation programs. *Applied Environmental Education and Communication*, *18*(3), 207–218. doi:10.1080/1533015X.2018.1451411

Zeidler, D. L., & Schafer, L. E. (1984). Identifying mediating factors of moral reasoning in science education. *Journal of Research in Science Teaching*, *21*(1), 1–15. doi:10.1002/tea.3660210102

Chapter 4
Ecological Sanitation (ECOSAN) in Oasis Communities:
Lessons From a Pilot Project in Five Villages of Draa Tafilalet, Morocco

Ahmed Karmaoui
https://orcid.org/0000-0003-3881-4029
Bioactives, Health and Environmental Lab, Faculty of Sciences and Techniques, University of Moulay Ismail, Morocco & Southern Center of Culture and Science, Morocco

Abdelkrim Ben Salem
https://orcid.org/0000-0002-2283-5928
Research Center BioBio, FS, Mohammed V University in Rabat, Morocco

Mohamed Yacoubi Khebiza
https://orcid.org/0000-0001-6579-1884
Faculty of Sciences Semlalia, Cadi Ayyad University, Morocco

Mohammed Messouli
Faculty of Sciences Semlalia, Cadi Ayyad University, Morocco

Jamie Fico
Syracuse University, USA

Ahmed El Aboudi
Laboratoire de Botanique, Mycologie, et Environnement, FSR, Mohammed V University, Morocco

Laila Rhazi
Laboratoire de Botanique, Mycologie, et Environnement, FSR, Mohammed V University, Morocco

Siham Zerouali
Southern Center for Culture and Science, Zagora, Morocco

ABSTRACT

The chapter highlights the importance of the concept of "ecological sanitation" (Ecosan) in environmental education. It also describes the implementation and management of the "Ecosan" project in several primary and middle schools and explores the lessons learned from reusing organic waste to promote sustainability in arid communities in the southeastern region of Morocco. The project planned and built Ecosan systems that recycle human wastes in five schools from five oasis villages of the Draa-Tafilalt region. The results suggest that the Ecosan system may be a sustainable option to support water, soil, and biodiversity in the vulnerable arid areas, but the scale-up of the project and adoption by the local communities will require more work and coordination with the government and local stakeholders.

DOI: 10.4018/978-1-7998-7512-3.ch004

INTRODUCTION

A global expansion in drylands by 3.1% in the period between 1980 and 2008 was recorded by Li et al., (2019) and this trend is projected to continue due to global warming (Koutroulis, 2019). Arid ecosystems and populations are threatened by climate change (CC), increasing human pressure (Liu et al., 2016), and water scarcity. The Moroccan arid oasis zones are particularly vulnerable. In fact, these zones are susceptible to anthropogenic and CC influences (Karmaoui et al., 2014, 2019). Morocco's agriculture sector is particularly threatened by rising temperatures and decreased precipitation in the coming years compared to other North Africa countries (Schilling et al. 2012). Increasing degradation of the oasis is accelerated by the local socio-economic vulnerability (high rate of poverty), which increases the dependence of local communities on the oasis ecosystem and the promotion of water-intensive cash crops such as watermelons (Karmaoui et al. 2014). The region is characterized by the high rate of population and poverty conjugated to high soil and water vulnerability. The socioeconomic fragility has impacted the sustainability of ecosystem services management, mainly water, and soil. Wells are dug deeper to reach receding water tables, which increases the salinity of the soil, harming crops. Composting human waste instead of using flush toilets is one way to conserve water resources in the region. In fact, Ecosan systems have a variety of benefits that include conserving water resources, enriching soil fertility, and improving food security and health (Langergraber & Muellegger 2004, 441).

Human waste systems such as septic tanks, latrines, or open defecation run the risk of contaminating food and water sources in socio-economically fragile regions. In fact, many studies have shown that feces can include various pathogenic species such as viruses, bacteria, protozoa, and worms. Bacterial (E. coli and enterococci) (Wanjugi et al., 2016). *Trichuris trichiura* and *Ascaris lumbricoides* infections associated with human excreta (Pham-Duc et al., 2013). Viruses, human adenoviruses, and polyomavirus were detected to be associated with human excreta (Dias et al., 2018). For example, enteric viruses are the pathogens that cause waterborne diseases such as hepatitis, meningitis, and gastroenteritis (Rodríguez-Lázaro et al., 2011). Septic tanks, which are used to store human waste in the study region, run the risk of seeping into water tables, contaminating drinking water and soil. In absence of a sanitation network along the Draa and Ziz river valleys (wadis), the dug septic tanks have largely contributed to the fecal contamination of the water and general bacteriological pollution (Messouli et al., 2008 & Ben Salem et al., 2011). In addition to the impact on health, human wastes have considerable repercussions on the environment including water, soil quality, and soil biodiversity.

Protecting the environment for human well-being requires awareness and change in our production, consumption, and waste disposal patterns. However, changing these behaviors depends first on education. Several approaches and techniques for protecting the environment can be taught and anchored in schools such as a sustainable development toolkit developed by teachers and implemented in a participatory manner by students, teachers, and the administration. For our case, we chose the implementation of the Ecosan technique, the abbreviation of the concept of ecological sanitation. This technique safely recycles nutrients from excreta to use in gardening and agriculture after disinfection. Simha & Ganesapillai (2017) define it as "a sustainable approach to promote closed-loop flows of resources and nutrients from sanitation to agriculture." Langergraber and Muellegger (2004) go further, explaining that Ecosan is a system as well as an attitude that views human excrement as a resource rather than a disposable waste. The Ecosan technique requires a conceptual shift in the dominant mentality of discarding human waste, to focus instead on its nutrient value and role in the ecological cycle. The technology is designed as a three-step process of containment, sanitization, and recycling (Rajbhandari 2008, 62) that separates the nutrients from harmful pathogens in the waste to use as a soil conditioner and fertilizer and support local food security.

Ecosan systems are used in both developing and developed countries. For development organizations such as the World Health Organization who note in 2017 that "2.3 billion people still lacked even a basic sanitation service and 892 million people worldwide still practiced open defecation" (WHO, 2017), Ecosan provides an ecologic and hygienic solution instead of digging latrines or open defecation. This technique was applied to several countries in Africa including the rural provinces of Kouritenga (2008), Boulkiemdé (in 2010), Oubritenga (2011) in Burkina Faso (Dickin et al., 2018), in the Hanahai and Paje villages in Botswana (in 2002), and Farchana refugee camp in Chad (von Münch & Ingle, 2012), and widely used in Uganda (Tumwebaze et al., 2011). Regardless of the context, Langergraber and Muellegger note, "conventional sanitation concepts, based on flush toilets, a water wasting technology, are neither an ecological nor economical solution in both industrialized and developing countries" (2005, 433). In their review study, Simha & Ganesapillai (2017) explore the implementation, adoption, and validation of this system reported that Ecosan studies extended over European countries in Netherlands, Denmark, Germany, and Sweden. They recorded also studies from Asia such as Indonesia, China, and India. Ecosan is especially beneficial in arid, water-scarce regions (Graham et all 2003). In the Draa-Tafilalt region of Morocco, the Ecosan system serves to conserve water resources, improve soil quality, and introduce a hygienic and ecological way to treat human waste. Ecosan conserves water resources by reducing the unnecessary use of water in flushing or sewage treatment (Tumwebaze et al., 2011; Werner et al. 2009, 396). Participants from similar projects in other areas have commented on the pesticide benefits of urine fertilizer (Rajbhandari 2008, 67). In regards to urine use as a liquid fertilizer, the Gethke et al., (2007) reported that the urine is stored even up to a period of eight months, while WHO (2016) advises that for crops the urine must be stored for a period greater than six months (T>20). Urine fertilizer has also significantly improved maize production at a school garden in Zimbabwe (Shangwa and Morgan 2009) while the solid waste serves as a soil conditioner adding nutrients to the earth. The Ecosan project in Draa-Tafilalt sought to protect groundwater and soil resources from contamination by human waste and the use of disinfected waste to fertilize the school garden. The implementation of this project provides several lessons from the concepts of the environment: the fight against the degradation of water, soil, and soil biodiversity, organic fertilization, recycling, purification, gardening, hygiene, sanitation, sustainable development, and the circular economy. Ecosan involves all the partners of the school and students to ensure the sustainable development of resources.

The project is part of the fight against the degradation of these two ecosystem services. The project contributes to improving sanitation in the two provinces of Zagora and Errachidia through reusing human waste in an agro-ecological way. The project was entitled "Contribution to the Fight Against Land Degradation, the Protection of Water Resources, and the Improvement of Living Conditions and Hygiene in the Provinces of Errachidia and Zagora, through the Protection of Water Points, Ecological Sanitation and the Reuse of Human Waste in an Agro-ecological Approach." This project was carried out by the Moroccan Association of Microbial Biotechnology and Protection of Natural Resources (MICROBIONA), which launched a call for a consultation for its evaluation and implementation. The vision of the project was to ensure 100% recycling of the school student waste at the partnering schools. In the toilets, the waste is collected to produce organic fertilizer that could be used in the school garden to improve the vegetation cover. The beneficiaries of the project are the students and staff of primary and middle schools. The concept was new for most of the primary school children, however, the university students at Maakat Al Bathaa college in Fezna Errachidia, the students were already familiar with the system, facilitating their integration with the system. This was due to the fact that dry latrines ("Ceikh dar" in Arabic) have been very popular in Tafilalet and farmers have widely used it to produce compost, passing on this practice from generation to generation. The present study seeks to use lessons learned from the Draa-Tafilalt project to improve future Ecosan projects, strengthen local food security, and sustain the environment through water and soil conservation.

MATERIALS AND METHODS

Project Area

The selected schools are located in two provinces: Errachidia (two schools) and Zagora (three schools). These two provinces are located in the southeastern zone of Morocco in the Draa-Tafilalet administrative region (Figure 1). The climate of this zone is arid to Saharan, where the high temperatures in the summer exceed 40 degrees Celsius and drop to less than 2 degrees Celsius on winter nights. The precipitation is sporadic, with the average annual rainfall of the two provinces is estimated in the interval of 80 to 140 mm. The terrain elevation is between 120- 2000m.a.s.l. The five project schools (inside the red border) are located in the Saharan climate as shown by the *bioclimatic stages of Emberger* (Figure 1).

Demographically, the two provinces include a population of 721,473 people (2.14% of the national population) (Table 1). The poverty and vulnerability rates at both sites are higher than national averages. The majority of houses are rural and the average occupancy rate in these two sites is higher than the national average. For the sewage disposal method, the sanitation network and the use of septic tanks are lower in this region compared to the national level. The same goes for household waste collection; data from the General Census of Population and Housing indicates that the collection rate at the study area level is lower compared to the national rate (Table 1). In rural areas, household waste such as ash from burning wood and dirt from sweeping can be recycled and used in the dry latrine. Mixing fecal matter with the ash benefits the compost process by helping the fecal matter dry.

The main economic activity for the people in this region is irrigated agriculture and livestock production. Increasingly, households are supplementing income with wage labor or remittances due to the decline of productivity and profits from agriculture in the region (Rössler et al., 2010). The majority of market-producing agricultural activities are carried on open fields, the main vegetal varieties are the date palm and cereals, and the main animal raised is the D'mman sheep (an adapted species to the arid region of Morocco). Many families still traditionally farm under the date palms growing small amounts of vegetables, barley and alfalfa watered by hand-dug irrigation canals, although Morocco's Green Plan has recently encouraged a shift to larger-scale, open fields using drip irrigation. The next section defines the methodological, theoretical, and geographical framework of the study.

PROJECT SUMMARY

This study was carried out in the Draa-Tafilalet region analyzing the implementation of Ecosan in five selected primary and middle schools. The use of these provinces was justified by their ecological and socioeconomic vulnerability. The project sought to recycle human wastes in five schools from five oasis villages of this region using sensitization workshops and installing Ecosan toilets. The construction

Figure 1. Localization of the Draa-Tafilalet region (red border, the area project including the five schools namely scholar group Moulay Youssef, Middle school Maarakat El Bateha, Aouriz School, Asrir School, and Tlat El Amal; and Bioclimatic stages of Emberger
Source: (Authors)

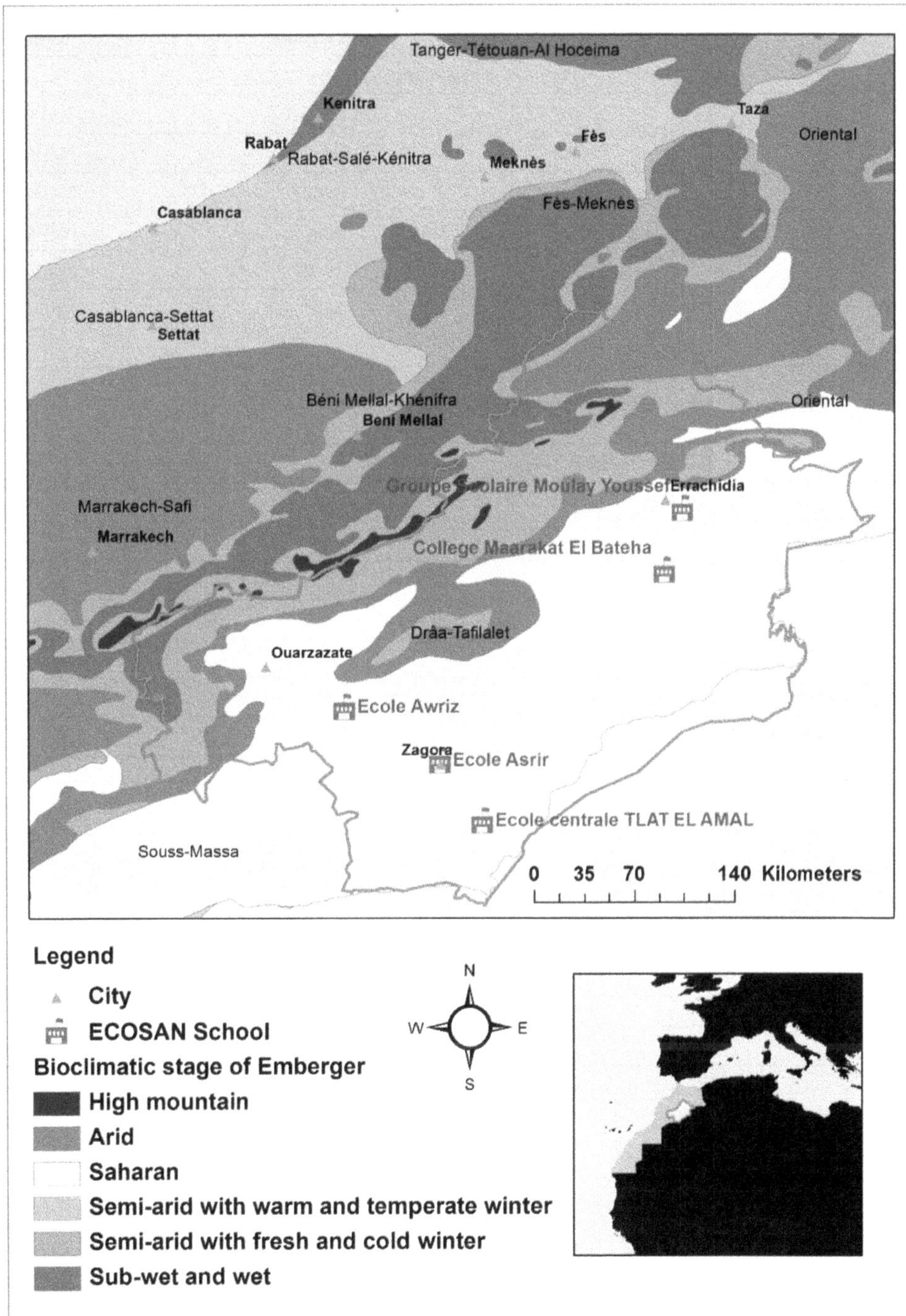

Table 1. Demographic, social, and Housing conditions of the two provinces: Zagora and Errachidia. Data was extracted and compiled from the last General Census of Population and Housing (2014) (http://rgphentableaux.hcp.ma/Default1/)

Indicator	Value		
Level	**Zagora**	**Errachidia**	**National**
Population and households			
Houses	38 067	75 264	7 313 806
Total population	305 510	415 963	33 610 084
Male (Urban)	24 885	97 455	10 034 689
Feminine (Urban)	25 783	94 466	10 249 747
Poverty rate*	20,1%	10,7%	5,87%
Vulnerability rate*	25,4%	18,9%	14,03%
Housing type (%)			
Villa	1.0	3.2	3.2
Apartment	1.4	1.2	11.6
Moroccan house	29.9	61.6	56.8
Rural houses	66.8	32.8	22.8
Occupancy rate	2.0	1.4	1.4
Sewage disposal method (%)			
Public network	14.0	41.7	58.9
Septic tank	42.9	26.0	23.2
Other	43.1	32.2	17.9
Household waste disposal method (%)			
Garbage bin of the municipality	17.1	16.7	46.1
Common or private truck	7.0	44.5	19.3
Other	75.9	38.9	34.6
Frequently used cooking mode (%)			
Gas	61.4	90.3	96.9
Electricity	1.4	1.5	4.8
Coal	0.9	0.4	1.5
Wood	55.7	30.5	10.2

* Data was extracted and compiled from the official website of the Moroccan High Commission for Planning: https://www.hcp.ma/Carte-de-pauvrete-monetaire-par-province-2014_a1944.html

of Ecosan in schools was carried out by an approved construction company, monitored by the project leader MICROBIONA, the administration of the schools, and the Administrative Affairs Department and financial, buildings, equipment and properties of the Provincial Delegation of National Education. The generated waste was stored until it was disinfected and used as compost and in the irrigation of a designed school garden with a drip irrigation system. Three units were implemented in Zagora province, while two units were implemented in Errachidia province. This pilot project was implemented through an agreement between the MICROBIONA and the Directorate of Education of both provinces with the contribution of the local association's networks. This collaboration contributed to common research and design activities and the participation of the schools' staff.

To achieve the objectives of the pilot project, five axes were considered:

- Participatory choice of schools with education directors from the provinces of the region;
- Information and awareness-raising activities for the stakeholders, environmental clubs for the students of beneficiary schools (natural resources, importance and vulnerability, and the Ecosan concept) and organization of information and awareness days;
- Capacity building for teachers and administrations (information on sanitation and Ecosan and information on the development of natural resources and the process of disinfecting excreta;
- Installation of the Ecosan system;
- Installation of pilot gardens and installation of the drip irrigation system;
- Evaluation of the project.

RESULTS

Participatory Choice of Schools with Education Directors from the Provinces of the Region

Prior to the start of the constructions, the project team organized meetings with the directorate of education and school staff to identify the appropriate schools and receive administrative permission.

Information and Awareness-Raising of Stakeholders, Environmental Clubs for Students of Beneficiary Schools (Natural Resources, Importance and Vulnerability, and the Ecosan Concept) and Organization of Information and Awareness Days

After the schools were identified, the students, parents' association and stakeholders need to understand the projects benefits and become invested in the implementation process. When the stakeholders (Regional Council, local communities, Directorate of Education, and Regional Directorate of Environment) understand the Ecosan concept and are engaged, they support the project. Figure 2 shows an example of some activities conducted in Asrit school in Zagora province. The most marked activities associated to the carrying out of the project are as follows:

- Public information about the implementation of the project in Errachidia province (Fezna, May 02, 2012).
- Dialogue and exchange with the NGO leaders or members of Zagora province (Asrir), May 12, 2014
- Raise awareness of schoolchildren on good water management, Zagora, November 12, 2014.
- Water Festival: Drawing competition in Tagounite (Zagora), November 10, 2015 and in Agdez in March 11, 2016.
- TV coverage and newspaper article about the project activities: Agdez, March 11, 2016.
- Project panel installed at the Awriz School in Agdez.

Figure 2. Examples of activities conducted in Zagora province (Asrir primary school)
Source: the authors

- Obtaining the "eco-school" label for the school of Asrir.

Capacity Building for Teachers and Administrations (Information on Sanitation and EcoSan and Information on the Development of Natural Resources and Sanitized Excreta)

- Inauguration celebration of ecological latrines in Zagora (Tagounite), May 02, 2015.
- Regional workshop on the state of the environment in the province of Zagora, May 15, 2015.
- Participation of decision-makers and managers in training in the proper management of natural resources, Agdez, January 12, 2017.
- Project closing workshop, Agdz, January 12, 2017.

Installation of the Ecosan System

The second part of the project of building ecological latrines. The below images are from this process in the Arsir school, Zagora. In the pictures, the toilets are in the process of being completed and some parallel activities related to the project are organized by Asrir School, MICROBIONA, and a group of stakeholders (Figure 3):

- A detailed presentation about the project of building ecological latrines for the students of local primary school and event guests.
- Visiting the Ecosan system to learn about the construction work
- Presentation on environmental protection and natural resources
- Introducing ECOSAN Project and raising awareness of its importance to the " Green Brigade of the Ecological School" students and to "Princess Lalla Hasna Foundation for Environmental Protection" representative.
- Drawing competition for the students of Asrir on the subject of the environment
- A sand stabilization workshop prepared by the Directorate of Water, Forests and Combating of Desertification of Zagora province.
- Participation of the students on the school environment club by preparing brochures documents and environmental documents.

Figure 3. Activities conducted in Asrir school in Zagora province
Source: the authors

Unit description (Figure 4 and 5)

The whole Ecosan system was built using concrete and iron. Each dry toilet unit is a double-vault toilet, consisting of two latrines that works by alternation and are connected to two tanks installed above the ground for excreta and to another tank that may store around 100 L of human urine.

In order to speed up and facilitate the compost process through the diversion and drying of organic matter, medium capacity containers were designed to avoid the accumulation of waste products. To

Figure 4.Simplified design of Ecosan system
Source: the authors

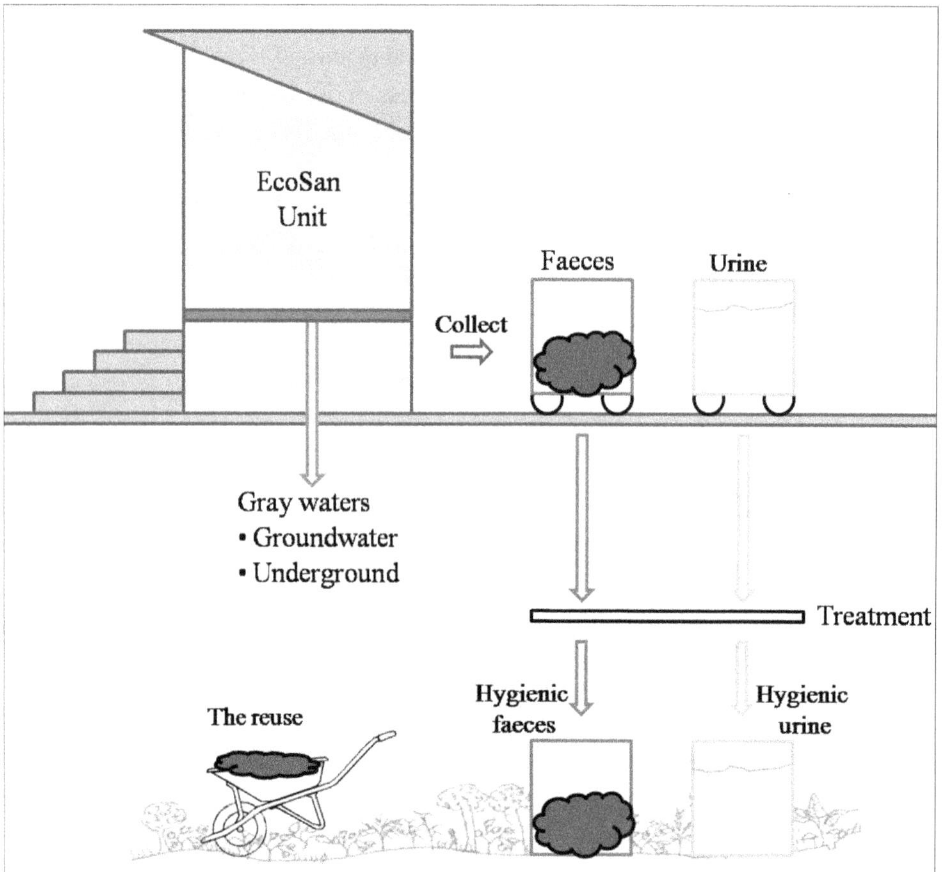

guarantee the continuity of Ecosan latrine, two openings were used for each toilet and two separate containers were installed to achieve the purpose of dry latrines. In addition, sensitization and monitoring activities performed ensure that the students were using the toilets correctly. Figure 6 provides an example of an Ecosan system.

Once construction is complete and the system is functional, students can use it as the regular toilet except that they must use one opening for soil waste and different one for liquid. This is because the system of the toilet is made up of two sections: one for urine and the second for feces to avoid that urine splash down into the dry area (See Figure 6 b).

After finishing, the user pours some ash on the feces to help with the compost process. As advanced by Heinonen-Tanski & van Wijk-Sijbesma, (2005) the urine and excreta may be supported with the ash and household organic waste to offer the necessary potassium and phosphorus for plants, to enhance soil structure, to reduce bad smells, and improve the hygiene of the pit content (Heinonen-Tanski & van Wijk-Sijbesma, 2005). It should be noted that in this context, any improper use of the excreta (such as the students using the wrong latrine or mixing urine with the solid waste) might threaten the children's health. As Tilley mentions, the urine-dividing toilet "is not intuitive or immediately obvious to some users. At first, users may be hesitant about using it, and mistakes made (e.g., feces in the urine bowl)

Figure 5. Ecosan system, the Design and Sizing. Source: adapted from Savadogo, (2008).

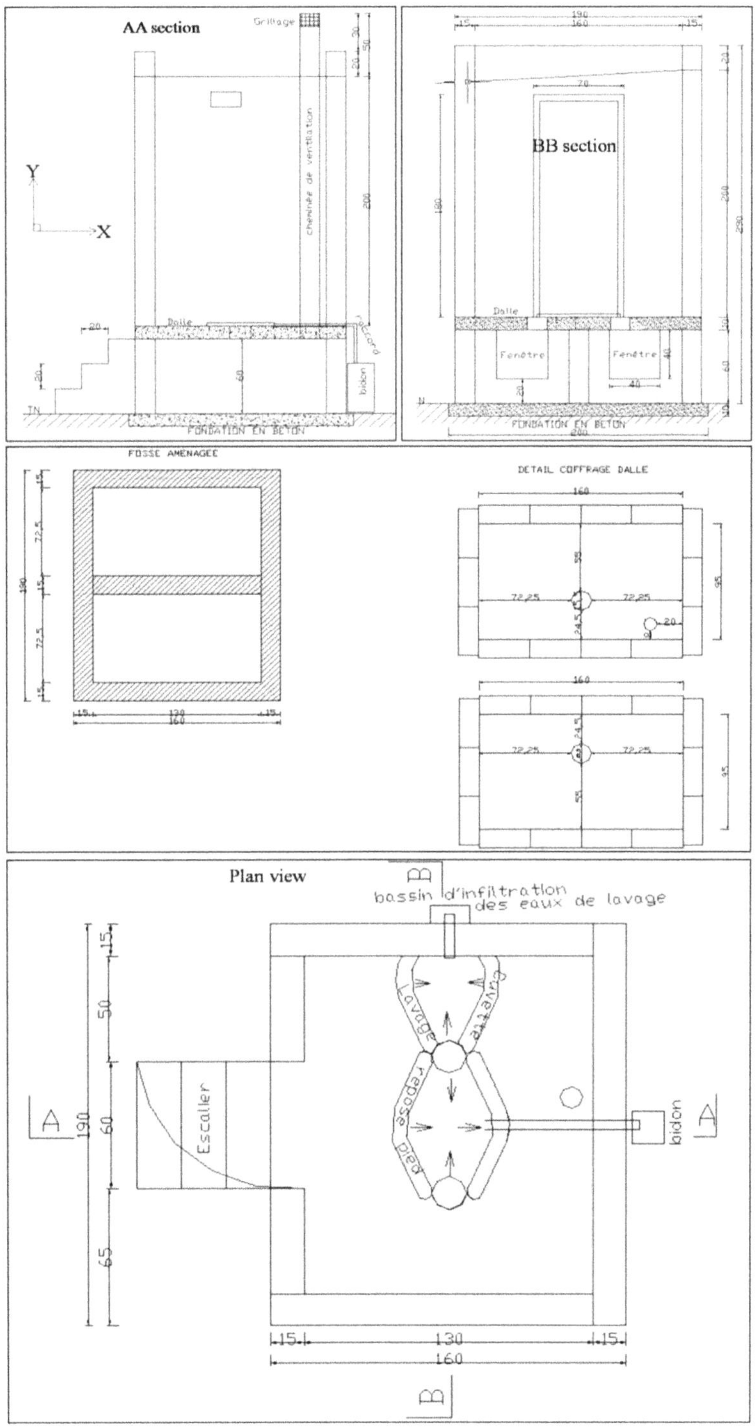

Figure 6. Ecosan system. a, the Ecosan units in process in Asrir school. b, the units ready to work in Aouriz Agdez, Zagora province
Source: the authors

may deter others from accepting this type of toilet as well. Demonstration projects and training are essential to achieve good acceptance with users" (2014, 46). This highlights the importance of continued training and awareness on the use of these dry toilets, especially for teachers who encourage the students to use the toilets correctly.

Installation of Pilot Gardens and Installation of the Drip System

The drip irrigation is a more efficient system to irrigate crops that reduces water evaporate and waste. This technique has been widely used in Morocco the last two decades adopted as a pillar in the Moroccan Green Plan. Its use in arid regions may help to decrease the soil salinization and the desertification process. This can help farmers to cope with the deleterious effects of CC. However, as Grafton et all note (2018), drip irrigation must be implemented with policies that control water usage in arid areas so that farmers do not keep expanding fields and using more water. Scientifically speaking, drip irrigation has an important impact on some parameters (climatic, soil, and vegetation) such as soil evaporation, crop transpiration, and Evapotranspiration (ET). In their study on these variables, Wang et al., (2020) demonstrated that the ET of maize under the drip irrigation was lower in drip irrigation compared to border irrigation (decrease of 40 mm per year), and the soil evaporation was decreased by 30 mm, and the growth days was shortened. Since the study area experiences frequent and prolonged droughts (Karmaoui, 2019), it is necessary to adopt a pilot drip irrigation model in the schools benefiting of the Ecosan system. The whole system is composed of irrigation distribution pipes made of PVC (Figure 7). These include water supply pipe main pipe, secondary pipe (Submain Line), collectors, polytube/lateral, ball valves, end stop, and drippers (Figure 7).

Sanitized Excreta Productions Estimation

The schools were able to recycle more than 50% of the generated human waste that was separated and stored using the Ecosan system. The organic fertilizer produced supports the vulnerable soil and protect groundwater and biodiversity. This project can be replicated in other schools as well as in urban sanitation.

Evaluation of the Project

The assessment of the project is based on the schedule of activities and on a set of project indicators reported by MICROBIONA. For this purpose, the NGO recruited a consultant to prepare a project appraisal, write an evaluation (sharing experiences and lessons), and carry out an in-depth analysis to evaluate to what extent the activities have achieved the objectives of the project. The project evaluation for the 3 schools in the Draa Valley region (not including the 2 Errachidia schools) reported that:

- 1,000 community members participated.
- Sustainable management for forests, water and agriculture was practiced on 200 hectares of land.
- 50 hectares of degraded land was restored and rehabilitated.
- 3 communities demonstrated sustainable land and forest management practices.
- 5 local or regional and 2 national policies were influenced.
- 10 days were increased for students going to school.
- A long-term estimate of $200 increase in purchasing power was achieved through the increase in yields of agricultural land fertilized by human manure (which reduces the cost of fertilizers).

While the project influenced a relatively small area of land (200 hectares), the impact extended to local and national policies, increased the amount of time students were learning in school, and saved a significant amount of money on fertilizers through recycling human waste. These results demonstrate

Figure 7. Simplified design of a drip irrigation system. a, Liquid tank including human urine; b, Hygienized liquid; c, Drip irrigation system, 1- Water supply pipe; 2- Main pipe; 3- Secondary pipe (Submain Line); 4- Collector; 5- Polytube/Lateral; 6- Ball Valves; 7- End Stop; 8- Drippers
Source: the authors

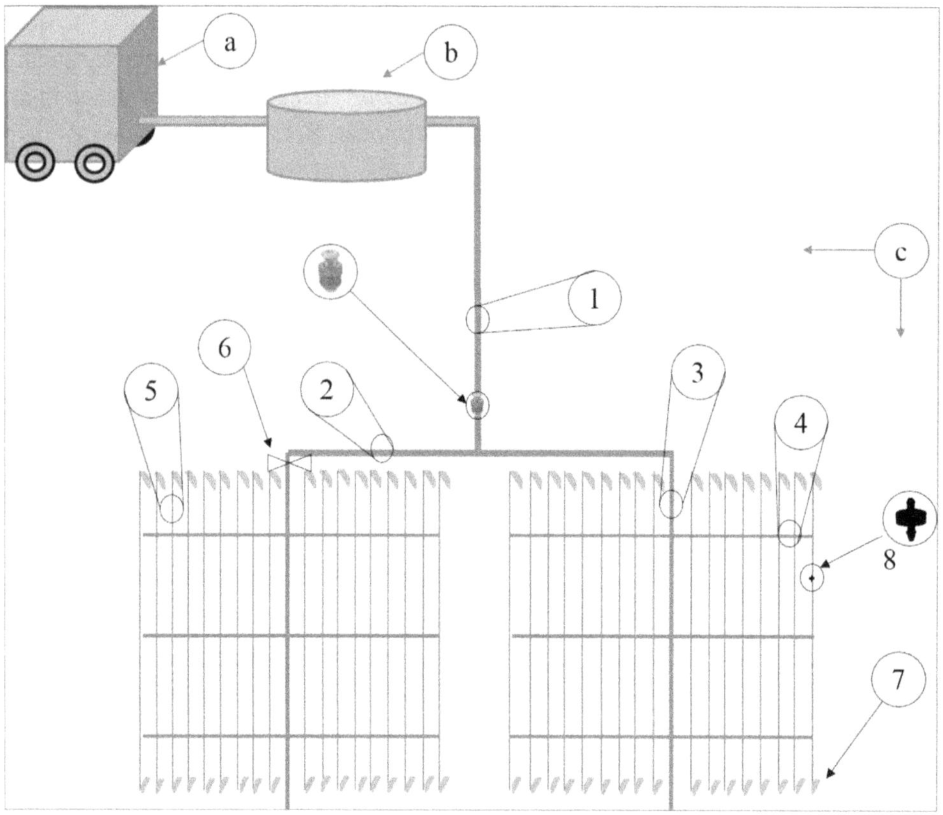

that if the Ecosan technique is adopted by more households in the region, the purchasing power of small-scale farmers can be greatly improved as well as the rehabilitation of land and biodiversity in the region. It also indicates that schools with functioning Ecosan toilets may increase the amount of time students are able to learn throughout the year.

Worldwide, Ecosan systems that recycle human waste are increasing, particularly in the global south. This is driven by water scarcity, degraded quality of soil, high cost of fertilizers, and rocky soil making it difficult to dig septic tanks or latrines. The high amount of nutrients in this human waste is another reason. However, this human waste contains a large number of chemical and biological pollutants that threaten the human health of an increased population. Parasitic diseases such as diarrhea, vector-borne diseases, and water-borne diseases are examples of the pathologies associated with human wastes. Ecosan mitigates this by separating liquid and solid waste and allowing an appropriate amount of containment time that kills parasites.

DISCUSSION

Drylands cover approximately one half of the Earth's surface and include 39% of the global population (Koutroulis, 2019). The aridity is expected to increase in the Mediterranean drylands and will negatively influence biodiversity (Wendt et al., 2020). While Muluneh, (2020) explored the impact of CC on soil water balance and maize production, Kannojia et al., (2019) reported the impact of CC on soil microorganisms through drought and elevated CO2. Therefore, the impacts of climate change may have considerable effects on soil productivity and water availability and consequently on the human well-being in these dry areas.

The human response must be focused on adaptation strategies based on new technologies and indigenous knowledge. Several techniques are used to protect soil and water resources such as composting, local irrigation, small dam construction, and recycling human wastes (solid and liquid), etc. This study focused on this later technic known as "ecological sanitation." As mentioned above, the Ecosan method plays several roles in water, soil, biodiversity, and human health in drylands. The current study applies the Ecosan concept to five systems in five schools of the Draa-Tafilalet region threatened by water scarcity. The project combined the implementation of the system with sensitization, hygiene, ecological activities, and gardening programs. Turning human waste into compost and fertilizer improves soil productivity and strengthens the fragile environment in the region. However, the Ecosan project faces scale- up challenges to becoming widely adopted by community members in their own homes.

Other Ecosan projects in developing countries have faced similar challenges to scale up the project and encourage community members to adopt Ecosan toilets. Nawab et all (2006) discuss cultural challenges of Ecosan adoption by Muslim populations in Pakistan who traditionally cleanse themselves after defecating with water. Tumwebaze et all (2011) note a similar lack of adoption by Muslim populations in Uganda and redesigned the project to accommodate these needs. Other challenges to the population's adoption of Ecosan toilets are the high cost, lack of local workers trained in their construction, and lack of understanding of the system. To combat these challenges, the Ugandan South-Western Towns Water and Sanitation Project (SWTSP) shared information about the toilets on the local radio and trained local masons in building the toilets (Tumwebaze et al., 2011, 298). Some residents were also able to reduce construction costs by using local, lower-cost materials (ibid). Thus, Ecosan projects represent an opportunity for local job creation when local people are provided the appropriate skills training. Additionally, the Ecosan design is flexible and can be adapted to fit local needs while producing environmental benefits. More recently, MICROBIONA incorporated a job aspect by training local masons in Ecosan construction for another project in the High Atlas region in 2018 (Microbiona 2018).

Reusing human excreta as fertilizer is not a new concept. Human excreta has been reused for soil fertility in Japan, Korea, more than 2,500 years ago in China, and in Nepal (Lüthi et al., 2010; Rajbhandari 2008, 66). Reusing excreta was also as endogenous technique in the province of Zagora (the study area). According to our survey, the traditional oasis agricultural society of Zagora had widely practiced this technique called locally *"Canib" or "Chikh Eddar"* for centuries in "Ksours" or "Qsours" (traditional southeastern Moroccan fortified castle). The oasis communities have long practiced adaptive strategies and resilience to cope with the harsh environmental conditions. However, with the transition of the construction from Ksours to concrete Moroccan houses, this type of sanitation was replaced by the *septic fossa*, an underground absorption well made without concrete through which domestic waste are accumulated but not reused. In this context, the current project constitutes an effective initiative to reincorporate traditional waste recycling practices using modern materials and supported by science.

Table 2.Ecosan system weaknesses and strengths

Weaknesses	Strengths
• Low local acceptance and application • Requires lot of sensitization • Requires monitoring of toilets • Possibility of mixing between solid and liquid • In some circumstances, the human urine can increase the soil salinity and pH. • Possible health disease due to children improper use • Long period of storage of urine	• Offer the necessary minerals and elements for plants • Enhance soil structure and moisture • Sanitation service • Support food security • Help to fight against desertification • Decrease water and soil pollution • Low cost of sanitation at long term sustainability • Less odor due to solid and liquid separation

In contrast, septic tanks run the risk of seeping into the groundwater table and contaminating the community's drinking water. The Ecosan initiative is considered as an ecologic solution that may help to reduce the vulnerability of oasis socio-ecosystems to climate change through:

- The fight against desertification;
- Providing low cost and long term sanitation service;
- Decreasing water and soil pollution;
- Offering organic fertilizer for plants;
- Improving hygiene and living conditions.

This project faced several limitations that impacted its effectiveness. Based on observations and preliminary analysis, the effectiveness is limited by climatic and socio-economic conditions which vary in time and space. Today, the installed units are not sufficiently used, not because they are unable to preserve natural resources, but because some students have stopped using them. This was due to the toilets requiring a lot of maintenance work and a lack of local labor, knowledge, and funding for their maintenance. In addition, a low acceptance by local population, the risk of improper use by children, possibility of mixing liquid and solid wastes, and long period of urine storage are other weaknesses (Table 2).

Moreover, Larsen et al., (2004) and Simha & Ganesapillai (2017) highlighted the issue of the 'precautionary principle' over application of this kind of wastes. The precautionary principle, proposed as a new guideline in environmental decision making, has four central components: taking preventive action in the face of uncertainty; shifting the burden of proof to the proponents of an activity; exploring a wide range of alternatives to possibly harmful actions; and increasing public participation in decision making. In this context, the intervention of the government and its services at the provincial level may help in informing and educating about the importance and fragility of the local resources and supporting the use of the ecological sanitation. Encouragement and monitoring of proper usage by teachers will also help the school to take ownership of the Ecosan system. The role of the local NGOs is essential in spreading health and environmental awareness and education for these stakeholders.

The team of this pilot project played a crucial role in the sensitization of all stakeholders. Several campaigns campaign to raise awareness among the population and local authorities continued after the installation of the Ecosan units. As shown in figure 8, an awareness day was organized including scientific communications about environmental resources and preservation presented by Prof. Yacoubi Mohammed and Prof. Karmaoui Ahmed. The day comprised of educative activities for students and field visit for the stakeholders. Additional sensitization activities were conducted by MICROBIONA such as

Figure 8. Awareness day including scientific presentations about environmental resources and preservation and the importance of the ecological sanitation, educative activities, and field visit (2017).
Source: the authors

a door-to-door campaign promoting the ecosan toilets in homes, some of which were adopted by tourist lodgings (France 24, 2015). Yet, further sensitization and participatory activities may be needed to further encourage the increased use and adoption of this system in the region. Other Ecosan programs have faced similar challenges due to the high cost of toilet installation and a lack of awareness of the toilet benefits. These can be overcome by subsidizing the cost of the toilets and continuing sensitization of the proper usage and benefits of the toilets in order to increase people adopting the system (Tumwebaze et all 2011).

There is still more work to increase the adoption of Ecosan systems and protect the world population from diseases linked to pollution by human waste. Communities that do not manage their human waste suffer from diseases like diarrhea, cholera, and malaria. This waste can contaminate drinking water and soil resources. If the project continues to sensitize the population, demonstrating the link between indigenous knowledge and the current technique, encourage proper use of the toilets and organic fertilizer application, and train local workers in toilet maintenance and construction, the Ecosan system could be adopted on a larger scale with local families constructing and utilizing their own Ecosan systems.

CONCLUSION

Waste management requires the introduction of agro-ecological approaches and the participation of the local community to combat the degradation of the ecosystem, specifically the water and soil. This effort supports the local population to adapt to climate change impacts at the local level as well as improve food security and work against diseases caused by sewer seepage or improper waste disposal. The approaches are numerous such as composting, biogas, organic pesticides, ecological sanitation (Ecosan). This later technique is based on several outcomes: decrease health risks, protect the groundwater and surface water against pollution, and produce organic fertilizers (GTZ, IWA, 2003) through recycling the human waste as an organic fertilizer for the local soil. In this study, this technique was applied through a pilot project in five schools in two provinces of southeastern Morocco. Ecosan is not just a concept, but a range of concrete actions. The commitment to an Ecosan project is multifold, encompassing human waste management, water conservation, soil rehabilitation, improved health, strengthened biodiversity, education, the involvement of the local community supporting the foundations of sustainable development. The major limitation of this project lies in the adoption and maintenance of the system by the local population and teachers which requires continued engagement and skill-transferring.

ACKNOWLEDGMENT

The study was based on the outputs of ECOSAN Project implemented in Draa-Tafilalet region, Morocco. The project and the associated activities were organized by the NGO "Association Marocaine de Protection des Ressources Naturelles (MICROBIONA), Marrakech and supported by Embassy of Finland in Rabat and in collaboration with the regional delegation of National Education in Zagora and Cadi Ayyad University. Authors are grateful to Microbiona Association for the study opportunity, the Cadi Ayyad University, Climate-leadership-program-AfriCLP-, and Ibn El Mahdi Eljirari Middle School for the technical assistance.

REFERENCES

Association Marocaine de Biotechnologie et de Protection des Ressources Naturelles (MICROBIONA). (2018, March). *Rapport Scientifique*. Author.

Ben Salem, A., Messouli, M., & Yacoubi-Khebiza, M. (2011). Developing an oasis-based water management tool: Ecohydrologic approach and weap software for a large arid catchment in Morocco. *International Journal of Water Resources and Arid Environments*.

Benmohammadi, A. (1998). Ensablement et désertification dans la vallée moyenne de l'oued Drâa: Le mythe de la fluctuation de la limite nord du désert. *Africa Geoscience Review*, *5*, 11–22.

Chelleri, L., Minucci, G., Ruiz, A., & Karmaoui, A. (2014). Responses to Drought and Desertification in the Moroccan Drâa Valley Region: Resilience at the Expense of Sustainability? *International Journal of Climate Change: Impacts & Responses*, *5*(2), 17–33. doi:10.18848/1835-7156/CGP/v05i02/37199

De Lange, H. J., Sala, S., Vighi, M., & Faber, J. H. (2010). Ecological vulnerability in risk assessment—A review and perspectives. *The Science of the Total Environment*, *408*(18), 3871–3879. doi:10.1016/j.scitotenv.2009.11.009 PMID:20004002

Dias, J., Pinto, R. N., Vieira, C. B., & de Abreu Corrêa, A. (2018). Detection and quantification of human adenovirus (HAdV), JC polyomavirus (JCPyV) and hepatitis A virus (HAV) in recreational waters of Niterói, Rio de Janeiro, Brazil. *Marine Pollution Bulletin*, *133*, 240–245. doi:10.1016/j.marpolbul.2018.05.031 PMID:30041311

France 24. (2015, October 13). *Morocco's desert oases now home to eco-friendly toilets*. The Observers France 24. Retrieved December 21, 2020, from https://observers.france24.com/en/20151013-morocco-oasis-toilet-waste-fertilizer

Gethke, K., Herbst, H., & Pinnekamp, J. (2007). Human urine-decomposition processes and nutrient recovery. *Gewasserschutz Wasser Abwasser*, *206*, 36.

Grafton, R. Q., Williams, J., Perry, C. J., Molle, F., Ringler, C., Steduto, P., Udall, B., Wheeler, S. A., Wang, Y., Garrick, D., & Allen, R. G. (2018). The paradox of irrigation efficiency. *Science*, *361*(6404), 748–750. doi:10.1126cience.aat9314 PMID:30139857

GTZ & IWA. (2003). Ecosan - closing the loop. In *Proceedings of the 2nd international symposium*. Deutsche GesellschaftfürTechnischeZusammenarbeit (GTZ) GmbH and International Water Association (IWA).

Heinonen-Tanski, H., & van Wijk-Sijbesma, C. (2005). Human excreta for plant production. *Bioresource Technology*, *96*(4), 403–411. doi:10.1016/j.biortech.2003.10.036 PMID:15491820

Kannojia, P., Sharma, P. K., & Sharma, K. (2019). Climate Change and Soil Dynamics: Effects on Soil Microbes and Fertility of Soil. In Climate Change and Agricultural Ecosystems (pp. 43-64). Woodhead Publishing. doi:10.1016/B978-0-12-816483-9.00003-7

Karmaoui, A., Messouli, M., Khebiza, Y. M., & Ifaadassan, I. (2014). Environmental Vulnerability to climate change and anthropogenic impacts in dryland (Pilot study: Middle Draa Valley, South Morocco). *J Earth Sci Clim Change*, S11. doi:10.4172/2157-7617.S11-002

Karmaoui, A., Minucci, G., Messouli, M., Khebiza, M. Y., Ifaadassan, I., & Babqiqi, A. (2019). Climate Change Impacts on Water Supply System of the Middle Draa Valley in South Morocco. In *Climate Change, Food Security and Natural Resource Management* (pp. 163–178). Springer. doi:10.1007/978-3-319-97091-2_8

Koutroulis, A. G. (2019). Dryland changes under different levels of global warming. *The Science of the Total Environment*, *655*, 482–511. doi:10.1016/j.scitotenv.2018.11.215

Kriebel, D., Tickner, J., Epstein, P., Lemons, J., Levins, R., Loechler, E. L., ... Stoto, M. (2001). The precautionary principle in environmental science. *Environmental Health Perspectives*, *109*(9), 871–876. doi:10.1289/ehp.01109871

Langergraber, G., & Muellegger, E. (2005). Ecological Sanitation—A way to solve global sanitation problems? *Environment International*, *31*(3), 433–444. doi:10.1016/j.envint.2004.08.006 PMID:15734195

Larsen, T. A., Lienert, J., Joss, A., & Siegrist, H. (2004). How to avoid pharmaceuticals in the aquatic environment. *Journal of Biotechnology*, *113*(1-3), 295–304. doi:10.1016/j.jbiotec.2004.03.033 PMID:15380662

Li, Y., Chen, Y., & Li, Z. (2019). Dry/wet pattern changes in global dryland areas over the past six decades. *Global and Planetary Change*, *178*, 184–192. doi:10.1016/j.gloplacha.2019.04.017

Liu, H. L., Willems, P., Bao, A. M., Wang, L., & Chen, X. (2016). Effect of climate change on the vulnerability of a socio-ecological system in an arid area. *Global and Planetary Change*, *137*, 1–9. doi:10.1016/j.gloplacha.2015.12.014

Lüthi, C., Panasar, A., Schütze, T., Norström, A., McConville, J., & Saywell, D. (2010). *Sustainable Sanitation in Cities: A Framework for Action.* Available on http://publications.lib.chalmers.se/records/fulltext/192469/local_192469.pdf

Messouli, M., Messana, G., Yacoubi-Khebiza, M., El Filali, A. E. A., Boughrous, A. A., & Boulanouar, M. (2008). The evolution of groundwater exploration methods in the Moroccan Oases through history, and managing ecological risk of their present pollution. In *Climatic Changes and Water Resources in the Middle East and North Africa* (pp. 267–288). Springer. doi:10.1007/978-3-540-85047-2_19

Minucci, G., & Karmaoui, A. (2017). Exploring the water-food-energy and climate nexus: insights from the Moroccan Draa Valley. In Peri-Urban Areas and Food-Energy-Water Nexus Sustainablity and Resilience Strategies in the Age of Climate Change. Springer International Publishing AG. doi:10.1007/978-3-319-41022-7_11

Muluneh, A. (2020). Impact of climate change on soil water balance, maize production, and potential adaptation measures in the Rift Valley drylands of Ethiopia. *Journal of Arid Environments*, *179*, 104195. doi:10.1016/j.jaridenv.2020.104195

Nawab, B., Nyborg, I. L., Esser, K. B., & Jenssen, P. D. (2006). Cultural preferences in designing ecological sanitation systems in North West Frontier Province, Pakistan. *Journal of Environmental Psychology*, *26*(3), 236–246. doi:10.1016/j.jenvp.2006.07.005

Pham-Duc, P., Nguyen-Viet, H., Hattendorf, J., Zinsstag, J., Phung-Dac, C., Zurbrügg, C., & Odermatt, P. (2013). Ascaris lumbricoides and Trichuris trichiura infections associated with wastewater and human excreta use in agriculture in Vietnam. *Parasitology International*, *62*(2), 172–180. doi:10.1016/j.parint.2012.12.007 PMID:23305757

Rajbhandari, K. (2008). Ecological sanitation latrines: The experience of Nepal. *Beyond Constructionuse by all: A collection of case studies from sanitaion and hygiene promotion practitioners in South Asia.*

Rössler, M., Kirscht, H., Rademacher, C., Platt, S., Kemmerling, B., & Linstädter, A. (2010). *Migration and resource management in the Drâa Valley, Southern Morocco. In Impacts of global change on the hydrological cycle in West and Northwest Africa*. Springer.

Savadogo. (2008). *Session Focat Ecosan; Design/Dimensionnement*. CREPA Siège.

Schilling, J., Freier, K. P., Hertig, E., & Scheffran, J. (2012). Climate change, vulnerability and adaptation in North Africa with focus on Morocco. *Agriculture, Ecosystems & Environment, 156*, 12–26. doi:10.1016/j.agee.2012.04.021

Shangwa, A., & Morgan, P. (2009). Teaching ecological sanitation in a school environment in Zimbabwe. *Waterlines, 28*(4), 327–332. doi:10.3362/1756-3488.2009.032

Simha, P., & Ganesapillai, M. (2017). Ecological sanitation and nutrient recovery from human urine: How far have we come? A review. *Sustainable Environment Research, 27*(3), 107–116. doi:10.1016/j.serj.2016.12.001

Tilley, E. (2014). *Compendium of sanitation systems and technologies*. Eawag.

Tumwebaze, I. K., Orach, C. G., Nakayaga, J. K., Karamagi, C., Luethi, C., & Niwagaba, C. (2011). Ecological sanitation coverage and factors affecting its uptake in Kabale municipality, western Uganda. *International Journal of Environmental Health Research, 21*(4), 294–305. doi:10.1080/09603123.2010.550036 PMID:21480021

Von Münch, E., & Ingle, R. (2012). *Compilation of 25 case studies on sustainable sanitation projects from Africa. In Sustainable Sanitation Alliance (SuSanA) and Deutsche Gesellschaft für Internationale Zusammenarbeit (GIZ)*. GmbH.

Wang, Y., Li, S., Qin, S., Guo, H., Yang, D., & Lam, H. M. (2020). How can drip irrigation save water and reduce evapotranspiration compared to border irrigation in arid regions in northwest China. *Agricultural Water Management, 239*, 106256. doi:10.1016/j.agwat.2020.106256

Wanjugi, P., Sivaganesan, M., Korajkic, A., Kelty, C. A., McMinn, B., Ulrich, R., Harwood, V. J., & Shanks, O. C. (2016). Differential decomposition of bacterial and viral fecal indicators in common human pollution types. *Water Research, 105*, 591–601. doi:10.1016/j.watres.2016.09.041 PMID:27693971

Wendt, C. F., Nunes, A., Verble, R., Santini, G., Boieiro, M., & Branquinho, C. (2020). Using a space-for-time approach to select the best biodiversity-based indicators to assess the effects of aridity on Mediterranean drylands. *Ecological Indicators, 113*, 106250. doi:10.1016/j.ecolind.2020.106250

Werner, C., Panesar, A., Rüd, S. B., & Olt, C. U. (2009). Ecological sanitation: Principles, technologies and project examples for sustainable wastewater and excreta management. *Desalination, 248*(1-3), 392–401. doi:10.1016/j.desal.2008.05.080

World Health Organization. (2016). *Sanitation Safety Planning: Manual for Safe Use and Disposal of Wastewater Greywater and Excreta*. WHO.

World Health Organization. (2017). Progress on drinking water, sanitation and hygiene: 2017 update and SDG baselines. Geneva: World Health Organization (WHO) and the United Nations Children's Fund (UNICEF).

Chapter 5
Demonstrating Wet Waste Composting as a Greening Society Initiative:
A Step Towards Climate Resilient Societies

Chinmai Kishor Hemani
https://orcid.org/0000-0002-4326-7456
RuChiNi ESP, India

ABSTRACT

The Indian Government's Smart Cities Mission intends to develop citizen-friendly, sustainable cities. Despite its laudatory objective, such a top-down scheme, doesn't reflect any grassroots level work to address water and waste issues or promote energy efficiency/saving at city-scale in which citizens also become participatory change agents. Out of overall food produced in the world, one-third is wasted. It's a matter of global concern with economic, social and environmental implications. Every tonne of food waste diverted from landfills saves 1.9tonnes CO2e (Carbon dioxide equivalent). This action research pilot project is a climate change mitigation initiative to treat wet waste at source through demonstration and techno-economic feasibility involving Rajkot citizens as participatory change agents. While currently awareness about such environmental issues and its environment-friendly solutions among Rajkot citizens is comparatively low. There is a potential of mass adoption of such initiatives, supplemented with government incentives making cities smart while also reducing Greenhouse Gas emissions. This paper describes successes and lessons from this pilot.

CITIES AND CLIMATE CHANGE ASPECT IN INDIAN CONTEXT

Cities are not only engines of growth for the economy but are also major contributors to climate change. Globally, cities produce about 70 percent of Greenhouse Gas (GHG) emissions. It is expected that the urban population will double by 2030 (World Bank, 2014), with 90 percent of that urban growth taking place in developing countries (World Bank, 2013). With regards to India 31 percent of its population

DOI: 10.4018/978-1-7998-7512-3.ch005

lives in cities contributing up to 63 percent of Gross Domestic Product (GDP); it is projected that by 2030 this urban population will grow to 40 percent with commensurate GDP growth to 75 percent by 2030 (Ministry of Urban Development, 2015). Increasing population, increasing purchasing power and lifestyle changes have led to an increase of 357 percent in India's GDP between 1990 to 2014, correspondingly leading to an increase of 180 percent increase in its GHG emissions. In 2014, India's total GHG emission were 3,202 million metric tonnes of carbon dioxide equivalent ($MtCO_2e$) (6.55 percent of global GHG emissions) with 68.7 percent GHG emissions from the energy sector, followed by agriculture, industrial processes, land-use change and forestry, and waste contributing 19.6, 6.0, 3.8 and 1.9 percent respectively (USAID, 2018)

The growing urban population implies an increase in the cost for cities from increased energy use to over strained public services. The stark reality of growing poverty, water and energy crisis calls out for increasing our efforts to mitigate and adapt to climate change and undertake the required shift to sustainable development. From being major contributors to climate change, cities should act as key players in low-carbon growth through innovative out of the box solutions and help their populations prepare for climate uncertainty and natural disasters.

Over the past decade, technological, economic and environmental changes including climate change, ageing populations, urban population growth and pressures on public finances have generated interest in smart cities. To improve the quality of living for an increasing urban population and to provide clean and sustainable environment through the application of innovative smart solutions, the Government of India (GoI) initiated Smart Cities Mission (SCM) in 2015. The SCM is an urban renewal and retrofitting program with the aim to make cities across the country citizen friendly and sustainable and in-turn develop them into smart cities.

The key approach of SCM is to target seven core sectors viz water, energy, waste, mobility, housing, education, healthcare and governance; collect and analyse vast quantities of data thus collected which could be used to improve quality of service, provide market signal feedback to users, automate processes and facilitate better decision-making (Ministry of Urban Development, 2015). However, the critical issue of waste management, its treatment and reduction are not one of the priority areas and has been much ignored.

A study by (Singh, Kumar, & Roy, 2018) suggests that 11 percent of all global methane (CH_4) emissions from Municipal Solid Waste (MSW) are generated by households and represent the third largest anthropogenic source of CH_4 emissions. India is the largest producer of MSW, where only 70–75 percent gets collected, of which only 20–25 percent of it is treated; the remaining waste goes to landfills untreated. MSW generated in India largely consists of 40–60 percent organic waste, 3–6 percent paper waste, and 30–40 percent ash and fine earth material waste, <1 percent of plastics, glass and metal wastes (Singh, Kumar, & Roy, 2018). A 2014 report by the Planning Gommission, GoI shows that the waste composition from 1996-2011 has changed rapidly with increasing proportion of high calorific value waste; paper and plastic waste increased by over 380 percent and 1650 percent respectively over the 15-year period (Planning Commission, 2014). This conspicious increase in consumption pattern indicates that a serious effort is needed to treat compostable and burnable waste by adopting both composting and waste to energy technologies.

Landfills are not only the third largest contributors to the total CH_4 emission in India, but also creates environment (air, water, soil) pollution, lead to leaching of ground water, and frequently catch fires. Such fires releases carbon monoxide, nitrogen oxide, sulphur dioxide, and carcinogenic hydrocarbons, apart from particulate matter into the air, resulting in India releasing 6 percent of CH_4 emissions only from garbage (compared to the global average of 3 percent) (Swaminathan, 2018)

To takle the increasing waste problem, GoI promulgated a revision to MSW law in 2016. This law intends to make the community aware of ways to reduce waste generation and involve them in the proper management of waste including technological innovations for recycling and energy recovery purposes. However solid waste management and its treatment is a huge challenge for Indian cities with segregation of waste the biggest problem as un-segregated waste reduces the treatment efficiency. For segregation of waste at source to become a way of life, the involvement of citizens as participatory agents, supported by an appropriate policy framework, needs to be in place. Such a citizen-centric participatory process, however can only succeed if communities are made aware of how apporpriate solid waste management (SWM) solutions can benefit their health as well as environment and should be the first step in SWM process.

To demonstrate the critical importance of community action for SWM, an action research project focusing on treating wet waste at source through stakeholders' involvement was conceived for Rajkot city in Gujarat state. Why Rajkot? In comparison to cities such as Bengaluru and Pune which have more environmentally-conscious residents, the awareness and adoption of environment friendly initiatives such as SWM initiatives is much lower in cities of Gujarat, including Rajkot. Thus, it was felt that the city would lend itself as an appropriate "pilot" for a bottoms-up initiative to address SWM, with the new lens of climate change.

The purpose of this pilot project was therefore, to provide science-based, objective and user-friendly information on the various aspects of wet waste management, planning and operation. The approach included awareness-building, sensitization and demonstration involving local citizens, supported with the right form of representation to the Rajkot Municipal Corporation (RMC) that would ensure broader acceptance and, in-turn, bridge the divide between top-down and bottom-up approaches.

The objectives of this project were:

- Create awareness and public acceptance for benefits (environmental, social, health) of waste segregation and wet waste composting through innovative techno-economic analysis.
- Establish technical, economic, social, health, environment feasibility through demonstration.
- Identify hurdles and barriers to implementation as the most important learnings and pave the way for advocacy and coalitions to bring about more relevant policy changes and contribute to the formation of knowledge and market networks
- Promote the pilot initiative as pro-environment action to combat climate change and enhance sustainable development practices leading to smart societies that would help address multiple Sustainable Development Goals (SDG) such as sustainable cities and communities – SDG-11, climate action SDG-13 etc.

THE NEED TO RETHINK OUR CURRENT APPROACH

With a population of over 1.5 million spread over 170 sq. km., Rajkot is the fourth largest city in the state of Gujarat, India. Despite having a proactive municipality and 115 registered "Smart Societies" across 23 wards, Rajkot has not been able to showcase substantial work by the municipality to manage waste. The city generates over 550 tonnes waste / day of which only about 15.7 percent is segregated (ICLEI,

Figure 1. Rajkot city Source: Google

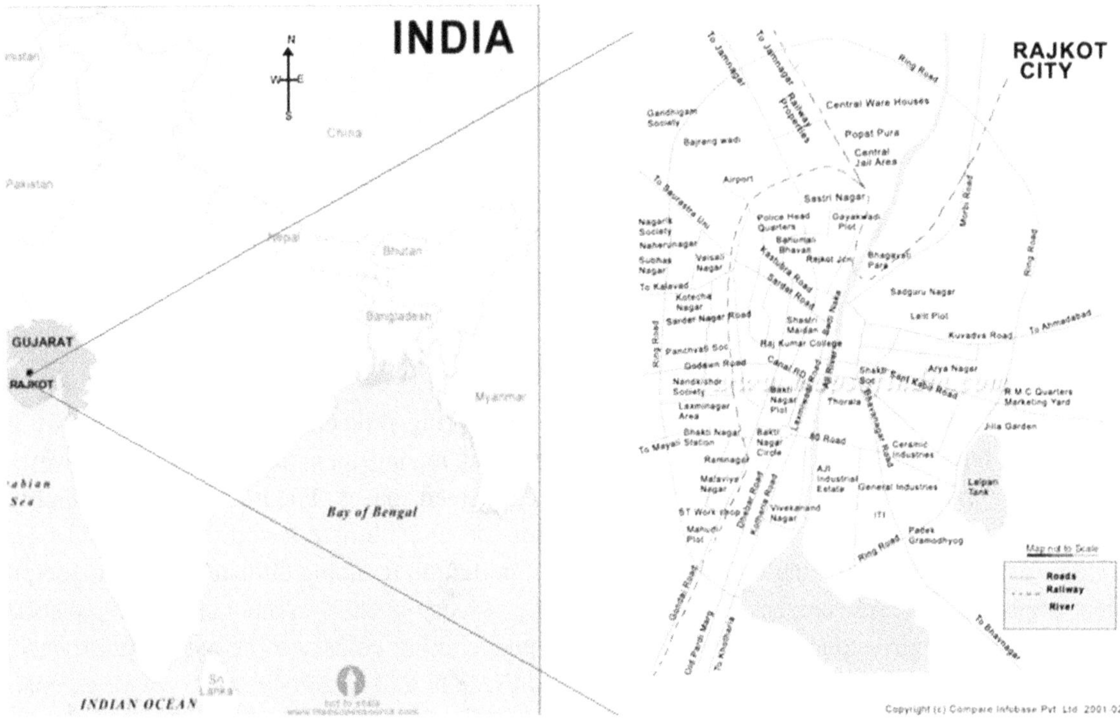

2016); the per capita waste is estimated at over 300 grams / day, out of which 40-50 percent is wet waste (RMC, 2020). Though door to door collection has been happening since 2015, the un-segregated waste continues to be dumped in three officially demarcated landfills of the city, viz

- Nakravadi (30 acres I 80000MT) which is already filled and overflowing since several years (NGT 2013) for which RMC has already been fined by Gujarat Pollution Control Board (GPCB) (WWFIndia, 2013),
- Manda Dungar (2.5 acres),
- Sokhda (11 acres) already closed for operations (RMC, 2020)

Despite the local government's wet and dry waste segregation campaign which has been in place since 2017, the waste collection tipper vans continue to pick up mixed waste due to the careless attitude of waste generators as well as the tipper van staff towards waste management. Lack of awareness on why segregation is important, how to treat wet waste at source, and hand-holding support for better planning and execution are some of the key factors that have impeded waste segregation at source. Moreover, the absence of strict penalties for not segregating waste at source, miniscule tax for waste collection, and a culture that looks down upon handling/treating waste have been deterrents to behavioural changes. GHG inventory of Rajkot city suggests that emission from residential and waste itself contributes to about 25 percent of total GHG emissions (ICLEI, 2016). Hence, a 5 percent increase of waste annually (Chary, 2013), with no scientifically built landfill to store the waste, would lead to not only air, water and land pollution but also avoidable CH_4 emissions from the anaerobic digestion of wet waste. The resultant health epidemic waiting to unfold, in addition to the increasing landfill requirement are two of the important fallouts of non-treatment of wet waste. CH_4 emissions are (56 times for 20 lifetime years and 21 times for 100 years) lethal than CO_2 emissions (UNFCCC, 2020).

Moreover, every ton of waste disposed of at a landfill costs the municipal corporation thousands of Indian Rupees (INR) per day for collection, transportation, segregation and its treatment. For Delhi Municipal Corporation (DMC) the cost is as high as INR 14500/tonne/day (USD 198/tonne/day; 1 USD = 73.19 INR as on 18th September, 2020) (Sambyal, 2018). Given that Delhi city is having population and cost to manage waste for DMC twice the cost of RMC, it would still cost RMC INR 7250/tonne/day) (USD 99/tonne/day) to manage the city's waste.

The Project

This pilot initiative - a citizen science and a social engineering project, promoted participatory approaches through workshop cum demonstration on wet waste management at the source by converting it into useful compost which, in-turn can be used to create green spaces. The initiative served multiple goals - in addition to engaging with citizens to help them develop climate-friendly skills such as dealing with waste at source, creating reduced environment pollution, reducing climate change impact; the project emphasized the role citizens can play to partially reduce the onus on the municipality to collect and treat waste, thereby reduced landfill needs while encouraging citizens to be participatory change agents in the Clean India Mission. GoI in 2014 launched Clean India Mission - a country-wide campaign initiated to eliminate open defecation and improve SWM. This pilot project with its emphasis on citizens' participation in environment issues, a first of its kind for Rajkot city, has the potential to be showcased as one of the climate mitigation strategies under both the Clean India and Smart City Missions.

The Approach

The project aimed to reach around 60 households by an open invitation through online and offline media, personal interaction as well as word of mouth with a target to capture at-least 10 percent of the participating individual households /society / institution ready for implementation. It was intended for intra-generational participants from different educational, professional backgrounds. Social inclusion was also ensured by inviting 10 students and 1 teacher from the city's deaf and dumb school.

In addition to hands-on learning, Information Communication and Technology (ICT) tools such as power point presentation as well as powerful video developed specifically about situation of solid waste in Rajkot city and how individuals can be change agents were used to motivate the participants. UNESCO Education strategy 2014-2021 states that vast amounts of data, information and knowledge provided from variety of sources makes critical thinking a key requirement for education and learning, along with ethical and responsive handling (UNESCO, 2014). Such demo-drives which use multiple strategies to share information about the subject in-turn leads to enhance critical thinking and raising concern about the issue of waste amongst the participants. This in-turn would help in creating environmental stewards who are committed to the environment, feel a sense of responsibility to improve it, and have the skills to take effective action.

In order to understand climate change and contribute to achieving SDG 13, through cognitive learning, participants understood current climate change as an anthropogenic phenomenon; they also learnt that human activities are adding most to this climate change, environment degradation, biodiversity loss as well all sorts of pollution. Socio-emotional learning happens along the way as the participant is able to understand his/her personal impact on local / national / global perspective. Participants also understood the interlinking of several socio-cultural – environmental - climate issues and resonate that we need to re-evaluate our world view and everyday behaviours. Participants were nudged to behavioural learning's as they are better able to anticipate, estimate and assess the impact of personal, local and national decisions or activities on other people and other living beings.

Figure 2. Images of overflowing Nakravadi landfill

Sustainable solutions such as these have the potential to gather momentum through people's participation, local Government recognition and corporate buy-in, centred around a social justice agenda and collective learning regarding its potential to transform society (Leicht, Heiss, & Byun (eds), 2018). Thus, in order to make this demonstration drive successful with lasting impacts collaborations from NGO, industry (as a part of their Corporate Social Responsibility (CSR)), Government were sought. The stakeholders for this project were:

- Housing societies/ individuals by way of open invitation
- NGO – Sister Nivedita Foundation provided the space for demonstration at token rate.
- Participating companies to support under CSR –Vasu Pharmaceuticals from Vadodara, RuChiNi ESP and Patel Brass works Ltd of Rajkot supported by partially funding
- Climate change department, Government of Gujarat (GoG) – which approved this project under Small Grants fund scheme of the Gujarat Energy Development Agency (GEDA) and provided partial funding under the grant.
- Vendors – Orbin supported the cause by providing the demonstration composter at their cost price
- Media: In addition to social media (WhatsApp, Facebook, Webpages), local newspaper gave good coverage to the demonstration drive.

The demo drive had a turnout of 55 participants of whom 10 percent adopted composting at home. This would amount to 1800 kg / year of wet waste avoided from being sent to landfill due to this initiative (considering average wet waste to be 1 kg / household/day). The overall impact of this bottom-up drive was that it reached to over 260-275 individuals considering 5 people / household on an average.

As (Leicht, Heiss, & Byun (eds), 2018) suggests, the participants empowered through this drive with the new learnings would also be able to support climate-friendly economic activities and public policies which would be discussed further as a way forward.

CASE STUDY

The project methodology consisted of primary research, secondary research, visit to various organisations treating wet waste and experimenting through purchase of low-cost solutions to determine the best and the most suitable mechanical composting technique and composter from range of available composters in the market.

Methodology of Implementation

In order to achieve its overall objectives, the pilot project was divided into two components:

1) Sensitization (Part A) and
2) Demonstration and participation (Part B)
   I. Study design

The methodology entailed a preparatory phase for data collection and analysis from secondary sources, followed by brainstorming with the stakeholders to develop a methodology for wet waste composting.

- Preparatory phase:
    - Desk study: involving a secondary research on already existing methods, indicators listed for sustainable green initiatives
    - Stakeholder consultation: Interaction with partners to brain storm and choose from various initiatives, to develop a plan for implementation, and agree upon the metrics for performance measurement

    II. Research to develop indicators for Measurements
- Understand various already available research options to find microbial culture which can decompose wet waste faster and its characteristics
- Developing metrics and indicators to measure performance of proposed green initiatives under the sensitization phase
- Conducting Baseline measurements and identify interested households through use of various media

  III. Implementation phase (included Awareness and Demonstration):
- Awareness phase involved:
    - Mapping of voluntary smart societies and their characteristic
    - Identifying households/individuals who are interested in implementation at home
- Awareness sessions and workshops on green low carbon and climate adaptive society. The venues included RMC community park / society hall / community science centre / colleges and public gathering places such as Yoga centres, Rotary clubs, running club etc. catering to a range of various audiences whoever agrees to partner with this project.

- Demonstration and Implementation:
    - Evaluating waste generation patterns. Creating targets for the various steps mentioned above for house-hold as well as community implementation in community spaces
    - Evaluating the cost of the initiatives chosen and adopted by the housing society, and identifying feasible cost-effective alternatives, as appropriate
    - Identifying of schemes and subsidies of the government which can be passed onto the people and society who buy in for suggested initiatives and changes.

    IV. Review, Monitoring and Evaluation Assessment
- Monitoring and evaluating of the adopted practices in terms of waste segregated and waste reduced, composting adopted including financial benefits accrued by individual households and RMC, if any
- Process review, followed by mid-course correction / action

Time frame to do this study was one year on a very stringent budget.

Figure 3. Methodology

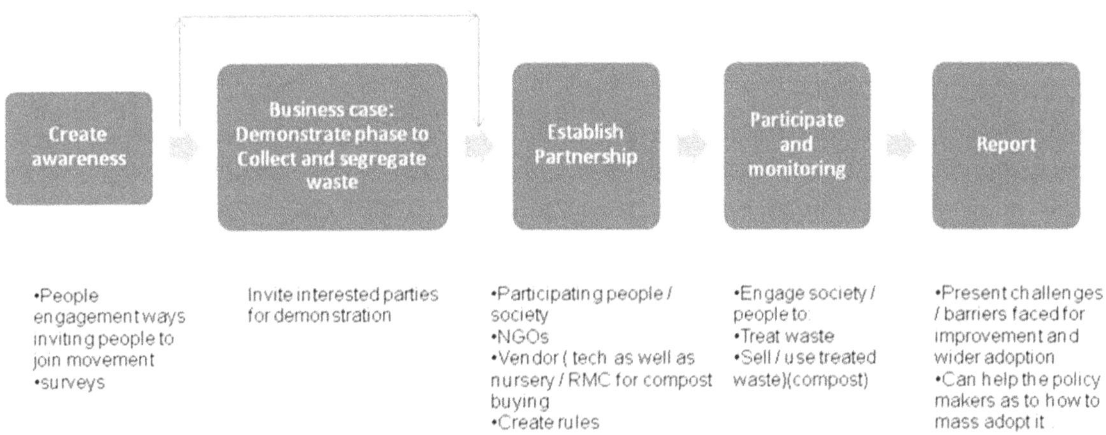

Desk Study and Stakeholder Consultation

This phase comprised of secondary research of already available methods for wet waste composting which would be most suitable for a city such as Rajkot city and its citizens. Based on interaction with different socio-economic strata of people of various societies, experts who have worked on the issue of waste, environment and study of bio-chemistry using Delphi method showed that the composting technology should be simple, cost effective and not time or effort intensive. This conclusion was in tune with author's understanding of local situation as a long-time resident of the city, considering the attitude of Rajkot residents towards greener environment, their willingness to be change agents and their appetite to spend for greener environment.

Choice of Technology

Both the primary and secondary research indicated that aerobic composting was the most appropriate option for a decentralized composting experience at source for Rajkot. Stress on using mechanical composters as choice of technology was mainly for the following reasons:

- citizens can take the ownership and responsibility of their waste and treat it and use the obtained compost for home garden use.
- such mechanical composters are reliant on natural process of decomposition without any other energy inputs unlike automatic and semi- automatic systems of composting needing external energy input which itself generates GHG emissions. The very purpose of treating wet waste at source is to reduce GHG emissions; semi-automatic and automatic versions are reliant on electricity to burn the waste rather than decomposing it and thus less effective and efficient.
- simple, cost effective or near zero budget solutions.

Moreover, the models studied emphasized on it being ergonomic, affordable and hassle free for better adoption rate.

Research on Indicators for Measurement of effectiveness of Composting

The indicators to help measure the effectiveness of the drive, viz reduction in food waste, carbon footprint related to food waste etc, were developed based on literature review, complemented by findings from stakeholder consultation through Delphi approach. It is envisaged that these indicators would also provide an insight into the feasibility of adoption of this practice on a larger scale in Rajkot city.

Pre-Survey Indicators

As a first step, a list of indicators (Table 1) was developed to understand the mindset of Rajkot citizens and to determine who would be interested in participating in such a demonstration drive and follow-up initiatives. These indicators captured the respondents' educational background, economic status, interest in working for a better environment, climate change, current knowledge on importance of waste segregation as well as treatment, per capita waste generation etc.

This pre-survey was extremely important as it led to understanding the motivation or barriers to adopting composting at source, helped determine the baseline measurements and contributed to identifying interested households through use of various media both, online as well as offline.

Recognising that a majority of the respondents were more comfortable responding in their mother tongue (Gujarati), the questionnaire was framed in both Gujarati as well as English.

Post Demonstration Drive Indicators

In order to get an understanding of the effectiveness of the demonstration drive several indicators were developed. These indicators would suggest the wet waste composting adoption rate, the direct and indirect benefits accrued relating to a better environment, etc. (Table 2).

Table 1. Pre-Survey Indicators

Theme	Indicators
Demographics	• Age • Gender • Education • Income range • Type of employment
Waste generation patterns	• Per capita waste generation • wet waste per capita • Nature / scale and origin of food waste
Knowledge about solid waste management	• Knowledge about waste segregation • adoption of waste segregation at source
Attitude towards waste and concern for environment	• Willingness to spend for acquiring new learning skill (Wet waste to compost), • Willingness to share this information with their family and friends, Mindset associated with Contributing towards better earth on adopting composting
Behaviour towards waste	• Willingness to buy composters

Table 2. Post Demonstration Indicators

Theme	Indicators
Registration response rate	• No. of registration response to emails – other personal engagements • No. of participant (actual)
Technology adoption rate	• Choice of wet waste composting technology opted • No. of compost bins / composters sold/bought
Waste treated	• Quantity of waste diverted from landfill (in tonnes)
GHG emissions avoided	• CO_2e emissions saved
Quantity and Quality of compost obtained	• Quantity of compost obtained (in kg) • Quality of compost (testing), Observance of pests, odour
Waste rules and metrics	• Framing of rules for waste management (penalty, team/ committee, responsibility etc.) • Developing metrics for comparative performance of societies

It is also envisaged that the above indicators and rules would help RMC to facilitate adoption and tracking of the effectiveness of composting at source underway in the Smart Societies on promoting to compost at source/premises. This exercise would also contribute to building a case for RMC to provide incentives to those housing societies that show better performance based on a comparative measurement of indicators.

AWARENESS PHASE

Interaction with stakeholders indicated that awareness on the issue of mixed waste needs to be highlighted through judicious use of both audio-visual documentation and dissemination of reference material/ handouts for greater effectiveness.

Thus, the awareness-generation stage entailed a range of activities such as technical research on city waste management, its strategies, visits to landfill site, identifying sensitive team of film maker to take videos at landfill site, conducting interviews with villagers and other stakeholders, writing up reference material, as well as the development and filling of survey forms (online and offline), etc.

A 30-minute video titled *"Where does my waste go?"* on waste management for Rajkot city was produced which graphically captured the ground reality of waste collected from Rajkot city being dumped unscientifically in landfill of Nakravadi.

Figure 4. Interaction with villagers turned to rag-picking at Nakravadi landfill site

Figure 5. Snippets from video ("Where does my waste go?") created on issues of mixed waste at Nakravadi landfill site

Completing the Survey Forms (Online/Offline)

Several online platforms such as WhatsApp, email as well as Facebook were used to spread the message as well as request interested individuals/housing societies to fill the online pre-survey form. More than 75 responses were received through online platform within a stipulated time frame.

However, to reach out to a wider audience, face-to-face sessions explaining the problem of waste in Rajkot city and the importance of segregating and treating waste at source were carried out at different colleges, yoga centres, running club, Rotary club, banking institutions etc. These interactive sessions also provided an opportunity for the participants to complete the pre-survey questionnaire; more than 220 responses were gathered offline, bringing the total respondents (online and offline) to 295.

Figure 6. Awareness session and filling of off-line Pre-survey form

PRE-SURVEY: KEY FINDINGS

The detailed pre-survey forms led to several findings. Identifying trends from data analysis were useful for creation of strategies to reach broader audience and invite them to attend the demo drive; it also helped in strategizing how to conduct the demo drive.

Analysis of Data from Pre-survey Questionnaire

Filling of pre-survey forms was voluntary; Out of surveyed respondents' 54 percent were female and 46 percent were male, and a majority (71 percent) were falling in age bracket of 21-50 years. It was also observed that male students were less interested to fill the forms as compared to their female counterparts when it came to offline filling of forms.

As reflected in Figure 9, a majority (94 percent) of the respondents resided in either apartments (31 percent), tenements (46 percent) or bungalows (17 percent), with the remaining 6 percent in other types of residence. The analysis also showed a trend towards nuclear families, given that around 70 per cent of the total respondents belonged to a family of 3-5 individuals, which was not a norm a decade ago. Younger, nuclear families more aware about environmental challenges tend to indicate a stronger possibility of adopting wet waste composting at their own premises when nudged in an appropiate way.

Figure 7. Age -Gender chart of Respondents

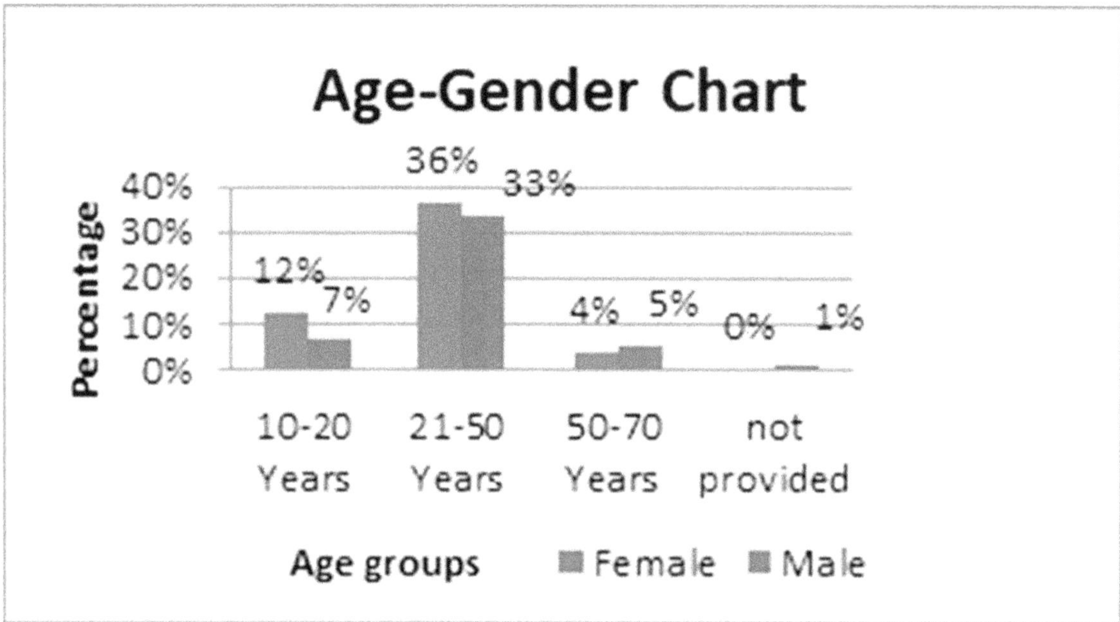

Figure 8. Occupation of Respondents

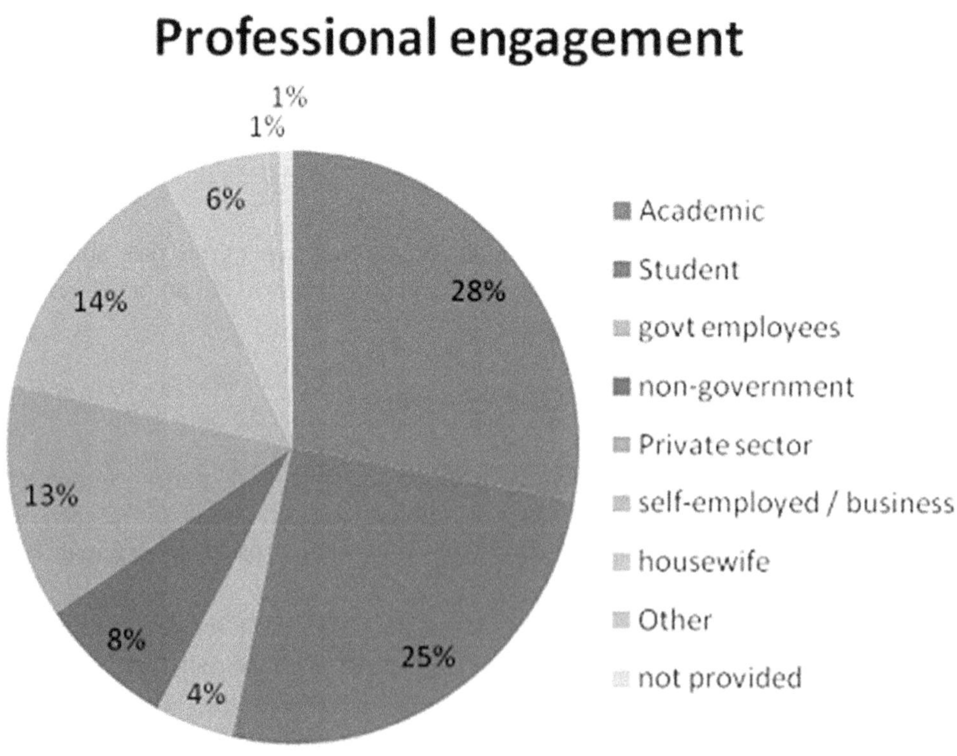

Figure 9. Respondents number of family members to type of residence they live in

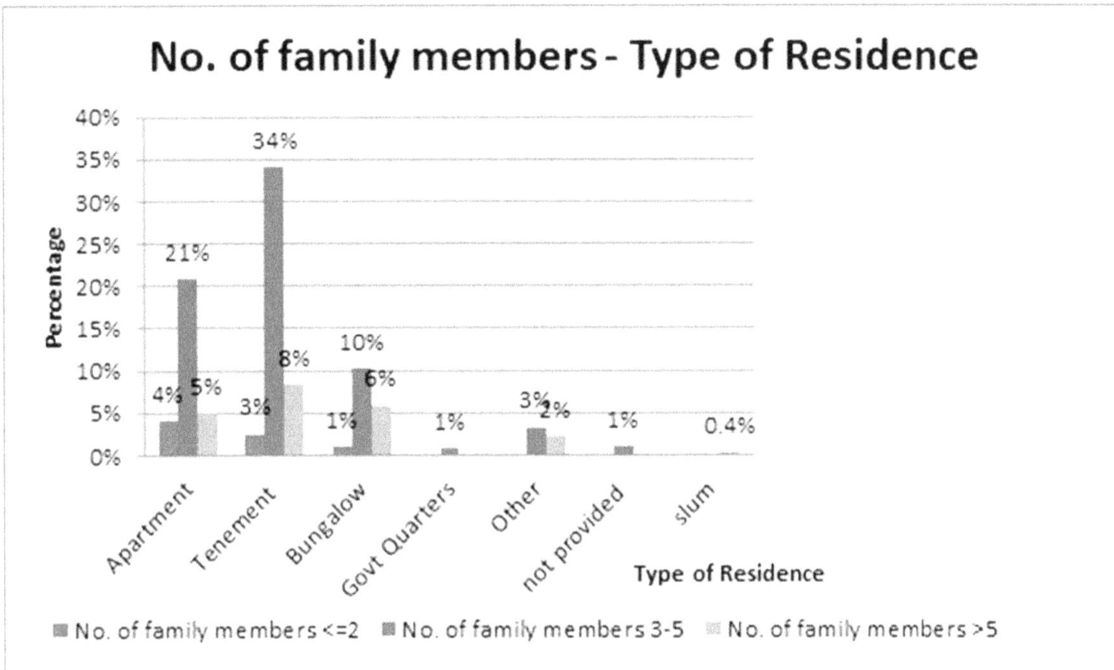

Figure 10. Respondents Income range

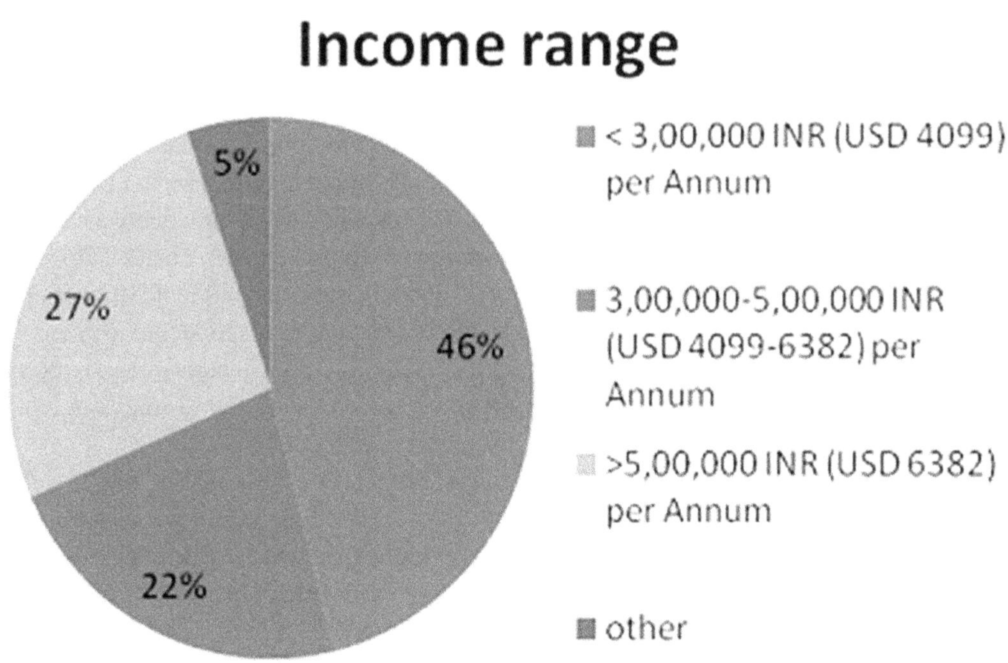

Figure 11. Respondents practice of waste segregation at source

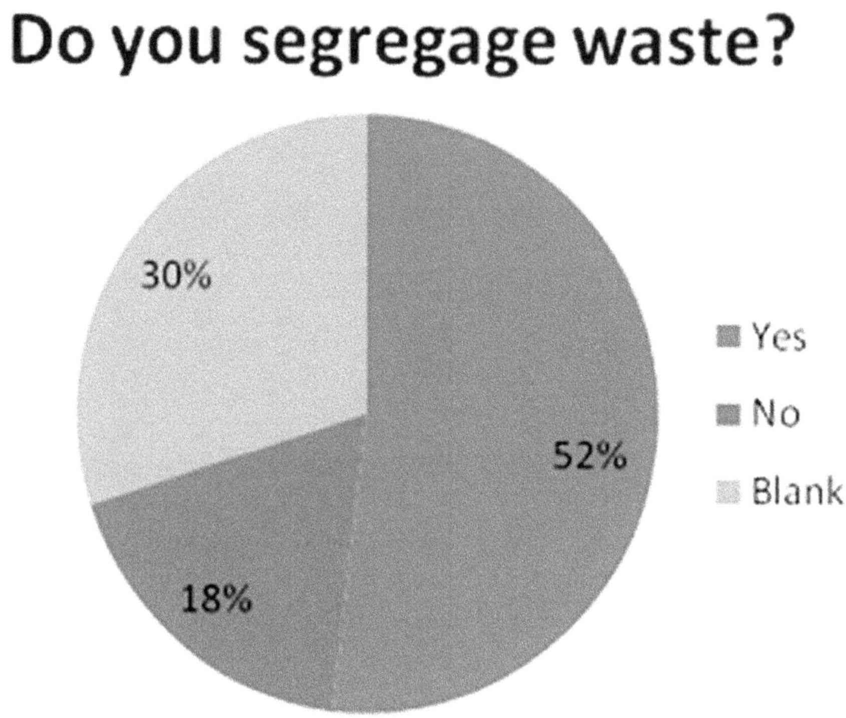

The analysis also showed that 46 percent respondents earned less than INR 3,00,000 (USD 4099) per annum while 22 percent earned between INR 3,00,000-5,00,000 (USD 4099 - USD 6382) per annum and 27 percent earned more than INR 5,00,000 (USD 6382) per annum. This data gives an indication of the spending potential, if there is will to shift to eco-friendly practice of treating wet waste at source.

It is noteworthy that more than half (52 percent) of the respondents do practice waste segregation (wet and dry) at their premises. If this 52 percent is motivated to move a step ahead and start treating their wet waste at home, it would reduce the quantity of waste being collected by RMC considerably.

Figure 12 shows a very interesting chart, it shows the percentage of respondents who are segregating waste at the source (primary y-axis), to the amount of wet waste they generate per day (x-axis); the cumulative quantity (in kg) of waste generated by these respondents who segregate waste is shown on secondary y-axis.

About 39 percent of those who segregated has wet waste generation within 0.5kg/day. If we calculate their cumulative amount, it is 21 kg / day wet waste which is depicted in the secondary y-axis. Similarly 43 percent of respondents who segregated waste at their premises generate wet waste between 0.5 kg to 1 kg, their cumulative wet waste comes to a starking 59 kg / day while 11 percent of respondents who segreagate waste generate between 1 – 2 kg waste / day generated 26 kg / day cumulatively as depicted in secondary y-axis. While though there were only 6 percent of respondents who generated more then 2 kg wet waste / day, their cumulative wet waste in kg was comparitively much higher and should not be ignored.

Figure 12. Respondents segregating waste and wet waste generation in Kg. with secondary axis showing cumulative wet waste in Kg.

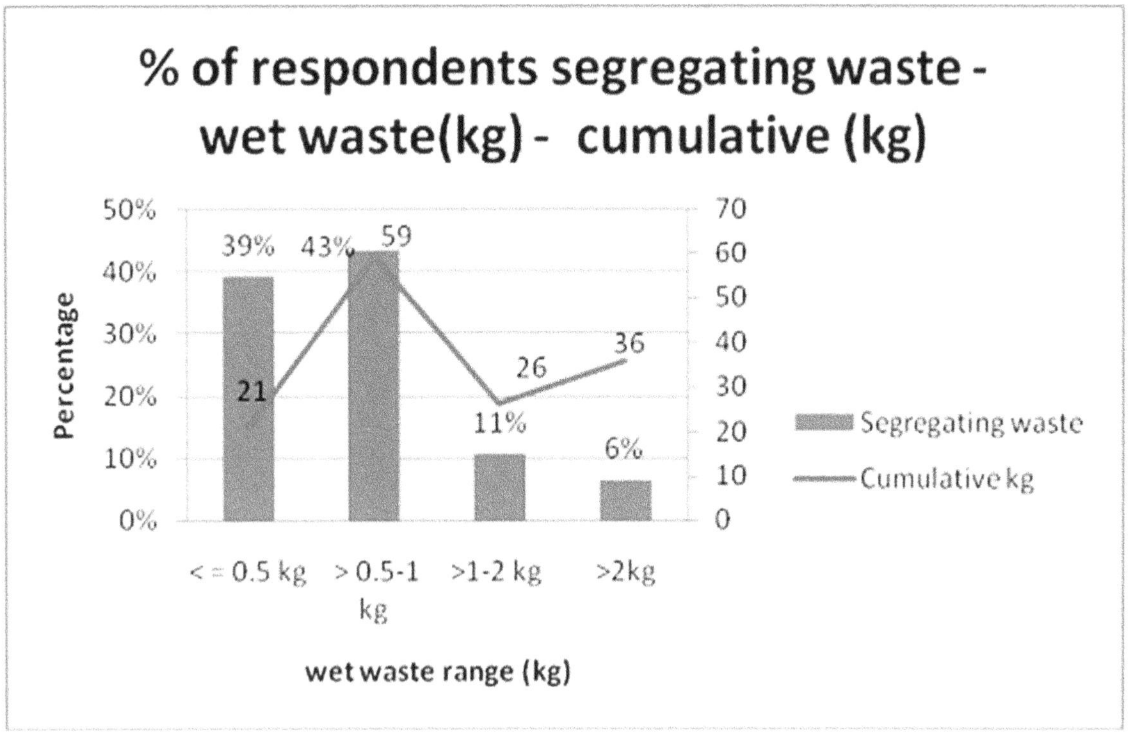

To extrapolate from the sample surveyed, if these households started treating their wet waste in-situ, a total of approximately 145 kg waste/day (i.e. 55,000 kg per year) would be reduced from the total waste collected/transported by the RMC. Given that every ton of waste disposed of at a landfill costs RMC USD 99/tonne/day, this action of treating waste at home could potentally lead to a saving of approximately USD 5240/year to RMC from this dataset alone. This shows that a much-needed potential nudge backed by a token incentive from the municipal corporation could see a significant shift in way the waste is managed in the city.

Climate change implication:

Landfill sites such as Nakravadi are overflowing with the waste and experiencing soil, air and water pollution in addition to CH_4 emissions due to anerobic deomposition of wet waste. Every tonne of food waste diverted from a landfill leads to a saving of 1.9ton CO_2e (Watch my waste, 2020). This implies that if 52 percent of respondents who are segregating waste start treating waste at the source would lead to a yearly reduction of 104.5 ton CO_2e emissions.

(Oakes, 2020) highlights that the GHG emissions from composting wet waste are just 14 percent of the same food dumped in a landfill. This approach, therefore, can become a low hanging fruit for the municipal corporation in its efforts to reduce climate change emissions.

To the question on their current knowledge about composting, around 63% indicated they were somewhat aware of what composting was all about primarily due to the awareness campaigns on various media platforms conducted by the Clean India Mission.

Figure 13. Knowledge about composting

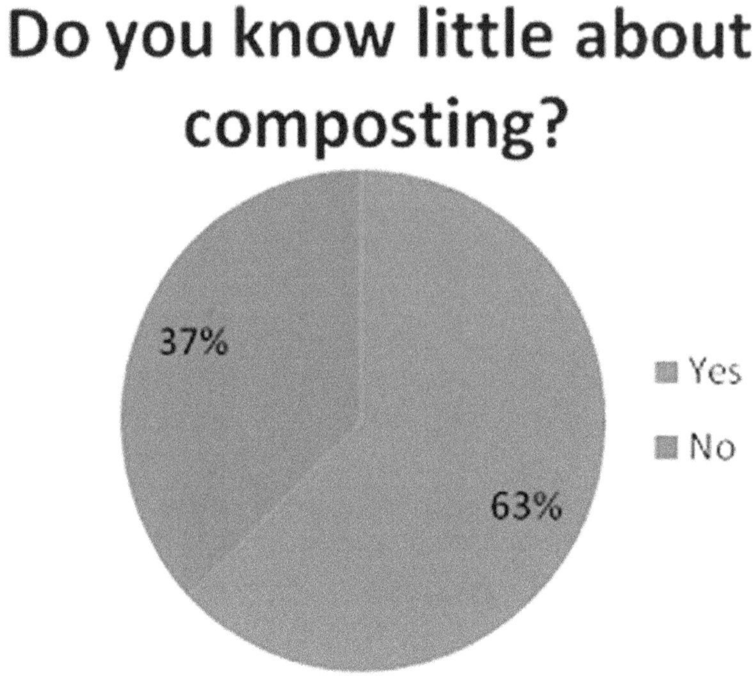

Figure 14. Respondents interest to learn composting

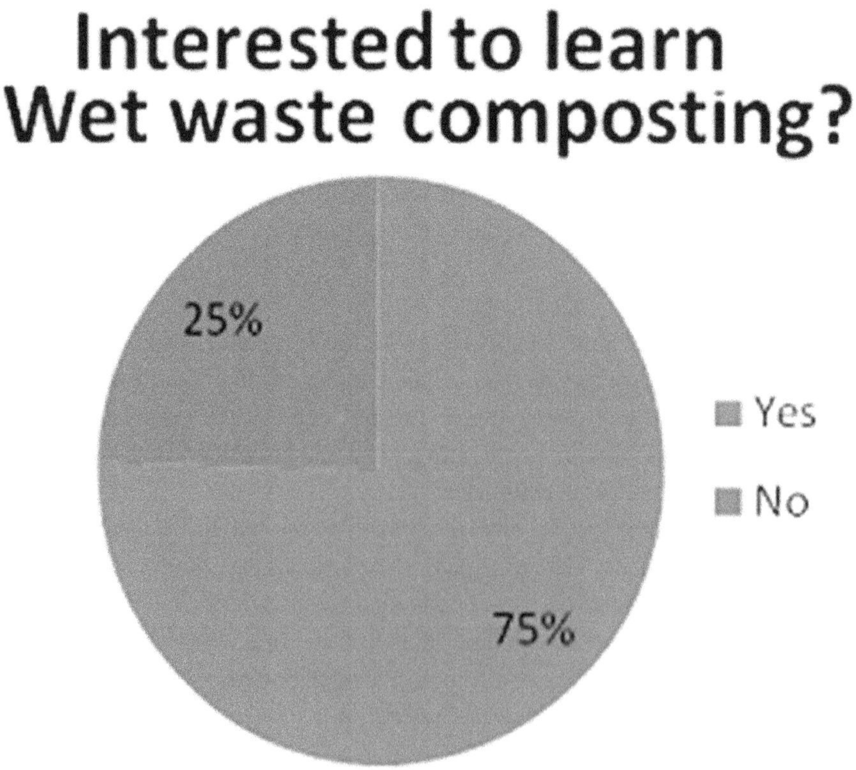

A significant 75% of the respondents indicated that they were interested to learn more about wet waste composting at source. However, when asked about their willingness to spend on learning a new skill, 65% of all the respondents showed their willingness to spend between INR 50 (USD 0.68) to INR 200 (USD 2.73) for this purpose (Figure 15), with 35% reluctant to spend on this activity.

Figure 15. Spending for learning new skill

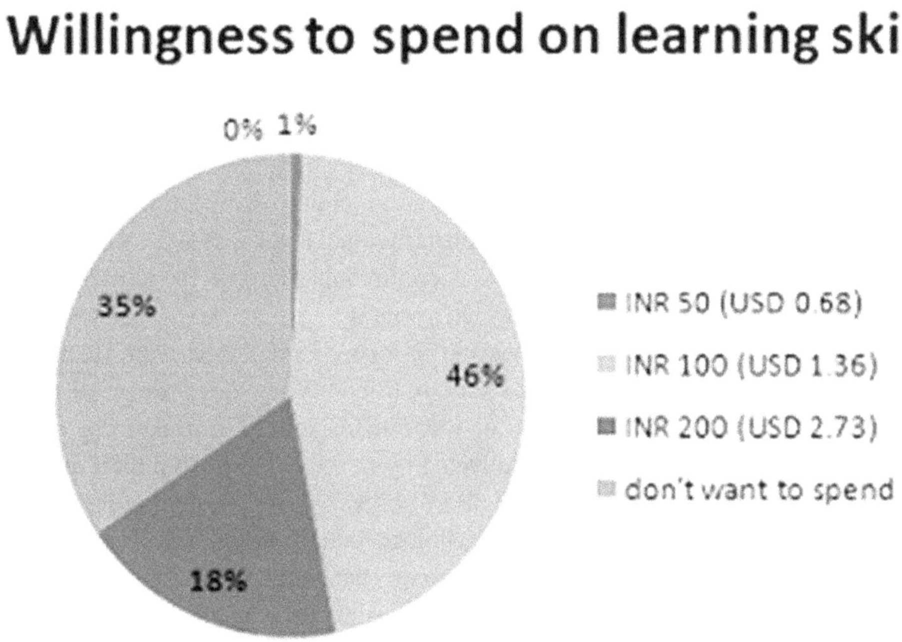

Figure 16. Contribution to better earth on adopting composting

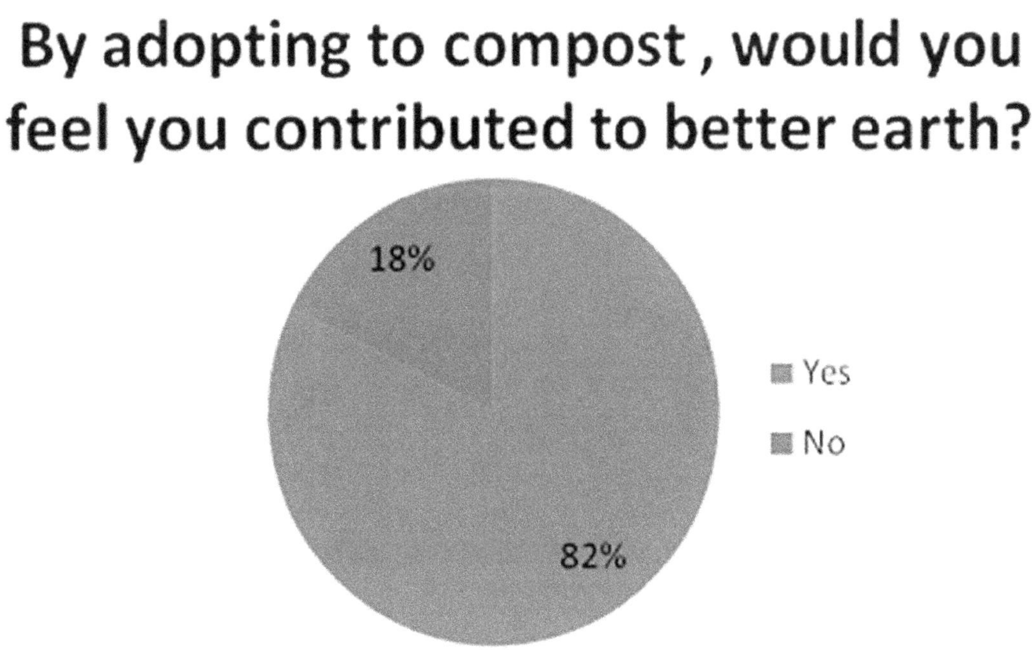

As indicated in Figure 16, 82 percent of the respondents also believed that if they adopted composting they would feel that they have contributed to better earth.

All this reconfirms that if potential nudge or small incentive is provided by the municipal corporation there would be a better adoption rate amongst citizens to treat wet waste at source.

PROMOTING CLIMATE AND ENVIRONMENTAL AWARENESS THROUGH DEMONSTRATION

The demonstration drive was a one-day exercise, with a nominal fee of INR100 (USD 1.36) per participant. It was felt that such an approach would ensure that only those genuinely interested in the issue and willing to replicate the process, would get enrolled.

The demonstration commenced with a brief presentation on why composting is an essential tool to combat climate change mitigation in a small way at the individual level, and as a collective best practice. The methodology was interactive to promote two-way communication with the participants, using tools such as quiz and giving award to those who answered correctly.

The presentation and question and answer session in form of small quiz were followed by screening of a short video on waste management in Rajkot city titled "Where does my waste go?" The video helped participants to connect with the process of waste management in Rajkot city and made them aware of what happens to the waste once it is collected by the tipper van from their area. This helped in perspective-building of the group that they are part of bigger waste problem of Rajkot city which is that of environment (Land, air, water pollution), causing social injustice in the lives of villagers living in the vicinity of the landfill site and are a party to climate change by not treating wet waste at source.

Power point presentations covered information on the salient aspects of composting and how it can be implemented. The idea was to keep the process as simple and affordable as possible. Several methods were discussed where participants were suggested options ranging from zero budget solutions such as using discarded earthen pots, plastic barrels, crates, to cheap composters like that of Daily dump,

Figure 17. Explanation of Wet waste composting using ICT and interactive approaches to participants including Deaf and Dumb students along-with their teacher

Figure 18. Participatory approaches and two-way communication during demo drive

to medium expensive ones such as Orbin. The live demo-drive motivated participants on how to do composting themselves. This was followed by a session resolving any queries and a feedback session.

Findings From the Demo Drive Feedback

As with the pre-survey questionnaire, completing the feedback form was voluntary. Of the total 55 participants, 27 per cent provided feedback; of these 100 per cent were (a) convinced that wet waste composting at their own premises was the best solution to climate and environment crises arising due to untreated food waste; (b) agreed to start composting at source; and (c) assured that they would recommend the workshop and spread the message to their family and friends. More over 100 percent of them were happy and satisfied with the demo workshop and were convinced that they would recommend this workshop to their family as well as friends.

An analysis of the feedback showed that 57 percent participants who gave feedback live in an apartment, with space restrictions. Of these apartment dwellers, 44 percent generated daily household wet waste of <1 kg and another 39 percent household had wet waste within 2 kg. Thus, the preference for composters which occupy less space and are sleek may be presumed.

Figure 19. Residence type

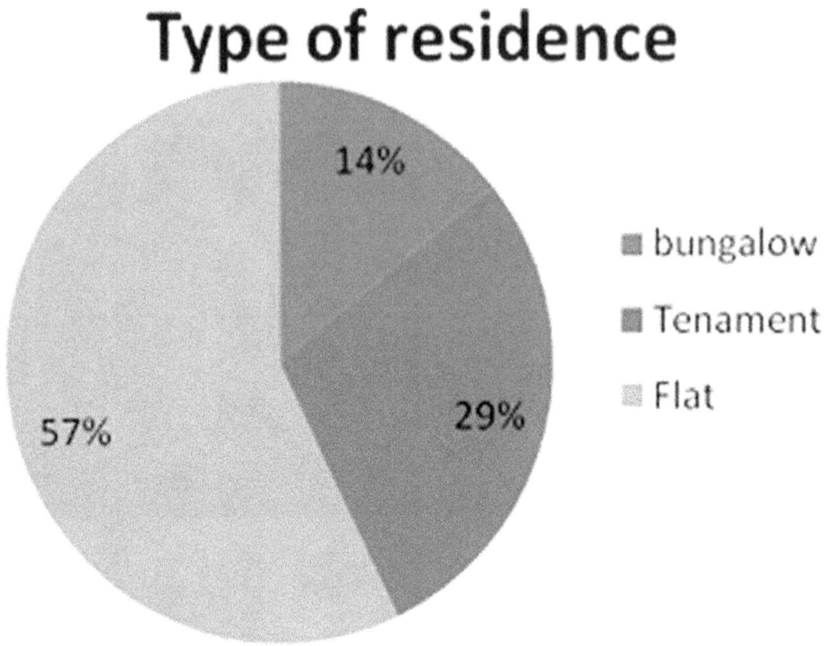

Figure 20. Avg. Wet waste / household

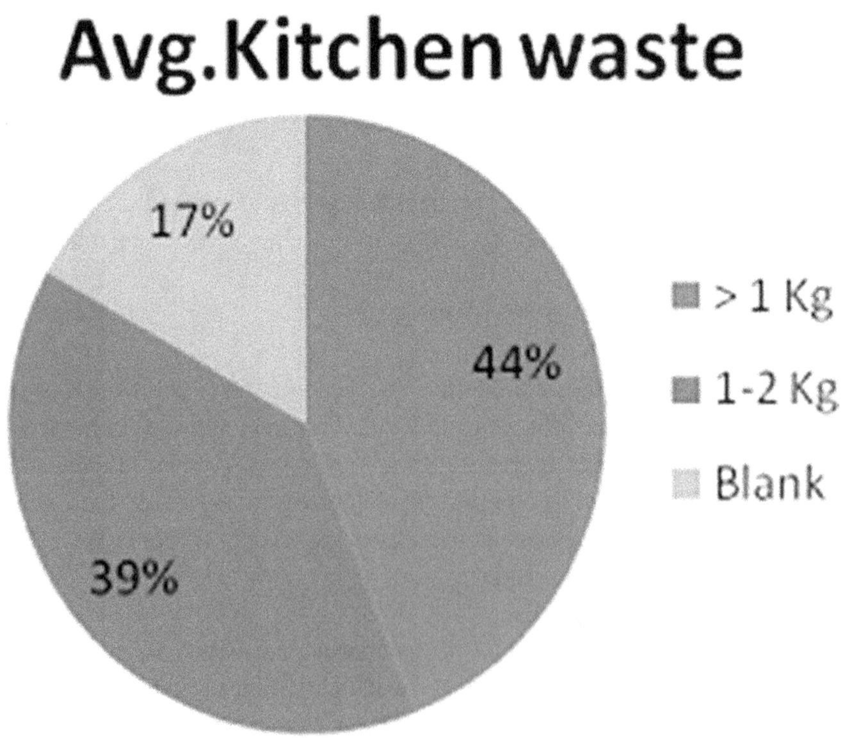

Figure 21. Inclination to buy composter

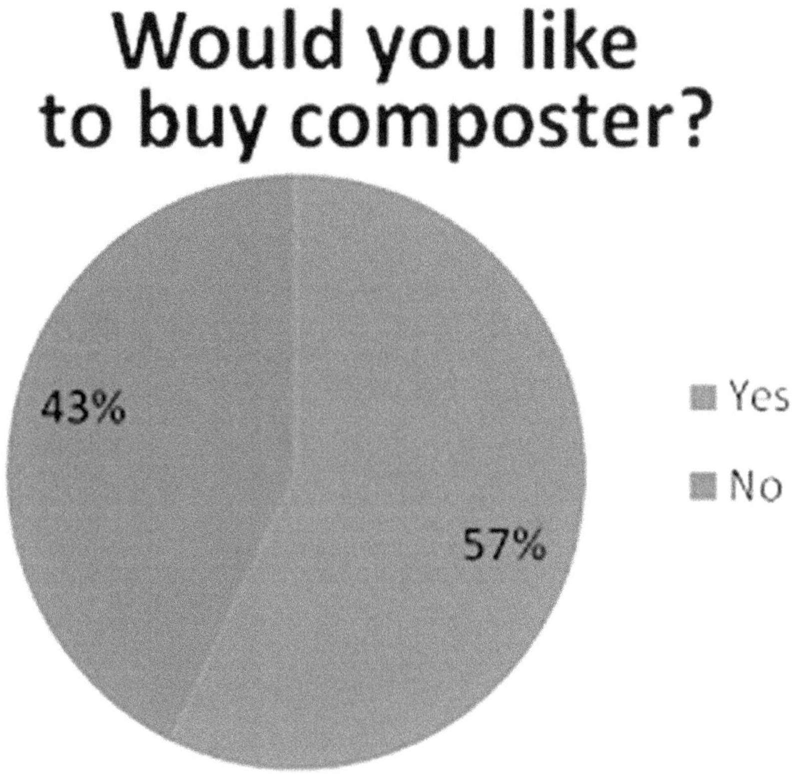

An interesting finding was that although 100 percent of participants who provided feedback responded that they were now convinced about composing at their premises; however, on further probing whether they would like to buy a composter to start composting revealed that 43 percent of them were not keen to buy a composter. This clearly shows how mindsets influence action – although most of the participants agreed that composting was useful, they were disinclined when it came to actually buying a composter.

While there can be many reasons to show less inclination to buy a composter, a leading reason could be demotivation to spend on buying one, or lack of space in their premises, another reason could be wanting to adopt zero-budget composters from discarded earthen pots, plastic bins/crates over other market options which were discussed and demonstrated in this demo drive.

A follow up with the participants showed that those who have adopted the composting at their premises have opted for zero-budget option by using discarded earthen wares or plastic drums for composting; only one participant actually bought a daily dump earthen pot composter.

The impact of the project can be summarised in table 3.

Table 3. Stage wise Demonstration drive Impact table

Input	Methods / tools used	Output	Impact
Awareness Phase	Secondary and Primary research, Key informant interviews, field visits	• Information gathering	Better sensitization of participants using audio visuals on journey of landfill and challenges faced by rag pickers
		Creation of video – "*Where does my waste go?*"	
		Prepared ready reckoner for wet waste composing as a reference tool	
	Pre-survey – online and offline	Interactive sessions on awareness on waste and composting. Completion of pre-survey questionnaire	• Close to 300 pre-survey forms filled online & offline • Reached out to > 10000 individuals through online & offline methods for sensitization
Demonstration phase	Media reach out for demo drive – online platform and newspaper media	Reached out to > 10000 individuals through online & offline methods	• >70 people registered • 55 participants participated
	Actual demonstration and hands on	Total Outreach: • 55 people total • 26 families, • about 300 people (~average 5 people / households)	• 5 households adopted • Virani Deaf & Dumb school adopted • Similar workshops to be organized for school students and doctors (based on feedback from a teacher and a doctor who attended) • Invite for conducting a workshop in Morbi city.

Feedback-cum-Review

To understand the effectiveness of the demo-drive a feedback was carried out with the participants in order to understand how many actually put into practice what they had learnt in the workshop, the challenges they faced, and the insights gained from their experience.

This feedback cum review process (a way of monitoring and evaluating this project) showed that issues such as presence of pests, as well as that of leachate due to improper wet waste treatment were faced by some who had adopted composting; these problems were addressed through handholding up to the 1st compost cycle. Many of the participants who attended the demo drive but did not adopt composting so far, are at least following the practice of segregation of waste before it is collected by the tipper van and have shown the inclination to start composting soon.

EXPERIENCES OF ACTION BASED LEARNING AND LESSONS LEARNT

The initiative covered a mixed batch of participants comprising of children, professionals, retired professionals, doctors, school teachers, homemakers and other sustainability enthusiasts. This varied group of individuals brought in several insights and ample learning that would contribute to replicating this initiative to cover a wider base of citizens in Rajkot city and beyond.

Figure 22. Participant feedback

Mrs Viditi Doshi – a business woman

"I am so happy that I can contribute whatever little to the environment as I can see nature working from waste to manure which I give to my plants. I am enjoying my composting experience which I learnt in the wet waste composting drive conducted by Ms. Chinmai. She has been immensely resourceful and helpful and has guided me whenever I was stuck or needed help. I would recommend this to my friends and family too"

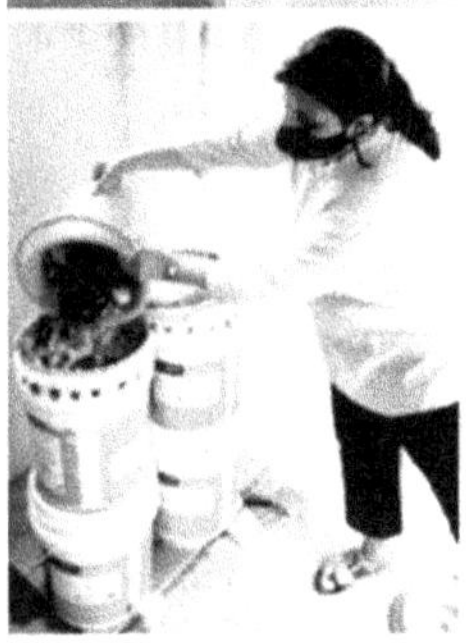

Mrs Puja Shah – A home maker

"We are happy with the outcome of good quality compost which we now use regularly in our garden. Now we don't need to buy external chemical laden fertilizers, this compost does the wonders. I have shared the positive results in our circle. Each one should take this up. I am glad I learnt it from Ms Hemani's workshop post which we have been following diligently and I urge all the urban people to participate and do their bit for better environment."

Dr Deepa Maniar – Renowned Gynaecologist

"Hi, myself Dr Deepa Maniar. I am happy to share my experience of composting. I live in an apartment where full parking lot is lined up with pots (herbs like tulsi, ardusi and flowers like rose, marigold) When I came to know about this composting drive through whatsapp group, I participated in it. At home I tried with help of my hospital staff. My 1st batch of compost which was ready by March was excellent but due to Covid further composting was restricted due to my busy schedule. I plan to reinitiate my composting experience now. Infact I had suggested Chinmaiji to take a similar session for us doctors however due to covid our priorities have shifted but will surely do in near future.

One good thing though that happened, my staff who is daughter of a traditional farmer, through this experience and her interest in this technique got her interested to learn more about organic farming. We directed her to get connected to Junagadh Agricultural University. She applied this composting knowledge and more information she learnt from university in her dad's farm. Thanks to Chinmaiji who is working really hard, to get us interested and to my staff who got a new direction. I am sure the good results will be seen in near future."

The key insights and learnings are summarized as follows:

1) At present, citizens of Rajkot are unaware of what happens to the waste and its treatment once it is collected from their premises. To address this, media such as video and power point presentations on local issues, followed by discussions and Q&A sessions, were some of the very useful interactive tools that were used during the workshop. This helped participants remain engaged, become aware, connect to the problem of waste and its treatment/management, take ownership and gradually evolve into change agents.
2) This kind of demo cum hands-on activity based on experiential learning focused on learners reflecting on their experience of doing hands-on, so as to gain conceptual insight as well as practical expertise.
3) Such demonstration and handholding for the participants who adopted composting at their premises led to their learning retain permanently.
4) The demo-drive empowered participants to make healthy life decisions by accepting personal responsibility for their actions and opting to start segregation and treatment of waste;

The demo-drive led to the identification of the following key factors that influenced its success and are seen as necessary for the roll-out of the program beyond the pilot:

- Participation of more than one family members, especially "mother – child duo" can reinforce the learning of the concept and make their involvement in the activity more fruitful
- Such team approach to practise composting and then using the compost in their garden/ premises and also to develop a kitchen garden provides long lasting results
- Nomination and participation of teachers, student leaders and/or Eco-club coordinator can contribute to the adoption of composting activities in school, thus reaching a wider base of possible practioners.
- The information and knowledge gained by the participating teachers would:
    - enhance their professional learning, and
    - help them develop innovative ways to link the hands-on activity of composting to the Science syllabus and other subjects such as Socially Useful Productive Work (SUPW), Home Science, etc.
- Such workshops provide life skill learning opportunity too.
- There is a critical need for follow-up planning that would help take the pilot initiative to scale. The broad elements of such a planning process would include appropriate technology options, costing, identification of funding sources, bringing on board government as well as non-government agencies and citizens groups, and generating wide-scale demand for wet waste management at source.

FUTURE RESEARCH DIRECTION

Considering the potential from this demonstration drive with 55 participants, what happens if majority of the citizens in a city start practicing composting at the source? Total solid waste generated by residents of Rajkot is about 0.3 kg / household daily out of which wet waste created by citizens amounts to approximately 40-50 percent i.e. about 0.15 Kg/household/day (RMC, 2011). Given the city's population of 1.5 million would result in a staggering 45000 kg / day! Thus wet waste composting at source has a great potential saving by RMC in terms of:

- monetary benefits due to reduced transporation trips by tipper vans and handling less waste;
- reduction in methane generation from the landfill due to reduced wet organic waste; this would help municipal corporation to meet its carbon goals;
- reduction in size of landfill apart from reduction in soil, water and land pollution.

Even if only 0.01 percent households of Rajkot city treat wet waste implies 15.6 ton CO_2e/year being removed from atmosphere. Thus the city will not only meet climate goals but also achieve several SDGs like Goal 6: Clean Water and Sanitation, Goal 10: Reduced Inequality, Goal 11: Sustainable Cities and Communities, Goal 12: Responsible Consumption and Production, Goal 13: Climate Action, Goal 14: Life Below Water, Goal 15: Life on Land.

Future Action Plan

With the objective of expanding the pilot to cover a wider citizen base in Rajkot city, engender a mass movement on waste management, promote greater adoption rate for composting options leading to an increase in climate benefits, it is proposed that RMC be approached with the following suggestions:

Replication and Up-Scaling

1. Share with RMC, the outcome and lessons learnt from this pilot project through analytical published reports showing the performance of different typologies of residential societies, successes with indicators for measurement, loop holes and failures generated and benefits accrued, so that similar models can be rolled out for other societies. If required, support could also be provided for the formulation of bye-laws or policies to be implementable at city level.
2. Support development of strategies on how slum dwellers / rag pickers can be involved in composting initiatives at community level, so that it becomes alternative source of income for them while they are trained to manage waste at community parks / spaces owned by the RMC.
3. Convince RMC to facilitate tie-ups between the participating societies and nurseries (Forest / State / Private) for compost procurement, thus creating a closed loop and market incentive to continue composting.

Costs, Financing and Incentives

1. RMC to explore the possibility of sharing information on this pilot initiative with all its registered Smart Societies. This would ensure that representatives from these societies participate in the initiative with full vigour.
2. It is also recommended that RMC provides space and other logistical/financial support to conduct these drives in community parks; such a step would encourage the participation of a larger number of households.
3. One of the recommendations of the pilot initiative was that a higher adoption rate would be feasible if RMC could provide some kind of an incentive to citizens if they purchase a composter. Some

options that are proposed could include (a) waiver of INR1,000/- (USD 13.70) for each composter purchased; (b) rebate on property tax or other municipal taxes. These incentives could act as motivation for adopting this practice by citizens or residential societies. As these have policy implications, support would be provided to RMC to assess the cost-benefit ratio and hand-hold the process until its fruition.

CONCLUSION

This pilot initiative has demonstrated that wet waste composting at source has multiple socio-economic and environmental benefits. It is a low hanging fruit that enables any urban local body to go clean, green and reduce its carbon emissions in a relatively cost-effective manner by investing on awareness generation cum action at the household level.

It is well recognised that the co-benefits of climate change mitigation and adaptation measures are most visible in cities as it is easier to link adaptation and mitigation actions with local development. Composting also has the potential for employment generation, can contribute to poverty reduction and result in improved sanitation and basic health. Moreover, by nudging to use compost in their own gardens to enhance soil fertility or inspire them to create their own green spaces and kitchen gardens where the compost could be used, has the potential to provide enhanced food security and nutrition.

ACKNOWLEDGMENT

This action research was selected in 2019 for funding under Government of Gujarat's Climate Change Department (CCD) and GEDA Small Grants Fund Programme with partial funding [Grant number: PRJ/10-2016/735712/B]. It was also supported by Vasu Pharmaceuticals Pvt Ltd and RuChiNi ESP as major partial sponsors along with other sponsor Patel Brass Works Pvt Ltd.

REFERENCES

ICLEI. (2016, March). *Rajkot, India Enhancing 'Livability' through Urban Low Emission Development.* Retrieved Sept 2020, from ICLEI: https://urban-leds.org/wp-content/uploads/2019/resources/case_studies/ICLEI_cs_175-Rajkot-UrbanLEDS_2016-web.pdf

Leicht, A., Heiss, J., & Byun, W. J. (Eds.). (2018). *Issues and trends in Education for Sustainable Development.* Retrieved Sept 2020, from UNESCO: https://bangkok.unesco.org/content/unesco-publication-issues-and-trends-education-sustainable-development

Ministry of Urban Development. (2015, June). *Smart cities Mission statement and guidelines.* Retrieved Sept 2020, from Ministry of Urban Development, Government of India: http://smartcities.gov.in/upload/uploadfiles/files/SmartCityGuidelines(1).pdf

Oakes, K. (2020, Feb 26). *How cutting your food waste can help the climate*. Retrieved Sept 2020, from BBC: https://www.bbc.com/future/article/20200224-how-cutting-your-food-waste-can-help-the-climate#:~:text=Onepercent20studypercent20estimatedpercent20thatpercent20the,frompercent20puttingpercent20itpercent20inpercent20landfill

Planning Commission. (2014, May 12). *Report of the Task Force on Waste to Energy (Volume9 I) (In the context of Integrated MSW Management).* Retrieved Sept 2020, from MNRE: https://mnre.gov.in/img/documents/uploads/7c1d4e00bb994cffbdd32d12cae627ce.pdf

RMC. (2020). *Existing infrastracture*. Retrieved Sept 2020, from Rajkot Municipal Corporation: http://www.rmc.gov.in/rmcwebsite/infrastructure.aspx

Sambyal, S. S. (2018, Jan 25). *Delhi's Solid Waste Management Bye-Laws focus on individual accountability*. Retrieved Sept 2020, from Downtoearth: https://www.downtoearth.org.in/news/governance/delhi-gets-it-new-byelaws-on-solid-waste-management-59559

Singh, C. K., Kumar, A., & Roy, S. S. (2018, Feb 13). Retrieved Sept 2020, from https://www.ncbi.nlm.nih.gov/pmc/articles/PMC5811459/

Steinig, S. Y., & Butts, D. M. (2009). *Generations Going Green: Intergenerational Programs Connecting Young and Old to Improve Our Environment*. Retrieved Sept 2020, from American Society on Aging: https://www.asaging.org/blog/generations-going-green-intergenerational-programs-connecting-young-and-old-improve-our

Swaminathan, M. (2018, Apr 20). *How Can India's Waste Problem See a Systemic Change?* Retrieved Sept 2020, from EPW: https://www.epw.in/engage/article/institutional-framework-implementing-solid-waste-management-india-macro-analysis

UNESCO. (2014). *UNESCO Education strategy 2014-2021*. Retrieved Sept 2020, from UNESCO: https://unesdoc.unesco.org/ark:/48223/pf0000231288

UNFCCC. (2020). *Global Warming Potentials (IPCC Second Assessment Report)*. Retrieved Sept 2020, from UNFCCC: https://unfccc.int/process/transparency-and-reporting/greenhouse-gas-data/greenhouse-gas-data-unfccc/global-warming-potentials

USAID. (2018, Sept). *Greenhouse emissions in India*. Retrieved Sept 2020, from USAID: https://www.climatelinks.org/sites/default/files/asset/document/India percent20GHG percent20Emissions percent20Factsheet percent20FINAL.pdf

Watch my waste. (2020). *Food Waste Greenhouse Gas Calculator*. Retrieved Sept 2020, from Watch my waste: https://watchmywaste.com.au/food-waste-greenhouse-gas-calculator/

World Bank. (2013, Jan 22). *New World Bank report offers policymakers roadmap for building resilient sustainable cities*. Retrieved Sept 2020, from The World Bank: https://www.worldbank.org/en/news/press-release/2013/01/22/new-world-bank-report-offers-policymakers-roadmap-building-resilient-sustainable-cities

World Bank. (2014, April 11). *Urban Development: Sector Results Profile*. Retrieved Sept 2020, from The World Bank: https://www.worldbank.org/en/results/2013/04/14/urban-development-results-profile

WWFIndia. (2013, Dec 20). *Paryavaran Mitra (Janvikas) and Ors. Vs Gujarat State Pollution Control Board and Ors.* Retrieved Sept 2020, from WWFIndia: https://www.wwfindia.org/?10463/Paryavaran-Mitra

ADDITIONAL READING

CHEEPO. (2020). *Manual on Municipal Solid Waste Management - 2016.* Retrieved Sept 2020, from Central Public Health and Environmental Engineering Organisation: http://cpheeo.gov.in/cms/manual-on-municipal-solid-waste-management-2016.php

Ministry of Housing and Urban Affairs. (2018, June). *Advisory on On-Site and Decentralized Composting of Municipal Organic Waste.* Retrieved Sept 2020, from Central Public Health and Environmental Engineering Organisation: http://164.100.228.143:8080/sbm/content/writereaddata/Advisory percent20on percent20decentralised percent20composting.pdf

Ministry of Housing and Urban affairs. (2020). *Swachh Bharat Misison Urban.* Retrieved Sept 2020, from SBM Urban: http://swachhbharaturban.gov.in/

Ministry of Urban Development. (2008). *National Urban Sanitation Policy.* Retrieved sept 2020, from Ministry of Urban Development: https://smartnet.niua.org/content/ce856338-d63d-4963-873f-1a97d9e30474#:~:text=The percent20policy percent20envisions percent20that percent20 percent22All,the percent20urban percent20poor percent20and percent20women percent22.

Ministry of Urban Development. (2015, June). *Smart cities Mission statement and guidelines.* Retrieved Sept 2020, from Ministry of Urban Development, Government of India: http://smartcities.gov.in/upload/uploadfiles/files/SmartCityGuidelines(1).pdf

Ministry of Urban Development. (2016). *Municipal Solid waste management manual Part II: The manual.* Retrieved from http://cpheeo.gov.in/upload/uploadfiles/files/Part2.pdf

NCDEQ. (2017, Oct). *Compost pilots and demos.* Retrieved Sept 2020, from NC Department of environmental quality division of waste management solid waste section: https://files.nc.gov/ncdeq/Waste percent20Management/DWM/SW/Composting percent20and percent20Land percent20Application/CompostPilotsDemoGuide11-6-17.pdf

Press Trust of India. (2018, July 26). *Gujarat announces Urban sanitaiton policy.* Retrieved Sept 2020, from Business Standard: Gujarat govt announces urban sanitation policy. Retrieved Aug 10, 2020, from Business Standard: https://www.business-standard.com/article/pti-stories/gujarat-govt-announces-urban-sanitation-policy-

RMC. (2006). *Detailed Project report for Strengthening Primary and Secondary solid waste management in the city.* Retrieved Sept 2020, from Swachh bharat urban: http://swachhbharaturban.gov.in/Auth/dsdocumentsfile.aspx?DOCTYPE=942&DOCID=103

KEY TERMS AND DEFINITIONS

Aerobic Composing: Is decomposition of organic matter using microorganisms that require oxygen. Heat, water, and carbon dioxide (CO_2) are by-products of aerobic digestion. Though CO_2 is a greenhouse gas, it's released as one of the outputs from the composting process is not counted in emissions inventory. Compost - the utilizable product as its output is earthy, dark and crumbly substance which enriches soil. Winrow composting, rapid composting including vermi-composting, using mechanical composter are several methods for aerobic composting.

Anaerobic Composting: Is a process of decomposition of organic matter using microorganisms which do not require oxygen to survive. An anaerobic system releases majority of the chemical energy contained within the starting material as methane. In the anaerobic process the utilizable product is methane gas (which can be used as bio-gas for home kitchen use) and sludge, which can be composted or used directly as manure.

Climate Change Adaptation: The process of adjusting of social, economic or ecological systems as a response to actual or expected climate change and its impacts.

Climate Change Mitigation: Involves reducing the flow of heat-trapping greenhouse gases into the atmosphere.

CO2e (Carbon Dioxide Equivalent): Our carbon footprints are measured through this standard unit.

Composter: Mechanical equipment, compost bins used for aerobic composting also called composters.

Composting: Composting is the decomposition of plant remains and other biodegradable organic matter. Biodegradable waste (organic - kitchen and garden waste) can be decomposed in two ways: Aerobic processes (in the presence of oxygen), and Anaerobic processes (in the absence of oxygen).

APPENDIX 1

Pre survey Form & Feedback to Online Pre-survey Forms

Survey No: Survey Questionnaire: Wet waste to compost / manure

This is a 2 minute survey of people interested in treating wet waste at their Homes / Colony at Rajkot

Greetings! I am Ms Chinmai Hemani, a Climate change and Sustainability professional from RuChiNi ESP. I plan to organize workshop cum demonstration of learning environment and climate friendly skills of wet waste management at the source and convert it to useful compost manure to be used to create green spaces. This initiative will not only create less burden on municipality to collect and treat wet waste and reduced landfill needs but also encourage citizens to be participatory agent in the Swatchhta Abhiyan. It can also be showcased as one of the climate mitigation strategies for Rajkot - Smart city. We plan to conduct 2 workshops open for all citizens targeting 70 - 100 people, We will contact you email or your phone to share further information of the workshop when it is planned if you have answered 'Yes' to question 10. (Please see: Composters are Compost bins used to treat wet waste). Thank You!

Name	
Phone no. -provide whatsapp no. too	Is this your whatsapp no.? Yes No
Email	
Gender: M / F / other	Age: 10 -20 years 21-50 years 51-70 years 70 & above
Profession (Tick mark applicable)	Academic Non-government Government Private-sector Self employed Restaurant/ café owner Student Housewife
No. of members in family	
What is your annual household income range?	< 3 lakhs/ annum >5 Lakhs / annum 3-5 Lakh / annum other
Name of Society, area you live in and full address	
What type of house do you live in? (Tick one)	Bungalow Apartment Government Quarters Others Tenement Slum Company accommodation
Do you segregate waste into wet and dry waste? Yes No	
How much kg waste (total) do you generate / day ? …… Kgs	
Out of it how much is wet waste …… Kgs	
Do you know about what is wet waste composting? Yes No	
Composting not only prevents all forms of air / water / land pollution but also helps prevent greenhouse gases which lead to climate change. By adopting to compost would you feel that you have contributed for better earth? Yes No	
Would you be interested to learn how to treat your wet waste you generate in kitchen or your garden into manure? Yes No	
How much would you be willing to spend for the workshop to get trained for treating wet waste into manure? • Rs 100 or Other ………… • Rs 200 (take reference material) • Rs 500 (Take away reference material handholding–1 visit/month upto 2 months (i.e. 2 compost cycles)	
Would you like to buy the composter? (Equipment to deal the wet waste) Yes No	
How much would you be willing to spend on Composters (For Household of 4-5 People), (For Community of 20 - 25 People) (Tick one) • Rs. 2000 (Household) - made from plastic Treats 1 kg waste / day or Other ………. • Rs. 3000 (Household) - made from clay Treats 1 Kg waste /day • Rs. 15000 (Household) - easy to use with wheels Treats 2 Kg waste / day • Rs. 80000 (Community - 20-25 households) • Rs. 150000 (Community - 40-50 households)	
Would you share this information to your family and friends? (We would be grateful if you do) through: Facebook Whatsapp Instagram Twitter Word of mouth Phone Email No I would not share Other If other - Please mention other	
If you would like to share name and Ph. Number/email of your family or friends who would be interested	
Remarks	

Chapter 6
Ecotourism as a Sustainable and Active Citizenship Approach

Ana Lúcia Almeida
ESHT, Polytechnic Institute of Porto, Portugal

Teresa Dieguez
https://orcid.org/0000-0002-4886-1446
Polytechnic Institute of Cávado and Ave, Portugal

ABSTRACT

Tourism offers both advantages and disadvantages to the involved countries. Ecotourism emerges as a tourism segment based on the principles of sustainable tourism, oriented to the preservation of nature, respect for local traditions and customs, interest in the history and culture of the destination, encouragement of intercultural communication, and aid to the local economy. This study aims to understand how students of tourism from the Porto Polytechnic Institute, in Portugal, know and define it. Additionally, it aims to understand how many of these students have participated in typology activities and the reasons that led them to do so. A quantitative research with 123 students (44% of the universe) was conducted and the results showed that 23% of students know the four basic principles of ecotourism and about half have experience in the practice of it. The main motivation is related to "contact with nature." Finally, it is discussed how ecotourism can be seen as a way of empowerment of students, regarding sustainability and the well-being of our environment and its inhabitants.

INTRODUCTION

The tourism industry is nowadays among the largest industries in the world, and nature-based tourism (NBT) is often said to be one of the fastest growing types of tourism. Its contribution to economic growth is relevant, but has negative impacts concerning environment and infrastructures, which makes it a subject of concern (Øian et al, 2018). The concept of sustainable tourism is complex and tends to bring out several paradoxes and dilemmas as the three dimensions of economic, social, and environmental concerns are difficult to balance (Gkoumas, 2019). Some short-term solutions may turn into unsustain-

DOI: 10.4018/978-1-7998-7512-3.ch006

able solutions in a longer time perspective (Sedarati, Santos & Pintassilgo, 2018). The United Nations World Tourism Organization (UNWTO) promotes tourism as a key to achieve sustainable development goals (SDGs) and countries with a more sustainable tourism sector are more likely to enjoy higher levels of positive peace in the future (WTTC, 2016). As the impact of sustainable development in the world's future is better understood, every industry is encouraged to get a solution for sustainability in its own field. Actions are conducted in a responsible way, through a set of principles and values, having in mind its impacts on economics, environment and society (Font & Mccabe, 2017; Dieguez, 2018).

Tourism is a very powerful social invention to deal with the plenty of desires of the modern industrialized world (Kunwar, 2018). It is "an arena in which many players interact and negotiate the construction of culture to different ends" (Salazar, 2006, p.329), and can be considered a "machine of happiness' (Wohlmuther & Winstersteiner, 2014, p .35), where the cultural "value" is "travelling to happiness" (Wohlmuther & Winstersteiner, 2014, p.35). As hallmark activities of the postmodern world, tourism and hospitality (TH) are "worldmaking" activities that have a creative and transformative role in the making of people and places and in the production of meanings, values and understandings about the past, present and future (Urry, 2003; Dredge, 2017). In other words, if TH is well managed within an integrated and sustainable approach, they can have significant impact on places and people, add value to cultural and environmental resources, and promote peace and understanding (Sharpley, 2009). They can raise social and cultural awareness and tolerance, deal with poverty, give power to communities and make them richer and more developed (Higgins-Desboilles, 2006, Dredge et al., 2012).

As Tourism is getting more importance in our society, knowledge production in tourism requires closer link with Higher Education Institutions (Castells, 2000, Portegies, De Haan, & Platenkamp, 2009). In fact, Universities play an important role as education is an important asset for people to positively and negatively compare things in life and strongly build a better future making smart decisions (Beier & Rittmayer, 2008). Also, researches and studies are critical to make available knowledge and understanding, in particular to students in learning how to control their behavior regarding the environment (Emanuel & Adams, 2011; Tan & Lau, 2009). Tourism can be considered as a form of education, whether it's by learning about the history of different places, about other gastronomic cultures or language, whether by learning to respect others peoples' space or our planet, by understanding it's our duty to live consciously. Many times, by leaving our usual place we understand how precious our planet is and how much we should care about it. Additionally, we also learn how diverse is our world and how respectfully we should be towards everyone. Tourism has the potential of teaching people how to take care of our planet and to respect and care about other communities (Dangi, 2018; Mondino, & Berry, 2019).

Among the innumerous typologies of tourism, Ecotourism is the one that matches the best with this kind of ideals. While it promotes nature conservation, Ecotourism, as an alternative tourism, involves visiting natural areas in order to learn, to study, or to carry out activities environmental friendly, that is, a tourism supported on the nature experience, which make possible the economic and social development of local communities (Kiper, 2013). It also aims to protect resources, in particular biological diversity, and maintain sustainable use of resources, which can take ecological experience to travellers, preserve the ecological environment and obtain economic benefit (Kiper, 2013). In other words, socially, historical, cultural destinations, local traditions and customs are promoted and preserved. Local products and businesses arise and local people are recruited to work. Economically, direct financial benefits appear for conservation and for local communities. People are more empower, more engaged, with legal power of decision. More active citizenship among the tourists are triggered sustainability is promoted.

In this sense, the more informed tourism students are about Ecotourism more empowered they are as future citizens, aware of their responsibilities towards the Earth and the people that inhabit in it (EEFEct, 2018). By both theoretical and practical skills about issues linked to Ecotourism, Higher Education Institutions can promote projects that aim offer a global vision of the possibilities offered by Ecotourism as a way to promote the economic development of rural and mountain areas, for example, a kind of tourism which doesn't compromise the ecosystem as well as activities to have fun by having healthy lifestyles.

This chapter will present a Case Study, a survey conducted in the Porto Polytechnic Institute, more specifically in the Hospitality and Tourism Higher School (ESHT-IPP), in Portugal. Its main goal is try to understand how conscious and informed the students from this Polytechnic are aware about Ecotourism and its impact on the present and future. A literature review about Ecotourism concept and basic principles will be presented, as well as the classification of ecotourists, the evolution of Ecotourism over the years and the importance of the involvement of the community. Then it will be presented the used methodology, the gathering data, the analysis and discussion of results. Finally limits of study are pointed, conclusions are made and further research is suggested.

LITERATURE REVIEW

Ecotourism

In Ecotourism, the major incentive is the surveillance and enjoyment of natural features and cultural related assets, although in adventure tourism it is somewhat the physical exercise and the testing situations in natural environments (Wood, 2002). In essence, Ecotourism must be planed and managed to successfully offer its key social and environmental objectives, which require specialized marketing, management skills, guiding and interpretation services, government policies and focused attention on local peoples, who must be given the right of prior informed consent, full participation and means and training to take advantage of this sustainable development option (Wood, 2002).

The study of Ecotourism is diminutive, across no more than 50 years. During 1960-1970, with the development of the environmental movement, people begin to care about tourism' environmental and social impacts and started exploring how to use the nature, while balancing tourism, environmental protection, and sustainable development (Fang et al, 2018). In 1981, Ceballos-Lascuráin, a Mexican architect, started using the Spanish phrase 'turismo ecológico' in order to express the Ecotourism approach (Wearing & Neil, 2009; Ceballos-Lascurain, 1987). Ecotourism can be seen as a substitute to mass tourism and took for the purpose of revising, enjoying, learning and experiencing the ecological environment in the lack of anthropogenic interfering or contamination, promoting local economic and ecological conservation (Ceballos-Lascuráin, 1987). A type of responsible tourism with the goal of admiring and enjoying nature (as well as some past or present cultural manifestations) which at the same time promotes the conservation, provides low impact by the visitors and positively contributing to the socio-economic involvement of the local community (Ceballos-Lascurain, 1996).

As a concept of environmental literacy and innovative business, Ecotourism since the 1990' includes adventure tourism, alternative tourism, environment-friendly travel, ethical tourism, green tourism, nature tourism, nature-oriented tourism, responsible tourism, soft tourism/tourisme doux, and sustainable tourism (Dolnicar, Crouch & Long, 2008; Xu & Fox, 2014). Ecotourism is a touristic segment ruled by the fundamental principles of sustainable tourism, following values such as nature preservation. Interest in the destination's history and culture, respect to the local traditions and customs, support of intercultural communication, assistance to the local economy by choosing local products and business, as well as interest on environmental events and on the reduction of non-renewable resources (Barkauskiene & Snieska, 2013).

In fact, Ecotourism is arguing for international recognition as a way to enhance sustainability and reduce the impact of tourism on the climate and environment. In this sense, it is related to responsible travel to natural areas, without negative impacts to local community's welfare and environment (TIES, 2018). Over the years, its range has expanded to incorporate environmental conservation, human rights and ethical issues, education, cultural preservation, economic development of local communities and social inclusion (Cobbinah, 2015). Examples of Ecotourism are ecolodges that minimize the environmental impact of the accommodation, as they protect the environment and provide experiences embedded in local culture. Often Ecotourism tours are organized that offer to small groups of people a well-planned interactive travel experience with new cultures and environments, with minimal environmental impact (UNEP, 2002).

The International Ecotourism Society (TIES) was launched by a team organized by Megan Epler Wood at a conference in Florida in 1989 as the world's first international non-profit dedicated to Ecotourism as a tool for conservation and sustainable development, having as long term goals the establishment of education programs (Jafari, 2000). The first board of directors and Chairman, David Western, only were recruited in 1990. Western and Epler Wood raise the first grants for the Society (TIES, 2020). Ecotourism is a accountable travel to natural areas in which the environment is conserved and the welfare of the community is encouraged. Therefore, this type of sustainable tourism focuses on two main aspects: the awareness regarding the environment preservation and the interest on the well-being of the visited location. (TIES, 2020).

Ecotourism is related to Ecotourism brings communities together, with preservation and travel, through memorable interpretative experiences to visitors and recognize the rights and divine philosophies of the natives and cooperates with them to create empowerment (TIES, 2020). It involves four essential elements namely concept of movement to areas as natural as possible, nature based through leisure, studies and volunteering, conservation's endorsement and educational role (Wearing & Neil, 2001). Ecotourism, according to TIES, divides itself into three specialty areas: conservation, communities and interpretation, as the figure below (Figure 1) shows:

A sustainable relationship with nature is required and Saraiva Santos (2017) presented some key concerns that should always be present, namely that:

Figure 1. Ecotourism specialty areas
Source: Adapted from TIES (2020)

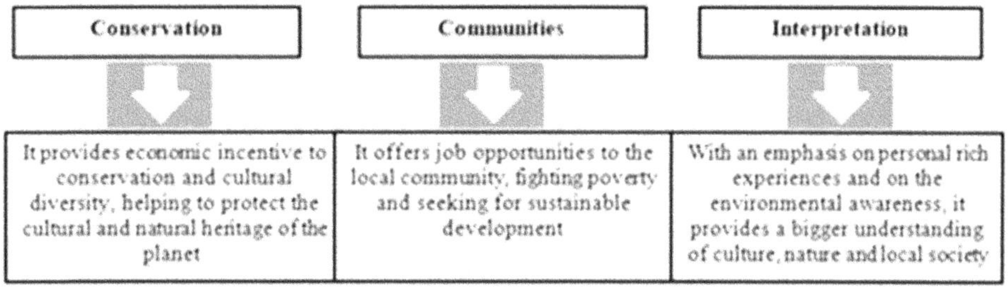

- nature's resources are not infinite, its reserves are limited and must be enjoyed in a rational way, avoiding waste and considering recycling;
- all species of the planet deserve our respect, being the conservation of its biodiversity essential to our survival;
- land-use planning in the urban and rural areas is essential, being needed to ensure decent living conditions, employment, transportation and leisure, with areas meant to food production and to the protection of natural resources.

People are increasingly aware of the protection of the planet, as recycling and public transports are popular and, due to these practices, among others, it's important to offer new touristic approaches, more eco-friendly (Kiper, 2013). The Ecotourism emerges among these new touristic approaches (WTO, 2013).

Ecotourist

Among tourism' researchers, ecotourists are defined in a more wide-ranging than consumers of Ecotourism (Anishchenko, 2016). While some studies define ecotourists as those who choose a specific travel experience and destination, in pristine natural environments (Eagles, 1992, p. 3), others argue that ecotourists should also interact with and benefit to the local community, what contradicts to pristine natural environment (Anishchenko, 2016). It is also possible to define ecotourists as those consumers of tourism, who prefer Ecotourism from all variety of tourism types, because of their natural inclination, preferences and convictions (Perkins & Grace, 2009).

Furthermore, ecotourists are the tourists who have interest in expanding knowledge about nature and are committed to sustainability and preservation of the environments they use as destination (Page & Dowling, 2002). For these reasons ecotourists would positively influence the intention, interest, and willingness to pay a higher price for Ecotourism products and services (Cheng, Gursoy & Chiappaet, 2016). More will to spend more will also increase the benefits to sustainable tourism destinations (Nickerson, Jorgenson & Boley, 2016).

However, there are different methods of classifying ecotourists such as site and activity-based, motivation-based, value-based, multi-criteria-based, and respondent-based (Hvenegaard, 2002).

Table 1. Concept of ecotourist

Eagles (1992, p. 3)	tourist "who select a certain travel experience and destination, that of nature-oriented experiences in pristine natural environments"
Page & Dowling (2002)	tourist who have interest in expanding knowledge about nature and are committed to sustainability and preservation of the environments they use as destination
Perkins & Grace (2009)	those consumers of tourism, who prefer Ecotourism from all variety of tourism types, because of their natural inclination, preferences and convictions

Source: Adapted from Anishchenko (2016)

- The method based on site and activity, considers ecotourists all the tourists visiting a particular place or site. From a motivacional perspective, it also considers all tourists visiting nature-based attractions (Weaver & Lawton, 2007; Lawton, 2009).
- The motivation-based method defines and categorizes ecotourists based on what stimulates them to travel being motivation more relevant criterion than visited places (Deng & Li, 2015).
- The value-based method identifies ecotourists based on their social or environmental values (Blamey & Braithwaite, 1997).
- Multi-criteria based method defines ecotourists based on few variables simultaneously, namely type of destination, type of activity tourists are engaged in and motivation to travel (Lee, 2007; Anishchenko, 2016).
- The respondent-based method permits interviewers to self-classify as ecotourists or non-ecotourists (Jinyang & Jian, 2015).

Table 2. Different classification of ecotourist according to its interest

occasional	natural monuments	interested	dedicated

Source: Adapted from Lindberg (1991

According to Lindberg (1991), ecotourists may be classified as occasional, natural monuments, interested and dedicated.

- Occasional ecotourist is the one that doesn't travel on purpose with Ecotourism as the main motivation, but accidentally, as part as a longer travel.
- Natural monuments ecotourist travels to visit natural attractions, but not as a habit. Not only is interested in natural environments but also in urban environments, craving for different experiences, either urban or not.
- Interested ecotourist travels looking for nature, has a special interest in subjects such as the environment, conservation and sustainability.
- Dedicated ecotourist travels motivated by scientific interests or because he/she has a strong environmental vocation. He/she seeks for learning and Ecotourism is a mean to achieve nature conservation.

Table 3. Different motivation of ecotourists

relaxation	adrenaline & enthusiasm	social interaction	self-esteem & development	Satisfaction

Source: Adapted from Holden & Sparrowhawk (2002)

The specifically focused on ecotourist motivations started to be conducted in 1990s and the accent was shifted to ecotourists' bio-centric attitudes (Perera, Vlosky & Sampath, 2012). However, the information is scarce and not clear in what concerns Ecotourism as a distinct segment supported on its particular motivational system (Anishchenko, 2016). Regarding the motivations of ecotourists, Holden & Sparrowhawk (2002) divided then into five main groups: relaxation, adrenaline and enthusiasm, social interaction, self-esteem and development and satisfaction.

- Relaxation is linked with the contact with nature and is the most important motivation for ecotourists.
- Adrenaline and enthusiasm related with sense of freedom. Social interaction has way to spend good times with friends.
- Self-esteem and development connected with the overcoming of capacities and the possibility of keeping in shape.
- Satisfaction with the proximity of nature.

Table 4. Different activities of Ecotourism

adventure activities	guided walks/tours	nature observation	travels with the intention of nature's conservation	tours with the local community

Source: Adapted from Iuliana, Cornelia, Ioan, & Anda (2017)

According to Iuliana, Cornelia, Ioan, & Anda (2017, p. 168) activities of Ecotourism may include. i) adventure activities (rafting, canoeing, equitation, bike trails, ski...), ii) guided walks/tours, iii) nature observation (fauna and flora), iv) travels with the intention of nature's conservation and v) tours with the local community (local attractions, traditional food, shopping local products).

Evolution and Importance

More and more tourism's supply and demand are getting autonomous, active, flexible and individual (Kontogeorgopoulos, 2004; Swarbrooke & Horner, 2007), which is interpreted either as the end of mass tourism (Poon, 1993) or its change into more multiple forms (Ioannides & Debbage, 1998; Torres, 2002). An increasing number of people from different social classes and from different places of the world travel to distant and exotic destinations.

During the 60's, mainly in the most developed countries, the public concern with the environment had increased. During that time, nature conservation organizations were created, with the purpose of pressuring the governments on creating places restricted to fauna, protecting ecosystems and offering a touristic experience with no interference in nature (Saraiva Santos, 2017). Therefore, the demand of places related to nature increased and ecotourists seek to be in contact with nature, gaining new experiences and knowledge. In this way, Ecotourism has been bringing the perspective of change towards a bigger environmental consciousness and has been gaining importance in several sectors, namely in hostels, leisure hotels, resorts and amusement parks (Santos Alves, 2009; Joyner, Lackey, & Bricker,2019).

The main destinations of this segment are mainly located in developed nations, but its recent growth promoted the development of destinations in more remote areas, such as villages in developing countries (Boyd & Butler, 1996). Regarding its growth, Ecotourism has been growing 10% to 12% per year, about three times more than the tourism industry as a whole (IES, *apud* Coria & Calfucura, 2012, p. 1) and has been considered as the market with the fastest growth in the tourism industry (Wight, 2001).

The involvement of the local community is one of the basic principles of Ecotourism, either by the use of local products or either by the interaction between residents and tourists. The participation of the community has been considered by stakeholders as an approach to the development of Ecotourism, since it promotes countless benefits, such as bigger conservation of natural and cultural resources, empowerment of host communities and improvement of their socioeconomic well-being (Mensah & Ernest, 2013). Additionally, regarding conservation and forest management, the involvement of the community in Ecotourism has become a tool together with the traditional means and may be considered as conservation and development tool (Mensah & Ernest, 2013). In a study conducted in Botswana (Okavango region), among indigenous communities, it was concluded that Ecotourism has become the main source of those communities livelihood and had replaced activities that harm the environment such hunting and cattle production (Mbaiwa & Stronza, 2010; Mensah & Ernest, 2013).

Following the above, all parts involved in the Ecotourism should be included, namely tourists, providers, local government, accommodation, tourist agencies, local communities, NGO's and ecologic groups (Dologlou, *apud* Diamantis, 2018). The population has an important role, since they know the place (Santos Alves, 2009) and according to the same author, the participation of the community in the process of development of local Ecotourism may be conducted in small locations through public policies. The policies should prevail ideals as social justice, income distribution, economic efficiency and nature conservation.

Prior research has indicated that High education level is a constant characteristic of eco-tourists (Fennell, 2008; Ye & Xue, 2008; Su & Wall, 2018). With high levels of education and rising environmental consciousness, university students represent potential future eco-tourists. Consideration about their insights of Ecotourism provides approaching into future Ecotourism demands as well as future development of Ecotourism. By exploring the perceptions of Ecotourism of university students in Portugal, this research adds value in what concerns empirical information on the topic from a Portuguese perspective.

EMPIRICAL EVIDENCE AND RESULTS

As general objectives, the study intends to understand how the tourism students of ES[1]HT-IPP: i) perceive the term Ecotourism, (ii) experience Ecotourism and iii) project the trend of this tourism niche.

As specific objectives it aims to: i) comprehend if the tourism students of ESHT-IPP know the basic principles of Ecotourism, ii) realize which Ecotourism activities were experienced by the tourism students of ESHT-IPP, iii) know the reasons that motivated the tourism students of ESHT-IPP to experience Ecotourism (if so) and finally iv) determine if the tourism students of ESHT-IPP have interest in activities of environmental conservation and social/local development.

Methodology

The research methodology consists of numerous steps (Saunders, Lewis & Thornhill, 2016) and according to him, research is something that people assume to find out things in a systematic way, thus raising their knowledge. Research "onion" is then classified into six stages: philosophy, approach, choice, strategy, time horizon and techniques & procedures. The research methodology can be seen as the holistic steps a researcher employ in embarking on a research work (Leedy & Ormrod 2001; Williams, 2011). It explains an issue or phenomenon through gathering data in numerical form and analyzing with the aid of mathematical methods, in particular statistics (Aliaga, & Gunderson (2002),

It is a convenience sampling, also known as haphazard sampling or accidental sampling, because the interviewed were chosen due their easy accessibility, availability at a given time and willingness to participate. Although commonly used in research, it is neither purposeful nor strategic (Palinkas et al., 2011). Some assumptions support the convenience sampling, namely that the members of the target population are homogeneous, which means that there would be no difference in the research results achieved from any sample gathered in some inaccessible part of the population (Etikan, Musa & Alkassim, 2015). One of obvious disadvantage of convenience sampling is that it is likely to be biased (Mackey & Gass, 2005), but the author of the present Thesis is aware of it and presents it as an exploratory study.

To answer the proposed research questions, a quantitative approach was used. It was chosen a Case study, probably one of the most criticized social research method (Yin, 2003), due the contemporary phenomenon, focusing on the dynamics of the case, within its real life context (Roth, 1999; Yin, 2003). A guide was drawn up for the elaboration of the questionnaire, which was evaluated by two experts in the field and two students (target audience), before being formally distributed by all students of Hospitality and Tourism studying in ESHT-IPP. The authors considered enough the comments of the two chosen experts to validate the content of the guide. Then an authorization was requested from the President of the School and the Course Directors to distribute it among students. Only after approval, the questionnaire was completed on classes, except for the second year master students who completed it by mail, as they do not had classroom classes. They were doing their Dissertations and only went punctually to the Faculty to meet their Tutors.

The questionnaire was drafted according to the proposed objectives and composed by three main areas, containing fourteen questions. The addressed issues were: i) related to demography, with the characterization of the student (gender, age, study cycle and study year), ii) perception of tourism students from ESHT about Ecotourism (knowledge and definition of the term) and iii) experience and reasons that trigger this type of choice (experience with Ecotourism, activities experienced, reasons of choice, willingness to experience activities of conservation and opinion on the importance of Ecotourism on environmental, social and economic sustainability).

Data Collection and Results

The study was carried out during the first semester of 2018, between March and June. It was conducted in ESHT-IPP, School of Hospitality and Tourism of Polytechnic Institute of Porto, in Portugal. For collecting data, the considered population was all the tourism students of ESHT-IPP, namely students from six classes of the Degree on Touristic Activities Management and two classes of the Master degree on Tourism Management.

Figure 2. Gender
Source: Own elaboration

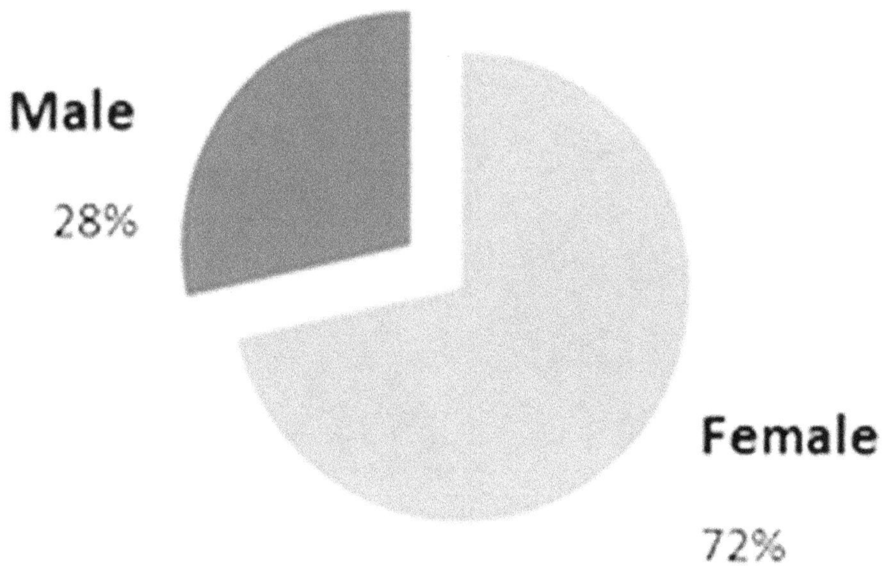

In total, 123 answers were collected (sample), representing 44% of the total amount of the students (283), which corresponds to an approximate error of around 0.055 in a 90% confidence interval. The data processing was made with the software IBM SPSS Statistics (IBM Corporation, 2010) and Microsoft Excel (Microsoft, 2010).

In an initial approach to the sample, a descriptive analysis was done (complementary to the sample characterization presented in the previous point), followed by a study of Cross-variables and Two Steps, in order to have a broader view of the results and to compare the variables gender and experience within Ecotourism.

Characterization of the Sample

The survey was analyzed based on three main fields: i) Demographic, ii) Perception on Ecotourism and iii) Experience and reasons for choosing this type of tourism.

In what concerns demographic and considering the gender, the sample is mainly constituted by females (72%), being the remaining 28% represented by males (Figure 2).

Regarding age (Figure 3), the sample is constituted mainly by students with the age between 18 and 24 years old (64%) and between 22 and 34 years old (33%). Only 2% have between 35 and 44 years old and only 1% has more than 44 years old.

Regarding the student's cycle of studies, the following figure (Figure 4) shows that the majority of the respondents is currently on the 1st cycle (degree), corresponding to 87%.

The more represented year of study (Figure 5) is the three years correspondent to the first cycle of studies, being 25% from the first year, 32% from the second year and 32% from the third year. The remaining percentages belong to the master students (4% from the first year and 7% from the second year).

Figure 3. Age
Source: Own elaboration

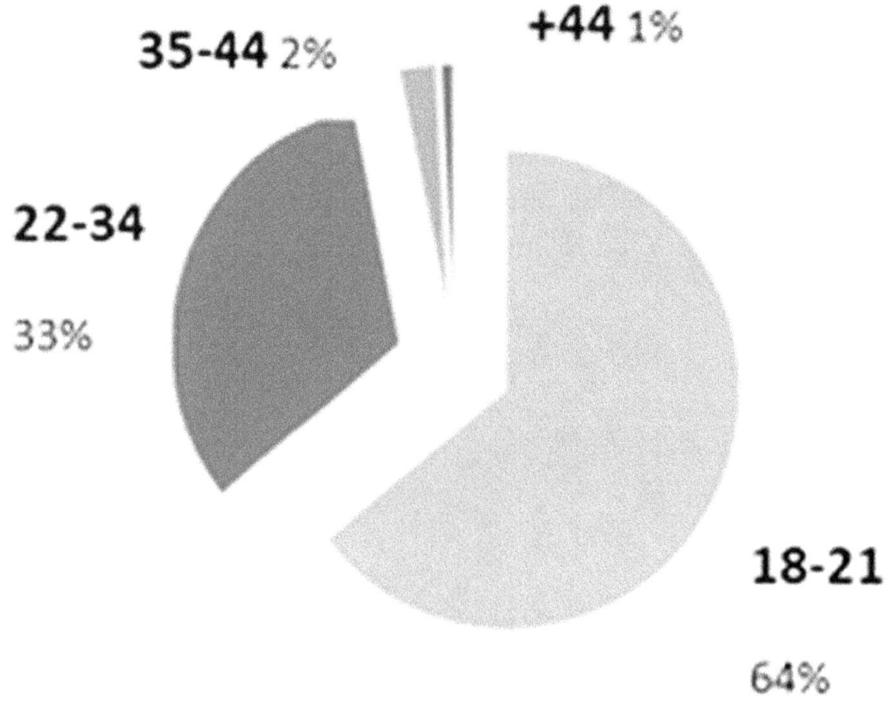

- *2.2 Results and Discussion*

Based on the results collected from the 123 questionnaires, in what concerns a descriptive analysis of the data, the following analysis consists on the results about Knowledge and definition of the term "Ecotourism", as well as Experience and reasons for choosing Ecotourism.

Data about Knowledge and Definition of the term "Ecotourism"

All of the students are familiar with the term "Ecotourism" (Figure 6) and these results seem to show the big familiarity that the tourism students from ESHT-IPP have with the mentioned term. This means that all students know the word and the concept, even if some of them have better knowledge of its broader concept.

Regarding the term's definition, the respondents' were divided by several options. Although 25% of the respondents have opted by the most right option ("All of the statements"), 35% of the sample chose the option "Tourism that aims for nature conservation" and 22% opted by the answer "Touristic activities that take place in nature". The two least chosen option were "Touristic activities that involve the local community" (11%) and "Touristic activities with an educative role" (9%), as we can see in Figure 7. These results seem to indicate that Ecotourism is a term that divides opinions and that two important characteristics are not of common knowledge yet ("Educative role" and "Local involvement"). However,

Figure 4. Students' Cycle of studies
Source: Own elaboration

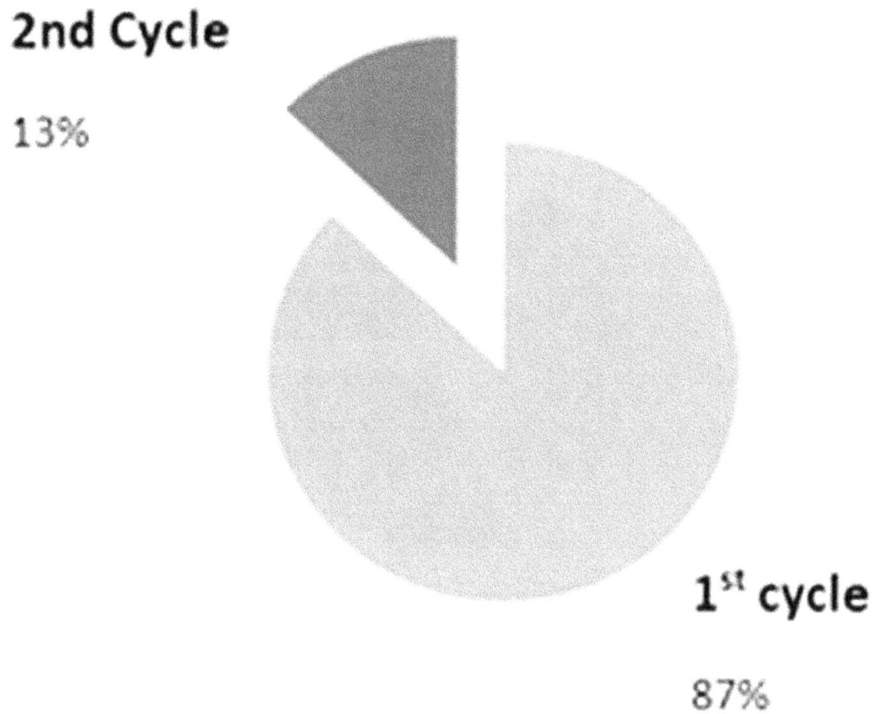

the fact that 35% of the students chose all of the statements presented, it's a positive sign regarding sustainable tourism.

Data about Experience and Reasons of Choice

Regarding the experience of the respondents with Ecotourism, the results divided themselves, since 56% answered "yes", while 46% answered "no" (Figure 8).

Regarding the activities experienced by the respondents, the four most chosen activities were "Adventure Activities" (29%), "Involvement with the local community" (20%), "Guided Tours" (18%) and "Wildlife observation" (16%). "The animals/nature conservation activities" represent 9% and "Volunteering" 8%, as we can see in the following figure (Figure 9).

In what concerns students that had never experienced Ecotourism, 91% answered that would like to do it (Figure 10) and the four most chosen activities were: volunteering (23%), animals/nature conservation activities (22%), wildlife observation (22%) and adventure activities (20%).

It's interesting to notice that the least chosen activity by the respondents that had already experienced Ecotourism is the most chosen activity by the ones who had never experienced it (volunteering: 8% *versus* 23%) (Figure 11).

Regarding the most chosen reason by the students to experience Ecotourism, the following figure (Figure 12) shows contact with nature (39%), social interaction (22%), adrenaline and enthusiasm (22%), and personal development (17%).

Figure 5. Year of study
Source: Own elaboration

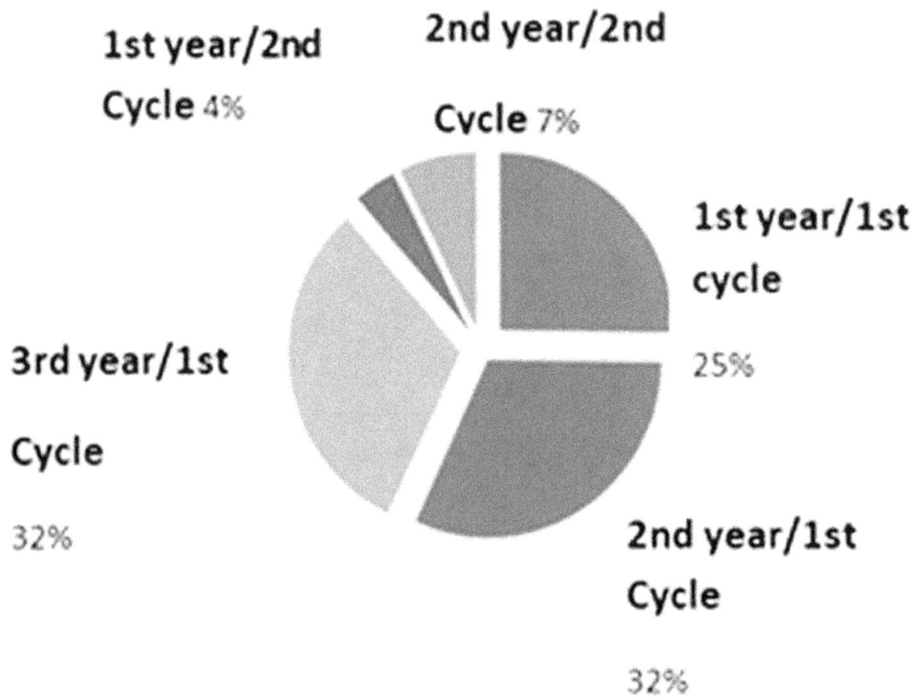

An interesting comparison is the fact that the option "Contact with nature" was the most chosen reason to do Ecotourism, while the option "Tourism that takes place in nature" was also the most chosen definition of the term Ecotourism. When asked if they would participate in activities of environmental conservation and of local/social development during their holidays, the majority of the students (87%) answered yes (Figure 13).

These results seem to point out that respondents have a great environmental consciousness, since usually during holidays they choose to have contact with nature and prefer tourism that takes places in nature. Therefore, these results show that ESHT-IPP' tourism students have interest in the preservation of the planet and that they would use their holidays to take part in sustainable activities.

To the question "If yes, why?", the majority of the students would like to enjoy their free time to take part in useful activities, that our planet needs and our consciousness demands, referring that tourism is a great tool to sustainable practices.

When asked about the importance of Ecotourism, the majority of the respondents (57%) considered Ecotourism indispensable to sustainability, 28% considered it important and the remaining (15%) as a competitive advantage in the future (Figure 14). None of the respondents considered that Ecotourism is not a priority to sustainability, which confirms that the majority of the tourism students of ESHT-IPP consider Ecotourism a tool to sustainability.

Figure 6. Knowledge about the term Ecotourism
Source: Own elaboration

Cross-Variables

In order to have a broader view of the results and to compare the variables, some of the variables were crossed, namely gender and experience within Ecotourism,

The following figure (Figure 15), show that when comparing the sample's gender with their experience with Ecotourism, the female respondents with experience within Ecotourism are in greater number when compared to male respondents (81,81%).

Regarding the relation between the age and the experience, all the older students (between 35-44 and over 44 years old) already have experienced Ecotourism. About the others age ranges the students between the ages of 18-21 years old show a great number of individuals who never experienced Eco-

Figure 7. Definition of Ecotourism
Source: Own elaboration

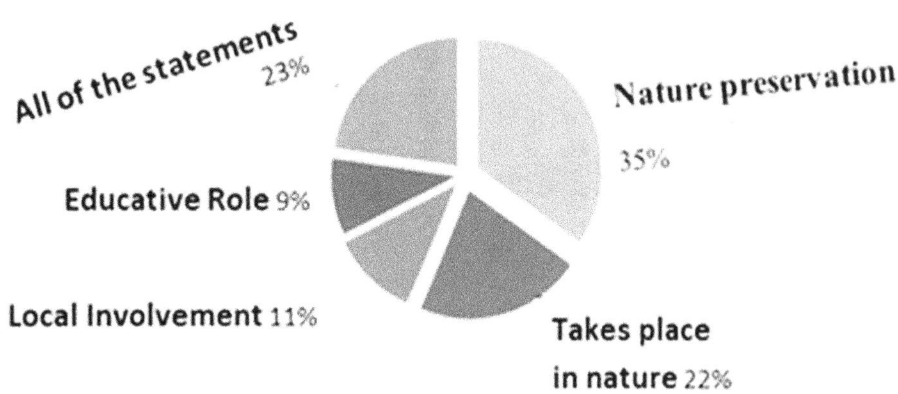

Figure 8. Experience of Ecotourism
Source: Own elaboration

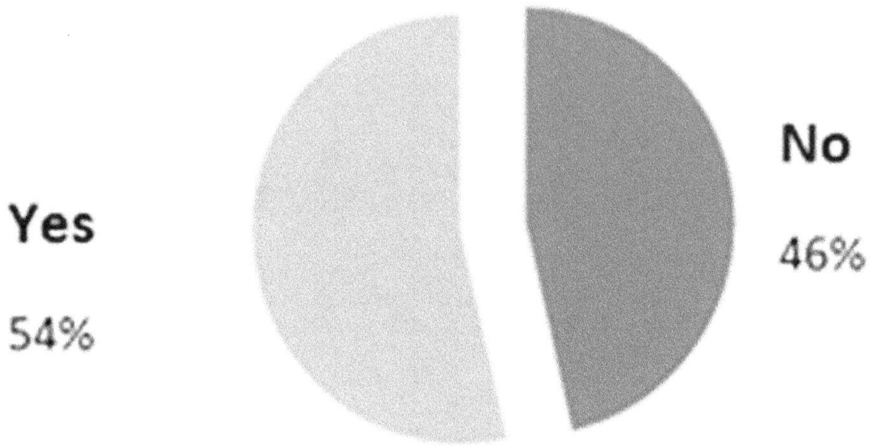

tourism and the students between the ages of 22-34 years old show a greater number of individuals with experience (Figure 15).

Regarding the relation between the student year of studies with the experience, the students from the second year of the 2nd Cycle are the ones who show bigger results (89%), followed by the students from the third year of the 1st Cycle (60%). At the end appears the students from the first year of the 1st Cycle (Figure 17).

Considering the relation between the participation in conservation activities *versus* experience, the results show that among the respondents that answered positively regarding the conservation activities during the holidays prevail the ones with experience on Ecotourism (Figure 18).

This information may mean that the students with greater experience with sustainable tourism are more likely to take part in other conservation activities. Regarding the relation between the opinions on

Figure 9. Activities experienced by the respondents
Source: Own elaboration

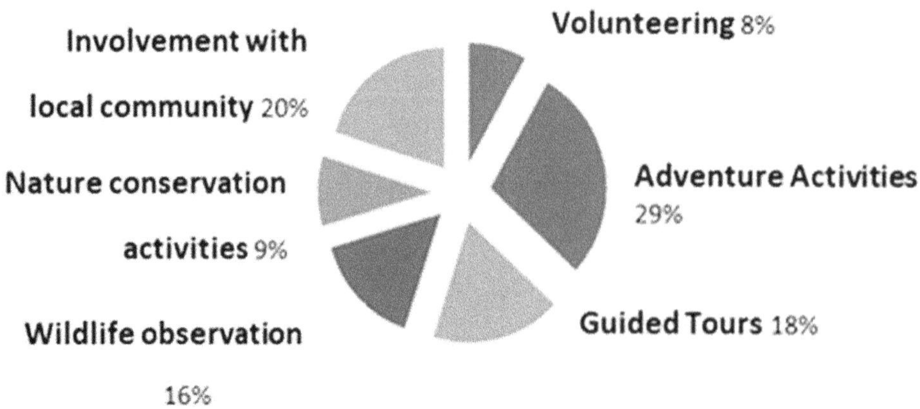

Figure 10. Percentage of respondents who have never experienced Ecotourism but would like to do it
Source: Own elaboration

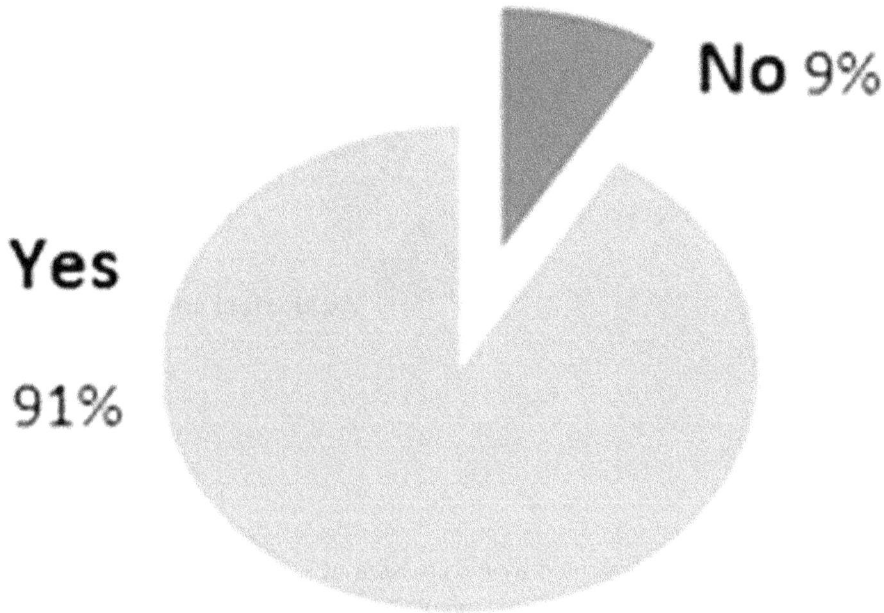

the importance of Ecotourism with their experience, Figure 19 allows us to see that the majority of the respondents that already have participated in Ecotourism (63%) consider it indispensable to sustainability. The same can be said about the students who never experienced Ecotourism (40%).

Figure 11. Activities that the respondents with no experience with Ecotourism would like to do
Source: Own elaboration

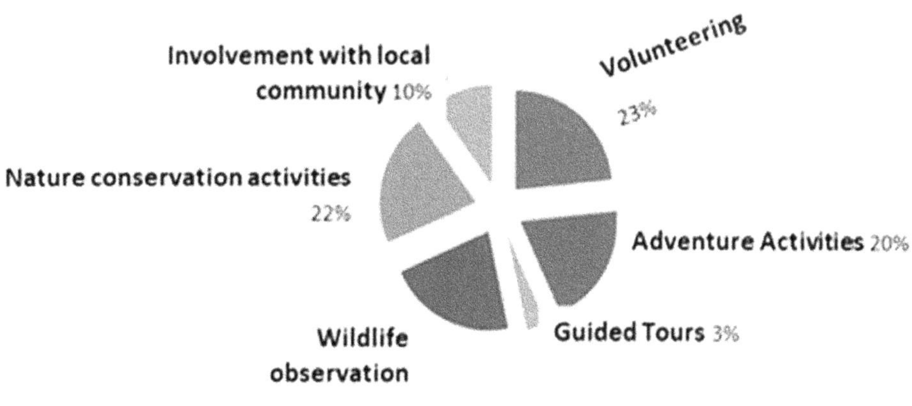

Figure 12. Reasons for experiencing Ecotourism
Source: Own elaboration

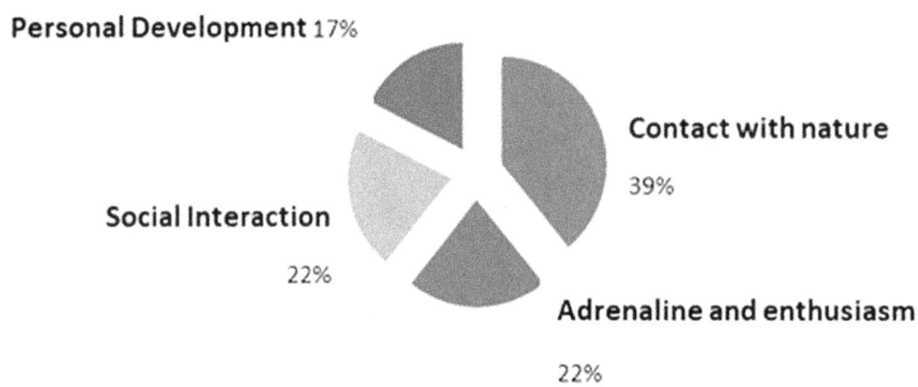

Two Steps

After a descriptive analysis, complemented by the crossing of variables, it was chosen to do a cluster analysis, as this type of analysis do not requires many assumptions about the structure of the data and can reveal interesting information. The results of this type of analysis of allies with the crossing of the

Figure 13. Percentage of students that would participate in activities of environmental conservation during their holidays
Source: Own elaboration

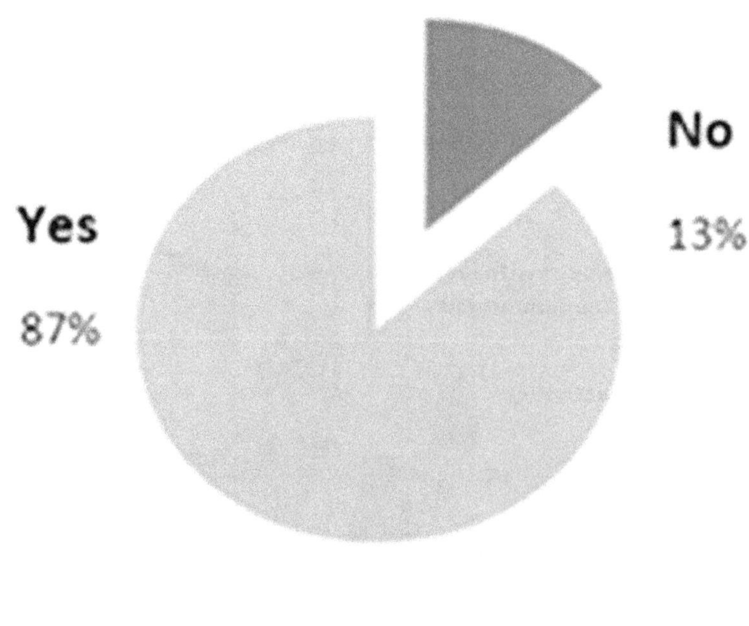

Figure 14. The importance of Ecotourism to sustainability (in the respondents' opinion)
Source: Own elaboration

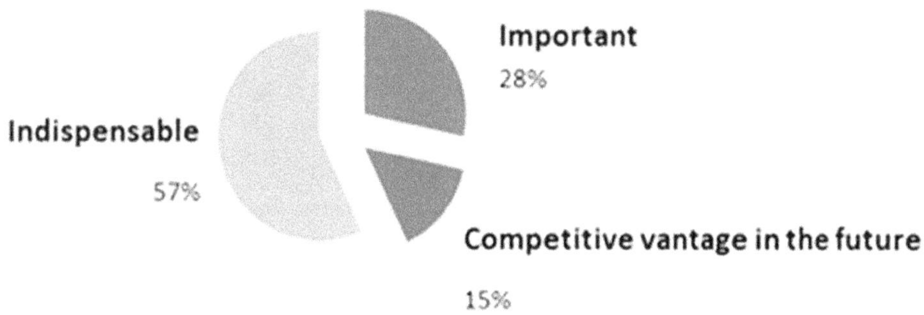

variables facilitate the interpretation of the obtained results. The two-step sorting method classifies the inquired into groups, based on common characteristics, assuming a maximum number of clusters to be built. In this case it was chosen 5 variables, namely: i) year of attendance (year of study of the Bachelor / Master); ii) experience (of students in Ecotourism activities); iii) definition (that students have about Ecotourism); iv) number of activities done (Ecotourism activities done by students) and v) age of students.

Within the chosen variables, it was tried to understand the importance that each variable assumes in the clusters. This information can be seen in the following figure (Figure 20), which highlights the Year, with 100%, followed by Experience, with 90%, Age and Definition, with around 70% and the Number of Activities Done at about 50%.

Figure 15. Gender versus Experience with Ecotourism
Source: Own elaboration

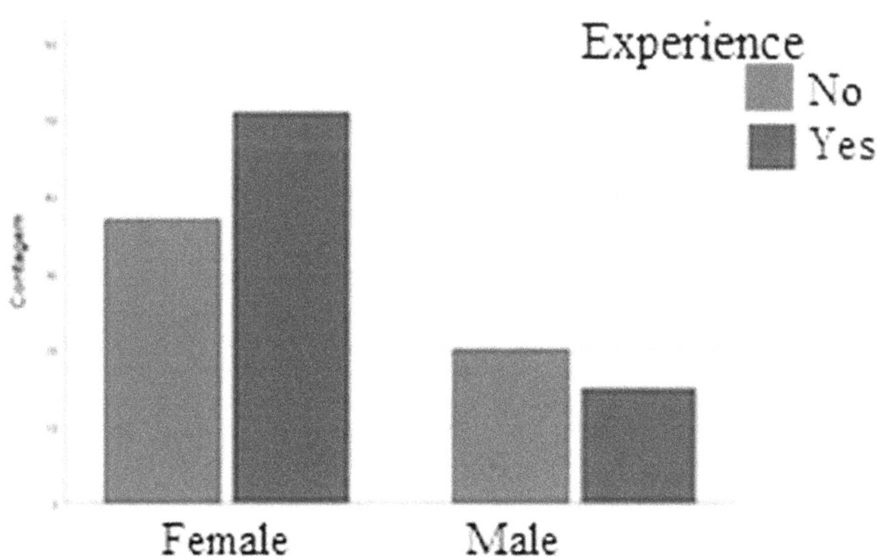

Figure 16. Ager versus Experience with Ecotourism
Source: Own elaboration

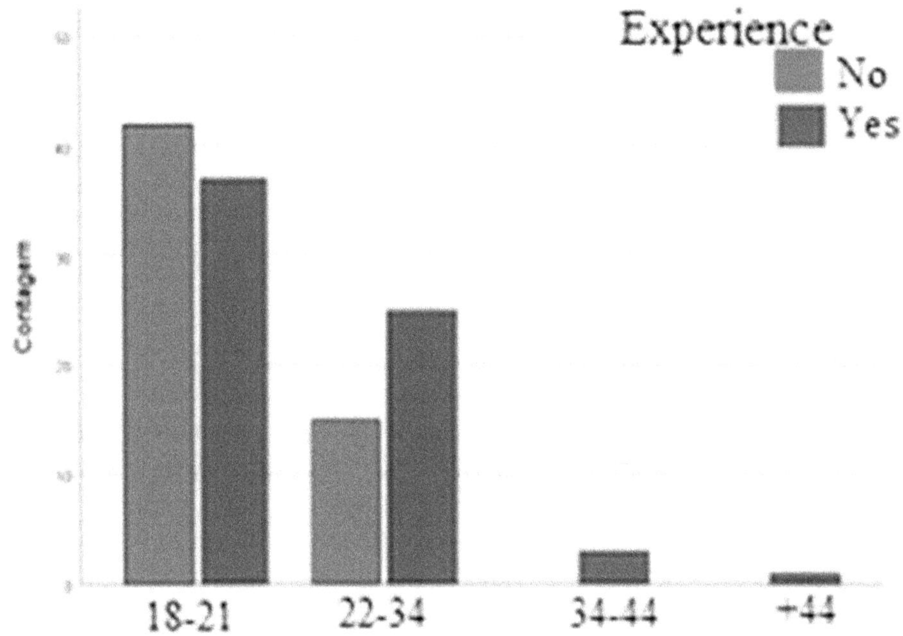

Figure 17. Year of studies versus Experience with Ecotourism
Source: Own elaboration

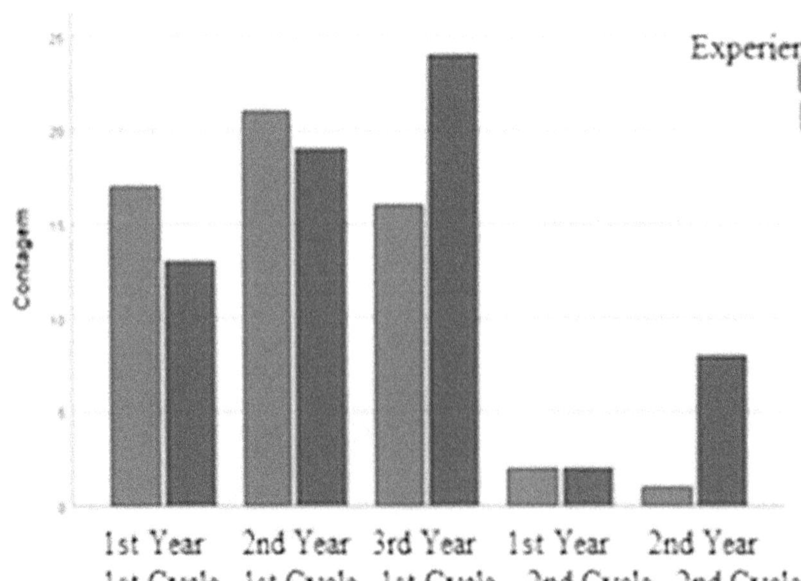

Figure 18. Conservation activities versus Experience with Ecotourism
Source: Own elaboration

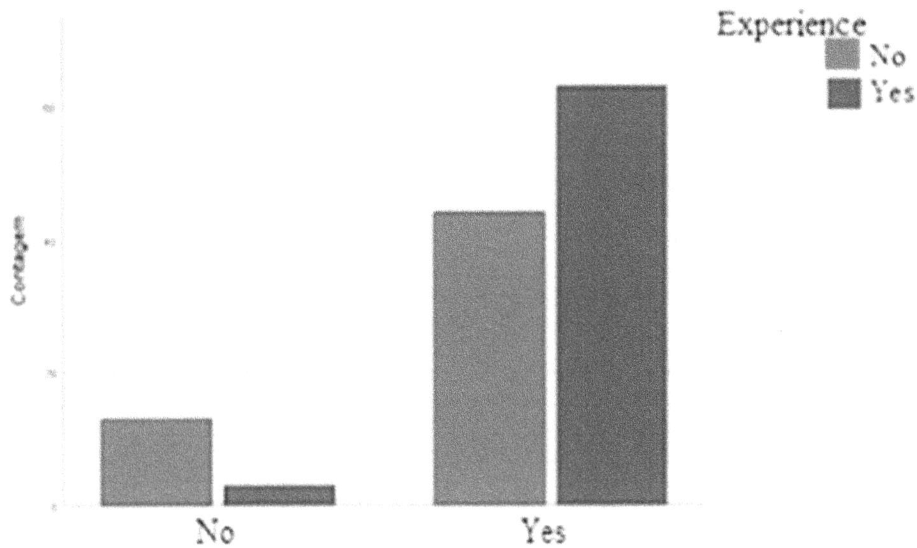

After identifying the importance of each variable in the construction of clusters, the quality of the cluster was determined, which allowed us to conclude that the quality is within a good precision range (0.5), as it can be seen in the Figure 21.

This step allowed the identification of 5 student profiles, represented on Table 5.

With the analysis of these 5 profiles it is possible to conclude that results are divided in what concerns "Definition of Ecotourism" and "Experience with Ecotourism". Most of the students know or believe

Figure 19. Opinion on Ecotourism versus Sustainability
Source: Own elaboration

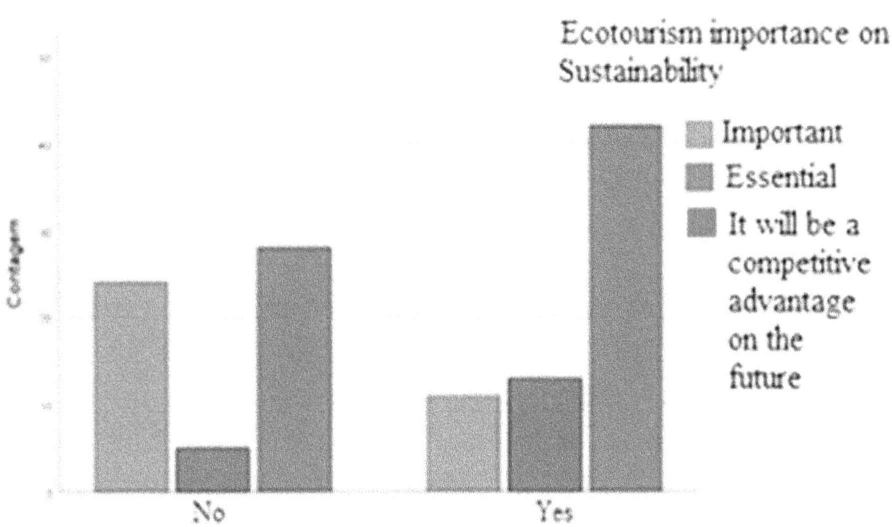

Figure 20. Importance of each variable in the construction of clusters
Source: Own elaboration

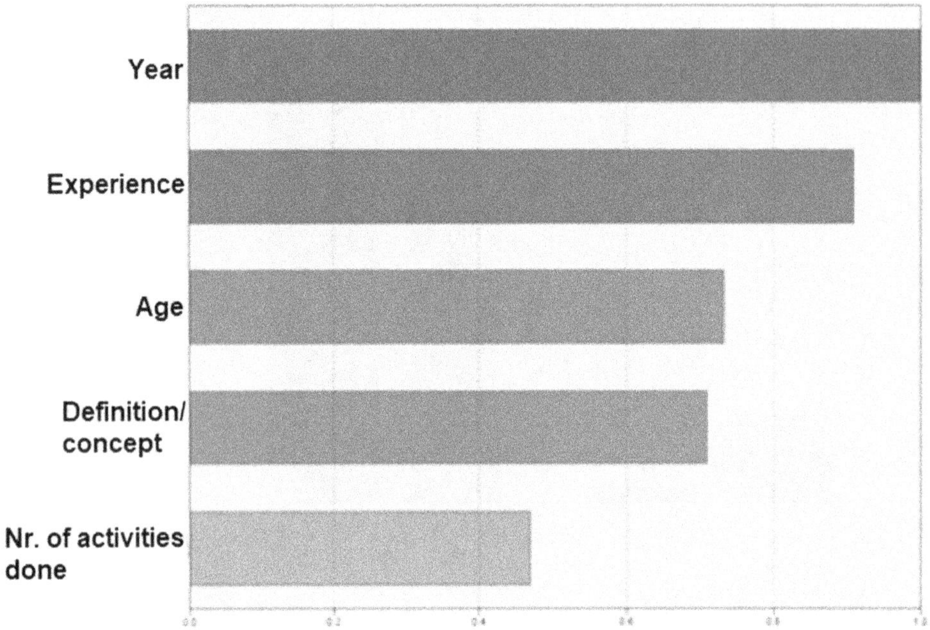

Figure 21. Cluster quality
Source: Own elaboration

Model Summary

Algorithm	TwoStep
Inputs	5
Clusters	5

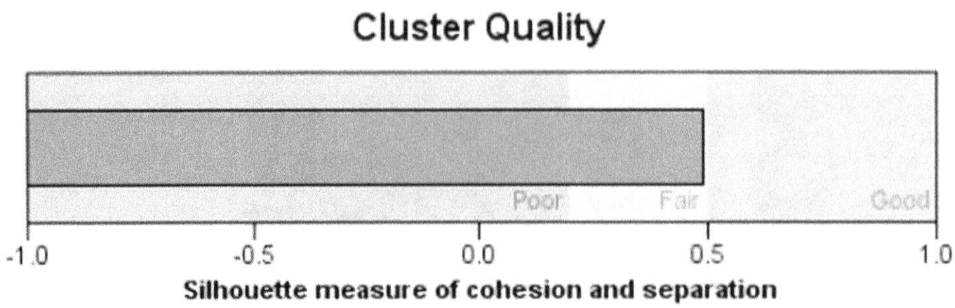

Table 5. Types of students generated by the clusters

Cluster 1 = 41 students
• Student who has never done Ecotourism but who knows or believes to know what Ecotourism is • Age between 18 and 21 years old • Most of the students attend the 2nd year of the Degree
Cluster 2 = 30 students
• Student who has done Ecotourism but does not know the inherent concept • Age between 18 and 21 years old • Most of the students attend the 2nd year of the Degree
Cluster 3 = 21 students
• Student who knows or believes that knows what Ecotourism is, but without experience in similar activities • Age between 22 and 34 years old • Most of the students attend the 3rd year of the Degree
Cluster 4 = 18 students
• All students in this cluster have experience in Ecotourism and all are aware of its concept • Age between 18 and 21 years old • All the students attend the 2nd year of the Degree
Cluster 5 = 13 students
• Most of the students have experience in Ecotourism activities (76,9%) and knowledge of its concept (61,5%) • Age between 18 and 21 years old • Most of the students attend the 2nd year of the Master's

Source: Own elaboration

to know what Ecotourism is. They also already have any kind of experience in Ecotourism. Of the 5 clusters, 4 point to a good notion of the concept and 3 clusters present positive results for the "Experience" variable.

As earlier mentioned, the research questions in this study are: "Does IPP Tourism students know about the definition of Ecotourism? " and" Does students of Tourism from IPP have experience in Ecotourism? ". Considering Figure 7 (Definition of Ecotourism) and analysis by clusters in Table 5, most of the tourism students at IPP are aware of what Ecotourism is. In fact, 23% of them gave the correct answer and 4 out of 5 clusters concentrate mainly respondents who know what Ecotourism is. However, it should be noted that two important aspects of Ecotourism were poorly chosen (involvement of the local community and Educational aspect). With regard to the experience of the sample in Ecotourism, it can be concluded that about half of the respondents have experience with it (54%), based on the Figure 8 and the construction of clusters (Table 5).

LIMITATIONS

Although the sample represents 44% of the total number of students from ESHT-IPP, School of Hospitality and Tourism of Polytechnic Institute of Porto, in Portugal, little demographic information about the total population is known, which can skew the results. Also some more specific questions could have been included in the survey, in order to better explain what students understand about sustainability and what they learn on ESHT-IPP about it. The Cross-Variables study is weak and could be better explored, through more integration between the identified variables. Also, the clusters analysis could be more complex as clusters analysis also allows to have a detailed view of other variables in comparison with the entire sample. For reasons of limitation of investigation time, it was decided by the authors not to do this type of analysis, although they are aware that it can open new perspectives, thus contributing with significant added value.

FUTURE RESEARCH DIRECTIONS

The present study provides pertinent information about ESHT-IPP tourism students, namely about how informed they are about Ecotourism and about their main motivations. In the future, studies may be carried out in relation to other areas of sustainability, not only with the students of the studied School (ESHT-IPP) but also with other IPP Schools, namely ISEP (Engineering), ISCAP (Management and Accounting), ESE (Education), ESS (Health Science), ESTG (Technology and Management) and ESMAE (Multimedia and Design), in order to better understand what students from the biggest Polytechnic Institute of Portugal think about Ecotourism and its impact on sustainability. Identical analogy would make sense with all other Higher Institutions, in order to compare the state-of-the-art and try to understand the reason that leads students to be sensitive about the destructive (negative impact) potential that tourism has at an environmental and socio-cultural levels.

Another suggestion is linked with appliance of regression analysis to data, in order to design predictive model that could explain the reason for students' choice for Ecotourism, which would allow the development of teaching methodologies to encourage the practice of Ecotourism in Schools. Finally, longitudinal studies are proposed to be strategically designed with the aim of being able to compare and evaluate the impact of education in these research areas. Use this present study, add some more information and during future periods collect new data and compare them.

SUMMARY AND CONCLUSION

Ecotourism has its own characteristics, as it is a segment that defends that the tourist who practices it should not leave any trace in the environment. On the contrary, it always should contribute with an educational perspective and with the inclusion of the local community. It is a growing segment of tourism and requires special attention from all of us.

As there is a common misunderstanding of the concept of Ecotourism, this study is a case study conducted in ESHT-IPP, the School of Hospitality and Tourism of Porto Polytechnic Institute, in Portugal, where it was intended to understand the level of knowledge of their students regarding Ecotourism as well as their experience with it.

After a literature review it was realized that the concept of sustainability is complex, used in many contexts and is not interpreted in the same way. The responsibility for sustainable tourism development lies with the responsibility of a number of stakeholders, including the State, tourist entrepreneurs, the local community and the tourist himself. Tourism always involves advantages and disadvantages, demanding a collective conscience. Education has an important role and proves to be essential to raise awareness and inform citizen/students about sustainability issues.

The study showed that student's perception of the meaning of the term were mainly divided between the options: "Tourism that aims to preserve the environment" (35%), "All statements" (23%) and "Tourist activities that occur in nature" (22%), with the least selected options being "Involvement with the local community" and "With an educational nature".

In what concerns the activities experienced by the respondents, the four most chosen activities were "Adventure Activities" (29%), "Involvement with the local community" (20%), "Guided Tours" (18%) and "Wildlife observation" (16%). "The animals/nature conservation activities" represented 9% and "Volunteering" 8%.

Most of the surveyed sample considered Ecotourism to be essential for Sustainability (57%), female experienced more Ecotourism activities than men and older students (more than 35 years old) already have practiced Ecotourism.

Students attending the 2nd year of 2nd Cycle (Master) are the ones with the highest ratio (89%) on experienced Ecotourism activities, followed by students in the 3rd year of 1st Cycle (Degree), with 60%. Lastly, appeared students from the 1st year of 1st Cycle, which may which may mean that ESHT-IPP works on the issue of sustainability in tourism in its training offer.

Students who had experience on Ecotourism are more concerned on develop preservation activities during its holidays, which may mean that students with more experience in sustainable tourism are more likely to carry out these kind of activities (environmental preservation) on their holidays.

These results show us that the inquired students are aware of Ecotourism and its importance. This analysis allows us to understand that these students are whiling to learn new approaches of tourism, towards more sustainable choices. With these results, it seems that the tourism students from this school are open to change their habbits, not only in tourism, but in general, in order to contribute to a healthier planet.

The students will certainly understand how Ecotourism represents a real resource for economic development as a "form of leisure that meets the needs of the tourists, the professionals of the tourism industry and the local community, without compromising the ability of future generations to meet their needs" (Brundtland definition). Therefore, this kind of projects will allow students to have fun by developing skills that may also open new job opportunities. Creativity, entrepreneurship and sense of initiative will continuously be stimulated thus promoting a real personal and professional growth of the students. Non-formal learning and learning by doing could also be used as main methodologies. Mutual approach among sending and hosting organisations will be the foundations for implementation of activities.

Moreover, Ecotourism was identified by the Youth and Environment Europe (YEE, 2018) as a current topic and a good way to promote active citizenship among young people. Ecotourism can attach diverse fractions of communities and could consequently be an innovative approach for encouraging youth participation and active citizenship. The study shows that Ecotourism can be a way of future citizens´ empowerment, highly contributing for the enhancing of the United Nations Sustainable Development Goals

REFERENCES

Aliaga, M., & Gunderson, B. (2002). *Interactive Statistics*. Sage Publications.

Anishchenko, M. (2016). *Ecotourists' motivation and its relation to the Ecotourism's principles* (Magister Thesis). Department of Social and Economic Geography, Umea University.

Barkauskiene, K., & Snieska, V. (2013). Ecotourism as an Integral part of sustainable tourism development. *Economics and Management*, *18*(3), 2029–2038. doi:10.5755/j01.em.18.3.4272

Beier, M. E., & Rittmayer, A. D. (2008). Literature Overview: Motivational factors in STEM. Interest and Self-Concept. In B. Bogue & E. Cady (Eds.), *Applying Research to Practice (ARP) Resources*. Retrieved from https://www.engr.psu.edu/AWE/ARPresources.aspx

Blamey, R. K., & Braithwaite, V. A. (1997). A social values segmentation of the potential Ecotourism market. *Journal of Sustainable Tourism*, *5*(1), 29–45. doi:10.1080/09669589708667274

Boyd, S. W., & Butler, R. W. (1996). Managing Ecotourism: An opportunity spectrum approach. *Tourism Management*, *17*(8), 557–566. doi:10.1016/S0261-5177(96)00076-3

Castells, M. (2000). A era da informação: economia, sociedade e cultura. In *A Sociedade em rede*. Paz e Terra.

Ceballos-Lascurain, H. (1987). The future of Ecotourism. *Mexico Journal*, *1*, 13–19.

Ceballos-Lascurain, H. (1996). *Tourism, Ecotourism and Protected Areas.* Reino Unido: IUCN/The World Conservation Union.

Cheng, A., & Gursoy, D. (2016). The influence of materialism on ecotourism attitudes and behaviors. *Journal of Travel Research*, *55*(2), 176–189. doi:10.1177/0047287514541005

Cobbinah, P. B. (2015). Contextualising the meaning of ecotourism. *Tourism Management Perspectives*, *16*, 179–189. doi:10.1016/j.tmp.2015.07.015

Coria, J., & Calfucura, E. (2012). Ecotourism and the development of indigenous communities: The good, the bad, and the ugly. *Ecological Economics*, *73*, 47–55. doi:10.1016/j.ecolecon.2011.10.024

Dangi, T. B. (2018). Exploring the Intersections of Emotional Solidarity and Ethic of Care: An Analysis of Their Synergistic Contributions to Sustainable Community Tourism Development. *Sustainability*, *10*(8), 2713. doi:10.3390u10082713

Deng, J., & Li, J. (2015). Self-identification of ecotourists. *Journal of Sustainable Tourism*, *23*(2), 255–279. doi:10.1080/09669582.2014.934374

Diamantis, D. (2018). Stakeholder Ecotourism management: Exchanges, coordination's and adaptations. *Journal of Ecotourism*, *17*(3), 203–205. doi:10.1080/14724049.2018.1502122

Dieguez, T. (2018). Growth or Development? A Sustainable Approach. *Journal of Business Management and Economic Research*, *2*(8), 38–46. doi:10.29226/TR1001.2018.56

Dolnicar, S., Crouch, G. I., & Long, P. (2008). Environment-friendly tourists: What do we really know about them? *Journal of Sustainable Tourism*, *16*(2), 197–210. doi:10.2167/jost738.0

Dredge, D. (2017). Responsibility and Care in the Collaborative economy. In Dredge & Gyimóthy (Eds.), Collaborative Economy and Tourism, Tourism on the Verge. Springer International Publishing. doi:10.1007/978-3-319-51799-5_4

Dredge, D., Benckendorff, P., Day, M., Gross, M. J., Walo, M., Weeks, P., & Whitelaw, P. (2012). The philosophic practitioner and the curriculum space. *Annals of Tourism Research*, *39*(4), 254–2176. doi:10.1016/j.annals.2012.07.017

Eagles, P. F. (1992). The Travel Motivations of Canadian Ecotourists. *Journal of Travel Research*, *31*(2), 3–7. doi:10.1177/004728759203100201

EEFEct. (2018). *Study on Ecotourism Development*. Erasmus+ EEFEct project, reference number: 2017-1- IT02-KA204-036713.

Emanuel, R., & Adams, J. N. (2011). College students' perceptions of campus sustainability. *International Journal of Sustainability in Higher Education*, *12*(1), 79–92. doi:10.1108/14676371111098320

Etikan, L., Musa, S. A., & Alkassim, R. S. (2015). Comparison of convenience sampling and purpose sampling. *American Journal of Theoretical and Applied Statistics*, 1–4.

Fang, W., Lien, C., Huang, Y., Han, G., Shyu, G., Chou, J., & Ng, E. (2018). Environmental Literacy on Ecotourism: A Study on Student Knowledge, Attitude, and Behavioral Intentions in China and Taiwan. *Sustainability*, *10*(6), 1886. doi:10.3390u10061886

Fennell, D. (2008). *Ecotourism* (3rd ed.). Routledge.

Font, X., & Mccabe, S. (2017). Sustainability and Marketing in Tourism: Its Contexts, Paradoxes, Approaches, Challenges and Potential. *Journal of Sustainable Tourism*, *25*(7), 869–883. doi:10.1080/09669582.2017.1301721

GKoumas. (2019). Evaluating a standard for sustainable tourism through the lenses of local industry. *Heliyon*, *5*, e027072.

Higgins-Desbiolles, F., Whyte, K. P., & Tedmanson, D. J. (2013). Tourism and environmental justice. In K. Schwab & D. Dustin (Eds.), Just leisure: Things that we believe. Academic Press.

Holden, A., & Sparrowhawk, J. (2002). Understanding the Motivations of Ecotourists: The case of trekkers in Annapurna, Nepal. *International Journal of Tourism Research*, *4*(6), 435–446. doi:10.1002/jtr.402

Hvenegaard, G. T. (2002). Using tourist typologies for Ecotourism research. *Journal of Ecotourism*, *1*(1), 70–18. doi:10.1080/14724040208668109

Ioannides, D., & Debbage, K. G. (1998). Neo-Fordism and Flexible Specialization. In D. Ioannides & K. G. Debbage (Eds.), *The Economic Geography of the Tourist Industry*. Routledge.

Iuliana, M., Cornelia, P., Ioan, B., & Anda, M. I. (2017). Issues Concerning the Classification of Ecotourism Destinations. *Lucrari Stiintifice*, *9*(1), 167–172.

Jafari, J. (2000). *Encyclopedia of Tourism*. Routledge.

Jinyang, D., & Jian, L. (2015). Self-identification of ecotourists. *Journal of Sustainable Tourism*, 255–279.

Joyner, L., Lackey, N. Q., & Bricker, K. (2019). Community Engagement: An Appreciative Inquiry Case Study with Theodore Roosevelt National Park Gateway Communities. *Sustainability*, *11*(24), 7147. doi:10.3390u11247147

Kiper, T. (2013). Role of Ecotourism in Sustainable Development. In M. Özyavuz (Ed.), *Advances in Landscape Architetures*. doi:10.5772/55749

Kisi, N. (2019). A Strategic Approach to Sustainable Tourism Development using the A'WOT Hybrid Method: A case study of Zonguldak, Turkey. *Sustainability*, *11*(4), 964. doi:10.3390u11040964

Kontogeorgopoulos, N. (2004). Ecotourism and Mass Tourism in Southern Thailand: Spatial Interdependence, Structural Connections, and Staged Authenticity. *GeoJournal*, *61*(1), 1–11. doi:10.100710708-005-8631-6

Kunwar, R. R. (2018). Tourism Education, Curriculum Spaces, Knowledge Production, and Disciplinary Pluralism. *The Gaze: Journal of Tourism and Hospitality*, *9*, 83–155. doi:10.3126/gaze.v9i0.19724

Lawton, L. (2009). Birding festivals, sustainability, and Ecotourism: An ambiguous relationship. *Journal of Travel Research*, *48*(2), 259–267. doi:10.1177/0047287509332330

Lee, T. H. (2007). An Ecotourism behavioural model of national forest recreation areas in Taiwan. *International Forestry Review*, *9*(3), 771–785. doi:10.1505/ifor.9.3.771

Leedy, P., & Ormrod, J. (2001). Practical research: Planning and design (7th ed.). Upper Saddle River, NJ: Academic Press.

Lindberg, K. (1991). *Policies for maximizing Nature Tourism's Ecological and Economic benefits*. World Resources Institute.

Mackey, A., & Gass, S. M. (2005). *Second language research: Methodology and design*. Lawrence Erlbaum Associates Publishers.

Mbaiwa, J., & Stronza, A. (2010). The effects of tourism development on rural livelihoods in the Okavango Del ta, Botswana. *Journal of Sustainable Tourism*, *18*(5), 635–656. doi:10.1080/09669581003653500

Mensah, I. & Ernest, A. (2013). Community Participation in Ecotourism: The Case of Bobiri Forest Reserve and Butterfly Sanctuary in Ashanti Region of Ghan. *American Journal of Tourism Management*, *2*(A), 34-42.

Mondino, E., & Beery, T. (2019). Ecotourism as a learning tool for sustainable development. The case of Monviso Transboundary Biosphere Reserve, Italy. *Journal of Ecotourism*, *18*(2), 107–121. doi:10.1080/14724049.2018.1462371

Nickerson, N. P., Jorgenson, J., & Boley, B. B. (2016). Are sustainable tourists a higher spending market? *Tourism Management*, *54*, 170–177. doi:10.1016/j.tourman.2015.11.009

Øian, H., Fredman, P., Sandell, K., Sæþórsdóttir, A. D., Tyrväinen, L., & Jensen, F. S. (2018). *Tourism, nature and sustainability. A review of policy instruments in the Nordic countries. Nordic Council of Ministers 2018*. Rosendahls.

Page, S. J., & Dowling, R. K. (2002). *Ecotourism*. Prentice Hall.

Palinkas, L. A., Holloway, I. W., Rice, E., Fuentes, D., Wu, Q., & Chamberlain, P. (2011). Social networks and implementation of evidence-based practices in public youth-serving systems: A mixed methods study. *Implementation Science; IS*, *6*(1), 113. doi:10.1186/1748-5908-6-113 PMID:21958674

Perera, P., Vlosky, R. P., & Sampath, B. W. (2012). Motivational and Behavioral Profiling of Visitors to Forest-based Recreational Destinations in Sri Lanka. *Asia Pacific Journal of Tourism Research, 17*(4), 451–467. doi:10.1080/10941665.2011.627353

Perkins, H., & Grace, D. A. (2009). Ecotourism: Supply of nature or tourist demand? *Journal of Ecotourism, 8*(3), 223–236. doi:10.1080/14724040802534992

Poon, A. (1993). *Tourism, Technology and Competitive Strategies.* CABI.

Portegies, A., De Haan, T., & Platenkamp, V. (2014). Embedded research: A pragmatic design for contextual learning - from fieldtrip to fieldwork to field research in Australasia. *Tourism Planning & Development, 15*(5), 600–613.

Roth, S. (1999). The state of Deign Research. *Design Issues, 15*(12), 18–26. doi:10.2307/1511839

Salazar, N. B. (2006). Building a "Culture of Peace" trough Tourism: Reflexive and analytical notes and queries. *Universitas Humanistica, 62*(Julio-Diciembre), 319–333.

Santos Alves, C. (2009). *A importância do ecoturismo no património arquitectónico de Cabo Verde: das potencialidades à prática.* Instituto Superior Técnico, Universidade Técnica de Lisboa.

Saraiva Santos, F. A. (2017). *Ecoturismo - desenvolvimento turístico e sustentabilidade social no parque natural da serra da estrela (Concelho de Seia - estudo de caso).* Tese de Doutoramento, Universidade de Coimbra.

Sedarati, P., Santos, S. P., & Pintassilgo, P. (2018). System Dynamics in Tourism Planning and Development. *Tourism Planning & Development.*

Sharpley, R. (2009). Shedding light on dark tourism: an introduction. In R. Sharpley & P. R. Stone (Eds.), *The darker side of travel: The theory and practice of dark tourism (3-22). Tonawanda: Channel View.*

Su, M. M., & Wall, G. (2018). Chinese university students' perceptions of ecotourism. *International Journal of Services Technology and Management, 24*(4), 356. doi:10.1504/IJSTM.2018.093345

Swarbrooke, J., & Horner, S. (2007). *Consumer Behavior in Tourism.* Butterworth Heinemann. doi:10.4324/9780080466958

Tan, B. C., & Lau, T. C. (2009). Examining sustainable consumption patterns of young consumer: Is there a cause for concern? *The Journal of International Social Research, 2*(9).

The International Ecotourism Society - TIES. (2018). Available on www.ecotourism.org

The International Ecotourism Society - TIES. (2020). *Principles of Ecotourism.* The International Ecotourism Society. Available on https://Ecotourism.org/

Torres, R. (2002). Cancun's Tourism Development from a Fordist Spectrum of Analysis. *Tourist Studies, 2*(1), 87–116. doi:10.1177/146879760200200l098

United Nations Environment Programme and World Tourism Organisation – UNEP. (2002). *The world ecotourism summit final report.* UNEP and WTO.

Urry, J. (2003). *Global Complexity.* Policy.

Wearing, S., & Neil, J. (2001). *Ecoturismo: impactos, potencialidades e possibilidades*. Manole.

Wearing, S., & Neil, J. (2009). *Ecotourism: Impacts, Potentials and Possibilities*. Butterworth-Heinemann. doi:10.4324/9780080940182

Weaver, D., & Lawton, L. (2007). Twenty years on: The state of contemporary ecotourism research. *Tourism Management*, *28*(5), 1168–1179. doi:10.1016/j.tourman.2007.03.004

Wight, P. (2001). Ecotourists: Not a homogeneous market segment. *The encyclopedia of Ecotourism*, 37-62.

Williams, C. (2011). Research methods. *Journal of Business & Economics Research*, *5*(3). Advance online publication. doi:10.19030/jber.v5i3.2532 PMID:21543382

Wohlmuther, C. & Winstersteiner, W. (2014). *International Handbook on Tourism and Peace*. Centre for peace Reasearch & UNWTO.

Wood, M. (2002). Ecotourism: principles, practices & policies for sustainability. UNEP, United Nations Environment Programme Division of Technology, Industry and Economics, Paris, France.

WTTC. (2016). *Tourism as a driver of peace. Quantitative analysis on the link between peace and Tourism, May 2016*. World Travel & Tourism Council and Institute for Economics and Peace.

Xu, F., & Fox, D. (2014). Modelling attitudes to nature, tourism and sustainable development in national parks: A survey of visitors in China and the UK. *Tourism Management*, *45*, 142–158. doi:10.1016/j.tourman.2014.03.005

Ye, W., & Xue, X. (2008). The differences in ecotourism between China and the West. *Current Issues in Tourism*, *11*(6), 567–586. doi:10.1080/13683500802475927

YEE. (2018). *Uniting Youth for ecotourism.Youth for Ecotourism*. Youth and Environment Europe.

Yin, R. K. (2003). *Case study research: Design and Methods*. Sage Publications.

KEY TERMS AND DEFINITIONS

Citizenship: Citizenship is a group of rights and responsibilities or duties: the rights are the citizen's entitlements from the nation, whereas the responsibilities are the duties that the nation is entitled to expect from the citizen; they are a set of legal obligations through which the citizen's status of belonging in the society is asserted. Although the concepts, rights and duties that compose citizenship may vary from one society to another, they do share specific mutual aspects. Perhaps the most important of these shared values is equality among all citizens in their citizenship rights and duties, facilitating mutual cooperation, strengthening their loyalty to their nation as well as helping them accept each other where there are differences of opinion.

Ecotourism: Ecotourism is a sustainable tourism where focus is on minimising negative impact while learning about the environment and contributing to environmental protection. It is an alternative to mass scale tourism.

Ecotourists: Ecotourists are defined as consumers of Ecotourism. This mean tourists who select a certain travel experience and destination, that of nature-oriented experiences in pristine natural environments, interacting with and benefit to the local community.

Local Community: Local community is a group of interacting people sharing an environment. Being a durable and organized human population that integrates a space of proximity, typically, their individuals share multiple ties: they are relatives, friends or simply acquaintances; they buy and sell among themselves, are in the same places and have common references. Like a state, the local community integrates population, government, and territory. Its durability, on the other hand, implies a high degree of self-sustainability - political, demographic, economic, social, and environmental.

Sustainable Development: Sustainable development relates to the systematizing standard for meeting human development goals in accordance to the preservation of environment and economic profits. It cares about future generations and its needs.

Sustainable Tourism: Sustainable Tourism is a conscious tourism, aware of its stakeholders needs and economic, social, and environmental impacts, in the present and future. The concept of sustainable tourism is complex and tends to bring about several paradoxes and dilemmas as the three dimensions of economic, social, and environmental concerns are difficult to balance.

Tourism: Tourism encompasses travels and stays outside the usual environment of the traveller, for more than one day and less than one consecutive year, for business, leisure, or other purposes.

Tradition: Tradition can be understood as a belief or behaviour (folk custom) that persists from the past to the present, accepted by those who receives it and who in turn will pass it on to the next generation. It has a symbolic meaning or special significance.

Chapter 7
Impact of Agriculture and Garden-Based Learning on Students of Secondary School

Sobha D. Namboodhiri
https://orcid.org/0000-0001-5905-8950
SRM University, Sonipat, India

Raghavendra G. Rao
SRM University, Sonipat, India

ABSTRACT

Agriculture and garden-based learning is the need of the hour as it helps in the cognitive and psychomotor development of the student. It provides skill-based education which not only ensures learning by doing but also helps in educating the child about growing various types of plants. This study is conducted by taking 500 students from the secondary level by providing them a self-prepared questionnaire by the research scholar in two schools of Panipat. Students are taken from fields of science along with maths, botany, and agriculture fields. Based on the questionnaire for plant grower, statistical analysis is done to find if there is any significant difference between learning agriculture in school and not learning it in school. On statistical analysis using the Anova test, it was found that learning agriculture was beneficial for children in holistic development. It also helps the schools in improving the nutritive value of midday meals, involving the students in some skill-based education which can help in solving the problems of malnutrition and providing jobs to millions of Indian youth.

BACKGROUND

R. Bowker et al 'Gardening as a learning environment: A study of children's perceptions and understanding of school gardens as part of an international project'(2007), learning environment research, Springer Link- Three countries selected for this project were England, Kenya and India. Sixty-seven schools were selected from these three countries and concept map was used to know the different ways

DOI: 10.4018/978-1-7998-7512-3.ch007

in which children utilized school gardening to understand the growth of crops. Based on this study it was found that school gardening helped children in learning about crop growth and they liked this activity.

D. Desmond et al., 'Revisiting garden-based learning in basic education' ResearchGate, January 2004- GBL (garden-based learning) has been especially beneficial in environmental education as well as in teaching scientific concepts. It allows developing GBL globally for researchers and educational administrators. Resources and ideas can be exchanged to promote GBL under all conditions.

According to L. Kanthimathi (2017)- 'Government schools raise kitchen gardens- Many Government Schools have grown kitchen gardens instead of ornamental plants to supplement food prepared under the noon-meal scheme.'

As per Bennur S.(2014) - 'Schools are growing vegetables, fruits in kitchen gardens to develop healthy ambiance.'

Figure 1. Students with teacher maintaining plants grown in a recycled plastic container
As shown in the website- www.schooleducation.kar.nic.in, SSA, Swasthplus.pdf

UNICEF and USAID during 2004-05 started the School Water and Sanitation towards Health and Hygiene (Swasth Plus) Programme in the Karnataka state of India in three districts which are Tumkur, Raichur, and Gulbarga. Ten blocks from these three districts were selected and this program gained momentum during 2005-06. It was a program aimed at an overall improvement in the quality by achieving universal enrolment, attendance, and retention in an environment conducive to learning. It helped to develop bio-intensive gardens in 360 schools and 369 teachers were trained in Bio intensive Gardening. The initiative was launched by UNICEF in 2005 in 120 schools of Gulbarga, 200 schools of Raichur, and 80 schools in Tumkur districts of Karnataka. School students have found a new reason to come to school. The students of the school would troop in early before regular classes begin so that they can tend to the plants sown by them. It resulted in improving attendance, mid-day meal, and extra surplus production for distribution.

According to Emily J. Ozer, (2008) -Learning in the domains of Academics, Social and Health can be improved by school gardens. This study provides a review of the impact of the curriculum of school gardening on the school as well as student's work, it tells us about how to use it and that further research is needed to improve this practice. There is a growing U.S. movement for the "greening" of the schoolyard."

As per the report of Food and Agricultural Organization of United Nations, FAO,(2010) 'School gardens are cultivated around or near to schools, tended at least partly by learners. School gardens are run for reasons like- promotion of good diet, nutrition education, the development of livelihood skills, together with the power to extend this learning beyond the school itself in a variety of ways.'

According to P. Hazra et al. (2010) "Garden for vegetable processing- Type of vegetable farming where vegetables are produced with the sole objective of supplying them to the processing factories.

Agriculture is an activity of man which is primarily aimed at the production of food, fiber, fuel, etc. by optimum use of terrestrial resources.

Composting is a biological process in which micro-organisms decompose the organic matter and lower the carbon-nitrogen ratio of the refuse and the final product of composting is well-rotted manure known as compost."

Based on the above findings in India and the U.S. following study is conducted to understand the impact of garden and agriculture-based learning on students of Secondary School.

INTRODUCTION

School education should be carried out for the holistic development of the individual. This education cannot be limiting. It has to include various spheres of knowledge which include the Cognitive and Psychomotor Domain of learning.

Mahatma Gandhi said 'Real education consists in drawing the best out of yourself.' It means Education should be imparted in such a manner that it brings out the best in the child. This was the philosophy of Father of Nation, Mahatma Gandhi.

Agriculture is the culture or growth of Agronomy based plants that have economic value while Garden is developed based on concepts of Horticulture which involves growing fruits and vegetables, flowering, and ornamental plants.

Both terms have differences based on their economical aspects, food value, and aesthetics. Agriculture involves growing cash crops while Horticulture includes growing flowers, fruits, and vegetables that have their cash value.

School students are made aware of their environment and nature using field trips particularly in subjects like Environmental Science as this brings them closer to nature and they can understand the various organisms around them. This approach can be improved by doing gardening and sowing some vegetables in the school.

Growing flowers helps in beautifying the school compound and these are liked by students. The fragrance and beauty of flowers help in attracting the small kids towards them and it helps in spreading a good environment in the school. As per Bhatt V.S.(1992) 'In India flowers are sanctified and are

Figure 2. Tagging and maintaining a school garden by students

commonly used in worshiping the deities. The use of flowers and garlands has become almost essential on all festive occasions.'

Secondary school students have to learn about various plants, their parts, uses importance, and floral structure. They also need to know various habitats and organisms in these habitats. All of these can be learned easily by maintaining a school garden that will attract birds of various types and small organisms like squirrels and monkeys.

There is a need to have partnership programs between agriculturists and educationists. Sustainable rural development is possible only when education and training strategies are utilized in various fields and disciplines of rural development.

Hence in the World Summit on Sustainable Development in Sept.2002, a program on educating the rural people was started by Directors-General of FAO and UNESCO to address the challenge of bringing together the practitioners and educationists- As per G. Hernes, Director, IIEP and E. Zulberti, Chief, Extension Education, and Communication Service, FAO.-D. Desmond, J. Grieshop, and A. Subramaniam Revisiting Garden-Based learning in basic education

Materials and Methods

Objectives

1. To study the difference in the learning of secondary school students regarding the inclusion of Agriculture in their curriculum.
2. To study the skill development of Children who have studied Agriculture in School and compare it with those who have not done it.

Hypotheses

1. There is no significant difference between children who learn agriculture and those who do not learn it.
2. There is no significant difference in skill development of children who have taken agriculture as a subject in school and those who have not taken it.

Delimitation

1. The study is delimited to two schools in Panipat District.
2. The study is delimited to Secondary School students.
3. The study is delimited to 500 students.
4. The study is delimited to the Panipat district of Haryana.

The reason for delimitation is the following-

1. This study is done in the Panipat District of Haryana as the government scheme of school gardening is in a preliminary stage and much work is not started in this area.
2. Secondary students are more active and willing to work in the garden, primary school students are not allowed to handle sharp tools of gardening and they are more prone to soil-borne pathogens due to less immunity in childhood.
3. Only 500 students are considered as those students who were willing to participate were considered after taking the consensus of their teachers and parents.
4. Two schools that have started the project of having agricultural gardens in their school compound were considered for the study, while other schools of the area are having only plans of starting the agriculture garden in school.
5. In the future, this study area will be expanded as all schools will promote garden-based learning.

Methodology

The study is conducted on Secondary School students of two schools of Panipat. Students are maintaining school gardens and tagging plants with botanical names. They are told to grow spinach and vegetable like tomato in the school garden. Each student is provided the responsibility of managing and taking care of one plant. They have to collect the measure of plant height, the product obtained and manure added to their plant.

Each student is asked some introductory questions related to their name, age, class, and interests in Science, Botany, and Agriculture. Then participating students are given a questionnaire of 50 questions, which they answer in a period of 1 hr. duration. These answers are analysed and graphs are prepared to know how many students are learning from the gardening done in school, is this an interesting activity, and how many other things they can learn from this activity like observing the plant growth, adding water, manures, and taking care of the plants.

Research Area

Delhi Public School Panipat and Shiv Deep Vidya Mandir Panipat.

Sampling

Systematic Sampling of students is done. A total of 500 students are selected from the two schools.

Method

A self-made questionnaire is given by the researcher to the students who are given a time of one hour to do the questions.

Participants involved in the garden work of their respective schools, then several plants grown by each student are counted, their involvement is measured by comparing the plant height and fruits obtained. All the participants were given the same time and plants were grown with the same amount of manure and the same environmental conditions. Plants were grown in recycled plastic containers and earthen pots.

Questionnaire

First introductory questions are asked to each participant with simple questions like-Name, Age, Class, and School Name. Then the questionnaire is provided after each participant is allowed to sit. The time limit of 1 hr. is provided for answering the questions.

Statistical Analysis

The questions are analysed, bar graphs are drawn for each answer, and the percentage of participants who agree, do not agree, totally disagree, totally agree, or are neutral is found out. The participants who agree or completely agree are compared to those who disagree or completely disagree, then these are compared with neutral ones. Their plant growth and production of fruits are also checked to justify their development of gardening skills. Then the ANOVA test is done in which variance between the samples and variance within the samples is calculated to find the F-ratio at a 5% level of significance.

Data Analysis and Interpretation

The calculated value of F-ratio is found to be greater than the F value at a 5% level of significance. hence the null hypothesis is rejected.

So it indicates that the difference between the sample means is significant.

Figure 3. Students discussing how to maintain the garden of the school

RESULT AND CONCLUSION

The test results and statistical interpretation gives the following results-

1. There is a significant difference between children who learn agriculture and those who do not learn it.
2. There is a significant difference in the skill development of children who learn agriculture and those who do not learn it.

The study shows that children who learn gardening can get many benefits from GBL (garden-based learning). Their overall development is holistic and they understand all botanical concepts, are affectionate for nature and organisms. They have a very good awareness of their surroundings. They can develop as self-independent individuals who are ready to start different sources of livelihood in the future.

CONCLUSION

The study indicates that Agriculture and Gardening Education is beneficial to the children of Secondary Level in Schools.

Educational Implications of Gardening and Agriculture in Schools

At the school level children from the primary level need to understand the plant parts, roots, stems and leaf functions, germination of seeds, floral parts, dispersal of seeds, use of fruits, flowers, vegetables, and grains.

At the secondary level children should be taught about sexual and asexual reproduction in plants, propagation, and pollination of plants. This education is imparted by providing the students practical knowledge through nursery visits,nature walks, and maintenance of the school garden. Students involve themselves from the sowing of the seed, understanding germination of plants, providing them fertilizers, and finally harvesting the crops. Due to their involvement schools are having good vegetable gardens and students are able to relate themselves to nature. This helps in the holistic development of the child as envisaged in the National Educational Policy 2020.

Also, children should be made to do practical applications of things learned in class like- taking out seeds, removal of weeds, asexual propagation of plants. This can be done in regular classes of Science or EVS. Also, there can be a hobby or Socially Useful Productive Work (SUPW) classes and Gardening Clubs in which the child should be able to do the practical application of things studied.

Children can be made to sow seeds, irrigate the ground, and remove weeds. This is a very enjoyable experience for the students as they can see the different growth stages of a plant. They can learn to use small tools like spade, trowel, and rake which help them to get knowledge of tillage, overturning of soil. They irrigate and harvest the crops which can be later used by them.

As per Verma V. (1994) "Children learn that plant growth is not only affected by a host of environmental factors but also several physiological processes."

Benefits Inclusion of Agriculture and Garden-Based learning has the following benefits for the students-

1. They remain closer to nature and learn how to protect it.
2. They can know how food comes on their plate. Hence they can give value to food.
3. They get to know the structure of soil, topography, and soil types.
4. They know the importance of water and how to wisely use it. They learn by doing as they can grow and nurture the plants in their school or community gardens themselves.
5. They learn to use various gardening tools. This also helps them in learning the skills and how to protect themselves from sharp tools.
6. The different stages of plant growth are visualized by them and it helps in imparting a lasting impression on their young minds.
7. They know the importance and uses of trees, their varieties, and which trees can be grown in their school or community garden.
8. They know that plants provide us clean and fresh air which we can get by growing evergreen plants and keeping them in our surroundings.
9. They can recognize the aesthetic value and other uses of flowers.
10. They learn sharing and caring by adopting each plant and providing it water and manures.
11. They learn different methods of asexual propagation and can propagate plants using these methods in their school gardens.
12. The students participate in the Van Mahotsav Programme and can grow plants and trees in their homes and colonies.
13. They can learn to identify different types of plants by their names.

Figure 4. Students sowing seeds and removing weeds in the school garden

Figure 5. Students growing ornamental plants like money plant in pots

14. They learn the herbal value of some herbal plants and can use them in their day to day life.
15. Students can also disseminate the information to their friends, parents, and grandparents who can help in saving the environment.
16. Gardening improves the aesthetic sense of students. They learn how to grow plants of various varieties with some special characters like- beautiful flowers, plant height, the essence of flowers, uses of plant parts, etc.
17. Students can differentiate between the various categories of plants and know about the useful qualities of these plants.
18. Students can be taught the processes of Composting, Vermicomposting by doing it practically in their school or community gardens.
19. They can be taken out for study tours to some Botanical Gardens which would help them to get more exposure to various types of plants, trees, and greenhouse plants.
20. Students also learn about the changing environment by observing the changes in climate, adaptations of plants, and variations in plant varieties due to some Anthropogenic activities.
21. They learn to prevent Soil erosion by doing afforestation.
22. History of Van Mahotsav, Chipko movements can be taught to students through the gardening club. This helps them to understand the sacrifices of our ancestors, the history of our country, and how our ancestors loved nature and its resources were taken care of by our ancestors.
23. They can become Eco-warriors through the gardening clubs in their schools to spread the message of afforestation, benefits of plants, growing indoor plants, herbal plants, etc.
24. They can help society to improve its air quality by growing indoor plants like- Aloe vera, Snake plant, Palms which improve the air by releasing oxygen and absorbing pollutants.
25. Millions of plant species in the world are treasure houses of medicines. Students can spread knowledge about these to their local population and this can build a society that can rely on natural medicine for curing their diseases.
26. Natural pest control systems can be utilized by the students in their gardens and about which they can spread a message to society.
27. Organic farming using natural fertilizers, natural pest control systems, herbicides are learned by students and they can spread the information in society.
28. Limited use of water resources can be learned by doing drip irrigation.
29. Land resources can be optimized by growing plants in small areas, pots, and baskets.
30. Soil can be replaced with coco peat and plants grown in them help to reduce the stress on land resources.
31. They can understand how trees can be grown in the boundaries of fields to absorb the pollutants and give this knowledge to Village Panchayat Head or farmers directly so that they can use this technique in their fields.
32. Farming can be promoted by developing its importance in young minds.
33. Learning about endangered species of plants, animals, and protection measures.
34. Learning about wild plants and animals, biodiversity, and its protection.
35. Herbarium of medicinal plants is made by students which helps in preservation and spreading knowledge about medicinal plants.

Figure 6. Visit to the local nursery to learn about various plants

36. According to Uberoi M. (2002) 'Herbaceous-Any plant which has soft stems rather than woody ones.'
37. Learning about pollinators, their habitat, and how to protect them.
38. Gardening helps the children to remain engaged with nature and if they develop it as a hobby, they can keep themselves in a pollution-free environment.
39. Gardening as a hobby helps the students to keep themselves away from mobiles and other internet-based games which occupy their time and makes them stressed.
40. Students can develop skills in poultry and dairy farming. They can make visits to these farms and understand how to do these types of farming.
41. Students can learn the difference between monoculture and multiple cropping.
42. Students can opt for some lesser-known career options like-Apiculture, Dairy farming, Poultry farming, Fisheries which can reap good benefits.
43. Students can go for start-ups in some fields which are new for society like- Hydroponics.
44. Gardening and Agriculture can teach students self-reliance as when they can give life to a plant or domestic animal they get to know about self-worth. It gives them confidence in building their lives.
45. Learning about Agriculture and Gardening helps in the holistic development of students.
46. Students understand the concept of biodiversity by seeing a variety of plants and arboreal animals, butterflies, honey bees, and birds.

Figure 7. Students collecting saplings of some trees for growing in the school garden

47. Students can organize exhibitions and learn how to be better than others by taking part in some competitions based on the good growth of their plants.
48. Growing food teaches them the economic value of plants, grains, vegetables, and fruits.
49. Students learn the importance of humus and natural fertilizers, how to use these wisely, and not to harm the topsoil.
50. According to Kulkarni M.N., (2015) Burning of crop residues should be avoided. Even weeds can be converted into good quality compost.
51. Soil types, pH value, porosity, stratification, and geology are easily understood.
52. The human history of civilization and ancestral behaviour can be easily understood by the students.
53. Gardening and Agriculture helps to know about Genetics and they can discuss whether Genetically Modified plants are good for them or not.
54. The problem of providing food, population explosion, climate change due to Global warming, and other similar ecological problems can be easily understood and young minds can be made to think about solving these important problems which can help in sustainable development.

Findings and Future Implications

Hence Garden and Agriculture based learning is the need of the hour. Every child should have these subjects in their school curriculum at the Secondary level.

Support System For starting this type of education the teachers and educators should be provided -

Figure 8. A layer farming model made by students in the school

1. Basic Guidelines
2. On-field experience
3. Interaction with Agricultural Scientists
4. Syllabus development training
5. Visits to Agricultural Farms and Gardens

Agriculture and Gardening can be also included in B.Ed. Training or other Teacher Education Programme which will help the teachers to be well trained for teaching these subjects.

Special Books should be published by NCERT (National Council of Educational Research and Training) which should be made available to students at a low cost so that children in Government Schools can also afford them. Other major Publishers can also start publishing affordable books for school children to give them exposure to various techniques of growing plants and managing gardens.

According to Sukhia S.P. (2005-06), 'Such facilities and equipment which are not possible for providing to each school should be arranged collectively for all the school complexes.'

HRD Ministry, Government of India, issued School Nutrition Kitchen Garden) Guidelines (2019)- "All schools to set up a kitchen garden and instructions have been issued by Centre to all States and Union Territories that Government Schools up-to Class Eight should include the produce from the Kitchen Gardens in the mid-day meal.

Aim This was aimed at improving nutrition in school children as the products of kitchen gardens can be used in mid-day meals. It will boost the nutritive value of midday meals and also encourage children to grow plants, vegetables, and fruits. The Government also wants to develop among children the skill of growing their vegetables and fruits in schools and homes.

Organizers Department of Agriculture and Horticulture, Krishi Vigyan Kendras, Food and Nutrition Board, State Agricultural Universities, and Forest Department provide the plantation requirements like technical and material assistance.

Mahatma Gandhi National Rural Employment Guarantee Act(MGNREGA) Scheme which employs rural poor people is utilized for leveling of land and making the boundary wall of the garden.

Allocation of Budget Ministry of Human Resource Development will allocate Rs 5,000 per year per school for this purpose and each school will work out its plan for considering the type of vegetables or fruits to be grown based on season and locality.

Workshop for Students UP Government has organized a workshop at Central Institute of Subtropical Horticulture (CISH) where teachers, staff, and students of schools were called to clear their doubts and were also taught gardening. They were also made aware of the benefits of seasonal vegetables.

Summary

The study conducted in two schools of Panipat, India can be extrapolated to all the students of the secondary level and it can be easily stated that inclusion of Gardening in the School Curriculum is the need of the hour and this has been realized by the Government of India as its HRD Ministry has given directions in this regard. Gardening and Agriculture trains the school students in a holistic manner which prepares them for their life as they develop the very important skill of growing and nurturing plants. It also allows them to develop sensitivity for nature and animals, gives them moral values, helps them support Sustainable Development by saving natural resources, and brings them closer to the real-world keeping them away from the virtual world. This also helps the schools in improving the nutritive value of mid-day meals involving the students in some skill-based education which can help in solving the problems of malnutrition and providing jobs to millions of Indian Youth.

REFERENCES

Bennur, S. (2014, May 29). Schools directed to grow vegetables, fruits in kitchen gardens. *The Hindu.*

Berleant, A. (2017, June 4). *Environmental Aesthetics.* Academic Press.

Bhatt, V. S. (1992). Handbook of Agriculture. Indian Council of Agricultural Research (ICAR).

Bowker, R., & Tearle, P. (2007). Gardening as a learning environment: A study of children's perceptions and understanding of school gardens as part of an international project. *Learning Environment Research.*

Desmond, Grieshop, & Subramaniam. (2004). *Revisiting Garden-Based learning in basic education.* ResearchGate.

Food and Agricultural Organization of United Nations. (2010). *A New Deal for School Gardens.* UN Report on School Gardens. FAO.

Hazra, P., Ghosh, S.K., Maity, T.K., Pandit, M.K., & Som, M.G. (2010). *Glossary of Horticulture.* Academic Press.

Kanthimathi, L. (2017, July 14). Government Schools raise kitchen gardens. *The Hindu.*

Kulkarni, M.N. (2015). Towards Sustainable Agricultural Systems. *Terra Green, 7*(11).

Miller, L. K. (1904). *Children's Garden for School and Home, A Manual of Cooperative Gardening*. School and Home.

Ministry, H. R. D. Government of India. (2019, May). *School Nutrition(Kitchen Garden) Guidelines*. http://www.schooleducation.kar.nic.in/SSA/pdfdocs/SwasthPlus.pdf

Sukhia, S.P. (2005-06). *Educational Administration, Organization and Health Education*. Academic Press.

Uberoi, M. (2002). *The Penguin Book of Gardening in India*. Academic Press.

Verma, V. (1994). *A Textbook of Plant Physiology*. Emkay Publications.

APPENDIX

Meaning of Some Terms

Agriculture- Growing plants for sale or growing plants for cash.
Anthropogenic activities- Activities which are done by man.
Asexual Propagation- Propagation or reproduction is done by a single parent using its parts which can be- Budding, Layering, Cutting, By roots, stem, or leaf.
Bio-intensive gardening - Using biological methods in good amounts in gardening.
Composting- Making compost using biodegradable waste materials like vegetable and fruit peels.
Educational Implications- Impact of education on students.
Fertilizer- Chemicals provided to plants for fulfilling their nutrient requirements. Nitrogen, Potassium, and Phosphorus are major nutrients needed for plants that can be provided by adding Urea, DAP, etc to the plants.
Garden- An enclosed area beautified with plant materials and other structures.
Genetically Modified Plants- GM crops are crops or plants which are grown after making some modifications in their genes in an artificial manner by human beings in their labs.
Germination- Development of a seed into a seedling by the growth of embryo under favourable conditions of air, water, and warmth.
Herbarium- A collection of plants or their parts dried and pasted in a book or diary to represent different varieties or some characteristic features like herbal or disease resistance in plants.
Horticulture- A branch of Agriculture that studies the growth of ornamental plants, flowering, or fruit-bearing.
Irrigation- Providing water to plants that are generally done artificially by a man using sprinkler, furrow, or basin methods. Nowadays to save water drip irrigation is a very popular method.
Mid-day meal-A free meal provided in the government schools to improve the attendance of students, their nutrition, health and to reduce child labor.
Monoculture- Growing plants of the same kind of variety.
Multiple cropping- Growing more than one crop in the same field.
School Garden- An enclosed garden of school which is generally present in the school premises.
Organic Farming- Farming using natural materials like-compost and which does not use any chemicals like fertilizers or insecticides.
Ornamental plants- Plants that are grown for decoration purposes and their cultivation is known as floriculture which is a major branch of horticulture.
Secondary School- School education from 6th standard to 10th standard.
Vermicomposting- Composting did by earthworms like- red worms.

QUESTIONNAIRE

A questionnaire was prepared based on the Likert Scale where answers range from-

Strongly agree, Agree, Not sure, Do not agree, Strongly disagree

The questions asked are (Questions related to name, age, class, and family are introductory questions asked to each participant at the beginning of questionnaire)-

1. It is good to have a school garden to improve the school environment, do you agree?
2. The gardening activity can be done on the school premises for providing cognitive learning.
3. The study of science is benefited by gardening as the various scientific methodology of growing plants is utilized.
4. Your family has an agricultural background and they want you to know the basic concepts of gardening.
5. Interest in the school garden is found more in students who like science.
6. The School garden project helps in providing vegetables for the Mid-day meal scheme.
7. Flowers grown in school gardens help to improve knowledge about flowering plants. Floriculture helps to improve the aesthetic sense of the children and to learn about other uses of flowers.
8. Growing vegetables help secondary children to know about fruiting plants.
9. The fruit development can be studied by the children when these grow in the schoolyard.
10. Learning by observation is promoted by studying plants.
11. Psychomotor learning is promoted by maintaining plants and carrying out the processes of gardening.
12. Students can identify different types of plants, know the herbal value of some plants, and use them in their day to day life.
13. The gardening project brings you closer to nature and helps in learning how to protect plants to keep yourself in a pollution-free environment.
14. Gardening helps to give value to food so you have reduced food wastage.
15. Students know the importance of water and how to wisely use it by watering plants in the school garden which also helps to learn by doing. Limited use of water resources can be learned by doing drip irrigation.
16. Students learn to use various gardening tools. This also helps them in learning psychomotor skills and to protect themselves from sharp tools.
17. The different stages of a plant's growth are visualized and students learn about caring by adopting each plant and providing it water and manures.
18. Students know that plants and trees provide us clean air, other uses of trees, their varieties which can be grown in their school or community garden.
19. Chipko movement and understanding the sacrifices of ancestors helps to improve historical knowledge.
20. Visiting botanical gardens and outdoor tours helps in learning plants of various varieties with some special characters like- beautiful flowers, plant height, the essence of flowers, uses of plant parts, etc.
21. Students can be taught the processes of Composting, Vermicomposting by doing it practically in their school garden.

22. Students also learn about the changing environment by observing the changes in climate, adaptations of plants, and variations in plant varieties due to some Anthropogenic activities.
23. Participating in Van Mahotsav, spreading the message of afforestation, benefits of plants, growing indoor plants, herbal plants make students eco-warriors.
24. They can help society to improve its air quality by growing indoor plants like- Aloe vera, Snake plant, Palms which improve the air by releasing oxygen and absorbing pollutants.
25. Organic farming using natural fertilizers, natural pest control systems, herbicides are learned by students and they can spread the information in society.
26. Land resources can be optimized by growing plants in small areas, pots, and baskets using coco peat.
27. Botanical study of plants, flowers, and fruits is easily done in the school compound which reduces efforts and learning happens in nature.
28. Integration of arts like poetry, painting, and flower organization with science helps to understand nature in a better way.
29. Students are motivated to learn botany and science by growing plants themselves.
30. Gardening activity is engaging and helps to impart knowledge to specially-abled children who enjoy doing this activity.
31. Learning about pollinators, their habitat, and how to protect them.
32. Learning about new methods of plant propagation like Hydroponics.
33. Gardening helps in holistic development as a child can know about natural surroundings in the school environment itself particularly its good for students who live in high-rise buildings with minimum vegetation.
34. Gardening helps in developing interpersonal skills as the child speaks to friends and learns to talk and ask elders for help while nurturing plants.
35. Gardening is one of the most important activities done which involves the students and keeps them engrossed.
36. Solution of problems of humanity is understood at a very young age which helps in developing a caring personality who understands the needs of the poor farmers and their hardships.
37. Learning about wild plants, biodiversity, and its protection helps to save endangered plants.
38. They learn different methods of asexual propagation and can propagate plants using these methods in their school gardens.
39. They can understand how trees can be grown in the boundaries of fields to absorb the pollutants and give this knowledge to Village Panchayat Head or farmers directly so that they can use this technique in their fields.
40. Students can learn the difference between monoculture and multiple cropping.
41. Farming can be promoted by developing its importance in young minds.
42. Learning about endangered species of plants, animals, and their protection measures.
43. Students can spread knowledge about herbal and medicinal plants to their local population by making herbarium and showcasing the importance of some herbal plants.
44. Gardening as a hobby helps the students to keep themselves away from mobiles and other internet-based games which occupy their time and make them stressed.
45. Students can opt for some lesser-known career options like-Apiculture, Dairy farming, Poultry farming, Fisheries which can reap good benefits.

46. Gardening and Agriculture can teach students self-reliance as when they can give life to a plant or domestic animal they get to know about self-worth and develop their confidence.
47. Organizing exhibitions and growing food teach the economic value of plants, grains, vegetables, and fruits.
48. Students learn the importance of humus and natural fertilizers, how to use these wisely, and not to harm the topsoil.
49. This helps to know the structure of the soil, topography, soil types, pH value, porosity, stratification, and geology.
50. Gardening and Agriculture help to know about Genetics and Genetically Modified plants.
51. Problem of providing food, population explosion, climate change due to Global warming, and other similar ecological problems can be easily understood and young minds can be made to think about solving these important problems which can help in sustainable development.

Section 2

Education and Sustainable Development

Chapter 8
Better Late Than Never: Elucidating Effective Education Methods for Sustainability Teaching

Eduardo Terán-Yépez
https://orcid.org/0000-0002-1260-2477
CIMEDES Research Center, University of Almeria, Spain

Gema M. Marín-Carrillo
CIMEDES Research Center, University of Almeria, Spain

María del Pilar Casado-Belmonte
University of Almeria, Spain

María de las Mercedes Capobianco-Uriarte
University of Almeria, Spain

ABSTRACT

This chapter, through an exploratory snapshot, aims to provide the first empirical work in the European context about the perceived effectiveness of sustainability education methods. Thus, based on a sample of 432 students from the Economics and Business Faculty of an Andalusian University (Southern Spain), this study identifies which methods are perceived as being most effective. The findings show that integrating sustainability into universities' strategic plans and implementing sustainability initiatives that actively involve the student through the development of strategic alliances with international universities and institutions, combined with the inclusion of field trips to learn about sustainability are perceived as the most relevant learning approaches.

DOI: 10.4018/978-1-7998-7512-3.ch008

INTRODUCTION

Acting sustainably requires that the core of any activity encompasses three goals: social, environmental, and economic development (Elkington, 1994; Schaltegger& Wagner, 2011). Accordingly, an increasing number of organizations are committed to developing sustainable strategies (Casado-Belmonte et al., 2019). Education plays a central role in sustainability-related awareness, providing students with the attitudes, values and behaviours needed to achieve sustainable development goals.

Mainly, it is the responsibility of higher-education institutions to embed learning approaches concerning sustainability into their education programs as students will later be responsible for making decisions on sustainability-related issues in their communities and professions (Sahin & Erkal, 2017). Even if the concept of sustainability is not new, sustainability in management education is a relatively novel area (Erskine & Johnson, 2012). Given that students are likely to be engaged in sustainable businesses, initiatives on the part of education institutions to integrate sustainability content into their current courses have been undertaken (Singhal et al, 2018). In this vein, the United Nations proposed the establishment of Principles of Responsible Management Education (PRME) in 2007 as *"an urgent call to modify business education in light of changing ideas about corporate citizenship, corporate social responsibility, and sustainability"* (Alcaraz & Thiruvattal, 2010, p. 542). The United Nations PRME include six principles, the first of those stating *"develop the capabilities of students to be future generators of sustainable value for business and society at large and to work for an inclusive and sustainable global economy"* (Erskine & Johnson, 2012; PRME, 2013).

Erskine and Johnson (2012) and Singhal et al. (2018) identified 37 unique learning approaches based on 125 publicly available PRME Sharing of Information on Progress reports, drawn from the self-reports of business schools reporting their pedagogical experiences as members of business schools that support PRME. The contribution of these studies is of great value as it enables the attainment of a plethora of education methods, which makes it possible to identify the most effective learning approaches. Effective learning approaches are thought to take into consideration the preference of students, based on the assumption of the robust relationship between motivation or willingness to learn and actual learning (Erskine & Johnson, 2012; Cole et al., 2004; Karns, 2005; Zocco, 2009). Therefore, the study of student perceptions about education methods on sustainability and the adequacy of these methods could heighten the chances of more successful learning on sustainability-related issues (Erskine & Johnson, 2012).

It should also be noted that according to Dyllick and Muff (2016), what society and the planet need in the next two decades is nothing less than a sustainability miracle, and companies which are in a position to generate that miracle. Therefore, the world needs entrepreneurs and businesspeople who are truly committed to sustainability and who can mitigate and even overcome the conflicts between economic and financial demands and the needs of society. Thus, there is still time to initiate change from the universities; however, to do so, it is necessary to consider what the most effective techniques for teaching sustainability issues are.

Previous research has been conducted in North America and Asia to understand student preferences when learning about sustainability, however to the best of our knowledge, there are no empirical studies that analyse this issue in the European context. To address the current research gap, and utilizing as a sample, students of the Economics and Business Faculty of an Andalusian university (southern Spain), the purpose of this study is to recognize effective learning approaches as perceived by higher-education students in the context of sustainability and suggest improvement guidelines for universities and business schools. The methodology to develop this study is based on a structured questionnaire that assesses the perceived importance for a total of 37 learning approaches.

Findings provide a categorization of the 37 learning approaches as perceived by the students and enable the identification of the most effective ones. In such a way, the most effective teaching methods for the entire Economics and Business Faculty as well as for each degree are obtained. Moreover, building on the most effective learning approaches, this study proposes some guidelines on how to facilitate the integration of sustainability in academia, with a special focus on business and management university degrees.

Therefore, this book chapter provides the first empirical work in the European context on the perception of university students about various learning approaches used in teaching sustainability issues, to identify which are the most effective education methods. Furthermore, based on these identified learning approaches, this study proposes some guidelines to set up these education methods. In summary, the findings help higher-education institutions, educators, and professors by providing information on the most effective learning approaches for teaching sustainability-related issues which could be used to transform and update policies, curriculum content, strategic planning, and networks of higher education institutions. The study first includes a theoretical background about the PRME, evolving methods in management and business education, and sustainability in management and business education. Thereafter the methodology, results and discussion, and finally, conclusions are presented.

LITERATURE REVIEW

Principles for Responsible Management Education (PRME)

Principles for Responsible Management Education (PRME) is a global call by the United Nations (UN) to integrate international values into business education such as those contained in the Global Compact on Human Rights, Labour, Anti-Corruption, and the Environment (Alcaraz & Thiruvattal, 2010). PRME's primary target is to spur a more sustainable and inclusive worldwide economy (Haertle et al., 2012). The six PRME are based on these values and aim to dramatically influence the learning and educational practices that take place in business schools to develop students' capabilities, incorporate the values of global social responsibility into academic activities, create adapted educational frameworks and materials, processes and environments, participate in conceptual and empirical research, interact with business managers, and facilitate the debate on sustainability issues among stakeholders and, in general society. The principles emerged in part as a response to criticisms made in companies and business schools in the first part of the twenty-first century (Erskine & Johnson, 2014). The UN in the PRME initiative addresses the responsibility of management education institutions to prepare the business professionals of today and tomorrow for the challenge of making business more responsible and sustainable (Godemann et al., 2014) In this book chapter we will try to shed light on how students are developing skills to be the future generators of sustainable value for business and society at large and work towards an inclusive and sustainable world economy.

What Is Sustainability?

Sustainability is mainly understood as "*satisfying the needs of the present generation without compromising the ability of future generations to meet their own needs*" (Brutland Commision, 1987, p. 8). While sustainability was born primarily out of concern for the environment, it pays equal attention to human and economic development (Erskine & Johnson, 2012). Thus, acting sustainably requires that the core of any activity encompasses three goals, social, environmental, and economic development (Elkington, 1994; Schaltegger & Wagner, 2011). At the organizational level, Rands (2009, p. 299) argues that "*based on the reality and severity of environmental threats, the actions from organizations should be simultaneously economically profitable (or at least break-even), environmentally sound, and socially just*". This tripartite approach is known as a triple bottom line (Elkington, 1998). At the educational level, the United Nations Environment Program (UNEP, 1972) in the Stockholm Conference in 1972 highlighted the relevance of education for the protection and preservation of the environment. Since then, numerous higher education institutions have shown an increasing awareness regarding sustainability, to the point of including sustainability issues throughout the academic period (Singhal et al., 2018).

Sustainability in Management and Business Education

In such a context, society faces the challenge of the adoption and implementation of sustainable behaviour through embedding sustainability in values (Nwagwu, 2020). In this regard, sustainability issues have drawn attention from both corporations and academic institutions. While the corporation context has witnessed the increasing awareness level of firms to respond to sustainability concerns and businesses' engagement with sustainability matters, academic institutions seem to be one step behind with a less than adequate exposure of students to sustainability matters (Jose, 2016).

Academic institutions, namely higher education institutions, are deemed to be the driving force for change in societies, contributing to building the capacity in future generations to face real concerns such as sustainability (Cole & Snider, 2019; Peterlin et al., 2018) and to foster lifelong learning skills for rapid adaptation (Erturgut, 2011). To achieve these purposes, educators should encourage students to acquire knowledge regarding the latest trends and research findings (Sohal, 2013).

In particular, this training gains special relevance in business education, where there exists a need to encourage and hasten the commitment to responsible management (Cole & Snider, 2019; Kelley & Nasher, 2014). In this vein, there is a call to change business education to prepare students, future leaders, and managers for responsible management (Nonet et al., 2016).

When talking about the connection between education and sustainable development, the first question is the lack of a clear understanding of what sustainability is and how it can be integrated into each part of the business to nourish sustainable development instead of just viewing it as a requirement to accomplish (Nwagwu, 2020).

Sustainability embedded in management/business higher education is vital for several reasons. First, the awareness of the premise that businesses go beyond profit-making. Second, the exposition to sustainability examples may increase global awareness and empathy in current and future managers (Cole & Snider, 2019). Third, the importance of soft skills is emphasized, the development of critical thinking, the comprehensive triple bottom line, and a shared vision for all stakeholders (None et al., 2016) along with broadening horizons with gender issue awareness and work-life balance (Peterlin et al., 2018).

Some authors (e.g., Jose, 2016) state that the difficulty in higher education is due to the lack of institutional capacity, motivation and incentives on the part of faculty members, lack of recruiter interest, and difficulty in access to high-quality resource material.

Although the inclusion of sustainability in management knowledge is pressing (Herrera-Cano, 2016), very little is known about how managers could be trained more effectively to embed corporate sustainability in their values (Hesselbarth & Schaltegger, 2014).

In this sense, awareness of student motivations/preferences could help educators to reshape the traditional pedagogical content to assume the responsibility of driving for sustainability and being challenged to educate future and current leaders.

The Relevance of Student Preferences

Several studies have shown that students' attitudes and preferences are key elements for the successful implementation of new pedagogical approaches (Erskine & Johnson, 2014; Barth & Timm, 2011; Martins & Kellermanns, 2004). According to Eastman et al. (2011), educators can propose new high value-added learning initiatives to achieve the curricular competences in a subject, but the learning process requires an important attentional and motivational deployment on the part of the student, since without his or her commitment and involvement the learning may not be significant.

Analysis of students' perceptions of what they consider to be the most effective ways of learning about sustainability can play a key role in designing learning objectives and activities in the subject curriculum content, as well as in defining other sustainability-oriented initiatives that can be offered outside the classroom (Erskine & Johnson, 2012). By contemplating their preferences, students can become more motivated, influencing their attitude, their learning and their behaviour (Cole et al. 2004; Karns, 2005). According to Entwistle (1988), student motivation makes the difference in terms of the amount of effort put into learning. Thus, to the extent that students have a more favourable attitude towards certain activities, they show a higher level of commitment and engagement. This greater involvement leads to better use of the learning experience, which contributes to more positive outcomes in their learning process (Clarke et al., 2001; Farias et al., 2010; Taylor et al., 2004).

The coherence between students' needs and preferences, in terms of sustainability, enables student satisfaction to be achieved more easily, and also the student's conviction of the usefulness and significance of the sustainability learning for their future jobs (Erskine & Johnson, 2012; Wingfield & Black, 2005; Kucharcikova et al., 2019).

METHODOLOGY

For this exploratory study, students from the Faculty of Economics and Business Sciences at an Andalusian university in southern Spain were taken as a sample. As stated before, Erskine and Johnson (2012) and Singhal et al. (2018) identify 37 unique education methods for sustainability education derived from 125 publicly available PRME Sharing of Information on Progress (SIP). Those 37 learning approaches are the basis for this research.

Procedure

To develop this analysis, and as suggested by Erskine and Johnson (2012) and Singhal et al. (2018), a questionnaire containing the 37 education methods was drawn up. The questionnaire allowed us to recognize which were the most effective learning approaches in the context of sustainability according to the perception of university students. Moreover, as recommended by Erskine and Johnson (2012) we also included some questions about the importance that students give to sustainability. The importance was measured through a seven-point Likert scale in which scaling is as follows (1= strongly disagree, 2 = disagree, 3 = somewhat disagree, 4 = neither agree nor disagree, 5 = somewhat agree, 6 = agree, 7 = strongly agree). As the original questionnaire developed by Erskine and Johnson (2012) and Singhal et al. (2018) was set in English, we translated it into Spanish with the help of an expert English-Spanish translator. Furthermore, we made the reverse translation of the questionnaire (Spanish-English) and no problems of translation and understanding were detected. Besides, the questionnaire was validated by a sample of 18 students, following which some changes were deemed necessary.

To clarify the concept of sustainability, we preceded the questionnaire with the following definition: "meeting the needs of the present generation without compromising the ability of future generations to meet their own needs".

Some of the more demanding challenges encountered during this stage relies on (1) the length of the questionnaire and (2) the fact that students are overexposed and burdened to filling several surveys in the university context such as pedagogic evaluation, academic studies and market research. Therefore, these issues could discourage students from completing it. However, we could overcome these difficulties due to the fact that the questionnaire was fulfilled during face-to-face class and we provided a little incentive once the questionnaire was completed. These "remedies" allow us to reach a vast number of questionnaires in the student population.

Sample

We recruited a convenience sample of 432 students of the Economics and Business Faculty of an Andalusian university. The majority of the students (82.4%) were between the ages of 19 and 23. The remaining 17.6% of respondents were 24 or older. Women comprised 58.8% of respondents while men comprised 41.2%. Regarding nationality, 87.3% of respondents were Spanish, whereas 12.7% were foreign nationals, particularly Chinese, Rumanian and Moroccan students.

Since our objective was to obtain a perception according to the various business and management degrees, we achieved a representative sample of the five degrees offered by the Economics and Business Faculty, so that the sample represents the demographic characteristics of the university student population. Thus, 46.8% of respondents were taking a Business Management and Administration Degree, 19.5% a Marketing Degree, 13.4% a Tourism Degree, 11.1% a Finance and Accounting Degree, and 9.5% an Economics Degree.

RESULTS

The perceived effectiveness of the 37 approaches to learning about sustainability from a student perspective is summarized in Table 1. The perceived effectiveness ranges from a maximum of 5.89 to a minimum of 4.00 in which a higher value indicates a higher perceived effectiveness of the learning approaches.

To offer a better interpretation of the results and to provide insight into the perceived relative importance of the learning approaches, these learning approaches were divided into five groups, using the mean as grouping criterion: least effective learning approaches (mean between 4.00 and 4.64), less effective learning approaches (between 4.65 and 5.08), midlevel effective learning approaches (between 5.09 and 5.23), more effective learning approaches (between 5.24 and 5.53) and the most effective learning approaches (over 5.54). The identified aggregate means for each group are noted in parentheses. A test comparing these aggregate means resulted in each aggregate mean being significantly different from every other aggregate mean at the < .005 level.

The *most effective learning approaches* for sustainability can be grouped into three categories. First, the importance is given to the integration of sustainability into the guiding principles of university management. In this vein, the incorporation of environmental improvement initiatives on campus is perceived by students as a way for the university to show its commitment to sustainability. Its management is expected to be consistent with its strategy of promoting sustainability issues in both teaching and research, under the assumption that a university that promotes sustainability must itself be sustainable. Second, it also highlights the importance of adapting the curriculum to integrate sustainability content into various subjects, i.e. searching for cross-sectional learning. Finally, students seem to give priority to initiatives that foster active experiences outside the university campus, such as international exchanges, internships in companies, or field trips.

Additionally, it is worth noting that in the group of the more effective learning approaches the integration of real business experiences in the classroom is highlighted. Even a cursory glance at the least effective approaches reveals the idea that activities unrelated to the academic curriculum, which are promoted by the student community themselves, lack student support.

Our results are partially consistent with the findings of previous studies such as that of Erskine and Johnson (2012) and Singhal et al. (2018). In the first study, Erskine and Johnson (2012) analyzed student attitudes towards the effectiveness of each approach from a sample of 183 business students at a large, public Midwestern university (U.S.). Four of the approaches identified in our study as being among the most effective are also detected in the study of Erskine and Johnson (2012), namely IM02, IM03, IM05, and IM06. Unlike our study, their results show that sustainability-related scholarships and business simulators are also considered by U.S. students as being part of the most effective approaches to achieving sustainability learning. U.S. students attach less importance to improving environmental initiatives on campus and field trips away from campus although they are considered effective. Moreover, incorporating sustainability into the university's mission is considered a midlevel effective learning approach.

Similarly, the study carried out by Singhal et al. (2018) collected the perceived effectiveness of the learning approaches for sustainability from a sample of 151 students of a top public management institute of India. In their results, as with ours, the most effective learning approaches include IM05, IM06, and IM07. Besides, this group includes guest lectures by experts in sustainability, conducting case studies in classes based on sustainability, a career fair about sustainability jobs, and business simulators. What is more, Indian students, in comparison with the Andalusian sample, assign less importance to promoting

Table 1. Perceived Effectiveness of Learning about Sustainability

Item	Perceived effectiveness of learning approaches	Mean	SD	n
	Most effective learning approaches (aggregate *M=5.72)*			
IM12	Enhanced environmentally friendly (green) initiatives on campus	5.89	1.252	426
IM07	A field trip away from campus to learn about sustainability	5.74	1.404	430
IM03	Promoting internships related to sustainable business practices	5.73	1.166	432
IM27	Sustainability incorporated into the university's mission	5.72	1.219	428
IM06	International opportunities to learn about sustainability	5.69	1.197	428
IM02	Sustainability integrated into various subjects	5.65	1.182	431
IM05	Implementation of sustainability practices at the university level	5.63	1.219	430
	More effective learning approaches (aggregate ***M=5.36)***			
IM11	Guest lectures by experts in sustainability (Master Class)	5.46	1.340	428
IM10	Guest speakers in classes to discuss sustainability	5.42	1.273	425
IM14	Incorporating sustainability learning goals within existing courses	5.39	1.283	430
IM13	Establishing a sustainable business centre at the university	5.38	1.300	429
IM09	Conducting case studies in classes based on sustainability	5.35	1.283	430
IM01	Providing sustainability-related scholarships	5.31	1.321	429
IM20	A career fair about sustainability jobs	5.31	1.429	429
IM37	Participation in sustainability rankings/ratings such as Princeton rankings	5.24	1.410	429
	Midlevel effective learning approaches (aggregate ***M=5.19)***			
IM19	Mentoring program to help students learn about sustainability	5.22	1.271	427
IM22	Conducting workshops on sustainability issues with teachers and business people	5.22	1.365	430
IM04	Use of sustainability-related business simulations	5.21	1.398	430
IM16	Floating elective subjects in sustainability for students	5.19	1.612	430
IM08	Additional points for students involved in sustainability projects	5.17	1.588	429
IM15	Promoting student team projects related to sustainability	5.16	1.319	426
IM26	Certification for students involved in innovative sustainability activities	5.15	1.458	427
	Less level effective learning approaches (aggregate ***M=4.89)***			
IM18	Promoting research-level studies on sustainability among students	5.07	1.262	432
IM21	Promote the creation of student organizations targeting sustainability projects	5.07	1.369	428
IM29	Organization of conferences focused on sustainability	4.92	1.459	426
IM23	Conducting call for sustainability study projects	4.89	1.366	427
IM35	A pledge taken by students to observe sustainability practices	4.87	1.421	429
IM36	Membership of international sustainability forums (such as PRME-Principles for Responsible Management Education)	4.81	1.450	429
IM32	Student network projects related to sustainability	4.79	1.395	428
IM24	Conducting student competitions in the classes based on sustainability issues	4.67	1.598	430
	Least level effective learning approaches (aggregate ***M=4.42)***			
IM25	Forming a student club focused on sustainability	4.61	1.566	428
IM31	Student panel discussion about sustainability	4.59	1.499	428
IM34	Online webinars related to sustainability	4.57	1.495	426
IM17	Starting a degree program in sustainability within the university	4.44	1.731	428
IM30	Holding a film series focused on sustainability issues	4.42	1.670	430
IM33	Social nights on-campus/off-campus for student groups to discuss sustainability issues	4.34	1.757	432
IM28	Compulsory sustainability course for students	4.00	1.924	429

Source: own elaboration

Table 2. Perceived Effectiveness of Learning about Sustainability by Major/ Bachelor Degree

Item	Perceived effectiveness of learning approaches	Mean	SD	n
	Top five: Business Management and Administration			
IM12	Enhanced environmentally friendly (green) initiatives on campus	5.88	1.232	200
IM07	A field trip away from campus to learn about sustainability	5.84	1.365	201
IM03	Promoting internships related to sustainable business practices	5.76	1.139	202
IM27	Sustainability incorporated into the university's mission	5.76	1.226	200
IM06	International opportunities to learn about sustainability	5.73	1.156	200
	Top five: Economics			
IM07	A field trip away from campus to learn about sustainability	6.02	1.060	41
IM06	International opportunities to learn about sustainability	5.83	1.160	41
IM12	Enhanced environmentally friendly (green) initiatives on campus	5.73	1.414	40
IM09	Conducting case studies in classes based on sustainability	5.68	1.404	41
IM03	Promoting internships related to sustainable business practices	5.68	1.474	41
	Top five: Finance/accounting			
IM12	Enhanced environmentally friendly (green) initiatives on campus	5.57	1.410	47
IM03	Promoting internships related to sustainable business practices	5.42	1.302	48
IM27	Sustainability incorporated into the university's mission	5.36	1.206	47
IM05	Implementation of sustainability practices at the university level	5.23	1.309	48
IM02	Sustainability integrated into various subjects	5.17	1.243	48
	Top five: Marketing			
IM07	A field trip away from campus to learn about sustainability	5.61	1.403	82
IM37	Participation in sustainability rankings/ratings such as Princeton rankings	5.50	1.125	82
IM11	Guestlectures by experts in sustainability (Master Class)	5.49	1.236	81
IM05	Implementation of sustainability practices at the university level	5.48	1.314	81
IM20	A career fair about sustainability jobs	5.46	1.363	82
	Top five: Tourism			
IM12	Enhanced environmentally friendly (green) initiatives on campus	6.21	1.005	58
IM07	A field trip away from campus to learn about sustainability	6.02	1.235	58
IM05	Implementation of sustainability practices at the university level	6.02	0.982	58
IM02	Sustainability integrated into various subjects	6.00	0.937	58
IM03	Promoting internships related to sustainable business practices	6.00	0.918	58

Source: own elaboration

internships related to sustainable practices, improving environmental initiatives on campus, integrating sustainability into various classes, and incorporating sustainability into the university's mission.

Table 2 shows the perceived effectiveness of the different learning approaches regarding the bachelor's degree of the students. Although common categories between degrees were not detected, three approaches stand out for being in three of the five degrees (Business Management and Administration, Economics and Tourism); namely items IM03, IM07, and IM12 show high values. The Finance and Accounting

Table 3.Attitudes Toward Sustainability

Sustainability attitudes	Mean	SD	n
Sustainability is an important topic in business	6.06	1.068	432
Sustainability is an important topic for me	5.69	1.251	432
I would prefer to work for a company that embraces sustainability as a core value	5.68	1.236	432
I am very knowledgeable about sustainability issues in business	3.85	1.364	432

Source: own elaboration

degree highlight two of the three (IM03, IM12) and the Marketing degree only indicates importance for IM07. To compare the importance of each approach, a weighting of the most effective approaches in each degree was established. In this vein, IM12 and IM07 appear for three degrees in the top position.

In comparing our results with Erskine and Jonhson's results (2012), only three degrees can be compared. In the Finance and Accounting degree, their results show coincidence with items IM02, IM03, and IM05; in the Business Management and Administration degree only IM03iscommon to both sets of results. Finally, in the Marketing Degree, there was no approach in common.

Table 3 shows the attitude toward sustainability. First, the results seem to indicate that students consider sustainability to be an important topic for both businesses (6.06) and their own value system (5.69). It is worth noting the key role given to businesses as relevant agents regarding sustainability issues. Second, the students' preferences to work for a company that embraces sustainability as a core value is revealed in their assessments (5.68). Finally, it is surprising the fact that students had the perception that they had insufficient knowledge about sustainability in business (3.85) despite studying Economics faculty degrees.

Previous studies that assess similar attitudes revealed interesting results. For the U.S. students (Erskine & Johnson, 2012), the attitude with the highest value is the preference to work for a company that embraced sustainability as a core value, while Indian students (Singhal et al., 2018) considered the importance of sustainability for businesses as the most highly assessed attitude, similar to the results in the Andalusian university. In Erskine & Johnson (2012) and Singhal et al. (2018), all items show high values, being the level of knowledge about sustainability the attitude with the lowest score. In this vein, the Indian students feel that they are more knowledgeable about sustainability than their counterparts in the U.S. and Spain. The Spanish students perceived their lack of knowledge in a stronger way.

The responses compiled on sustainability attitudes (see Table 4) show that more than 70% of students feel that sustainability is a crucial issue in businesses, showing agreement or strong agreement. The

Table 4. Attitudes Toward Sustainability (Proportion of respondents agreeing)

Sustainability attitudes	% Agree	%Strongly agree	%Agree & strongly agree
Sustainability is an important topic in business	29.7	44.1	73.8
Sustainability is an important topic for me	31.3	31.1	62.4
I would prefer to work for a company that embraces sustainability as a core value	29.2	31.6	60.8
I am very knowledgeable about sustainability issues in business	7.9	1.9	9.8

Source: own elaboration

percentage of respondents that agree and strongly agree with the importance for themselves is lower (62.4%), and the strong preference for working in a company with sustainability as a core value reaches 60.2%. Notwithstanding the importance given to the topic in business, less than 10% of respondents rate themselves knowledgeable about sustainability issues in business.

The comparison with the attitudes of students in Indian and American samples reveals that Indian students reflect every attitude more strongly. The importance of sustainability for businesses is agreed and strongly agreed by more than three-quarters of the respondents: India (89.6%) and U.S. (78.0%), whereas in the Andalusian sample this reaches 73.8%. It is worth noting that American students stand out with the lowest rate of respondents that agree and strongly agree with the statement of the importance of sustainability as a topic for themselves (50.0%), despite their preference for working in a company with sustainability as a core value (71.0%). Finally, in all samples the percentage of students that feel they are knowledgeable about sustainability topics is under half, India (43.7%) and U.S. (35.0%), nevertheless these values are far above the result in the Andalusian university of 9.8%.

Furthermore, Table 5 reveals findings on attitudes toward sustainability by degree. As seen, Economics students are more likely to feel that sustainability is a relevant theme in business (80.8%) and to agree that they are the most knowledgeable about sustainability issues (22.0%); however, Tourism students are the ones that are more likely to feel that sustainability is an important topic for themselves. Marketing students stand out in reporting attitudes towards working in companies that embrace sustainability (61.4%), despite considering themselves the least knowledgeable (6.0%).

In comparison with the results obtained by Erskine and Johnson (2012), U.S. Finance/Accounting and Marketing students award more importance for sustainability as a topic in business, more willingness to work in a company that encompasses sustainability as a core value, and more knowledge about sustainability issues in business, while Spanish students indicate that sustainability is more relevant for themselves. In the case of Business Management and Administration students, the Andalusian sample considers sustainability to be an important topic for both businesses and their own values more than the American sample, while the American students report more knowledge and willingness to work in businesses that embrace sustainability as a core value.

CONCLUSION

From an overall perspective, this book chapter contributes to existing research regarding sustainability learning approaches. This study offers clarity to the understanding of the most and least important higher-education learning techniques when it comes to sustainability context. In such a way, it is offered initial empirical evidence in the European context concerning which are the most effective learning approaches for sustainability teaching. Therefore, it is an exploratory snapshot to establish preliminary grounds on which to try to improve the techniques that are currently being used by higher-education institutions.

The findings show that two approaches stand out as being very relevant since they are coincident in previous studies and our results. Namely, the approaches dubbed the *Implementation of sustainability practices at the university level* and *International opportunities to learn about sustainability*. Accordingly, it can be said that these techniques could be used globally. However, taking into consideration the results in the Spanish sample, two approaches stand out as the most effective. Specifically, *Enhanced environmentally friendly (green) initiatives on campus* and *A field trip away from campus to learn about sustainability*. Thus, these approaches would appear to be relevant in the local context.

Table 5. Attitudes Toward Sustainability by Degree

Sustainability attitudes	% Agree	%Strongly agree	%Agree & strongly agree
Business Management and Administration			
Sustainability is an important topic in business	29.9	45.8	75.7
Sustainability is an important topic for me	30.3	30.8	61.2
I would prefer to work for a company that embraces sustainability as a core value	26.9	33.3	60.2
I am very knowledgeable about sustainability issues in business	6.0	2.5	8.5
Economics			
Sustainability is an important topic in business	22.3	58.5	80.8
Sustainability is an important topic for me	29.3	36.6	65.9
I would prefer to work for a company that embraces sustainability as a core value	24.4	34.1	58.5
I am very knowledgeable about sustainability issues in business	17.1	4.9	22.0
Finance/accounting			
Sustainability is an important topic in business	31.3	25.0	56.3
Sustainability is an important topic for me	27.1	20.8	47.9
I would prefer to work for a company that embraces sustainability as a core value	29.2	14.6	43.8
I am very knowledgeable about sustainability issues in business	6.3	2.1	8.4
Marketing			
Sustainability is an important topic in business	32.5	42.2	74.7
Sustainability is an important topic for me	34.9	26.5	61.4
I would prefer to work for a company that embraces sustainability as a core value	34.9	26.5	61.4
I am very knowledgeable about sustainability issues in business	6.0	0	6.0
Tourism			
Sustainability is an important topic for me	34.5	43.1	77.6
Sustainability is an important topic in business	29.3	46.6	75.9
I would prefer to work for a company that embraces sustainability as a core value	12.1	32.8	44.9
I am very knowledgeable about sustainability issues in business	12.1	0	12.1

Source: own elaboration

It is worth mentioning the lack of knowledge about sustainability revealed by the Spanish students, despite belonging to the Economics and Business Faculty. This fact indicates the need for proper learning approaches about sustainability and it shows that there is room for further effort in this area. Thus, our results may help universities to reshape their learning techniques regarding sustainability issues. Therefore, by taking into account student preferences, we find that to facilitate the integration of sustainability in academia, it is necessary to integrate sustainability into universities' strategic plans and implement sustainability initiatives that actively involve the student through the development of strategic alliances with international universities and institutions, and the inclusion of field trips to learn about sustainability.

As any other research study, this book chapter is not exempt from limitations. First of all, it should be acknowledged that our sample is based on the student population of a solely faculty on a Spanish university. In this vein, scholars should be cautious since our results may not be generalizable to other contexts. As another limitation, it should be highlighted that our sample is heterogeneous in terms of the number of students of each degree which could make it difficult to compare them. However, it should be noted that our sample is faithful and depicts the real composition of the faculty students' population.

Thus, this study opens the door to future research that examines student preferences for learning about sustainability issues in different European countries. Furthermore, an IPA analysis can shed more light since this type of analysis not only enables the comparison of the importance of such learning approaches but also the evaluation of the students' perceived satisfaction with their sustainability education.

REFERENCES

Alcaraz, J. M., & Thiruvattal, E. (2010). An interview with Manuel Escudero. The United Nations' principles for responsible management education: A global call for sustainability. *Academy of Management Learning & Education*, *9*(3), 542–550. doi:10.5465/amle.9.3.zqr542

Barth, M., & Timm, J. (2011). Higher education for sustainable development: Students' perspectives on an innovative approach to educational change. *Journal of Social Science*, *7*(1), 13–23. https://ssrn.com/abstract=1978184

Brundtland Commission. (1987). *Our common future: Report of the World Commission on Environment and Development*. Oxford University Press.

Casado-Belmonte, M. P., Martínez-Victoria, M. C., Martínez-Romero, M. J., Ramos-Requena, J. P., & Martínez-Alonso, R. (2019). *Students' Perceptions of Ethical Development in Business Education Programs. In Proceedings of ICERI2019*. IATED.

Clarke, I. III, Flaherty, T. B., & Mottner, S. (2001). Student perceptions of educational technology tools. *Journal of Marketing Education*, *23*(3), 169–177. doi:10.1177/0273475301233002

Cole, M. S., Feild, H. S., & Harris, S. G. (2004). Student learning motivation and psychological hardiness: Interactive effects on students' reactions to a management class. *Academy of Management Learning & Education*, *3*(1), 64–85. doi:10.5465/amle.2004.12436819

Cole, R., & Snider, B. (2019). Managing in turbulent times: The impact of sustainability in management education on current and future business leaders. *Journal of Cleaner Production*, *210*, 1622–1634. doi:10.1016/j.jclepro.2018.11.135

Dyllick, T., & Muff, K. (2016). What does sustainability for business really mean? And when is a business truly sustainable? In S. Jeanrenaud, J. Gosling, & J. P. Jeanrenaud (Eds.), *Sustainable Business: A One Planet Approach* (pp. 381–408). Wiley.

Eastman, J. K., Iyer, R., & Eastman, K. L. (2011). Business students' perceptions, attitudes, and satisfaction with interactive technology: An exploratory study. *Journal of Education for Business, 86*(1), 36-43. doi:10.1080/08832321003774756

Elkington, J. (1994). Towards the Sustainable Corporation: Win-Win-Win Business Strategies for Sustainable Development. *California Management Review*, *36*(2), 90–100. doi:10.2307/41165746

Elkington, J. (1998). *Cannibals with forks: The triple bottom line of 21st Century business*. New Society.

Entwistle, N. (1988). Motivational factors in students' approaches to learning. In *Learning strategies and learning styles* (pp. 21–51). Springer. doi:10.1007/978-1-4899-2118-5_2

Erskine, L., & Johnson, S. D. (2012). Effective learning approaches for sustainability: A student perspective. *Journal of Education for Business*, *87*(4), 198–205. doi:10.1080/08832323.2011.590162

Erturgut, R. (2011). Increasing demand for logistics technician in business world and rising trend of logistics programs in higher vocational schools: Turkey case. *Procedia: Social and Behavioral Sciences*, *15*, 2776–2780. doi:10.1016/j.sbspro.2011.04.187

Farias, G., Farias, C. M., & Fairfield, K. D. (2010). Teacher as judge or partner: The dilemma of grades versus learning. *Journal of Education for Business*, *85*(6), 336–342. doi:10.1080/08832321003604961

Godemann, J., Haertle, J., Herzig, C., & Moon, J. (2014). United Nations supported principles for responsible management education: Purpose, progress and prospects. *Journal of Cleaner Production*, *62*, 16–23. doi:10.1016/j.jclepro.2013.07.033

Haertle, J., Parkes, C., Murray, A., & Hayes, R. (2017). PRME: Building a global movement on responsible management education. *International Journal of Management Education*, *15*(2), 66–72. doi:10.1016/j.ijme.2017.05.002

Herrera-Cano, C. (2016). Disaster risk management in business education entrepreneurial formation for corporate sustainability. *AD-Minister*, (28), 33–48. doi:10.17230/ad-minister.28.2

Hesselbarth, C., & Schaltegger, S. (2014). Educating change agents for sustainability - Learnings from the first sustainability management master of business administration. *Journal of Cleaner Production*, *62*, 24–36. doi:10.1016/j.jclepro.2013.03.042

Jose, P. (2016). Sustainability education in indian business schools: A status review. *AD-Minister*, *28*(28), 255–272. doi:10.17230/ad-minister.28.13

Karns, G. L. (2005). An update of marketing student perceptions of learning activities: Structure, preferences, and effectiveness. *Journal of Marketing Education, 27*(2), 163-171. doi:10.1177/0273475305276641

Kelley, S., & Nahser, R. (2014). Developing Sustainable Strategies: Foundations, Method, and Pedagogy. *Journal of Business Ethics*, *123*(4), 631–644. doi:10.100710551-013-2014-6

Kucharcikova, A., Miciak, M., Malichova, E., Durisova, M., & Tokarcikova, E. (2019). The Motivation of Students at Universities as a Prerequisite of the Education's Sustainability within the Business Value Generation Context. *Sustainability*, *11*(20), 5577. doi:10.3390u11205577

Martins, L. L., & Kellermanns, F. W. (2004). A model of business school students' acceptance of a web-based course management system. *Academy of Management Learning & Education*, *3*(1), 7–26. doi:10.5465/amle.2004.12436815

Nonet, G., Kassel, K., & Meijs, L. (2016). Understanding Responsible Management: Emerging Themes and Variations from European Business School Programs. *Journal of Business Ethics*, *139*(4), 717–736. doi:10.100710551-016-3149-z

Nwagwu, D. I. (2020). Driving sustainable banking in Nigeria through responsible management education: The case of Lagos Business School. *International Journal of Management Education*, *18*(1), 100332. doi:10.1016/j.ijme.2019.100332

Peterlin, J., Dimovski, V., Tvaronavičienė, M., Grah, B., & Kaklauskas, A. (2018). The strategic process of developing social aspects of sustainability through the vision reflection in business education. *Technological and Economic Development of Economy*, *24*(4), 1718–1736. doi:10.3846/tede.2018.5198

PRME. (2013). *Principles for Responsible Management Education, an initiative supported by the United Nations*. Retrieved from https://www.unprme.org/index.php

Rands, G. P. (2009). A principle-attribute matrix for environmentally sustainable management education and its application. *Journal of Management Education, 33*(3), 296-322. doi:10.1177/1052562908323191

Şahin, H., & Erkal, S. (2017). An Investigation of University Students' Attitudes toward Environmental Sustainability. *European Journal of Sustainable Development*, *6*(4), 147–154. doi:10.14207/ejsd.2017.v6n4p147

Schaltegger, S., & Wagner, M. (2011). Sustainable entrepreneurship and sustainability innovation: Categories and interactions. *Business Strategy and the Environment*, *20*(4), 222–237. doi:10.1002/bse.682

Singhal, N., Gupta, H., & Mittal, G. (2018). Importance-performance Analysis to Identify Effective Learning Approaches for Sustainability in an Indian Business School. *Vision: The Journal of Business Perspective*, *22*(3), 276–283. doi:10.1177/0972262918785963

Sohal, A. S. (2013). Developing competencies of supply chain professionals in Australia: Collaboration between businesses, universities and industry associations. *Supply Chain Management*, *18*(4), 429–439. doi:10.1108/SCM-07-2012-0228

Taylor, S. A., Humphreys, M., Singley, R., & Hunter, G. L. (2004). Business student preferences: Exploring the relative importance of web management in course design. *Journal of Marketing Education*, *26*(1), 42–49. doi:10.1177/0273475303262352

UNEP. (1972). *Declaration of the United Nations Conference on the Human Environment.* Retrieved from https://www.unep.org/Documents.Multilingual/Default.asp

Wingfield, S. S., & Black, G. (2005). Active versus passive course designs: The impact on student outcomes. *Journal of Education for Business*, *81*(2), 119–123. doi:10.3200/JOEB.81.2.119-128

Zocco, D. (2009). Risk theory and student course selection. *Research in Higher Education*, *3*, 1–29.

KEY TERMS AND DEFINITIONS

Andalusia: A geographical and cultural region of southern Spain. The population in the media is portrayed as cheerful and open-minded.

Learning Approaches: Different pedagogical methods undertaken by the education institution.

Management Education: A discipline of higher education by which students are instructed to be able to work in the broad spectrum of business and management.

PRME: An upheld activity by the United Nations for raising the profile of sustainability in education far and wide.

Students Preferences: Students' most valuable choices of type of classroom forms, regardless of the collaboration with mates.

Sustainability: Acting in the way that focuses on the satisfaction of present needs' population without compromising future generations' needs.

Sustainable Business: Type of company whose business model combines obtaining profits in line with a positive impact on the environment and society.

Sustainable Development: The process of improving people's quality of life and economic well-being without compromising the resources of future generations.

Chapter 9
Environmental Education and Its Effects on Environmental Sustainability

Joan Nyika
https://orcid.org/0000-0001-8300-6990
Technical University of Kenya, Kenya

Fredrick M. Mwema
https://orcid.org/0000-0001-6116-5587
Dedan Kimathin University of Technology, Kenya

ABSTRACT

Environmental education (EE) for sustainable development remains a valuable subject of contemporary society, which is characterized with environmental issues such as climate change, pollution, loss of biodiversity, and resource degradation. The delivery of EE is based on the North American Association for environmental education values of knowledge, dispositions, competencies, and responsible behavior towards the environment. EE is a transformative tool to learners since it prepares learners with skills, attitudes, knowledge, and values to resolve environmental problems. It promotes environmental activism and action-oriented resolution of environmental issues. The full benefits of EE are challenged by limited human capacity, questionable professionalism, limited resources, and poor transformation of knowledge to practice. These challenges however can be alleviated through community engagement in formulating EE programs, multidisciplinary engagements, and research on EE delivery and quality.

INTRODUCTION

Initiatives focusing on the need for environmental protection began two centuries ago with the motive of rescuing endangered species. However, the growing indicators of environmental degradation including global warming, climate variation and change, ozone layer thinning, soil and water pollution have imposed the need for environmental protection and diversified education initiatives focused on this subject (Boca

DOI: 10.4018/978-1-7998-7512-3.ch009

& Saracli, 2019). It is from these predispositions that the awareness of the importance to act responsibly towards nature at individual and collective level has grown among humankind. Consequently, mutual collaborations at all administrative levels to protect nature are emerging in the field of environmental education (EE). According to Thomas (2005), EE is the acquisition of knowledge aimed at behavioural and action modification by students and is skewed to wise use of environmental resources at individual and group level. Similarly, Grosseck et al. (2019) pointed out that recent research tendencies focused on understanding EE promote sustainable development (SD). A research by Choudhary et al. (2019) also indicated that EE is an opportunity to acquire skills, attitudes and knowledge essential in ecological improvement. The Rio de Janeiro earth summit that promulgated the Agenda 21 emphasized the role of education in environmental sustainability (Boeve-de Pauw et al., 2015). Other efforts towards EE for SD include the Johannesburg world summit of 2002 that declared 2005 to 2014 as the period, which this initiative was to be implemented and the campaigns by UNESCO aimed at reshaping curriculum to inculcate EE (Grosseck et al., 2019).

Although the initiatives aimed at promoting EE are existent, the actions to address environmental issues and participate in actual initiatives is low among students. Boca and Saracli (2019) highlighted this issue observing that there is a rift between the lack of involvement in practical environmental protection despite the rich ecological knowledge among communities. This tendency could be a flaw of existing curriculums whose attention does not specify methods aimed at practical civic and moral education towards environmental sustainability. Therefore, EE aimed at SD does not only disseminate knowledge on environmental issues but also enriches the value associated with the ecology to instil problem-solving skills and change the attitudes and perceptions of ecosystems for the better (Liu & Guo, 2018). Authors such as Maurer and Bogner (2019) assessed the role of EE and SD and highlighted their inter-relatedness and overlap. Environmental education, "resonates with the model which refers to the environment in a holistic, human-oriented approach as interacting biophysical (organisms and life support systems), social (people living together), economic (livelihood, money and services) and political (power, policy and decisions) dimensions" (Goldman et al., 2013, p. 517). These are some of the tenets of SD, which can be interrelated with EE. The purpose of this book chapter will be to highlight the chronological development of EE; discuss its delivery determinants and challenges involved in its dissemination with the aim of providing a way forward. Additionally, it will discuss the effects of EE in relation to learner outcomes using named case studies of Botswana and Colombia. The case studies were chosen because EE in these two countries is compulsory unlike other developing countries where it is voluntary or offered on ad hoc basis. The methodology applied in the chapter is an examination of existent literature on the topic.

The History of EE

EE is traceable from the early educational reformers from Europe such as Humboldt, Goethe, Pestalozzi, Rousseau and Comenius prior to its definition and integration of this concept to mainstream education (Rosaleen, 2012). In the 1970s, an international working meeting on environmental education was held in the USA and came up with an EE definition (IUCN, 1970). In the meeting, EE was defined as a process of developing values and understanding concepts to cultivate attitudes and skills essential to appreciate the interrelationships among humans, their culture and biophysical features. Furthermore, EE was a tool useful in making decisions, coding behaviours and best practices on environmental quality issues. Following this development, the United Nations (UN) in a conference on human environment held at

Stockholm recognized EE formally (UNESCO, 1972). The conference lead to the conceptualisation of EE as Hollweg and others stated claiming that this concept served,

"to develop a world population that is aware of, and concerned about, the environment and its associated problems, and which has the knowledge, skills, attitudes, motivations and commitment to work individually and collectively toward solutions of current problems and the prevention of new ones" (2011, p.2).

This move triggered a number of related publications and most notable of these was the, "Silent Spring' book by Carson that detailed the harmful effects of using pesticides such as Dichlorodiphenyltrichloroethane (DDT) on the environment (Maurer & Bogner, 2019). The book's claims were used to enforce a global ban for the pesticide use years later after environmentalists lobbied against DDT use. Later in the 20th century, the global population's awareness on environmental issues increased due to the prevalence of ozone layer declines, air pollution and acid rain. These trends were in line with EE efforts of inculcating commitment and motivation to eradicate environmental protection (UNESCO-UNEP, 1978; UN, 1992). The Tbilisi declaration and the Belgrade charter further reinforced these efforts (UN, 1992). A broader viewpoint of EE emanated in the 1990s at the Rio conference through the Agenda 21 aimed at a global action plan towards sustainable development, SD (UN, 1992). These initiatives resulted to the concept of environmental SD, which focuses on ecological development and is one of the three components of SD including social and economic development (Reilly, 2008).

There is consensus from the education history that EE promotes eco-friendly lifestyles among growing children and learners apart from acquisition of knowledge on their natural bio-physical environment (Kaiser et al., 2008; Boeve-de et al., 2015; Choudhary et al., 2019). In the 1960s, physical systems and species were used in EE (Maurer & Bogner, 2019). In the 1970s, EE advanced to practical models to incorporate urban studies, outdoor adventures while the 1980s saw the application of diversified knowledge dissemination and acquisition approaches including inquiry and discovery learning, case-based, project-based, problem-based and just in-time teaching (Palmer, 1998; Prince & Felder, 2006). In the 1990s, EE was recognized globally and its integration in affective, psychomotoric and cognitive learning aspects was formalized as De Kock et al. (2004) observed. The diversified learning and teaching approaches do not automatically translate to success especially if learners' behaviour, attitudes and knowledge on the environment is limited (Rickinson, 2001). Grosseck et al. (2019) supported this ideology suggesting that apart from factual knowledge, action-related knowledge is essential for effective EE. The focus of this chapter is to assess the factors that influence EE, the challenges encountered during the process and the resultant effects of disseminating the concept based on literature.

The Need for Environmental Education

Previously, the needs of man considering the scarce population of humankind were limited and had no interference on nature. However, in the contemporary society, population rise of humankind has constrained man's needs and consequently affected the environment negatively to the extent of threatening the sustainability of land and water (natural) resources. Indiscriminate and unplanned use of the resources by humans has led to their scarcity. This situation is further exacerbated by modernization, deforestation, pollution and rapid industrialization. Therefore the need to device strategies aimed at introducing and enhancing EE among people to save existent natural resources and the entire earth is indispensable (Choudhary et al., 2019). Introducing EE in schools increases awareness on the surroundings among young learners, which assists them to understand and acquire knowledge prerequisite in developing appropriate skills and attitudes necessary in solving the environmental challenges they encounter daily through responsible acting and scientific thinking (Edsand & Broich, 2020). Although through EE, pro-environmental behaviour cannot be acquired, emotional involvement and pro-environmental consciousness can be enhanced (Kollmuss & Agyeman, 2010).

The Domain of Environmental Education

EE is underpinned on a framework that was initiated by the North American association for environmental education (NAAEE) (Hollweg et al., 2011). The framework describes environmental literate individuals as those who make wise decisions on environmental issues and are willing to take action that will influence the welfare of individual, group and world's environment positively during their civic life. Therefore, environmental literate persons should possess a number of characteristics according to Sharma (2016). These include 1) having a high standard of conduct to apply understanding and knowledge when making decisions regarding varied environmental contexts, 2) cognitive abilities and skills, 3) affective and cognitive dispositions and 4) knowledge and understanding of various environmental issues, problems and concepts. A summary of the framework and its sub-domains is provided in Table 1 and discussed in the following sub-titles.

Table 1. Components and sub-components of the EE domain according to Hollweg et al. (2011)

Domain Components	Sub-components
Knowledge What do you know?	❖ Ecological and physical systems ❖ Political, cultural and social systems ❖ Environmental issues ❖ Action strategies and citizen participation ❖ Many solutions to environmental problems
Dispositions How do you respond to environmental problems?	❖ Concerns and attitudes to the environment ❖ Sensitivity ❖ Personal responsibility assumption ❖ Concerns and attitudes to the environment ❖ Self-efficacy ❖ Intent and motivation to take action
Competencies What skills and abilities do you possess? When and how do you apply them?	❖ Identify environmental problems ❖ Question about environmental problems under different environmental conditions ❖ Analyse environmental problems ❖ Use primary and secondary sources to investigate the science and social aspects of environmental problems ❖ Assess and make judgements on environmental problems considering the socio-political systems ❖ Use evidence and knowledge-based facts to propose solutions to the problems ❖ Monitor and evaluate plans to solve environmental problems at various levels and scales
Responsible behaviour towards the environment	Participate in habitual and deliberate behaviours at individual and group level to solve current environmental problems sustainably and prevent re-emergence of new ones.

Knowledge

Literacy on physical and ecological systems; environmental issues; socio-cultural and political processes; citizen involvement and action planning and a variety of solutions to environmental problems constitute the five types of knowledge. Researchers on EE as early as 1960s and 70s identified these sub-domains of knowledge (Disinger, 1983; Hart, 1981; Harvey, 1977). Documents prepared by NAAEE (2004) and UNESCO (1978) included these as some of the objectives of EE explicitly. Knowledge on physical and ecological systems focuses on earth sciences, atmospheric, oceanic and geological systems relevant to environmental literacy. These systems are interdependent and link the environment to humans as earth system and ecosystems' main variables (Coyle, 2005). According to Berkowtiz et al. (2005), the area of knowledge also evaluates the role of humans as protective agents and their associated activities as influencers of these systems using evidence-based research. Knowledge on socio-cultural and political processes such as legal, economic, transportation, kinship and agricultural systems as well as their spatial and temporal variations in development is another domain of knowledge (Rickinson, 2001). The area considers practices and beliefs encountered during civic participation and problem solving of environmental issues. Citizen involvement in various environmental issues is key to collaborative solution seeking through effective political action and communication strategies. Several technical, scientific, educational and regulatory efforts are and have been underway to address various environmental problems and form part of EE. The knowledge on environmental issues that arise from human conflicts and biophysical changes of the natural world are key in environmental literacy.

Dispositions

Dispositions are negative or positive determinants of behaviours towards the environment. These behavioural determinants influence the willingness of EE students to recognize and adapt to various values while resolving environmental problems (Sharma, 2016). The motivation to participate and act during problem resolution forms part of the predispositions. According to Dada et al. (2017), affective dispositions focus on the sensitivity, attitudes, locus-of-control, worldview, values and personal responsibility of learners regarding environmental issues. Dispositions influence competency on environmental issues and are influenced by levels of environmental knowledge (Igbokwe, 2016).

Competencies

Competencies are a group of abilities and skills that are required and expressed for a specific environmental issue. They enhance the capacity of learners to identify, analyse, investigate, question and evaluate environmental problems to make a personal judgement regarding them and informed by critical thinking and evidence-based research (Hollweg et al., 2011). EE therefore advances competences that enhance the aptitude to deal with real life environmental problems (Roczen et al., 2014). Pro-environmental competencies are cultural beliefs, environmental perceptions, personal motives and skill that promote ecological matters. Through these abilities, learners can assume responsibility in addressing environmental matters for the betterment of communities (Sharma, 2016).

Responsible Behaviour to the Environment

This is an expression of competencies, disposition and knowledge regarding environmental issues. By being responsible, EE learners actively participate in solving issues related to the environment. Hsu (2004) observed this predisposition among college students of Taiwan whose environmental responsibility, perceived knowledge on such issues, intention to act and skills towards resolving environmental problems was enhanced following EE.

Influence of School Education in Inculcating Environmental Literacy

In the science curriculum, social scientific themes and topics are essential in equipping learners with skills to become responsible citizens, manage society sustainably as well as participate in social life (Wan & Bi, 2019). This aspect has been proven by a number of empirical studies on the effects of environmental education and schools. Bradley et al (1999) assessed the relationship between EE and environmental altitudes of high school learners of Texas, USA. The study reported a positive correlation between higher scores in EE and enhanced environmental attitudes among students. In Mexico, an empirical study using qualitative data correlated EE to ecological knowledge positively among middle school learners (Ruiz-Mallen et al., 2009). In New Haven, Connecticut, third and fifth grade students showed enhanced environmental awareness after an open space EE program was initiated according to Fisman (2005). Elementary school student-teachers of Turkey showed improved attitude towards the four dimensions of EE (knowledge of environmental challenges individual responsibility to the environment, general attitudes towards environmental solution and problems) following enhanced awareness on the concept (Ozden, 2008).

Despite these positive reports on the correlation of EE and environmental awareness, a number of studies did not establish any relationship between the two aspects. A study by DeChano (2006) investigating EE and environmental knowledge among high schoolers of USA, Switzerland, England and Chile did not find any significant relationship between EE and resultant environmental attitudes. In Israel, there was no significant correlation between EE and environmental behaviour (Negev et al., 2008). However, the impact of EE programs examined by Israeli out-of-school environmental associations established an enhanced sensitivity to human-environment interrelationships with limited cognitive domain on environmental issues (Goldman et al., 2013).

Schools have been evaluated for their role in promoting environmental awareness and attitudes. A study by Coertjens et al. (2010) in Flanders region of Belgium is a case example. The study concluded that, "schools in which science is taught in a more hands-on manner are associated with higher student environmental awareness whilst environmental learning activities are associated with more pro-environmental attitudes amongst students" (Coertjens et al., 2010, p. 497). Environmental responsibility towards a sustainable environment was enhanced according to Lin and Shi (2014) in a study correlating school and individual-related environmental literacy determinants in USA and Canada. The authors emphasized that a deeper sense of responsibility to the environment is inculcated following EE at school. Schools cultivate creative problem solving skills among learners through EE, which prepares them for the current environmental challenges they face in their daily life. Huang et al. (2019) as well as Karpudewan and Roth (2018) confirmed this supposition among primary learners since EE improved and advanced their reasoning capacity and ability to make evidence-based decisions on various environmental issues.

Most studies evaluating the effects of school EE and environmental literacy are conducted in developed countries as observed by Edsand and Broich (2019). There are limited empirical studies investigating this aspect in developing countries and emerging economies. A case example is a study by Kim and Hamdan Alghamdi (2019), which examined the science education reforms among secondary school girls of Saudi Arabia that have incorporated EE. The authors concluded that EE has the capacity to reform education such that the metacognitive learning orientation, knowledge and attitude of the girls in science education and practice is improved. Similarly, extra-curricular environmental learning in primary schools of Cote d'Ivoire enhanced environmental awareness and attitude towards environmental issues among learners according to Borchers et al. (2014). An Indonesian study established a positive correlation between EE and student knowledge but not their resultant environmental awareness and attitudes (Haryono et al., 2014). Children from rural Madagascar who had prior exposure EE showed enhanced knowledge and positive attitudes on the matter compared to those who did not (Rakotomamonjy et al., 2015). In this chapter, the gap between this aspect in developed and developing nations will be evaluated by exploring the challenges encountered during EE delivery.

These discussed empirical studies show the existent of a conflict between EE as a tool to enhance knowledge and attitudes on the environment as well as the resultant behaviour to this subject. A number of documentations support the ideology that EE raises awareness and concerns on environmental matters but these do not necessarily translate to behaviour change (Bamberg & Moser, 2007; Zsoka et al., 2013). Other researchers have supposed that values and attitudes are drivers for action though their effectiveness to altering behaviour is questionable (Marjaine et al., 2011). There is however consensus that pro-environmental attitudes and EE are interrelated and strengthen each other by motivating information-seeking on environmental problems. Among learners at young and adult age, EE was a crucial stimulant to awareness of environmental probes in Germany and Finland (Asunta, 2004) as well as in Canada (Michalos et al., 2009). In addition to internal factors of environmental literacy defined by its domain components (Table 1), external factors such as traditions, pressures and norms transmitted by the social environment also influence pro-environmental behaviours and actions (Zsoka et al., 2013). These divergent dispositions of EE and its resultant impact present a challenge in its implementation as discussed in the following subsection.

CHALLENGES OF EE DELIVERY

Lack of Professionalism

A number of challenges emerge during the delivery of EE. One of the commonest bottleneck is professionalism flaws that are either implicit or explicit. It is argued that EE programs should be well delivered and conceived, that experts working in them should specialize in pedagogy and that programming and professional development of the program should be improved using evidence-based research (Marcinkowski, 2009). Although the milestones realised in developing EE since the 1990s have enhanced professional programming and development, the growth is not yet optimal. For a long time, EE has been criticized as being advocacy oriented and presenting messages laced with emotionalism, despair, misinformation and hype as Marcinkowski (2009) highlighted. These criticisms raise the uncertainty of EE particularly in the United States. An Independent Commission on Environmental Education (ICEE, 1997) criticized EE as weak in economics and science as well as controversial since it did not adequately equip learners with skills to understand the trade-offs when dealing with contentious environmental problems. The criticism questioned the credibility of ill-prepared EE providers, the use of superficial and inaccurate materials, the program's advocacy practices and limited programs. In justification of this view, Salmon (2000) claimed that the ICEE report was an unbiased viewpoint on its guides and materials but did not assess the broadside of EE. The lack of professionalism in delivering EE resulted to collaboration among groups of experts such as the NAAEE, Environmental Education and Training Partnership (EETAP) and US Environmental Protection Agency (EPA) to develop a set of guidelines for such programs (NAAEE, 2004). These guidelines are not reconciled up-to-date and remain contentious among EE experts (Zsoka, 2013). In Nigeria for instance, EE is still associated with activism by non-governmental and international organizations and its implementation in mainstream educational sector remains non-objective despite its good intentions (Erhabor & Don, 2016).

A Poor Interconnection of EE and SD

A second challenge is the complex relationship between EE and education for sustainable development (ESD). The initial guidelines on EE included economics and sustainability as content areas. However, ESD scope is intricate since it is coupled with its complex interrelationships with technological, economic, social and environmental demands and issues. Additionally, the involvement of multidisciplinary experts in addressing education for sustainability has made it difficult to agree on a way forward in reconciling these concepts. The recognition of a document titled, "Environmental education for sustainable societies and global responsibility" by the NAAEE resulted to an emphasis on sustainability within the UN (UNESCO) rather than EE (Marcinkowski, 2009). It is from this imbalance that it has been difficult to interrelate EE and ESD. Case studies of Denmark and Sweden by Breiting and Wickenberg (2010) pointed the difficulty in integration EE methodologies to the three dimensions of SD especially during pedagogical training. Some authors argue that ESD has its roots from EE (Disinger, 1983) while others claim that the two serve in a similar capacity (Sato, 2006). Although EE has a longer history compared to ESD that emerged with the wave of SD, the latter has a broader outlook encompassing social, economic and technological goals while EE is limited to environmental issues (Marcinkowski, 2009). As such, the challenges that emerge in linking the two aspects can be summarised into three. These include: 1) to what level does EE follow the ESD concepts and seek to address issues related to basic literacy, healthcare and poverty and 2) to what level do proponents of the two concepts believe that ESD is superior compared to EE. Lastly, 3) to what extent do supporters of ESD agree that its focus is broad and addresses issues of biodiversity, environmental quality and natural resource better that EE. These issues remain contentious and are major hindrance to optimal delivery of EE programs.

The Climate Change Challenge

A third challenge in the delivery of EE is the climate change issue. The national evaluation of professional needs in environmental literacy recommended that climate change is an essential content area. However, climate change concepts have exceptional pedagogical and methodological challenges to EE learners as Marcinkowski (2009) highlighted. Climate change on one hand is the predominant environmental crisis that needs societal attention and education following the increased loss of biodiversity, ozone depletion, population growth, pollution and resource depletion. On the other hand, EE serves to streamline behavioural patterns and lifestyles through training and educational programmes to overcome the aforementioned environmental challenges. The extent to which EE is applicable in climate change adaptation and mitigation is therefore the challenge. For instance, East African countries of Kenya, Uganda and Tanzania have registered progress in inducing EE in the region but the challenge of harsh weather conditions exacerbated by climate change have reversed the gains (Johnson-Pynn & Johnson, 2005). This supposition is because of the multifaceted, large scale and interrelated nature of climate change, which poses the most potent environmental threat of interest in EE compared to the existent ones. Therefore, three issues can be deduced from this challenge

1. To which level are environmentalists willing to cooperate with EE experts towards common goals for climate change adaptation and mitigation?
2. To which level are EE experts ready to engage collectively or individually with other experts that have a common goal of addressing climate change and its resultant effects?
3. To what extent are governing and policy formulating agencies who have an appreciation of EE willing to incorporate it during decision-making on responses to climate change?

To address these issues, cooperation of EE experts and climate change advocates in multiple disciplines is essential coupled with knowledge sharing and participative decision-making (Darnton, 2008; Marcinkowski, 2009).

The Effects of EE on Environmental Sustainability

Stanford University experts (2013) have documented the impacts of EE in relation to environmental sustainability extensively in a review. Apart from enhanced awareness of the environment, a number of positive effects resulted from EE. These are as summarized in Figure 1. The positive effects including improving academic excellence enhanced critical thinking and reasoning skills, life building skills and personal growth. Additionally, learners were shown to be better civic engagers with good environmental behaviours. Ardoin et al. (2018) in a review study provided evidence that alluded to EE learning as a transformative learning opportunity to promote environmental sustainability. Similarly, Stern et al. (2014) demonstrated that EE learning instils confidence, skills, behaviour, motivation and better academic performance among learners towards sustainable environmental practice. Stern et al. (2011) observed that EE enhances interest and passion to resolve environmental issues, which is the mainstay of environmental sustainability. Learners with EE skills were more likely to engage in community clean-ups, waste recycling and water reuse activities compared to those without (Harness & Drossman, 2011).

Apart from equipping learners with knowledge to deal with environmental issues, EE learners can synthesize complex information through decision-making and critical thinking to produce long-life problem solvers and active community participants (Ernst & Monroe, 2004). Additionally, it offers motivation and personal responsibility to address current environmental issues such as water management (Stern et al., 2011) pollution and climate change (Jennings et al., 2005). The following section discusses the effects of EE on environmental sustainability using named case studies.

Figure 1. Effects of environmental education to learners in relationship to environmental sustainability. Source: the authors

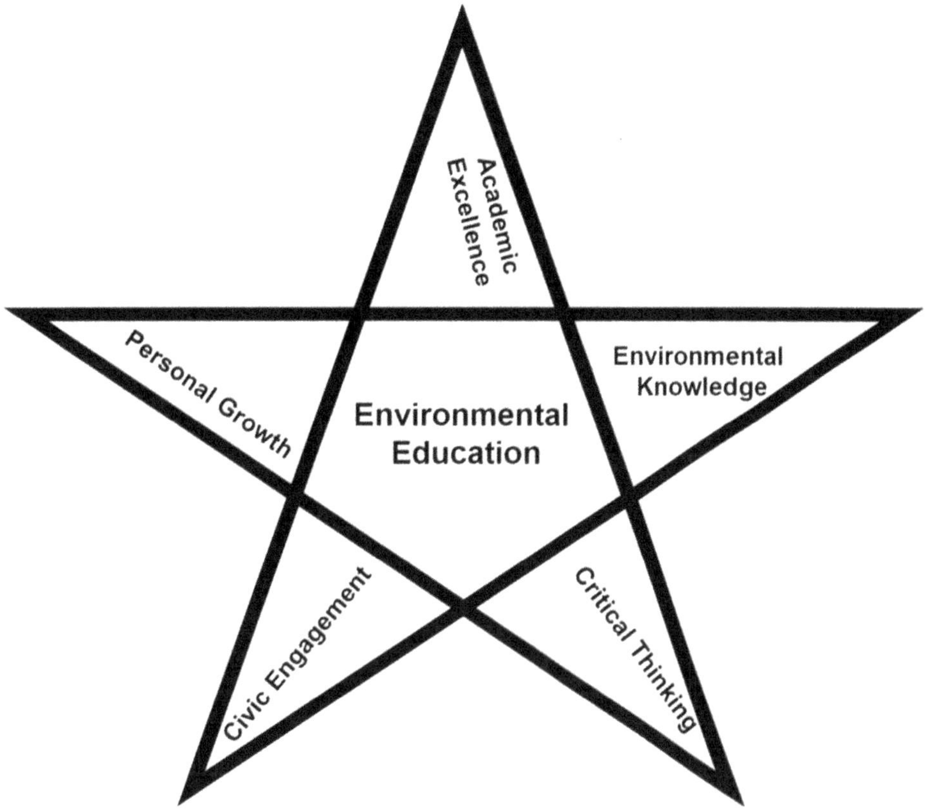

CASE STUDIES

Effects of EE in Botswana

The Botswana government made EE compulsory in secondary schools in 1994 following the revision of the country's policy of education (Botswana Government, 1994). The move was in response to fighting environmental degradation that has worsened and is under pressure because of improved economic development and growth of the country's population (Silo, 2015). From this initiative, EE has enabled transmission of skills, knowledge and values among the youth, which has resulted to a protective attitude in their use of environmental resources. According to Kanene (2016), EE has enhanced personal responsibility and commitment to conservation practices as motivated action to enable ethics and values of environmental sustainability. An empirical study by Sundar (2010) in Botswana showed that through EE, learners have better awareness of their environment in addition to positive valueS and attitudes that are requisite in ESD. Mogensen and Schnack (2010) explained the role of EE in Botswana as, *"the action competence approach points to democratic, participatory and action-oriented teaching–learning that can help students develop their ability, motivation and desire to play an active role in finding democratic solutions to problems and issues connected to sustainable development that may even consist of the aforementioned tendencies, ideas and agendas" (p. 62).*

Their sentiments are also shared by Kanene (2016) who reported improved perceptions of environmental issues and a positive attitude towards their solutions after the introduction of EE in secondary schools of Botswana. The use of civic clubs to promote EE in Botswana has also been shown to promote environmental stewardship by transforming attitudes and knowledge towards resolving environmental issues (Ajiboye & Silo, 2008).

Although these effects have been realized and are supported by policies, integration of EE in the mainstream education system has faced its challenges in Botswana. According to Velempini (2016), integration of EE failed to engage community who are key stakeholders and have valuable contribution. Consequently, the acceptance of EE learners by and in involvement of community environmental issues remains a contentious issue characterized by tension, as both parties perceive each other as less knowledgeable. A case study by Velempini et al. (2017) in secondary schools of Okavango Delta region of Botswana raised similar issues citing that indigenous knowledge through community participation was not incorporated in EE and if this can be done to what extent and of what relevance would it be. The authors also cited limited education resources such as the lack of transport means to cover outdoor learning activities stipulated in the EE curriculum and the translation of skills learnt to hands on practice. Educators particularly teachers who deliver EE curriculum are insufficiently equipped and the use of English language for communication of environmental issues was also a limiting factor in optimizing the effectiveness of EE in rural Botswana as noted by Velempini et al. (2017). Despite these challenges in greening the curriculum, Botswana has made great strides in EE adoption compared to other developing countries of sub-Saharan Africa (Kanene, 2016; Velempini et al., 2017).

Effects of EE in Colombia

Compulsory education for children aged between 5 and 15 years was introduced in Colombia in 1991. This move was aimed at; 1) expanding public service provision, 2) raising the enrolment figures in secondary schools and 3) expanding learning capacities in schools (Edsand & Broich, 2020). The efforts resulted to increased enrolment in schools by 20% in a period of two decades though wide discrepancies in the quality of education occurred in private and public schools as well as rural and urban areas (Faguet & Sanchez, 2014). A number of national efforts have supported the incorporation of EE in the education system of Colombia. An example is the recognition of EE as a tool to protect the environment and gain knowledge on its related issues by the Colombian national constitution since the early 1990s (Sauvé et al., 2005). A collaboration by the ministries of education and environment have outlined the goals and funding alternatives to facilitate EE delivery. Regionally initiatives such as the school environmental projects (PRAE), the inter-institutional technical committees of EE (CIDEA) and citizen projects for EE (PROCEDA) enable the delivery of EE at municipality level (Faguet & Sanchez, 2014).

Through these initiatives, a number of impacts have been realized among learners. According to Faguet and Sanchez (2014), EE has enhanced awareness on environmental problems among learners, instilled responsible environmental behaviour towards SD including the use of renewable energy and application of EE knowledge to resolve environmental issues. Another study by Youkhana et al. (2018)

highlighted that EE cultivates the creation of environmental activists who enforce ESD ideologies through community participation and action-oriented problem-solving skills. Jefferson et al. (2016) observed that EE resulted to positive attitudes and perceptions of the environment towards SD and hence the motivation to use ecosystems and protect them sustainably. These achievements have been realized despite the challenges of questionable quality of EE in the country, spatial inequities in EE delivery, the non-involvement of community members in streamlining the curriculum and transformation of acquired knowledge to practice. These problems are attributable to the sluggish development of EE in the country and its undervalued impact towards ESD as Youkhana et al. (2018) noted.

CONCLUSION

This book chapter explores the history, challenges of delivery and impacts of EE towards SD. It is evident that EE is necessary in contemporary society that faces a myriad of environmental issues including pollution, land degradation, climate change and biodiversity loss, which are worsened by the growth of population and increased economic practices. EE plays a role in promoting awareness, knowledge, positive values and attitudes towards the environmental and environmental issues. Ultimately, the novelty in this chapter revolves around how the EE concept is a powerhouse to SD since it equips learners to be action-oriented and responsible environmentalists. These positive effects have been proven in a number of empirical studies and in case studies of Botswana and Colombia in the chapter. Issues of limited human capacity, questionable professionalism, limited resources and poor transformation of knowledge to practice hinder optimal benefits of EE. To alleviate these issues, the need for more support of EE programs through extensive research, funding and involvement of many disciplines is essential to reconcile contentious issues. Additionally, the involvement of community among other stakeholders during integration of EE programs is requisite to enhance the acceptance of learners while practicing environmental activism and projects within the society. Therefore, the book chapter emphasizes the need to raise awareness on EE and implement it as part of the mainstream education just as is the case in Botswana and Colombia. This will enhance its application in resolving SD related problems and in specific, those associated with environmental conservation.

REFERENCES

Ajiboye, J., & Silo, N. (2008). Enhancing Botswana children's environmental knowledge, attitudes and practices through the school civic clubs. *International Journal of Environmental and Science Education*, *3*(3), 105–114.

Ardoin, N., Bowers, A., Roth, W., & Holthuis, N. (2018). Environmental education and K-12 student outcomes: A review and analysis of research. *The Journal of Environmental Education*, *49*(1), 1–17. doi:10.1080/00958964.2017.1366155

Asunta, T. (2004). Knowledge sources, attitudes and self-reported behaviour of secondary-level science students concerning environmental topics. In A. Laine, J. Lavonen, & V. Meisalo (Eds.), *Current Research on Mathematics and Science Education. University of Helsinki. Research Report 253.*

Bamberg, S. & Möser, G. (2007). Twenty years after Hines, Hungerford, and Tomera: a new meta-analysis of psycho-social determinants of pro-environmental behaviour. *Journal of Environmental Psychology, 27*, 14-25.

Berkowitz, A., Ford, M., & Brewer, C. (2005). A framework for integrating ecological literacy, civics literacy, and environmental citizenship in environmental education. In E. A. Johnson & M. J. Mappin (Eds.), *Environmental education and advocacy* (pp. 227–266). Cambridge University Press.

Boca, G., & Saracli, S. (2019). Environmental education and student's perception for sustainability. *Sustainability, 11*(6), 1553. doi:10.3390u11061553

Boeve-de Pauw, J., Gericke, N., Olsson, D., & Berglund, T. (2015). The Effectiveness of Education for Sustainable Development. *Sustainability, 7*(11), 15693–15717. doi:10.3390u71115693

Borchers, C., Boesch, C., Riedel, J., Guilahoux, H., Ouattara, D., & Randler, C. (2014). Environmental education in Côte D'Ivoire/West Africa: Extra-curricular primary school teaching shows positive impact on environmental knowledge and attitudes. *International Journal of Science Education, 4*(3), 240–259.

Botswana Government. (1994). *The revised National Policy on Education*. Government Printers.

Bradley, J. C., Waliczek, T. M., & Zajicek, J. M. (1999). Relationship between environmental knowledge and environmental attitude of high school students. *The Journal of Environmental Education, 30*(3), 17–21. doi:10.1080/00958969909601873

Breiting, S., & Wickenberg, P. (2010). The progressive development of environmental education in Sweden and Denmark. *Environmental Education Research, 16*(1), 9–37. doi:10.1080/13504620903533221

Choudhary, S., Saha, A. R., & Tiwary, N. K. (2019). The role of compulsory environmental education in higher learning: A study in the University of Delhi. *Applied Environmental Education and Communication*, 1–13. doi:10.1080/1533015X.2019.1605946

Coertjens, L., Boeve-de Pauw, J., De Maeyer, S., & Van Petegem, P. (2010). Do schools make a difference in their students' environmental attitudes and awareness? Evidence from Pisa 2006. *International Journal of Science and Mathematics Education, 8*(3), 497–522. doi:10.100710763-010-9200-0

Coyle, K. (2005). *Environmental literacy in America: What ten years of NEETF/Roper research and related studies say about environmental literacy in the U.S.* The National Environmental Education and Training Foundation.

Dada, D., Eames, C., & Calder, N. (2017). Impact of environmental education on beginning preservice teachers' environmental literacy. *Australian Journal of Environmental Education, 33*(3), 201–222. doi:10.1017/aee.2017.27

Darnton, A. (2008). *Reference report: An overview of behaviour change models and their uses. A GSR behaviour change knowledge review*. Centre for Sustainable Development, University of Westminster.

De Kock, A., Sleegers, P., & Voeten, M. (2004). New Learning and the Classification of Learning. *Review of Educational Research, 74*(2), 141–170. doi:10.3102/00346543074002141

DeChano, L. M. (2006). A multi-country examination of the relationship between environmental knowledge and attitudes. *International Research in Geographical and Environmental Education, 15*(1), 15–28. doi:10.2167/irgee/184.0

Disinger, J. (1981). Environmental education in the K-12 schools: A national survey. In A. Sacks & ... (Eds.), *Current issues VII: The yearbook of environmental education and environmental studies* (pp. 141–156). ERIC/SMEAC.

Edsand, H., & Broich, T. (2020). The impact of environmental education on environmental and renewable energy technology awareness: Empirical evidence from Colombia. *International Journal of Science and Mathematics Education, 18*(4), 611–634. doi:10.100710763-019-09988-x

Edsand, H., & Broich, T. (2020). The impact of environmental education on environmental and renewable energy technology awareness: Empirical evidence from Colombia. *International Journal of Science and Mathematics Education, 18*(4), 611–634. doi:10.100710763-019-09988-x

Erhabor, N., & Don, J. (2016). Impact of environmental education on the knowledge and attitude of students towards the environment. *International Journal of Environmental and Science Education, 11*(12), 5367–5375.

Ernst, J., & Monroe, M. (2004). The effects of environment-based education on students' critical thinking skills and disposition toward critical thinking. *Environmental Education Research, 10*(4), 507–522. doi:10.1080/1350462042000291038

Faguet, J., & Sanchez, F. (2014). Decentralization and access to social services in Colombia. *Public Choice, 160*(1-2), 227–249. doi:10.100711127-013-0077-7

Fisman, L. (2005). The Effects of Local Learning on Environmental Awareness in Children: An Empirical Investigation. *The Journal of Environmental Education, 36*(3), 39–50. doi:10.3200/JOEE.36.3.39-50

Goldman, D., Assaraf, O., & Shaharabani, D. (2013). Influence of a non-formal environmental education programme on junior high-school students' environmental literacy. *International Journal of Science Education, 35*(3), 515–545. doi:10.1080/09500693.2012.749545

Grosseck, G., Tiru, L., & Bran, R. (2019). Education for sustainable development evolution and perspectives: A bibliometric review of research, 1992-2018. *Sustainability, 11*(21), 6136. doi:10.3390u11216136

Harness, H., & Drossman, H. (2011). The environmental education through filmmaking project. *Environmental Education Research, 17*(6), 829–849. doi:10.1080/13504622.2011.618626

Hart, E. (1981). Identification of key characteristics of environmental education. *The Journal of Environmental Education, 13*(1), 12–16. doi:10.1080/00958964.1981.9943018

Harvey, G. (1977). A conceptualization of environmental education. In J. Aldrich, A. Blackburn, & G. Abel (Eds.), *A report on the North American regional seminar on environmental education* (pp. 66-72). Columbus, OH: ERIC Clearinghouse for Science, Mathematics and Environmental Education.

Haryono, A., Soemarno, Djati, M. S., & Setyoleksono, A. (2014). Learning attitude and awareness against students in cultured environmental success in Probolinggo. *Journal of Environment and Earth Science, 4*(16), 72–81.

Hollweg, K. S., Taylor, J. R., Bybee, R. W., Marcinkowski, T. J., McBeth, W. C., & Zoido, P. (2011). *Developing a framework for assessing environmental literacy*. North American Association for Environmental Education.

Hsu, S. (2004). The Effects of an Environmental Education Program on Responsible Environmental Behavior and Associated Environmental Literacy Variables in Taiwanese College Students. *The Journal of Environmental Education*, *35*(2), 37–48. doi:10.3200/JOEE.35.2.37-48

Huang, T. C., Lin, W., & Yueh, H. P. (2019). How to Cultivate an Environmentally Responsible Maker? A CPS Approach to a Comprehensive Maker Education Model. *International Journal of Science and Mathematics Education*, *17*(S1), 1–16. doi:10.100710763-019-09959-2

Igbokwe, B. (2016). *Environmental literacy assessment: assessing the strength of an environmental education program (EcoSchools) in Ontario secondary schools for environmental literacy acquisition* (PhD Thesis). University of Windsor.

Independent Commission on Environmental Education. (1997). *Are we building environmental literacy?* Author.

International Union for Conservation of Nature (IUCN). (1970). *International Working Meeting on Environmental Education in the School Curriculum, Final Report*. Gland, Switzerland: IUCN.

Jefferson, G. M., Ciro, P. M., & María Andrea, M. S. (2016). Environmental education and the Bogotá River: An intervention to be carried out in Cundinamarca (Colombia). *International Research in Geographical and Environmental Education*, *26*(4), 281–296. doi:10.1080/10382046.2016.1262510

Jennings, N., Swidler, S., & Koliba, C. (2005). Place-based education in the standards-based reform era—Conflict or complement? *American Journal of Education*, *112*(1), 44–65. doi:10.1086/444522

Johnson-Pynn, S., & Johnson, L. (2005). Successes and challenges in East African conservation education. *The Journal of Environmental Education*, *36*(2), 25–39. doi:10.3200/JOEE.36.2.25-39

Kanene, K. (2016). The impact of environmental education on the environmental perceptions/attitudes of students in selected secondary schools of Botswana. *European Journal of Alternative Education Studies*, *1*(2), 1–19.

Karpudewan, M., & Roth, W. M. (2018). Changes in primary students' informal reasoning during an environment-related curriculum on socio-scientific issues. *International Journal of Science and Mathematics Education*, *16*(3), 401–419. doi:10.100710763-016-9787-x

Kim, S. Y., & Hamdan Alghamdi, A. K. (2019). Female secondary students' and their teachers' perceptions of science learning environments within the context of science education reform in Saudi Arabia. *International Journal of Science and Mathematics Education*, *17*(8), 1–23. doi:10.100710763-018-09946-z

Kollmuss, A., & Agyeman, J. (2002). Mind the Gap: Why do people act environmentally and what are the barriers to pro-environmental behaviour? Mind the Gap: why do people act environmentally and what are the barriers to. *Environmental Education Research*, *8*(3), 239–260. doi:10.1080/13504620220145401

Lin, E., & Shi, Q. (2014). Exploring individual and school-related factors and environmental literacy: Comparing U.S. and Canada using PISA 2006. *International Journal of Science and Mathematics Education, 12*(1), 73–97. doi:10.100710763-012-9396-2

Liu, S., & Guo, L. (2018). Based on Environmental Education to Study the Correlation between Environmental Knowledge and Environmental Value. *Eurasia Journal of Mathematics, Science and Technology Education, 14*(7), 3311–3319. doi:10.29333/ejmste/91246

Marcinkowski, T. (2009). Contemporary challenges and opportunities in environmental education: Where are we headed and what deserves our attention? *The Journal of Environmental Education, 41*(1), 34–54. doi:10.1080/00958960903210015

Marjaine, S., Zsoka, A. & Széchy, A. (2011). Consumer behaviour and lifestyle patterns of Hungarian students with regard to environmental awareness. *Society and Economy, 33*(1), 89-110.

Maurer, M., & Bogner, F. (2019). How freshmen perceive environmental education (EE) and education for sustainable development (ESD). *PLoS One, 14*(1), e0208910. doi:10.1371/journal.pone.0208910 PMID:30640908

Michalos, C., Creech, H., McDonald, C., & Hatch Kahlke, P. M. (2009). *Measuring Knowledge, Attitudes and Behaviours towards Sustainable Development: two Exploratory Studies*. International Institute for Sustainable Development.

Mogensen, F., & Schnack, K. (2010). The Action Competence Approach and the "New" Discourses of Education for Sustainable Development, Competence and Quality Criteria. *Environmental Education Research, 16*(1), 59–74. doi:10.1080/13504620903504032

NAAEE. (2004). *Guidelines for the preparation and professional development of environmental educators*. NAAEE.

Negev, M., Sagy, G., Garb, Y., Salzberg, A., & Tal, A. (2008). Evaluating the environmental literacy of Israeli elementary and high school students. *The Journal of Environmental Education, 39*(2), 3–20. doi:10.3200/JOEE.39.2.3-20

Ozden, M. (2008). Environmental Awareness and Attitudes of Student Teachers: An Empirical Research. *International Research in Geographical and Environmental Education, 17*(1), 40–55. doi:10.2167/irgee227.0

Palmer, J. (1998). Environmental Education. In *The 21ST Century Theory, practice, progress and promise*. Routledge Publishers.

Prince, M., & Felder, R. (2006). Inductive teaching and learning methods: Definitions, comparisons, and research bases. *Journal of Engineering Education, 95*(2), 123–138. doi:10.1002/j.2168-9830.2006.tb00884.x

Rakotomamonjy, S. N., Jones, J. P. G., Razafimanahaka, J. H., Ramamonjisoa, B., & Williams, S. J. (2015). The effects of environmental education on children's and parents' knowledge and attitudes towards lemurs in rural Madagascar. *Animal Conservation, 18*(2), 157–166. doi:10.1111/acv.12153

Reilly, S, Petrillo, D. & Demchik, M. (2008). Environmental Education's Role in Sustainable Development: Three Case Studies from India, South Africa & the United States. *International Journal of Human Resources Management*.

Rickinson, M. (2001). Special Issue: Learners and learning in environmental education: A critical review of the evidence. *Environmental Education Research*, *7*(3), 208–320. doi:10.1080/13504620120065230

Rickinson, M. (2001). Learners and Learning in Environmental Education: A Critical Review of the Evidence. *Environmental Education*, *7*(3), 207–317. doi:10.1080/13504620120065230

Roczen, N., Kaiser, F., Bogner, F., & Wilson, M. (2014). A competence model for environmental education. *Environment and Behavior*, *46*(8), 972–992. doi:10.1177/0013916513492416

Rosaleen, J. (2012). *Outdoor learning: past and present: Past and Present*. Open University Press.

Ruiz-Mallen, I., Barraza, L., Bodenhorn, B., & Reyes-García, V. (2009). Evaluating the impact of an environmental education programme: An empirical study in Mexico. *Environmental Education Research*, *15*(3), 371–387. doi:10.1080/13504620902906766

Salmon, J. (2000). Are we building environmental literacy? *The Journal of Environmental Education*, *31*(4), 4–10. doi:10.1080/00958960009598645

Sauvé, L., Brunelle, R., & Berryman, T. (2005). Influence of the globalized and globalizing sustainable development framework on national policies related to environmental education. *Policy Futures in Education*, *3*(3), 271–283. doi:10.2304/pfie.2005.3.3.5

Sharma, P. (2016). *Learning outcomes for environmental literacy: A guide for tectbook writers, teachers and researchers*. Centre for Environmental Education.

Silo, N. (2015). Using Science as a Tool for infusing Environmental Education in the School Curriculum–a Barrier to Education for Sustainable Development. *European Journal of Educational Studies*, *7*(1), 11–22.

Stern, M., Powell, B., & Hill, D. (2014). Environmental education program evaluation in the new millennium: What do we measure and what have we learned? *Environmental Education Research*, *20*(5), 581–611. doi:10.1080/13504622.2013.838749

Stern, M. J., Powell, R. B., & Ardoin, N. M. (2011). Evaluating a constructivist and culturally responsive approach to environmental education for diverse audiences. *The Journal of Environmental Education*, *42*(2), 109–122. doi:10.1080/00958961003796849

Sundar, I. (2010). Teaching & Learning Methods in Environmental Education. Sarup Book Publisher.

Thomas, G. (2005). Facilitation in Education for the Environment. *Australian Journal of Environmental Education*, *21*, 107–116. doi:10.1017/S0814062600000999

UNESCO-United Nations Environmental Program (UNEP). (1978). *The Tbilisi Declaration: Final report intergovernmental Conference on Environmental Education*. Paris, France: UNESCO ED/MD/49.

United Nations Educational Scientific and Cultural Organisation (UNESCO). (1972). *The Stockholm Declaration*. UNESCO.

United Nations (UN). (1992). *Rio Declaration on Environment and Development—Preamble*. UN.

Velempini, K., Martin, B., Smucker, T., Ward Randolph, A., & Henning, J. (2017). Environmental education in southern Africa: A case study of a secondary school in the Okavango Delta of Botswana. *Environmental Education Research*, *24*(7), 1000–1016. doi:10.1080/13504622.2017.1377158

Velmpini, K. (2016). *The integration of environmental education in the secondary school curriculum: A case study of a 10th grade junior secondary school curriculum in the Okavango Delta, Botswana* (PhD Thesis). University of Ohio.

Wan, Y., & Bi, H. (2019). What major "socio-scientific topics" should the science curriculum focused on? A Delphi study of the expert community in China. *International Journal of Science and Mathematics Education*.

Youkhana, E., Leifkes, C., & Leon-Sicard, T. (2018). Epistemic marginality, higher and environmental education in Colombia. *Gestion y Ambiente*, *21*(2), 15–29. doi:10.15446/ga.v21n2supl.77752

Zsoka, A., Szerenyi, Z., Szechy, A., & Kocsis, T. (2013). Greening due to environmental education? Environmental knowledge, attitudes, consumer behaviour and everyday pro-environmental activities of Hungarian high school and university students. *Journal of Cleaner Production*, *48*, 126–138. doi:10.1016/j.jclepro.2012.11.030

Chapter 10
Training Strategies for Organic Agriculture as a Pathway to Achieve the SDGs

Miguel de Porras
FiBL Europe, Belgium

Lisa Haller
FiBL Europe, Belgium

Sigrid Alexander
Beratung Nachhaltige Lebensmittel, Germany

ABSTRACT

The current agricultural and food system needs to become more sustainable. Scientific evidence shows that organic agriculture is particularly environmental beneficial and can contribute to the achievement of the SDGs. Yet, it is a knowledge intensive practice and all actors along the value chain need to be accordingly trained. Focusing on farmers, this chapter assesses available training initiatives in the sector. In order to become more effective, public bodies should engage more in capacity building and training on organics. Moreover, additional training initiatives should focus on specific crops or agricultural systems, which also addresses the training need of farmers in conversion. The use of ICT has the potential to facilitate the development of more efficient training interventions. The further development of training initiatives is a vehicle to promote organic agriculture and ultimately to contribute to achieving the SDGs and climate mitigation and adaptation in particular.

DOI: 10.4018/978-1-7998-7512-3.ch010

INTRODUCTION

In order to tackle the global environmental, social and economic challenges of the 21st century, the United Nations Member States have compromised to the 17 Sustainable Development Goals (SDGs) under the commitments of the Agenda 2030. The SDGs address climate change and the preservation of nature, oceans and forests, while at the same time set goals to end poverty and other deprivations, improve health and education, reduce inequality and foster economic growth (UN 2020a).

Land and land use, including agricultural production, are the basis for our livelihoods. In the context of a changing climate and environmental degradation, agriculture and food systems need to provide enough nutritious and high quality food for a growing population. The intensification of agriculture in the last century has led to a significant increase in production and food security globally. However, current agricultural practices are also a major driver for environmental problems(e.g. biodiversity loss, pollution, soil degradation)including climate change. According to the IPCC (2019), 21-37% of the total GHG emissions can be attributed to the food system, including land use, agriculture, storage, transport, processing and packaging as well as retail and consumption. While land use is responsible for approximately a quarter of the total direct and indirect GHG emissions (10-12 Gt CO_2-equivalents), half of the emissions derive from agriculture (5-5,8 Gt CO_2-equivalents; IPCC 2014). These agricultural emissions are mainly emissions of methane (CH4) from enteric fermentation of livestock (mainly cattle); methane emissions from rice fields; and nitrous oxide/nitrous oxide (N_2O) emissions from agricultural soils (Tubiello et al. 2013; Smith et al. 2014).Regarding emissions from livestock, however, the balance of intensive, so-called industrial production that is highly dependent on concentrated feed, is different from species-appropriate grazing with limited input of concentrated feed. Crop production and animal husbandry account for 4-14% of the GHG emissions and land use and land use change is estimated to be the cause of 5-14% of the GHG emissions. If "business as usual" continues and current trends in the food system are pursued, it is estimated that the food system will account for 30-40% of the total GHG emissions in 2050 (Shukla et al. 2019).

At the same time, the food system, and particularly the agricultural sector, are highly vulnerable to a changing climate. Extreme weather events disrupting the food chain are expected to increase in the future, both in terms of intensity and in terms of frequency. Desertification and land degradation caused by climate change further adversely impact food security. Moreover, higher CO_2 levels in the atmosphere may lower the nutritional quality of crops (IPCC 2019).

Therefore, it is evident that an environmentally sound and sustainable transformation of the whole food system is needed.According to the FAO (2020) "to be sustainable, agriculture must meet the needs of present and future generations, while ensuring profitability, environmental health, and social and economic equity". There are three dimensions to sustainability – environmental, social and economic, that all have to be met, even though there might be trade-offs between them.

The transformation has to take place both on the production (e.g. increase of the share of sustainable agricultural practices such as organic agriculture, see below) as well as on the consumption side. Modelling efforts at a global scale show that an overall more environmentally sustainable agriculture can only be achieved if both the production and consumption of animal products as well as food waste and losses are significantly reduced (Schader et al. 2015; Müller et al. 2017).

The Contribution of Organic Agriculture to the SDGs

In order to achieve the necessary transformation towards more sustainability and fight against the most relevant environmental challenges that our food systems are facing, there are several alternative approaches to agriculture that can make a significant contribution. For this chapter we have selected organic agriculture, as one of the systems that can support this transition. There is sound scientific evidence on the positive environmental performance of organic agriculture and its potential contribution to achieving the SDGs.

It has proven to be beneficial to a number of environmental parameters, such as biodiversity, soil fertility, water protection or resource efficiency (Sanders and Hess 2019, Haller et al. 2020).

Organic agriculture is, therefore, a production system that can contribute to the accomplishment of multiple SDGs (cf. Figure 1), such as climate action (SDG 13), responsible production & consumption (SDG 12), life below water & on land (SDG 14 & 15), zero hunger (SDG 2), clean water (SDG 6), no poverty &decent work and economic growth (SDG 1 &SDG 8) and good health & wellbeing (SDG3). In this context, organic agriculture can not only reduce the negative impacts of agricultural production but also increase its positive impact. For the latter, increased carbon sequestration, higher biodiversity on organic farms or superiror water holding capacity are only a few examples of this increased positive impact. At the same time, organic agriculture can help mitigating the negative impact of agricultural production, e.g. due to the ban of agrochemicals, which have adverse effects on biodiversity, soils, health (De Schaetzen 2019). In addition, organic agriculture can contribute to innovation (SDG 9) and sustainable communities and cities (SDG 11). Innovative approaches in organic agriculture can also act as a driver for innovations applied to conventional agriculture. E.g., the limited availability of permitted inputs fuels the research and innovation potential to find alternatives (cf. Horizon 2020 RELACS project to phase out the dependency on and use of inputs considered contentious in organic farming systems). In the context of global challenges in the food system, the local level and short supply chains gain increasingly importance. Particularly in urban areas, a transformation of the food system needs to take place and shorter supply chain strengthened. Organic with its higher demand in cities and shorter supply chains can be a key driver (cf. Moschitz and Frick 2020, Dubelling et al. 2015).

Furthermore, agriculture can play an important role for climate change mitigation. Globally, carbon sequestration is considered to have the greatest potential (Smith et al. 2008). Measures to increase the sequestration potential include the cultivation of perennial fodder legumes and the recycling of farm manure (Gattinger et al. 2012). In conventional agriculture, the use of nitrogen fertilizers leads to both direct and indirect N_2O emissions, direct N_2O emissions derive from fertilized soils and indirect emissions from the transport of reactive nitrogen compounds (ammonia and nitrate) to soils. In addition, emissions arise in the course of the production and transport of nitrogen mineral fertilizers as well as through the consumption of fossil resources. Yet, in climate reporting, these indirect emissions are not attributed to agriculture (UNFCCC 2019). The climate protection potentials of organic agriculture are, therefore, (1) the renunciation of mineral fertilizers and synthetic pesticides, (2) the lower purchase of animal feed, (3) the limited use of peat-based substrates and (4) the focus on building up humus through the cultivation of humus-producing crops (Gattinger et al. 2012; Freibauer et al. 2004). Studies show higher CO_2 sequestration and lower N_2O emissions per hectare in organic crop production compared to conventional crop production (Weckenbrock et al. 2019).

Regarding the potential for climate change adaptation, soil fertility plays a crucial role (Levin et al. 2019). At the same time, high soil fertility positively affects flood control and erosion protection, which

Figure 1. Organic Agriculture and SDGs; Source: UN (2020b), own adaptation

will gain importance in the future due to a predicted increase of drought periods and heavy precipitation events (Brasseur et al. 2017). Moreover, a high infiltration potential of soils can facilitate adaptation to dry periods, since a high seasonal water retention can stabilize the regional water regime (Levin et al. 2019). Jung and Schmidtke (2019) showed in their extensive study that organic agriculture can havea positive effect on soil fertility. Key factors are higher soil organic carbon(SOC) contents, more diverse crop rotations, longer soil cover, reduced use of pesticides and reduced soil management.

Yet, the complexity of the agricultural systems and their sustainability is high. In spite of the positive environmental performance mentioned above, organic agriculture, as any other agricultural production system is quite diverse. Within this diversity, organic farmers with a higher degree of intensive production and increased off-farm inputs and therefore a potentially less positive environmental performance, can be identified (De Wit and Verhoog 2007, Rosset and Altieri 1997). These "intensive" organic farming systems might lead to critical developments such as separation of plant and animal production, monoculture structures and other examples of over-specialization,challenging or even conflicting with the principles of organic agriculture (especially ecology and health). In addition, the globalization of organic markets and the increased reliance on global value chains (cf. off-farm inputs) might have negative impacts on the social and economic performance of this kind of agriculture (Guthman 2004; Skalidou and Oya 2017).

Organic Agriculture, the State of the Art

Nevertheless, organic agriculturecan play an important role in achieving the SDGs. It is a highly dynamic sector and it represents one of the biggest drivers for the necessary transition towards more sustainable agricultural and food systems. Due to the still relevant premium prices of organic produce and overall reduced production costs (Crowder & Reganold 2015) this sector represents a real incentive to change farmers' behavior and shift current practices towards more sustainable agricultural ones. Over the past decade, the global organic retail sales have almost tripled. In 2018, organic retail sales reached almost 97 billion € worldwide, yet, the demand for organic products remains concentrated in Northern Europe and Europe, with 42% and 39% of global retail sales respectively (Willer et al. 2020). To serve this market, the number of organic farmers is also quickly growing, reaching 2.8 million farmers in 2018. In total, they farmed 71.5 million hectares of farmland in 186 countries of the world, representing 1.5% of total agricultural land globally. More than 66% of organic agricultural land is made up by grassland and almost 19% is arable land. Additionally, over 35.1 million hectares of wild collection and beekeeping areas, which are also regulated by organic laws, were reported for 2018.Over a 10-year period, the worldwide growth of organic farmland was 97.2% and of organic producers 55%, making it one of the fastest growing agricultural sectors (Willer et al. 2020). Ninety percent of the 2.8 million organic producers are based in Asia, Africa and Europe, with 2.4 million farmers (86% of all organic producers) from countries on the list of the Development Assistance Committee (DAC; Willer et al. 2020). India, with 1.15 million has by far the biggest number or organic producers. With the blooming growth and based on the trend of an increasing organic market over the past decade, it can be assumed that this trend will continue in the coming years, positioning organic agriculture as a key driver of sustainability at global level. Figure 2 gives an overview of the current state of organic agriculture worldwide.

Figure 2. Organic agriculture worldwide in 2018. Source: Willer et al. (2020)

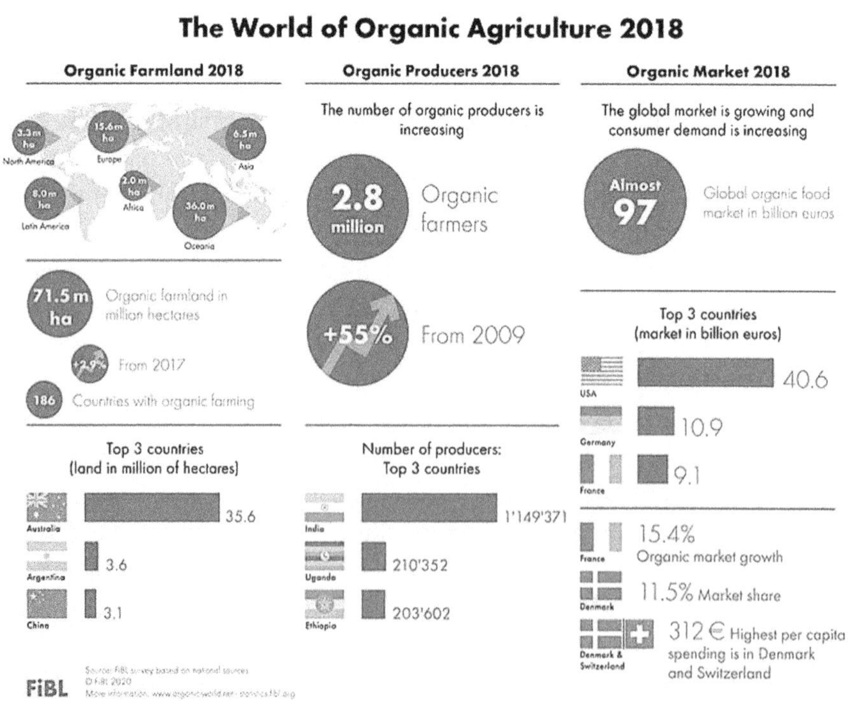

Organic Agriculture – Current Needs and Bottlenecks for Training

Organic production is a particularly knowledge-intensive production system that often depends on context-specific solutions adapted to the agro-climatic conditions (Morgan and Murdoch 2000). In addition, existing technologies and innovations are not always disseminated and known among organic producers from different countries or continents with similar challenges. Access to knowledge and capacity building programs for organic production is often a challenge due the dispersion of producers, and the massive dominance of conventional approaches in rural extension and capacity building programs (Kroma 2006). In other cases, there might be a certain reluctance within some actors of the organic sector to take up new techniques or innovations produced in academic or research contexts. Furthermore, being the only clearly codified sustainable production system with a legal regulatory framework and set-out standards, certified organic agricultural products are one of the most controlled goods. In 2019, 84 counties had organic standards in place, while additional 17 countries were in the process of drafting a respective legislation (Willer et al. 2020). Therefore, complying with all regulatory requirements is an additional layer of complexity that organic operators and their partners in the supply chain and the control system have to deal with to access the organic market.

In order to maximize the growth of the sector and to guarantee the long-term positive impact of organic agriculture, knowledge and capacity building, as well as the development of skills particularly focused on organic production are crucial. Currently, the lack of the needed information, knowledge and skills of the actors involved is one of the key barriers for the expansion of the organic sector (Wheeler 2008) and hence its successful contribution to achieving the SDGs. A case study in Jordan identified special training workshops on organic agriculture farming next to financial support for organic farmers as necessary key drivers for the sector (Altarawneh 2016).Due to the niche dimension of the organic market, so far systematic training on organic agriculture has been one of the most important gaps in agricultural capacity building programs globally, at all different levels, from most of the public education institutions, advisory services, or agricultural extensions programs, etc. Yet, a strong push to move the organic market out of the niche can be observed. As part of its Green Deal, the European Commission recently published its Farm to Fork strategy. One of the ambitious objectives set out is to reach at least 25% of EU's agricultural land under organic farming by 2030, which will go hand in hand with the aim to make consumption and diets more sustainable and healthy (European Commission 2020). As the EU is the biggest importer and exporter of agri-food products (European Commission 2020), the EU also intends to raise standards globally and to avoid negative environmental and social impacts in countries of production outside the EU (European Commission 2020). This ambitious framework is one of the most relevant globally, but not the only one, as there have been several initiatives in other regionsduring the last years.In India, the full transition of the Sikkim State to organic agriculture was performed at the end of 2015, which is pushing the adoption of this agricultural method in the whole country (IFOAM 2017).Therefore, while in the past only a limited number of institutions and stakeholders have been working on training focusing particularly on organic basics and production methods, more ambitious strategies to further upscale the sector globally need to be developed. In addition, the requirement of certification and the implementation of strict controls in organic agriculture makes the market access financially and administratively more difficult and costly especially for small-scale farmers (El-Hage Scialabba 2012). For this reason, training, which includes the regulatory aspects of organic agriculture are specially needed to grant access to organic markets to small-farmers.

In a desk study, we explore different forms and modalities of resources available for training and education in organic agriculture. Based on this first overview, we present four case studies of existing training initiatives around the globe. This will allow to better understand the current status of available training initiatives and their potentials.

Additionally, we assess the potential of Information and Communication Technologies (ICT) to develop more efficient training interventions and tools to support organic agriculture. These technologies have an enormous potential for capacity building and knowledge sharing interventions that can be key for future interventions in the organic sector worldwide. These technologies make knowledge sharing tools at global level possible, with the potential to allow interventions oriented to closing the huge skills gaps identified in the organic sector (cf. Aiyelaagbe et al. 2015).

Therefore, this chapter allows for a better understanding of the role that different training interventions can have and already have on the development of the organic sector. The further development of training initiatives is a vehicle to promote organic agriculture and ultimately to contribute to achieving the SDGs in general and climate mitigation and adaptation in particular.

TRAINING IN ORGANIC AGRICULTURE – CURRENT FORMS AND MODALITIES

To address the complexity of the food system in general and the connected agricultural practices in particular, a transdisciplinary whole-systems approach that includes an assessment of long-term impacts is needed (Francis 2011). While all actors along the value chain are potential target groups for training, education and advisory work, we focus on farmers, addressing the full spectrum from small-scale to large-scale farmers. They are at the core of agricultural production, not only producing agricultural produce but also providing public goods in a wider sense (e.g. cultural landscape). The applied practices have direct influence on the sustainability of the agricultural system and, therefore, particular attention has to be paid to maintain its function, e.g. soil fertility, animal welfare, quality and variety of regional produced staple food. Knowledge and skills are crucial in this context. Moreover, based on our research, there is more (publicly available) training material available for farmers than for other actors along the value chain, facilitating a comprehensive comparison. Yet, there are other training initiatives targeting specific actors along the value chain, e.g. EU`s Better Training for Safer Food (BTSF) focusing on organic agriculture is specifically tailored for representatives of Control Authorities and Control Bodies. At the same time, training initiatives targeting processors are still underrepresented and should get more attention in the future in order to ensure processing that fits the purpose of organic products in a particular and guarantees a transition of the whole food system towards more sustainability in general (Moschitz et al. 2021).

Despite the urgent need for more training and advisory work on organic agriculture as described above, there are already various training initiatives using different modalities and approaches established. We performed a desk study to identify the different training materials available in organic agriculture. In our desk study, we mapped various existing training strategies at regional, national and global level. There is a vast variety of training initiatives already in existence and our aim was to identify the different approaches used, rather than deriving a complete list. Overall, we took more the 40 different training initiatives into account, of which 24 addressed farmers, operating on different geographical levels reaching from regional to global scale. For our analysis, in a first step, we grouped the different modalities applied by the examined initiatives in the following four categories:

1. **Training materials:** This category includes all training materials giving practical information that are either addressed directly to farmers for individual self-learning or trainers as supporting materials. Training materials can reach from written materials, to videos or infographics. Written materials include concise practice abstracts targeting particular topics (e.g. the best treatment for a particular disease) or comprehensive manuals informing about the principles of organic production, soil fertility, plant health etc. Short didactical videos or infographics are widely used to impart information addressing specific topics. Videos are easy-to-access tools that can convey knowledge and showcase practical skills that is accessible for a broad audience.
2. **Training courses:** There is a wide range of different training courses available, from online courses (incl. e.g. live online trainings or tutorials,webinars)in modules, courses combining online tools and classroom or field modules to entirely face-to-face learning. Also regarding the duration there is a whole spectrum ranching from one-day courses to courses with a duration of up to a year.
3. **Peer-to-peer:** The role of farmers as innovators is particularly central in organic agriculture and their knowledge of crucial importance. Initiatives that facilitate the knowledge exchange among farmers using a peer-to-peer approach are, therefore, a common training practice widely used in organic agriculture.
4. **On-site demonstrations/farms:** Demonstration farms and on-site trainings also play an important role in organic agriculture training, as they are an important means to teach practical knowledge and skills on-site.

However, it has to be noted that the majority of initiatives do not fall under only one specific category but uses a blended approach, combining the above-mentioned modalities.

As there are little national programs for training in organic agriculture, the majority of initiatives are private, initiated by farmers associations, research institutions and universities or publicly funded (e.g. in the framework of projects funded be the European Union).Also downstream actors of the value chain (traders, processors, retailers etc.) offer trainings for farmers or finance advisors to support farmers in better implementation (cf. OTA 2020). Regarding the topics addressed by the different initiatives, they cover the full spectrum from crop production (including pest and weed management as well as soil fertility), and animal husbandry to the structuring of farms, certification procedure and marketing as well as regulatory issues in general. Yet, only a few of them include also training on social or economic aspects. Table 1 gives an overview of the selected initiatives specifically targeting farmers.

TRAINING AND EDUCATION IN ORGANIC AGRICULTURE – CASE STUDIES

To illustrate the different approaches commonly used for training in organic agriculture, we made a selection of four case studies based on three qualitative criteria. We reflected the different modalities of initiatives identified in the categorization mentioned above. We also targeted different geographical regions in order to show the ongoing activities globally. Finally, the case studies reflect the diversity of initiating bodies, from the organic agricultural movement itself to research institutions and universities.

Table 1. Overview of selected initiatives specifically targeting farmers.

Nr.	Initiative	Geographical scope	Institution	Target group	Category	Content
	Global					
1	Organic Academy by IFOAM Organics International	global	IFOAM	all actors involved in organic sector	course	business leadership or regenerative agriculture
3	Organic Leadership courses	global	IFOAM	farmers, all actors involved in organic	course	managing organic businesses, increasing sustainability of value chains
4	FAO Training manual for organic agriculture	global	FAO, IFOAM, FiBL	farmers, advisors	Training material	conversion, organic production, water management, nutrient,pest and weed management, soil
5	IFOAM Training Manual for Organic Agriculture	Tropics, Arid and Semi-Arid tropics	FiBL, IFOAM	farmers, trainers	Training material	guidelines for main crops in the topics, marketing strategies
	Africa					
6	African Organic Training manual	Africa	FiBL	farmers	Training material	conversion, organic production, water management, nutrient,pest and weed management, soil, marketing
	America					
7	Basics of organic farming	USA	OFRF, UC SAREP, Cal Poly	farmers, newcomers	course	soil health & weed management
8	Organic Farmer Training Programme	USA	Michigan State University	farmers	course, on-site	organic production, business development
9	Soil Health and Organic farming Webinars series	USA	OFRF, eOrganic	farmers	course, Training material	soil health
	Asia					
10	Azerbaijan State programme on organic agriculture	Azerbaijan	Azerbaijan Government	farmers, stakeholders in the organic sector	course	organic production
11	BOKK II	Democratic People's Republic of Korea	IFOAM, PIINTEC	farmers, academia, policy makers	on-site	capacity building, mainstream organic agriculture to address food and nutrition security
	Europe					
12	Organic Growers Alliance courses	UK	Organic Growers Alliance	farmers	course	organic production
13	SEAE training courses	Spain	SEAE	farmers, advisors	course	organic production
14	LFI training courses	Austria	LFI	farmers, advisors	course	organic (and conventional) production, inputs
15	TYFU Specialist advice and training	UK (Wales)	TYFU CYMRU	farmers (horticulture)	course	farm management, marketing
16	Jornadas de transformacion	Spain	Red de inovacion rural	farmers (in conversion)	course	milk and meat production, wine, cereals, horticulture
17	Future Growers	UK	Soil Association	farmers (newcomers)	course, on-site	horticulture
18	Commercial growers course	UK	Kindling trust	farmers (newcomers)	course, on-site	marketing & management, soil fertility, vegetable production
19	Bio Austria training courses	Austria	Bio Austria	farmers	course, on-site	organic production
20	Trailer rides	UK	Abbey Home Farm	growers and consumers	on-site	awareness
21	Organic Farm Knowledge Platform	Europe	FiBL, ICROFS, IFOAM EU	farmers	Training material, peer-to-peer	crop production, animal husbandry, seeds and breeding
22	DigiPlanAlp	Europe (alpine regions)	Agridea, FiBL, Agroscope, Büro Alpe	farmers (alpine)	Training material	problem plants in the Alps
23	Bionet	Austria	FiBL, LFI, LKÖ, Bio Austria, Raumberg-Gumpenstein, BOKU	farmers, advisors	Training material, course	transfer/link between research and practice, crop and vegetable production
	Oceania					
24	Hortcourses	Australia	ACD Distance Education	farmers	course	organic principles, farm management, organic production

Organic Farm Knowledge Platform (No. 21)

The online platform aims at providing easy-access to practical tools and materials to enhance organic production. Resources and tools include practice abstracts and guidelines but also calculation tools, online courses or videos. Currently the resources are developed by seven partner organizations from all over Europe (FiBL, ICROFS, IFOAM Organics Europe, AIAB, Bioland, CIHEAM Bari, ORC, ITAB) and the website is translated into 14 EU languages. In addition, the developed resources and tools are available in different language versions depending on the respective issuing partner organization. The platform serves as a virtual meeting place that allows for cross-border exchange and learning (Organic Farm Knowledge 2020).

The platform was developed as a result of two multi-actor projects funded under the framework of Horizon 2020 by the European Union, OK-Net Arable and OK-Net Eco Feed. The geographical focus of the platform is on Europe and it offers easy-accessible resources to farmers and advisors free of charge. It is a "living platform" meaning that additional resources and tools are added on a regular basis.

When searching the toolbox, one can search, among others, by theme, language or keyword. For each available tool, a brief introduction on the problem addressed, solution offered and description is given facilitating the targeted search for users. Themes addressed include crop production, animal husbandry as well as seeds and breeding. To ensure a more specific search, the themes are further divided into sub-topics (e.g. soil quality and fertility, arable crops, nutrient management, pest and disease control, cropping systems, animal health, feeding and rotation planning, seed production, seed quality, plant breeding and variety trails, economics and regulation of organic seeds, etc.). The available resources can be directly accessed via the platform, be watched (videos) or downloaded (leaflets and practice abstracts). Currently almost 300 up-to-date tools and resources are available. In addition to accessing the tools, users have the possibility to rate and comment on them (Organic Farm Knowledge 2020).

African Organic Agriculture Training Manual (No. 6)

The African Organic Agriculture Training Manual is a tool available online free of charge that provides material on organic agriculture with the objective of promoting organic and other sustainable agricultural practices in Africa (African Manual 2020). It uses graphical illustrations, easy language and practical examples. Its content is particularly adapted to farming circumstances in Africa. It includes both materials for trainers as well as handouts for farmers, with practical tips, presentations that trainers can use and booklets for farmers. The materials for trainers include background information on the specific topics as well as ready-to-use presentation slides they can use in courses and workshops with farmers, including suggestions for discussion points and group exercises to make it as interactive as possible and additional readings. Additional resources are constantly added. Currently it is available in English and partly in French and Swahili, with the intention to provide further translations. The manual is divided into 10 modules covering a broad range of topics related to organic farm management, including definition and benefits of organic agriculture, soil fertility, pest and weed management, animal husbandry, farm management marketing and trade, conversion to organic farming, crop management, and animal species. The modules are further sub-divided into sub-themes like soil and water conservation or post-harvest management and storage. The manual was developed by the Research Institute of Organic Agriculture (FiBL, African Manual 2020).

Organic Academy by IFOAM Organics International (No. 1)

The International Federations of Organic Agriculture Movements (IFOAM Organics International) is the biggest global member-based organization representing the organic sector (not only farmers). Founded in 1972, IFOAM has become a relevant player in the agricultural sector worldwide, connecting farmers, companies and NGOs in a social movement pushing for the transition towards sustainable agriculture. In 2012, IFOAM Organics International has established a training and advisory program that operates globally, the IFOAM Academy. Under this training scheme, IFOAM offers different formats, including webinars, residential sessions or topic specific one-day trainings. The fees vary according to format and duration of the training. The aim is not only to offer training to actors of the value chain but also to build and foster the organic network (IFOAM 2020).

One of the flagship elements of the IFOAM academy is the organic leadership course oriented to capacity building and network development of the organic movement. These courses are an interesting training initiative due to their global focus and the main focus on strengthening the interconnection among organic players. Farming and organic production methods are only a limited part of these courses but they are offered in every continent and they are an excellent platform for facilitating the interconnection of organic farmers. In addition, the Organic Academy offers training and tailored formative courses, such as the Nutrition in Mountain Agroecosystems course or the Organic for Extension Agents, with a bigger farming component, resulting in a relevant platform for the sector. In addition, IFOAM also offers trainings and consultancies on Participatory Guarantee Scheme (PGS) and on the development and improvement of policies that support the development of the organic sector (IFOAM 2020).

This case is especially relevant since it is a training program developed by the organic movement for target audiences interested in increasing and strengthen it. Social movements often have limited resources to provide training and build the capacities of their participants. The IFOAM Organic Academy can, therefore, be considered as a success story of a social movement that has gained enough resources to develop independently their own training and capacity building program at global scale.

Organic Farmer Training Program (OTPF) (No. 8)

The Michigan State University offers an 8-months training program on organic agriculture to 23 participants per year who intend to start their own organic business. Connected to the training program is a study-farm, giving the opportunity to gain practical knowledge. The program follows a mixed approach including lectures, demonstrations and practical activities, with a 50/50 mix of field and classroom activities.

In the field, participants are engaged in hands-on activities, farm walks and equipment demonstrations. The classroom activities consist of small group activities, full group discussions, guest speakers' presentations and participant presentations. In addition to the study-farm, participants visit various farms in the region. The focus is both on agricultural production and on business development creating knowledge that is necessary to operate a small farm. Topics covered include vegetable production, harvest and post-harvest handling, farm ecology, seasonal planning and crop rotation, livestock management, farm financial management and direct and wholesale marketing. Outside of class, video lectures, readings and practical assignments, like writing a Farm Business Plan connect learning themes from the class to participants' own farm goals (MSU 2020).

The selected cases reflect the broad range of the available training approaches. It shows that context specific training initiatives exist, both in terms of content and geographical scope. This is particularly important for organic agriculture as a one-fits-it-all approach is not applicable to organic farming. Local conditions have to be taken into account and vary depending on the area, climatic zone and landscape structure. The cases also show that there are training offers available for both newcomers to the organic sector (e.g. Organic Farmer Training Program) as well as experienced farmers (e.g. Organic Farm Knowledge).The institutions behind the development of the selected cases are fundamentally research centers involved in organic agriculture (FiBL and ICROFS), organic social movements (IFOAM) and one University (Michigan State University). This selection also reflects the main typology of actors putting efforts in offering organic training to farmers worldwide. This chapter does not offer a comprehensive list of the training initiatives in organic agriculture worldwide, however with the data collected it becomes evident that currently the presence of private institutions (including research centers and NGOs) is quite significant in comparison to public institutions. Involvement of public bodies (especially Universities) is quite relevant but we can consider that their weight is much lower than in other agricultural sectors. As regards of the training contents, the collected information offers a wide variety of focus showing a considerable diversity, from business and leadership to organic horticulture. However, considering the selected cases as well as the courses identified in our desk research, we observe, despite the existence of many specific courses there is only a limited number of crop- or animal husbandry specific trainings available. Many of the organic farming courses are rather general, focusing in organic production and while lacking a focus on a specific crop or system.

The four cases have been highlighted as examples of best practices in organic training provision. Organic Farm Knowledge and the African Organic Training Manual represent two of the best examples of training materials including a peer-to-peer knowledge exchange available. Easy to access, with a simplified language far from scientific standards and a quite specific content directly targeted to specific farmers' and advisors' needs these two cases can be presented as best cases of materials accessible through Information and Communication Technologies (cf. section on ICT technologies below).

The Organic Academy and the Organic Farming Training Program are best practices of face-to-face training in organic agriculture. The first one organized by a global movement is a key instrument to build the global network of organic actors including also farmers, increasing their impact on policies for supporting the sector. The Organic Farming Training Program is an excellent course provided by a University, which combines theoretical and on-farm teaching to train successful organic farmers. The inclusion of basic financial topics are also key for the development of competitive organic farms.

Even though the majority of the considered training initiatives are easy to access, a constraint might be that farmers need to proactively engage to participate. The use of digital and online resources improves the access to training tools. Yet, while organic farmers tend to be quite open for additional training due to the knowledge intensive practices applied, the lack of public training schemes and programmes might result in a limited range of the initiatives.

As organic farming is operating within a clearly defined legal framework, with a strict certification system in place, it is particularly important for farmers to be aware of the applicable rules and standards. To guarantee the market access for farmers (and processors) especially from third countries, they have to be trained continuously also on these updated requirements for a better understanding of the legal requirements and needs to facilitate a more efficient and harmonized implementation.

Moreover, despite the important contribution of organic agriculture to the accomplishment of the SDGs this tends to be addressed only in a limited manner in the training initiatives under consideration.

Farmers are usually well aware of the functioning of ecosystems and on how the applied practices influence them. However, to foster their role as multipliers a stronger focus on the positive impacts of organic agriculture and the underlying reasons might be useful.

ORGANIC AGRICULTURE TRAINING AND ICT

Information and Communication Technologies (ICT) have opened a new scale to training and capacity building programs, especially in those related to agriculture where sometimes access to qualitative extension programs and advisory services are limited. Even if the use of these technologies is far from being exhausted, partly due to lacking infrastructure (limited connectivity in rural areas) and partly due to social constrains (lack of basic ICT education, cultural barriers, etc.), several initiatives have been developed globally quite quickly with very positive impact. This shows to the world how agricultural producers and rural areas are ready to exploit the potential of ICT for accessing agricultural knowledge fully.

Several institutions at global level are investing and supporting the deployment of these technologies in agriculture trough different initiatives.

Two key publications assessing the adoption of Digital Agriculture, including ICT technologies, are the report “The Digitalization of African Agriculture report” (CTA, 2019) focusing on the African continent and another one referred to Europe, summarized in the report “New Skills for digital Farming” (EIP-AGRI, 2020). These reports present the picture that ICT technologies are already a quite significant part of digital agriculture, since they are becoming the main vehicle for knowledge and data sharing for all actors in the agri-food value chain. From the assessment of these reports, we can derive that ICT are the most widespread digital technologies in agriculture. This is connected to the fact that other technologies (e.g. robotics) might require higher investments, hence being only scalable in developed economies, such as Europe, but not the rest of the world. ICT technologies are characterized by lower investments and easy access for farmers, normally linked to overall household connectivity not specifically connected to their applications to agriculture. For this reason, ICT technologies have reached a higher uptake among farmers all over the world. Despite these positive aspects certain controversial aspects need to be discussed, such as data ownership, considering that IT companies become owners of the knowledge produced by farmers and other actors. Another critical aspect is the concentration of market share as IT companies and service providers are becoming highly concentrated. The dependency on certain services combined with the lack of competition within these markets might lead to economic inefficiencies which might impact in farmer income. Therefore, the complexity of the deployment and use of new technologies, such as ICT, need always to be taken into account.

Specifically, for organic agriculture these technologies have opened the door in many countries to bypass several structural lock-ins for knowledge sharing among producers interested in more sustainable agricultural methods. Even though the focus of the application of these technologies is often not limited to organic certified agriculture, their key objectives are usually to increase the efficiency of agricultural production while reducing its environmental impacts, objectives that match those of the organic certified production systems and can be easily applied to them.

On the one hand, ICT technologies can provide access to many kinds of trainings and teaching materials for farmers. At the same time, they are also the mean for the creation of different typologies of innovative process in organic agriculture (Knierim, 2020). Between organic farmers, these technologies allow the creation and development of peer-to-peer networks and communities of practice (Wenger, 1998).

Furthermore, ICT technologies can have an essential role in supporting the development of Innovation Platforms (Kilelu, Klerkx, and Leeuwis, 2013) between organic farmers and other actors of the organic value chain, including control bodies, traders, processors and retailers.

Peer-to-peer networks and communities of practice are key drivers for the development of organic agriculture. Indeed these kind of instruments are essential to overcome the geographical dispersion of certified operators, allowing cross-regional sharing of knowledge and experience about organic crop production. ICT here are of an uttermost importance for the horizontal integration of farmers and for making available to them digital resources and teaching material that could support them when developing similar production. In this way, farmers have the tool not only to gain knowledge from other actors, mostly from research and extension services, but also to exchange and share their own knowledge in a co-creation process to solve their production problems and build collectively the needed capacities for organic production.

Innovation Platforms are the vehicle for the vertical integration of organic farmers, supporting their business to be better integrated and increase their bargaining power in the agri-food value chain. Organic operators, meaning retailers, processors and traders, represent a minor part of the global agri-food business and, therefore, having an integrated value chain from the production of the raw material to the market shelf is not an easy task. There is a considerable number of mismatches between the different segments of the value chain that can create many problems and inefficiencies in the development of the organic sector. These mismatches, such as lack or excess of raw materials, transport and timing problems, etc. can generate economic loses to all actors of the chain but often are transferred to farmers. Here, ICT are also key to increase the integration of all steps of the value chain, putting in contact and sharing the data of certified operators, traders, processors and retailers to avoid mismatches and ensure an efficient and fair distribution of the value created.

These two examples of innovative processes are essential to understand that the potential of ICT for the organic sector are not only limited to make accessible training programs and teaching materials, but also are key to ensure a co-creation process of organic farming knowledge. By ensuring this, these technologies should be considered as very relevant tools for solving some of the abovementioned structural lock-ins of the organic sector.

DISCUSSION

This chapter has presented information on how organic agriculture has the potential to contribute to achieving the SDGs. In this context we have highlight the importance of knowledge exchange and capacity building for the development of this sector. It became evident that organic agriculture needs to be strengthened with the support of suitable and adapted trainings for all actors, but especially for farmers (cf. Moschitz et al. 2021, Altarawneh 2016). We have gathered information about existing training materials and resources available for different geographical scopes. We have classified them in four different typologies trying to summarize the diversity of the different examples of which four cases have been assessed in more detail.

The current political momentum putting organic agriculture high on the political agenda (cf. Farm to Fork Strategy by the European Commission) suggests that the sector will grow, and with it the need for tailor made training materials and resources. As mentioned before, our desk research has identified more than 40 initiatives worldwide. This is, however, not an exhaustive list of existing initiatives and additional materials might be identified in different languages and channels. However, our assessment provides an overview of the existing offer. There are materials available worldwide, but, considering the increasing relevance of organic agriculture, they are still far from fulfilling the potential demand at global and regional level incl. covering offers in local languages. In addition, the initiatives and materials assessed, have, despite being quite accessible, still a limited connection with the existing formal educational and training systems in the field of agriculture. Thus, to facilitate the development of effective education and training systems tailored for the dynamic and growing organic sector, we consider that some key gaps and limitations presented below need to be urgently addressed.

Organic agriculture has a demonstrated great potential to support the achievement of the SDGs globally as well as clear environmental benefits (cf. (De Schaetzen 2019, Sanders and Hess 2019). For this reason, it is important, as it is happening in certain regions of the world, to support the transition to organic agriculture with different public policies and political commitments (cf. European Commission 2020, Sikkim State in India). In order to be successful, the transition of the production system and its linked overall agricultural and development policies need to go hand in hand with training and educational policies. This is the only way to ensure direct impact on the food systems. Policymakers need also to consider education and training as a cornerstone dimension of the policy coherence in the field of organic agriculture. Any political ambition in this field needs to be accompanied by equally ambitious educational and training policies in order to ensure an effective implementation. To facilitate the transition of farmers towards organic agriculture and increase their skills on a regular basis it is not enough to plan objectives and interventions in the field of agricultural policies, but educational and training strategies need to be also addressed. From the assessment of the selected initiatives and materials, we have identified two key issues that need to be addressed to overcome the bottlenecks currently existing in organic agriculture and allow the scaling up of this agricultural practice.

The underrepresentation of public bodies as driving institutions for training materials on organic agriculture assessed is quite relevant. This issue needs to be addressed in order to scale up organic agriculture and increase its positive impact on the SDGs. Despite some exceptions and the quick development of new training initiatives, which involve public bodies, the public sector is lagging behind in comparison with the private sector. This fact is difficult to justify considering the participation of public bodies on conventional agriculture training, and the current push for organic agriculture. We also believe that this engagement is key to increase the recognition of organic farming and to move the sector out of the niche, in which it is still "trapped" in most countries (cf. Willer et al. 2020).

In line with the finding by Aiyeleegabe et al. (2015), we conclude that organic standard development and certification, organic livestock production, organic aquaculture management, and organic crop production are the key areas that need further training efforts. For the latter, specifically the provision of crop-specific or system-specific training is an additional gap identified in the training offer identified. We believe that this gap needs to be addressed since this will better support the specific needs of organic farmers who have to deal with their specific crops. However, we believe that the biggest advantage of increasing this training offer is to support the conversion of conventional farmers. The availability of crop-specific training might rise the interest of conventional farmers and create appropriate incentives to experiment with organic production. To invest in these crop-specific trainings might support the development of community of practice and farmer networks that have a fundamental role for supporting the transition towards more organic agriculture and hence to achieve the SDGs.

An important lever for effective training is the pedagogical approach. Many training courses focus on the amount of content: most courses should cover as many relevant technical aspects of the topic as possible. This bears the risk that all those involved are overtaxed and the training can be of less use to the person who uses them later. If, however, e.g. findings from applied brain research were to be included, this would increase the efficiency of the training courses commissioned. This approach then leads to a different design of the trainings: materials can then be developed in an adapted way and further methodology can be applied. And the trainers themselves should also be sufficiently trained to be able to apply didactic principles in the teaching in addition to their typical specialist knowledge. Live online training, exploiting the possibilities of ICTs, will play an important part in this. Shorter, but more frequent and targeted training sessions can increase the impact of training activities, reaching a bigger audience.

FUTURE RESEARCH DIRECTIONS

This chapter presents the results of a collection effort of existing material and initiatives in organic agriculture training. Trying to identify synergies and increase the impact of the existing initiatives that are often distributed in a too scattered and isolated landscape. As stated before, one of the biggest risks in the future of organic agriculture training is the risk of "reinventing the wheel", maintaining parallel disconnected efforts to achieve the same goal. We have to acknowledge that the strong growth perspectives of organic farming globally is directly related with the growth of the demand organic farming training.

However, the possibility of having a "reinventing the wheel" scenario, with a lot of local disconnected activities without a common learning is still quite likely. The sector is still lacking enough investments with global perspective and relies on local, regional and national funding. This will jeopardize an efficient and synergistic use of the resources allocated for organic training development which could represent a huge drawback in the development of the sector.

To overcome this scenario, research, innovation and training initiatives in organic farming need to learn and exchange from the existing materials and initiatives. Future organic training need to avoid the temptation of developing ever new products. Itis rather key to ensure a synergistic learning from the initiatives developed in different parts of the world.

The future of research and innovation in organic training should aim to develop harmonized methods, tools and processes that can be applied in different contexts. At the same time they need to ensure the maximum mutual learning and knowledge exchange as well as identify and exchange best practices.

CONCLUSION

Based on the above mentioned issues we can draft the following recommendations for training and educational policies that can support the achievement of the SDGs.

1. It is key that public bodies offering agricultural training start or increase their offer of organic agriculture programs. Universities, vocational training institutes, extension and advisory services, even basic educational programs with agricultural components need to progressively integrate organic agriculture in their curricula. Governments need to invest resources of the public sector in educational and training interventions with an organic agriculture component. By increasing the participation of the public sector in organic agriculture training, this agricultural system will increase its political weight and its importance for the rest of society, increasing exponentially its

visibility. Due to the scientific evidence on the positive impact of organic agriculture, the increase of the organic agriculture sector will also support policymakers in the ambition to achieve of the SDGs.

2. To develop crop-specific training programs is a key task for public and private training providers in the field of organic agriculture. Addressing this identified gap is especially relevant for attracting conventional farmers to organic agriculture systems. The necessary transition to more sustainability needs to be developed by a big number of farmers who are used to manage highly specialized farms in conventional systems. Therefore, we believe that crop-specific trainings can become an effective intervention to attract conventional farmers to an organic production system. Moreover, the development of crop-specific training materials is also key for organic farmers. This will also help to better connect them.
3. The best practices presented in this chapter reflect some key aspects in organic training we consider very relevant for the replication in further training and educational initiatives worldwide. It is essential to ensure a process of knowledge exchange and mutual learning between existing organic farming training initiatives for the development of new organic training programs. In order to achieve effective and efficient training strategies, it is essential to build on the existing experiences in order to avoid duplication of efforts. Being traditionally excluded from mainstream agricultural training systems there is a considerable risk that new organic training providers (in)tend to "re-invent the wheel" if the above mentioned process of mutual learning is not granted. Hence, ensuring this learning process is fundamental for increasing the efficiency of investments in organic agriculture training.
4. Finally, from our assessment we derive that any organic farming training and educational policy need to also consider the potential of ICT for their deployment. In spite of the risks and potential challenges, we believe that its integration in organic training programs, in combination with traditional training strategies, can increase their efficiency as well as their accessibility. Yet, the positive impact of the integration of ICT should not be taken for granted and needs to be carefully planned to maximize it. However, the integration of these technologies with traditional training, together with public and private actors is a must for any forward looking training and educational policy in organic agriculture.

Organic farming is a dynamic sector with great potential to facilitate the transformation of the food system and to support the achievement of the SDGs. Due to the importance of the food system to pressing global challenges, such as climate change, it is of uttermost importance to give it visibility and priority in policy making at global level in order to increase the sustainability of our societies in general. This chapter has focused on the training and educational dimension of this transition and in bottlenecks that need to be addressed. Ambitious policies in this field need to be developed to contribute to getting our societies closer to achieving the SDGs.

REFERENCES

African Manual. (2020). *African Organic Agriculture Training Manual.* Available at: https://www.organic-africa.net/training-manual.html

Aiyelaagbe, I., Harris, P., & Olowe, V. (2015). Skills gaps in organic agriculture and SWOT analysis in higher educational institutions (HEIs) in Anglophone West Africa. *Organic Agriculture*, *6*(2), 109–118. Advance online publication. doi:10.100713165-015-0119-1

Altarawneh, M. (2016). Determine the barriers of organic agriculture implementation in Jordan. *Bulgarian Journal of Agricultural Science*, *22*, 10–15.

BMEL. (2019). Ökologischer Landbau in Deutschland. Bundesministerium für Ernährung und Landwirtschaft (BMEL).

Brasseur, G. P., Jacob, D., & Schuck-Zöller, S. (2017). *Klimawandel in Deutschland: Entwicklung, Folgen, Risiken und Perspektiven.* Springer. doi:10.1007/978-3-662-50397-3

Crowder, D., & Reganold, J. (2015). Financial competitiveness of organic agriculture on a global scale. *Proceedings of the National Academy of Sciences of the United States of America*, *112*(24), 7611–7616. doi:10.1073/pnas.1423674112 PMID:26034271

CTA. (2019). *The Digitalization of African Agriculture Report, 2018-2019*. CTA.

De Schaetzen, S. (2019). *Organic Agriculture and the Sustainable Development Goals – Part of the Solution*. EOSTA.

De Wit, J., & Verhoog, H. (2007). Organic values and the conventionalization of organic agriculture. *NJAS - Wageningen Journal of Life Sciences, 54*(4), 449-462. doi:10.1016/S1573-5214(07)80015-7

Dubelling, M., Wiskerke, H., & Moschitz, H. (2015). *Nachhaltige Lebensmittelversorgung in der Stadt.* RUAF Fondation und SUPURBFOOD-Projekt.

EIP-AGRI. (2020). *New Skills for Digital Farming.* Final Report. Available at: https://ec.europa.eu/eip/agriculture/en/publications/eip-agri-seminar-new-skills-digital-farming-final

El-Hage Scialabba, N. (2012). *Current situation of organic market access. In: Let the good products flow! Global organic market access in 2012 and beyond.* Available at: http://www.fao.org/3/an455e/an455e00.pdf

European Commission (2020). *Communication from the Commission to the European Parliament, the Council, the European Economic and Social Committee and the Committee of the Regions - A Farm to Fork Strategy for a fair, healthy and environmentally-friendly food system* (COM/2020/381).

FAO. (2017). *Greenhouse gas emissions from aquaculture – A life cycle assessment of three Asian Systems.* FAO Fisheries and Aquaculture Technical Paper 609.

FAO. (2020). *Sustainable Food and Agriculture.* Available at: http://www.fao.org/sustainability/en/

Francis, C. A., Jordan, N., Porter, P., Breland, T. A., Lieblein, G., Salomonsson, L., Sriskandarajah, N., Wiedenhoeft, M., DeHaan, R., Braden, I., & Langer, V. (2011). Innovative Education in Agroecology: Experiential Learning for a Sustainable Agriculture. *Critical Reviews in Plant Sciences*, *30*(1-2), 226–237. doi:10.1080/07352689.2011.554497

Freibauer, A., Rounsevell, M. D. A., Smith, P., & Verhagen, J. (2004). Carbon sequestration in the agricultural soils of Europe. *Geoderma*, *122*(1), 1–23. doi:10.1016/j.geoderma.2004.01.021

Gattinger, A., Müller, A., Haeni, M., Skinner, C., Fliessbach, A., Buchmann, N., Mäder, P., Stolze, M., Smith, P., Scialabba, N. H., & Niggli, U. (2012). Enhanced top soil carbon stocks under organic farming. *Proceedings of the National Academy of Sciences of the United States of America*, *109*(44), 18226–18231. doi:10.1073/pnas.1209429109 PMID:23071312

Guthman, J. (2004). *Agrarian Dreams: The Paradox of Organic Farming in California*. University of California Press. Retrieved November 19, 2020, from http://www.jstor.org/stable/10.1525/j.ctt1pnn9k.10

Haller, L., Moakes, S., Niggli, U., Riedel, J., Stolze, M., & Thompson, M. (2020). *Entwicklungsperspektiven der ökologischen Landwirtschaft in Deutschland.* UBA Texte 32/2020. https://www.umweltbundesamt.de/publikationen/entwicklungsperspektiven-der-oekologischen

Hülsbergen, K.-J., & Rahmann, G. (2015). Klimawirkungen und Nachhaltigkeit ökologischer und konventioneller Betriebssysteme - Untersuchungen in einem Netzwerk von Pilotbetrieben: Forschungsergebnisse 2013 - 2014. Johann Heinrich von Thünen-Institut.

IFOAM. (2017).*Guidelines for public support to organic agriculture.* Available at: https://www.ifoam.bio/sites/default/files/2020-05/policy_toolkit_main_report.pdf

IFOAM. (2020). *Organic Academy by IFOAM Organics International.* Available at: https://www.ifoam.bio/our-work/how/training

IPCC. (2014). *Mitigation of Climate Change. Contribution of Working Group III to the Fifth Assessment Report of the Intergovernmental Panel on Climate Change.* Cambridge University Press.

IPCC. (2019). *Special Report on Climate Change and Land – Summary for Policy Makers.* Available at: https://www.ipcc.ch/srccl/chapter/summary-for-policymakers/

Jung, R., & Schmidtke, K. (2019). Bodenfruchtbarkeit. In Leistungen des ökologischen Landbaus für Umwelt und Gesellschaft (Vol. 65). Braunschweig: Johann Heinrich von Thünen-Institut.

Kilelu, C. W., Klerkx, L., & Leeuwis, C. (2013). Unravelling the role of innovation platforms in supporting co-evolution of innovation: Contributions and tensions in a smallholder dairy development programme. *Agricultural Systems*, *118*, 65–77. doi:10.1016/j.agsy.2013.03.003

Knierim, A., Garster-Bentaya, M., Mekonnen Birke, F., Bae, S., & Kelly, T. (2020). *Innovation advisors for interactive innovation process: Conceptual grounds and common understandings*. i2connect project. https://i2connect-h2020.eu/resources/library/

Kroma, M. (2006). Organic Farmer Networks: Facilitating Learning and Innovation for Sustainable Agriculture. *Journal of Sustainable Agriculture*, *28*(4), 5–28. doi:10.1300/J064v28n04_03

Levin, K., Brandhuber, R., Freibauer, A., & Wiesinger, K. (2019). Klimaanpassung. In Leistungen des ökologischen Landbaus für Umwelt und Gesellschaft (Vol. 65). Braunschweig: Johann Heinrich von Thünen-Institut.

Morgan, K., & Murdoch, J. (2000). Organic vs. conventional agriculture: Knowledge, power and innovation in the food chain. *Geoforum*, *31*(2), 159–173. doi:10.1016/S0016-7185(99)00029-9

Moschitz, H., & Frick, R. (2020). City food flow analysis. A new method to study local consumption. *Renewable Agriculture and Food Systems*, *35*, 1–13. doi:10.1017/S1742170520000150

Moschitz, H., Muller, A., Kretzschmar, U., Haller, L., de Porras, M., Pfeifer, C., Oehen, B., Willer, H. & Stolz H. (2021). *How can the EU Farm to Fork strategy deliver on its organic promises? Some critical reflections.* (unpublished)

MSU. (2020). *Organic Farmer Training Program.* Available at: http://www.msuorganicfarm.org/organic-farmer-training-program.html

Muller, A., Schader, Ch., Scialabba, N. E.-H., Brüggemann, J., Isensee, A., Erb, K.-H., Smith, P., Klocke, P., Leiber, F., & Stolze, M. (2017). Strategies for feeding the world more sustainably with organic agriculture. *Nature Communications*, *8*(1), 1290. doi:10.103841467-017-01410-w PMID:29138387

Organic Farm Knowledge. (2020). *Organic Farm Knowledge Platform.* Available at: https://organic-farmknowledge.org/

OTA. (2020). *Organic Education Webinars.* Available at: https://ota.com/programs-events/organic-education-webinars

Ponisio, L. C., M'Gonigle, L. K., Mace, K. C., Palomino, J., de Valpine, P., & Kremen, C. (2015). Diversification practices reduce organic to conventional yield gap. *Proceedings of the Royal Society B: Biological Sciences, 282*(1799). 10.1098/rspb.2014.1396

Rosset, P. M., & Altieri, M. A. (1997). Agroecology versus input substitution: A fundamental contradiction of sustainable agriculture. *Society & Natural Resources*, *10*(3), 283–295. doi:10.1080/08941929709381027

Sanders, J., & Heß, J. (2019). Leistungen des ökologischen Landbaus für Umwelt und Gesellschaft (Vol. 65). Braunschweig: Johann Heinrich von Thünen-Institut.

Schader, C., Muller, A., Scialabba, N. E., Hecht, J., Isensee, A., Erb, K. H., Smith, P., Makkar, H. P. S., Klocke, P., Leiber, F., Schwegler, P., Stolze, M., & Niggli, U. (2015). Impacts of feeding less food-competing feedstuffs to livestock on global food system sustainability. *Journal of the Royal Society, Interface*, *12*(113), 20150891. doi:10.1098/rsif.2015.0891 PMID:26674194

Shukla, P. R., Skea, J., Slade, R., van Diemen, R., Haughey, E., Malley, J., Pathak, M., & Portugal Pereira, J. (2019). Technical Summary, 2019. In Climate Change and Land: an IPCC special report on climate change, desertification, land degradation, sustainable land management, food security, and greenhouse gas fluxes in terrestrial ecosystems. Academic Press.

Skalidou, D., & Oya, C. (2018). The challenges of screening and synthesising qualitative research in a mixed-methods systematic review. The case of the impact of agricultural certification schemes. *Journal of Development Effectiveness*, *10*(1), 39–60. doi:10.1080/19439342.2018.1438495

Smith, P., Martino, D., Cai, Z., Gwary, D., Janzen, H., Kumar, P., McCarl, B., Ogle, S., O'Mara, F., Rice, C., Scholes, B., Sirotenko, O., Howden, M., McAllister, T., Pan, G., Romanenkov, V., Schneider, U., Towprayoon, S., Wattenbach, M., & Smith, J. (2008). Greenhouse gas mitigation in agriculture. *Philosophical Transactions of the Royal Society of London. Series B, Biological Sciences*, *363*(1492), 789–813. doi:10.1098/rstb.2007.2184 PMID:17827109

Tubiello, F. N., Salvatore, M., Rossi, S., Ferrara, A., Fitton, N., & Smith, P. (2013). The FAOSTAT database of greenhouse gas emissions from agriculture. *Environmental Research Letters, 8*. doi:10.1088/1748-9326/8/1/015009

UN. (2020a). *Sustainable Development Goals.* Available at: https://sustainabledevelopment.un.org/sdgs

UN. (2020b). *Communication material Sustainable Development Goals.* Available at: https://www.un.org/sustainabledevelopment/news/communications-material/

UNFCCC. (2019). *Climate Change Threatens Progress Across Sustainable Development.* https://unfccc.int/news/climate-change-threatens-progress-across-sustainable-development-warns-new-un-report

Weckenbrock, P., Sanchez-Gellert, H. S., & Gattinger, A. (2019). Klimaschutz. In Leistungen des ökologischen Landbaus für Umwelt und Gesellschaft (Vol. 65). Braunschweig: Johann Heinrich von Thünen-Institut.

Wenger, E. (1998). *Communities of practice: Learning as a social system* (Vol. 9). Cambridge University Press. doi:10.1017/CBO9780511803932

Wheeler, S. (2008). The barriers to further adoption of organic farming and genetic engineering in Australia: Views of agricultural professionals and their information sources. *Renewable Agriculture and Food Systems*, *23*(2), 161–170. doi:10.1017/S1742170507002128

Willer, H., Schlatter, B., Travnicek, J., Kemper, L., & Lernoud, J. (2020). The World of Organic Agriculture. Statistics and Emerging Trends 2020. Research Institute of Organic Agriculture (FiBL), Frick, and IFOAM – Organics International, Bonn.

KEY TERMS AND DEFINITIONS

Capacity Building: Capacity-building is defined as developing and strengthening skills and competencies, not only at an individual level but also at institutional and systemic level. It maximized the ownership and included behavioral change and ability to adapt to changes and participation.

Digital Learning: Digital learning is a practice of learning using technologies in an effective way, combining different elements such as blended or virtual learning using e.g. mobile technologies or e-learning. It requires a combination of technology, digital content, and instruction.

ICT: Information and communications technology (ICT) is commonly defined as an extensional term for information technology (IT) stressing the integration of telecommunications as well as adequate software and audiovisual systems. With these technologies, information can be accessed, stored, and transmitted.

Innovation Platform: An innovation platform can be defined as a multi-actor configuration with the purpose to facilitate and undertake various activities around identified innovation challenges and opportunities.

Peer-to-Peer Network: Peer-to-peer networks involve individuals sharing knowledge and experience with each other. They focus on the fact that learning takes place between individuals, facilitating and encouraging interpersonal exchange.

Sustainable Development Goals: The Sustainable Development Goals (SDGs), were formulated and adopted by the United Nations Member States in 2015. They set 17 integrated goals to end poverty, protect the planet and ensure that all people enjoy peace and prosperity by 2030. At the core is that any development must balance the three pillars of sustainability, social, economic, and environmental sustainability.

Section 3
Teaching Climate Change

Chapter 11
Students' Understanding of Climate Change in Prone Areas Exposed to the Impact of Climate Change:
Case Study of Lembata and North Timor Tengah Districts, Indonesia

Bertha Anggraini Safitri
IPB University, Indonesia

Perdinan
IPB University, Indonesia

Ika Rachmawati Suratno
Deutsche Gesellschaft für Internationale Zusammenarbeit, Indonesia

ABSTRACT

Information about climate change is useful for children to anticipate the potential impacts of climate change. However, until 2016 in Lembata and North Timor Tengah Districts, Indonesia, which is the vulnerable area exposed to the effects of climate change, it was found that children's understanding is still low, even though 98% claim that they can feel some changes in their environment. The result of this research were obtained by collecting data from student from both regions and were proceeded to find climate change awareness index (CAI). It showed that most students (89%) have more knowledge about climate change, and the rest (11%) do not understand. This condition is influenced by facilities and infrastructure factors, access to books and internet, teacher quality, and gender. It is also known that information about climate change in Indonesian textbooks have not reached 50% according to UNICEF recommendations to provide students with good understanding of climate change.

DOI: 10.4018/978-1-7998-7512-3.ch011

INTRODUCTION

The science of climate change has been growing in recent years. Research conducted based on direct observations, climate models, and historical sources has concluded that human activity contributes to changing climate conditions on earth that have an impact on future generations, ranging from rising sea levels, rising temperatures, and increasing the frequency of natural disaster events. According to NOAA (2016), 2015 was the hottest year since 1880, when global temperature increases were known to begin to occur (Currie & Deschenes, 2016). Global temperatures continue to increase in the 20th century along with greenhouse gas concentration exceed the level of natural climate system variability (Crowley, 2000).

According to Oppenheimer and Anttila-Hughes (2016), most parties believe that humans are the main cause of climate change. Humans emit 35 billion metric tons of carbon per year. While naturally, oceans and land, including organic and inorganic materials, produce 20 times that amount, but the increase tends to be constant because it is naturally absorbed by the ocean and through photosynthesis. Human interference makes the carbon cycle that occurs becomes unbalanced.

In fact, greenhouse effect is one of the important factors that support life on earth. One of the most talked about greenhouse gases is carbon dioxide gas. Svante Arrhenius in 1886 showed that increasing concentrations of carbon dioxide in the atmosphere resulted in higher temperatures on earth known as the greenhouse effect. The temperature increase was considered beneficial than a problem at that time (Oppenheimer & Anttila-Hughes, 2016). At present, the increase in carbon dioxide has reached 40 percent from the preindustrial period due to mining and fossil fuels burning, deforestation, agricultural activities that emit greenhouse gases, and industrial waste (Oppenheimer & Anttila-Hughes, 2016).

Climate change events have different impacts in different places. However, the changes will affect all humans and life on earth. The carbon dioxide produced in one country is different from other countries, but the longevity of carbon dioxide in the atmosphere makes it able to spread and increase the concentration of carbon dioxide globally so that it can contribute to changing the climate in other countries and the impact is global. The changes that humans will feel include an increase in average temperature and increasing frequency of extreme temperatures. This causes more cases of death and illness, reduces the children ability to learnn, and affects the adults mentality(Givin & Shrader, 2016). The phenomenon of climate change also affects children's rights related to health, education, and security (Gibbons, 2014).

This research was conducted to find out the rate of student's understanding about climate change phenomena and how much information about climate change conveyed precisely in their school handbooks.

BACKGROUND

Climate change is now being the one of environmental issues that almost discussed in any forum. The impacts of climate change are now becoming everyone's attention because it can affect the life sustainability on Earth. Some methods have been discussed to reduce the impacts of climate change. Children are physically more vulnerable than adults to the direct impacts of natural disasters caused by climate change (Currie & Deschenes, 2016). In addition, the indirect impacts of climate change on children can hold them to reach their maximum potency due to some diseases caused natural disaster or conflict.

Based on research of Leiserowitz *et al.* (2011), children's knowledge about climate change in United States still lower than adults. Less than 1/5 children get some good information about climate change. 73% children in Junior and Senior High School surf the internet to get information about climate change and only 27% get the information about climate change, specifically in global warming, and deep learning in formal education in their school.

Students may get some knowledge about climate change from formal and nonformal education. Formal education that mustbe taken by Indonesian children has a main role to give proper information about climate change. But the reality is that educations about climate change cannot be delivered to them caused by many factors. One of the factors that can support them to increase their understanding about climate change are some handbooks that can help teacher to give information. Teachers should be helped by giving them number of climate change handbooks so they can provide climate change information, such as the definition, causes, impacts, and example of mitigation and adaptation actions.

Chen (2017) on his research found that people have lack understanding aboutclimate change phenomena because they know climate change just as an object. Meanwhile, to have a good understanding, people should know climate change both as an object and as a process. This information can be applied to some school subjects based on formal education curriculum in Indonesia.

East Nusa Tenggara Province is one of regions with low rate of information dissemination. Meanwhile, its geographic profile with many small islands and borders directly with Indian Ocean, making it has more vulnerability to the impacts of climate change, so that the population of this Province must better understand about climate change. Based on the data of Direktorat Jenderal Pembangunan Daerah Tertinggal (2016), Lembata is one district in East Nusa Tenggara Province in Indonesia and an island in a group islands named Solor Islands. Lembata District is located at 8° 10' - 8° 11' S dan 123° 12' - 123° 57' E. The geographical conditions of Lembata consists of coastal, hills, and mountain areas with the high peak on Labalekang Mountain (1.613 meters above sea level), Uyelewun Mountain (1.513 meters above sea level), and Ile Ape Mountain (1.450 meters above sea level). The slope varies between 5% to 40%. Lembata District has a tropical climate and generally as a dry area with rainfall 500-1200 mm. Total population in Lembata in 2012 is 124,912 in an area of 1266.48 km^2.

North Timor Tengah located between 9° 02' 48" - 9° 37' 36" S and 124° 04' 02" - 124° 46' 00" E. North Timor Tengah classified with D type of climate based on Schmidt-Ferguson climate classification. Based on Koppen classification, North Timor Tengah has type A climate. Total population in North Timor Tengah was about 229,000 at 2010 in an area of 2,669.7 km^2 in the form of land with 50 km coastline based on the data of Direktorat Jenderal Pembangunan Daerah Tertinggal (2016).

METHOD

This research was conducted in Lembata and North Timor Tengah Districts. Both districts were selected because of their main economic sectors. Their main economic sector in Lembata is fisheries while in North Timor Tengah is agriculture which are sensitive to the impacts of climate change. This research and data processing was carried out from February until May 2017 at the Climatology Laboratory in Geophysics and Meteorology Department of Bogor Agricultural University and PI Area Bogor.

Data were collected directly with giving questionnaire to 57 students in Lembata and 76 students in North Timor Tengah which selected randomly. The data covers student's knowledge about the definitions of climate change in general, the causes of climate change, the impacts of climate change, and some actions that they can do with climate change phenomena.

Analysis for Student's Understanding to Climate Change Phenomena

Every question in questionnaire has their value with the score of answer between 0 until 5. The value of Climate Change Awareness Index (CAI) is counted by this formula (Rahman *et al.*, 2014):

$$\text{CAI}(\%) = \frac{\sum_{i=1}^{n}(R_i \times W_i)}{\sum_{i=1}^{n}(R_m \times W_i)} \times 100 \cdot$$

R_i is the score of answer for question i, W_i is the value for question i, and R_m is the maximum score for question i (Rahman *et al.* 2014). Final score of CAI is between 0-100% that shows the percentage of understanding with 0% shows the minimum of understanding and 100% is the maximum rate of understanding. The value of questions in questionnaire is based on the research of Rahman *et al.* (2014) with some modifications and adjustments by the questions.

Analysis of The Effect of Social-Economic Factors to Student's Understanding

The score of CAI analysed by 2-sample t-test to view the average differentiation between two groups of independent variables. Independent sample t-test or 2-sample t-test is one of statistical test that used to compare the average from two unpaired groups, which mean the analysis is conducted for two sample subjects or different independent variables.

Table 1. Value of P interpretations

No.	Nilai P > α	Nilai P < α
1.	Accept H_0	Decline H_0
2.	Sample variation is evident	Sample variation is not evident
3.	μ_1 is not different significantly with μ_0	μ_1 is different significantly with μ_0

Source: Sugiyono, 2007

Zero Hypothesis (H_0) which used in this research is that there is no significant different between the average of CAI at the different social-economic condition with value of $\alpha = 0.05$. While the Alternative Hypothesis (H_1) that there is some significant different between the average of CAI. And μ represent to the average of CAI.

CAI score also were analysed with double regression method to find theeffect of social-economic factors to student's understanding. Social economics that are being analysed are parent's occupation, distance from domicile to capital city, academic level, and gender.

Simple regressions are based on to the cause-impact relation between dependent variable and independent variable with the simple formula:

$\hat{Y} = a + bX$. $\hat{Y}$.is the dependent variable, a .is intersect $\hat{Y}$.when $X=0$. b is the coefficient of regression, and X .is the independent variable. The value of *a* and *b* can be found with this formula:

$$a=\frac{(\Sigma Y_i)(\Sigma X_i^2)-(\Sigma X_i)(\Sigma X_iY_i)}{n\Sigma X_i^2-(\Sigma X_i)^2}\ .\ b=\frac{n\Sigma X_iY_i-(\Sigma X_i)(\Sigma Y_i)}{n\Sigma X_i^2-(\Sigma X_i)^2}\ .$$

Regression formula for 4 predictors are below:

$\hat{Y} = a + b_1X_1 + b_2X_2 + b_3X_3 + b_4X_4$ .The value b_1, b_2, b_3, and b_4 can be found by this formula:

$$\Sigma X_1Y=b_1\Sigma X_1^2+b_2\Sigma X_1X_2+b_3\Sigma X_1X_3+b_4\Sigma X_1X_4 \quad . \quad 2 \quad .$$
$$\Sigma X_2Y=b_1\Sigma X_1X_2+b_2\Sigma X_2^2+b_3\Sigma X_2X_3+b_4\Sigma X_2X_4 \quad . \quad 3 \quad .$$
$$\Sigma X_3Y=b_1\Sigma X_3X_1+b_2\Sigma X_2X_3+b_3\Sigma X_3^2+b_4\Sigma X_3X_4 \quad . \quad 4 \quad .$$
$$\Sigma X_4Y=b_1\Sigma X_4X_1+b_2\Sigma X_4X_2+b_3\Sigma X_3X_4+b_4\Sigma X_4^2$$

.And a can be found by this formula: 1.

$$a=\bar{Y}-b_1\overline{X_1}-b_2\overline{X_2}-b_3\overline{X_3}-b_4\overline{X_4}\ .$$

Parent's occupation and gender variables are qualitative data, so to do the regression, we use dummy variable with 0 and 1. For gender, 0 value for female students and 1 for male students. For parent's occupation, 0 is for entrepreneurial occupation such as the farmers and fisherman, and 1 is for non-entrepreneurs such as government employees and general employees.

RESULT

Student's Understanding to Climate Change Phenomena

Research conducted by Oppenheimer and Anttila-Hughes (2016) shows that children will face the extreme temperature in the future. Higher temperature has a direct impact on their health. It also has indirect impact by affecting agricultural system, inflation, and the increasingfood scarcity. Climate change is also predicted to increase the frequency of natural disasters which have dangerous impact for children. Not only the physical impact, it also has long-term indirect impact for children such as ecosystem damage. Based on this theory, children's understanding for climate phenomena can be important for them to face the future impacts.

Research conducted by Perdinan *et al.* (2016) in Lembata and North Timor Tengah Districts shows that most of Junior High School and Senior High School in these regions are familiar with and understand about climate change, 37% are familiar but do not quite understand climate change, and 3% are not familiar to climate change. The proportion shows in Figure 1. Meanwhile, almost all students can feel the changes of their environment conditions and only 2% cannot feel the change. However, 24% of them claim that they do not know that it is related to climate change, and 2% claim that it is not related with climate change. It shows that there are some students with low level knowledge about climate change even have incorrect understanding.

Student's understanding about climate change which measured by Climate Change Awareness Index (CAI) shows that respondent's understanding is still low in Lembata and North Timor Tengah Districts

Figure 1. Distribution of children's knowledge to climate change distribution (Perdinan et al., 2016)
Source: Perdinan et al., 2016

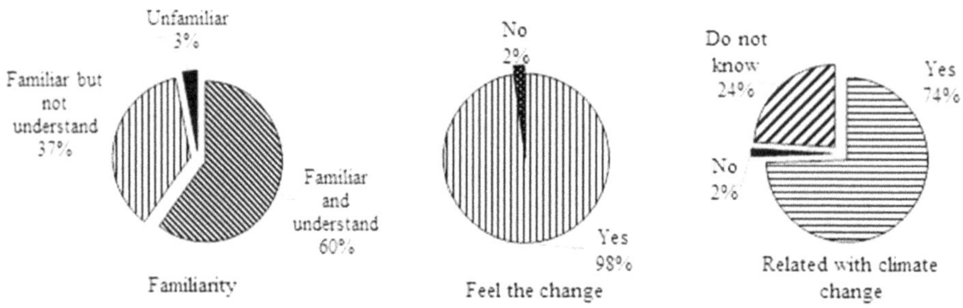

as shown in Table 2. 11% of respondents have CAI score between 0 until 40 which classified by Ochieng & Koske (2013) as a low understanding level. Meanwhile, there are no respondents who have a good level of understanding and 89% or most respondents have an intermediate level of understanding. It shows that student's understanding about climate change are onlyhave general information. Student's understanding in Lembata and North Timor Tengah Districts should be improved to face climate change impact in futureto do climate change adaptation, mitigation, and prevention actions.

The Influences of Social-Economic Factors

Social-economic factors like gender, academic level, and family's income can influence student's understanding to climate change phenomena based on the research of Rahman *et al.* (2014) and Pitpitunge (2013). T-test that have been done shows that student's understanding will be different in the different socioeconomic condition. Figure 2 compares the CAI score distribution of respondents in the tworegions. Students in Lembata District have a bigger variation of distribution with the smallest score is 29% and the biggest at 80%, meanwhile in North Timor Tengah have smaller variation between 50-80%. The P-value for regional factor is 0.000, below $\alpha = 0.05$. It shows that it has a significant difference in the average of CAI score between the regions with the average at North Timor Tengah is bigger than Lembata District.

Regression based on student's data shows the relation of parent's occupation, academic level, distance to capital city, and gender with CAI are following the formula bellow:

CAI = 80,1 + 0,01 parent's occupation + 1,14 academic level - 0,144 distance to capital city - 9,18 gender

With the following graphic and the significance:

Table 2. Distribution of respondents' CAI scores

Level of Understanding	CAI	Frequency	Percentage (%)
Good	80 - 100	0	0
Middle	41 - 79	118	89
Low	0 - 40	15	11
Total		**133**	**100**

Figure 2. Residual plot CAI as the result of Minitab analysis

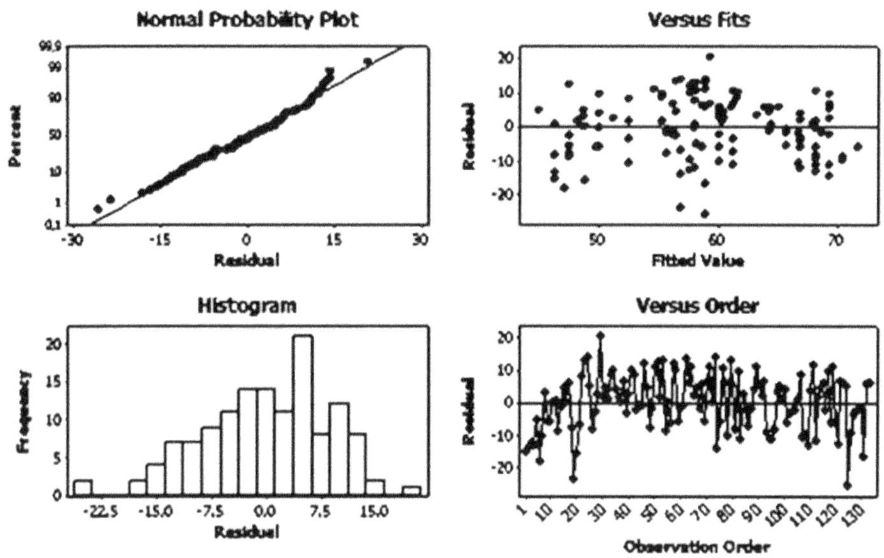

The Histogram and Normal Probability Plot at Figure 2 show that data are distributed following normal distribution. Versus Fits shows that data distributed above and below 0 without following a certain pattern. VIF value of all variables are under 10 are shown inTable 3. All of analysis above are the important thing to do the regression and its fulfilled at this regression.

Parent's occupation factor has coefficient of 0.01. the positive value means that students with parents as entrepreneurs have a higher understanding level than students with parents who work in non-entrepreneur sectors. Figure 3 shows the difference of CAI scores based on parent's occupation. Parent's occupation factor has been used to represent student's welfare as one of the factors that can influence their access to climate change information. Students with parents who work in non-entrepreneur sectors are considered to have prosperity with steady income. But the result for this factor should be reviewed in future because of the significant difference at the quantity of respondent's parents who work as entrepreneur and as non-entrepreneur with the P-value is 0.273, above of $\alpha = 0.05$ and P-value of regression in 0.998 shows that parent's occupations are not significantly influence student's understanding.

Based on academic level, Senior High School students have higher CAI score than Junior High School with significant difference shown by P-value = 0.000 in Figure 4. It shows that Senior High School

Table 3. Relation and the significance of independent variable in regression formula

Independent Variable	P-value	Significance	VIF	Multicollinearity
Parent's occupation	0.998	Non-significant	1.11	VIF < 10 No Multicollinearity
Distance to capital city	0.000	Significant	1.05	
Academic level	0.018	Significant	1.17	
Gender	0.000	Significant	1.01	

Figure 3. The difference of CAI distribution by respondent with parents who work as entrepreneur (a) and non-entrepreneur (b)

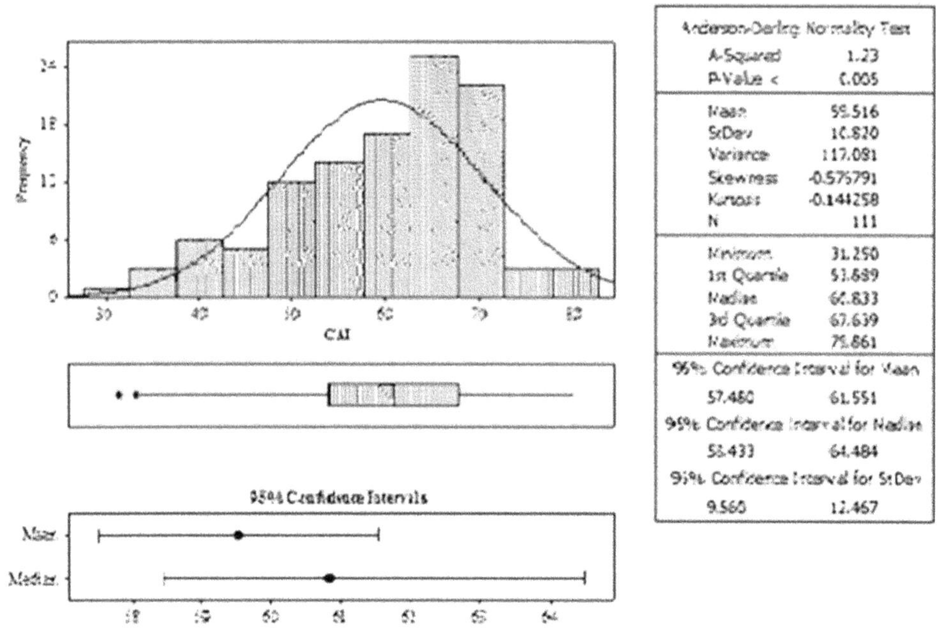

(a)

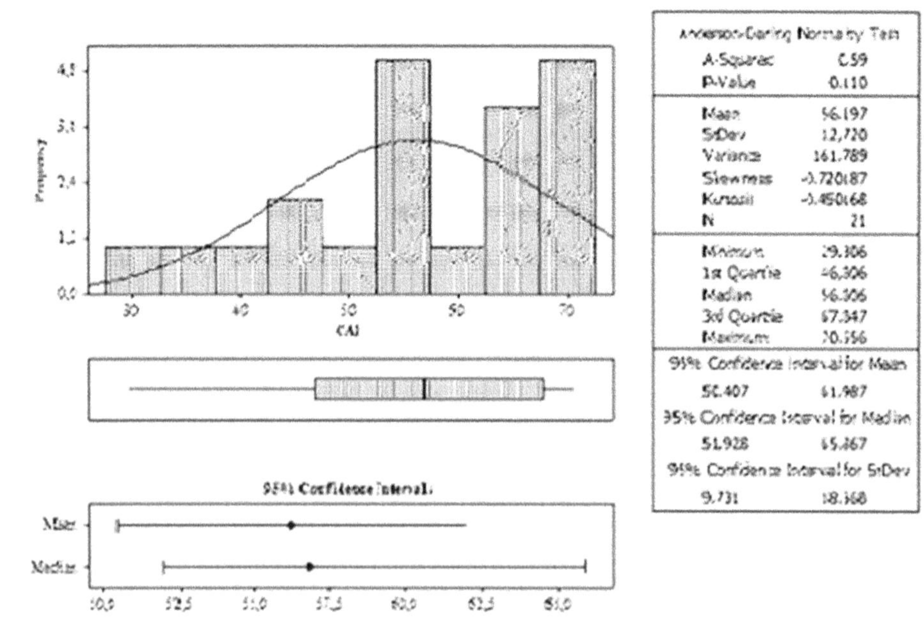

(b)

students have better knowledge of climate change than Junior High School Students in accordance with amount of climate change information that delivered at each level.. CAI scores of Junior High School students distributed at 30-70% while Senior High School students has scores at 35-80%. The Histogram and Boxplot at Figure 4 show more Senior High School students have higher CAI scores with less distribution. It means that their understanding of climate change is better and spread more evenly than Junior High School students.

North Timor Tengah District has a closer distance to Kupang City as the capital city of East Nusa Tenggara Province. Meanwhile, Lembata is further and separated by Sawu Sea. Relation of the distance to capital city produce the coefficient of -0.144. The negative value means that CAI scores decreases by the farther distance. The distribution of CAI score in both districts by Figure 5 shows less variation means student's understanding in North Timor Tengah are more evenly distributed with higher level of understanding than Lembata District. It indicates that the distribution of information about climate change are not evenly distributed by region. Students with farther region from capital city have less access to the information. The differences in facilities and infrastructure, access to books and internet, or teacher's quality are some factors should be improved to increase student's understanding.

Another socio-economic factor that influences student's understanding to climate change is gender. T-test have been carried out to show that the average of CAI score between male and female students is significantly different with P-value of 0.000. CAI score for female students shown in Figure 6 are higher than male's score. It is related to the research conducted by Ghazvini and Khajehpour (2011) with the result that female students are better in concept and knowledge than male. Meanwhile, male students easier to understand logical things and mathematics. The difference caused by gender influence in locus of control, self-concept, and learning strategy. Therefore, the right delivery method needs to be improved to increase student's understanding to climate change to anticipate the impact of climate change at the future.

P-value at Table 3 shows that variable of academic level, the distance to capital city, and gender can influence student's understanding to climate change with the value under critical point at 0.05. The coefficient of determination (Adjusted R^2) from the regression formula is 36.8% which means 36.8% of CAI score is influenced by the factors above and it has 63.2% influenced by the other factors which not discussed in this research.

Introduction to Climate Change for Children

As a generation that will face the impact of climate change in the future, children should be given the right information and knowledge about climate change. Figure 7 shows that most of students in Lembata and North Timor Tengah get information and knowledge about climate change from media and school. It means that formal educations are one big source to distribute the information and knowledge about climate change for children. Figure 8 shows 22% students state that no issued in climate change delivered in lessons at school, 43% do not know, and 35% students state that there are some subjects discussing about climate change issues.

NOAA (2016) reveals that introduction to climate change for children must be adjusted to the age of children so they do not become overwhelmed due to too much information. UNICEF (2011) states that elements of climate change knowledge at formal subjects for children be adjusted by classifying some subjects below.

Figure 4. The difference of CAI scores distribution for Junior High School (a) and Senior High School (b) students

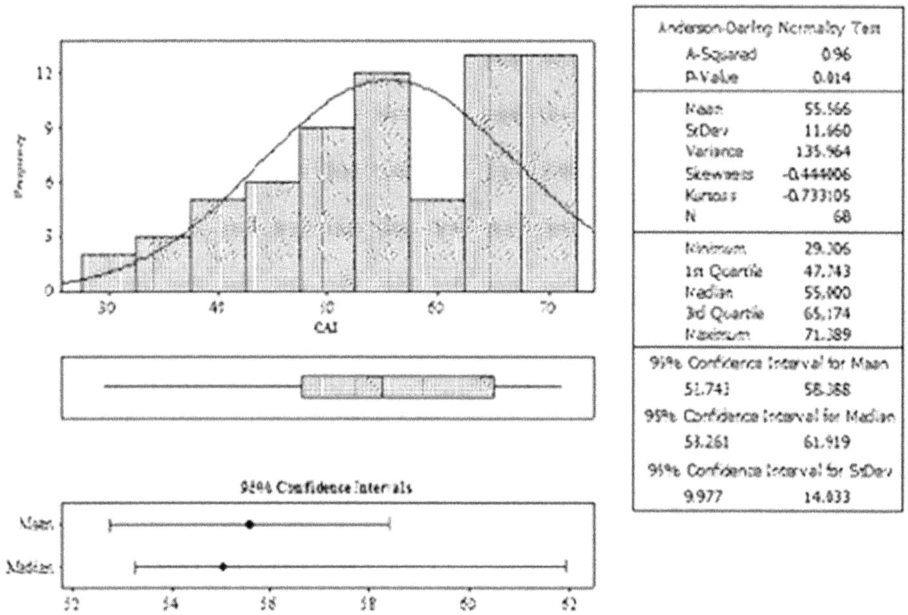

(a)

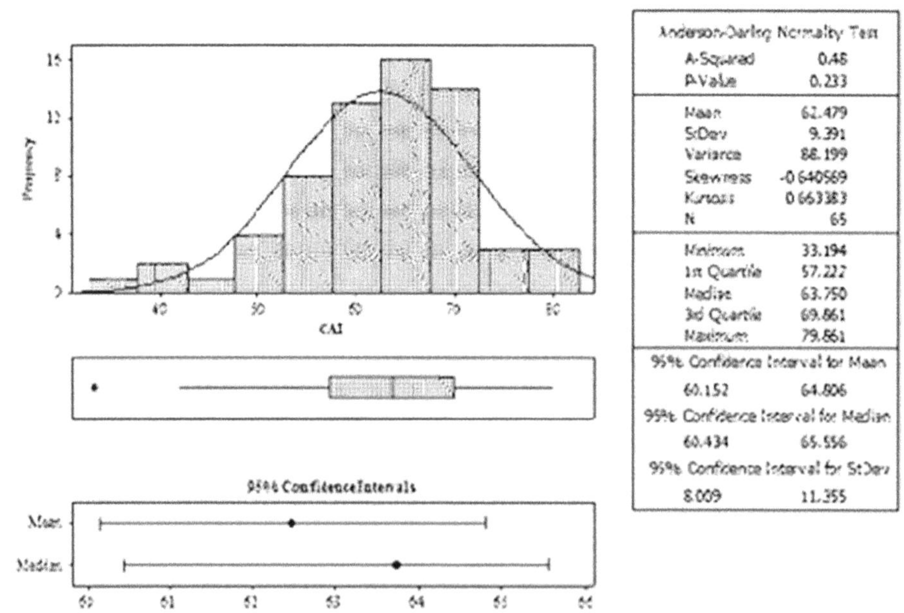

(b)

Figure 5. The distribution of CAI scores in Lembata (a) and North Timor Tengah (b)

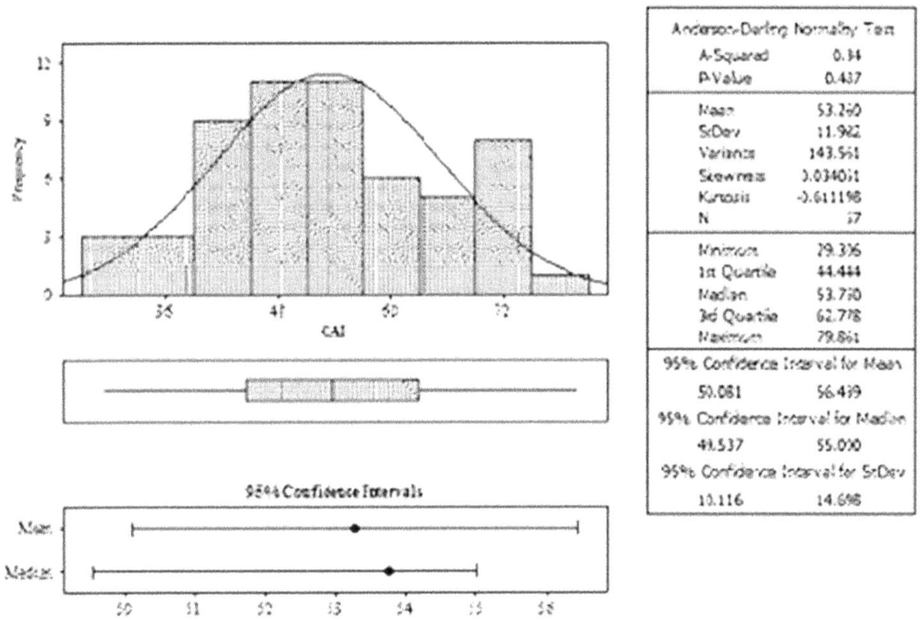

(a)

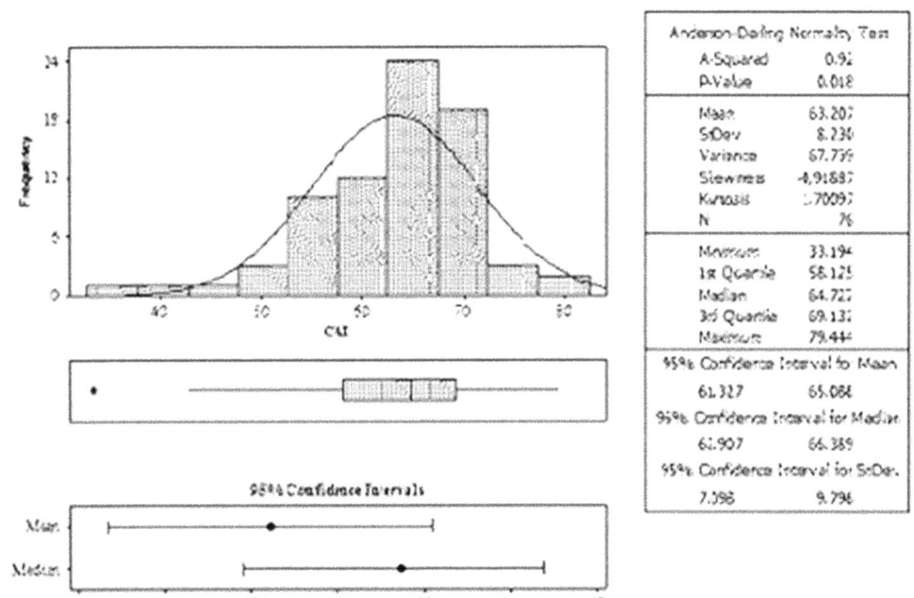

(b)

Figure 6. The difference of CAI scores distribution at male (a) and female (b) students

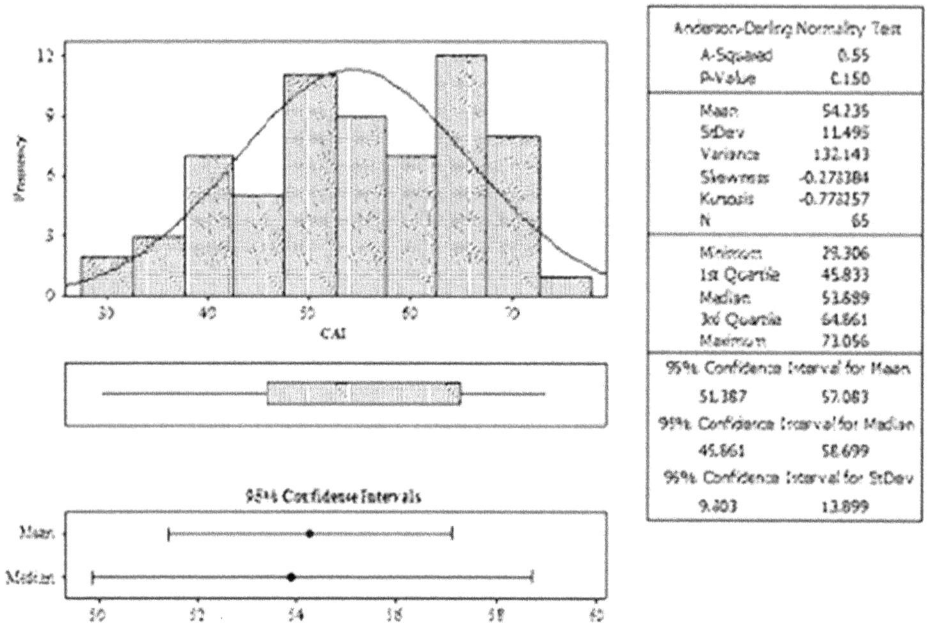

(a)

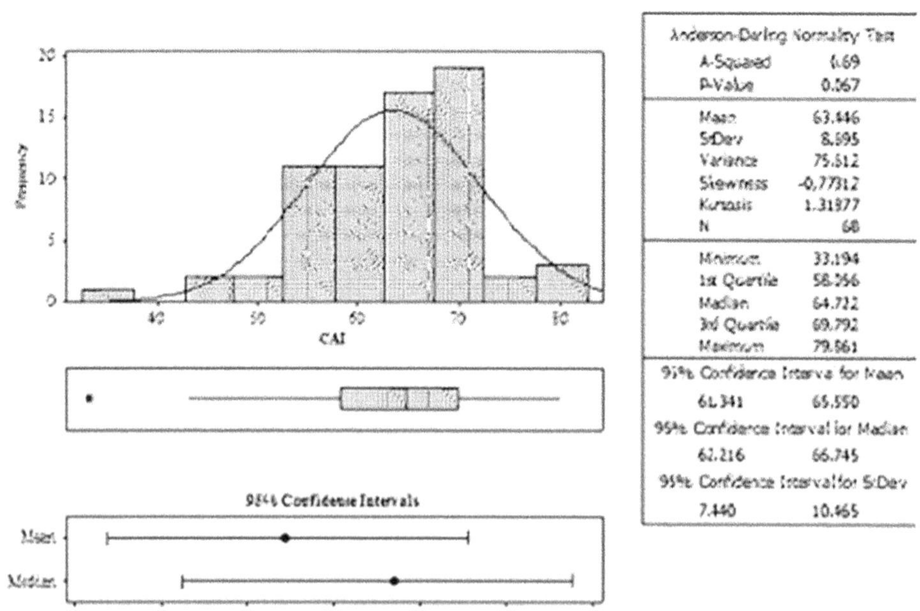

(b)

Figure 7. Percentage of source of climate change information distribution (Perdinan et al., 2016)
Source: Perdinan et al., 2016

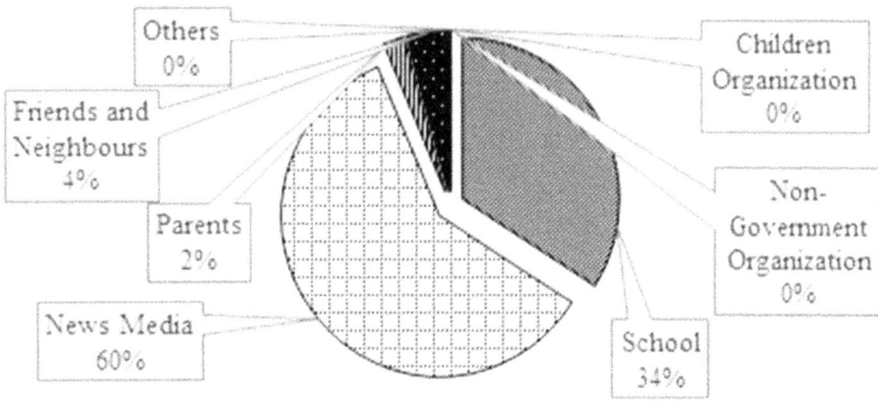

Junior High School subjects in the 2013 Curriculum are divided into two groups, namely Group A and Group B. Group A consists of subjects of Religious Education and Character, Pancasila and Citizenship Education, Indonesian Language, Mathematics, Natural Sciences, Social Sciences, and English which emphasize cognitive and affective aspects. Meanwhile, Group B consists of subjects in Culture and Art, Physical Education, Sports and Health, and Workshops, including local content, which places

Figure 8. Information about climate change in curriculum (Perdinan et al., 2016)
Source: Perdinan et al., 2016

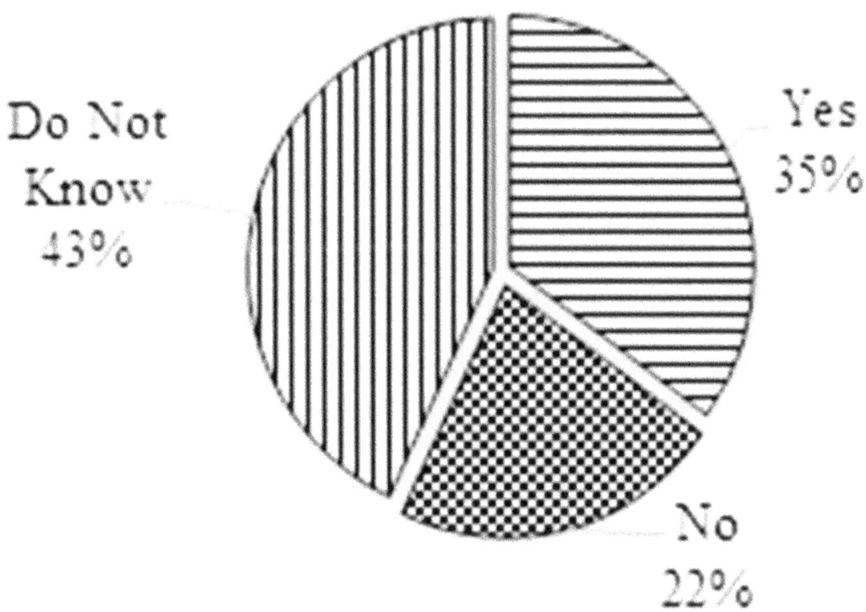

Table 4. Introduction to climate change at school subjects

Subjects	Climate Change Elements Delivered
Language	The definitions, causes, effects, and action can be done with reading and writing activities
Natural Science	The definition, causes, and effects to the environment
Geography	The impacts of climate change to the world, population with biggest impact, and sources of the main causes of climate change
Civic Education	Children's role as the agent of change for the world in future

Source: UNICEF, 2011

more emphasis on affective and psychomotor aspects. Natural Sciences and Social Sciences courses in the 2013 curriculum were developed as integrative science and integrative social studies subjects, not as disciplinary education. Both focus more on the application of science, the development of thinking skills, learning abilities, curiosity, and the development of caring and responsible attitudes towards the social and natural environment.

High school subjects are divided into compulsory subject groups, specialization subjects, cross-interest elective subjects, and deepening subjects. Subject groups must be divided into two groups, namely Group A and Group B. Group A consists of Religious Education and Character, Pancasila and Citizenship Education, Indonesian Language, Mathematics, Indonesian History, and English. While Group B consists of Arts, Culture, Physical Education, Sports, and Health, as well as Workshops and Entrepreneurship. Group A is a subject that gives more competency orientation to cognitive and affective aspects while Group B places more emphasis on affective and psychomotor aspects. Specialization in high school consists of three groups, namely Specialization in Mathematics and Science, Social Specialization, and Specialization in Language. The subjects studied in this study focus on compulsory subject groups, specialization in Mathematics and Science (consist of Mathematics, Biology, Physics, and Chemistry) and Social Specialization (consist of Geography, History, Sociology and Anthropology, and Economics subjects).The matrix of student's handbooks with National Curriculum 2013, there are some subjects with information about climate change in some topics shown in Table 5 with checklist marked means that the elements of climate change information are conveyed in the handbooks. It shows that CAI score for grade 9 at Junior High School is 39%. The conformity with UNICEF recommendation at the handbook for grade 9 at Senior High School in Natural Science is smaller with score of 22%. Handbooks for grade 12 Senior High School in Social Science have the highest conformity, but still only 44%. It shows that the handbooks does not comply with UNICEF recommendation.

The conformity of climate change elements in school handbooks with UNICEF recommendation amounted by adjusted CAI. Environment and climate issues discussed in Junior High School handbooks at Indonesian Language, Natural Science, and Social Science subjects. Environmental damage, human activities that cause environmental damage, impacts, until the action that must be done to minimize and prevent environmental damage described both directly and indirectly, but it less introducing about climate change. By the Indonesian Language subjects, environmental knowledge most describe directly with reading ang analysing some text. Several types of environmental damage also directly explained by Natural Sciences subjects in associated with human life and activities, both causes and impacts, also the action they must take. Furthermore, global warming information can be found at the inserted text, although its only as additional text beside the main text, the information is quite informative and can

Table 5. Elements of climate change in school handbooks

Grade	Subjects	Environmental Damage				Climate and Climate Change				CAI
		Definition	Cause	Impact	Action	Definition	Cause	Impact	Action	
9 Junior High School	Indonesian	v	v	v	v					39
	Natural Science		v	v	v					
	Social Science	v			v	v				
12 Senior High School (Natural Science)	Indonesian	v			v					22
	English					v				
	Biology					v				
	Physics								V	
12 Senior High School (Social Science)	Indonesian	v			v					44
	English					v				
	Geography					v				
	Sociology					v	v	V	V	

bring students curiosity to climate change in general and especially about global warming. Meanwhile, Social Sciences subjects most introducing climate commonly.

Indonesian Language subjects at grade 12 of Senior High School only explain the action to protect the environment, while in English subjects examine meteorological hazards on the topic "Read the Nature", such as storm, tornado, and elements such as clouds, wind, lightning, and air pressure. Language subjects in this grade still not enough to discuss climate change which must be maximized with includingsome text in the handbook. Students with a specialization in Natural Science get some topics with climate elements can influence growth and development in Biology and the action that they can do to reducecarbon dioxide emission in Physics. Meanwhile, students with specialization in Social Sciences get most climate change information in Geography and Sociology. Climate knowledge can be found in the Spatial Interaction of Villages and Cities topics in Geography subjects. Sociology subjects examine the environment and its influence the human life. There are also topic with climate change information in causes, impacts, and the actions they can take. Unfortunately,here are no climate change or environmental damage information at Civic Education, both in grade 9 and grade 12. The information in the handbooks still a general information and has not be applied to the subject types based on UNICEF recommendations. Therefore, it must be improved and enriched with some detailed information include the application in daily activities besides the basic information such as the different between climate and weather (NOAA, 2016) that children must know. It can help students to absorb the information because it relates to their daily activities. The next important thing is to introduce children to take some actions to delay global warming and minimize its impacts. Complex information such as greenhouse effect can be informed at the higher grade. The important point is that children get some information that some gas can trap the heat and causes global warming. They also need to know the source of that gas. The information must be adjusted with subjects and the curriculum.

Insufficient understanding about climate change for the students leads to differences between CAI studentin 26-76% and CAI handbook in 21-44%. It can occur due to several probabilities. First, some students use other handbooks beside the main books from Ministry of Education and Culture published in 2015. Second, students get the climate change information besides the formal education, such as internet and other media which are explained in Figure 6 because climate change information from the handbooks are not enough. Lack of student's understanding can be caused by the insufficient information delivered at both regions, also the information from formal education which one source that must be accessible and reliable. In addition, student's knowledge will increase along with the increase in information received at the previous grade.

CONCLUSION

Student's understanding of climate change phenomenon in Lembata and North Timor Tengah Districts is not good enough based on their CAI scores which is under 80%. Most of student's understanding is only in the intermediate level with the general knowledge of climate change.Children in both districts are not only must have good awareness and understand the phenomena of climate change to anticipate the impacts in future, but they should also have the climate change preventive knowledge.

Social-economic condition/ factors of students can influence student's understanding in climate change. It shows with the significant difference at the average of CAI at some social-economics condition. Social-economic factors such as the distance of their domicile to the capital city can influence their access to climate change information. Significant difference in CAI scores between students that living close to the capital city with the farther shows that information distributions are not spread evenly. Psychology factors based on gender also causes understanding level in climate change between male and female students. Academic level and source of information also can influence student's understanding level. School's role as the main source of information make the formal education has responsibilities to the student's understanding, especially in climate change which have the effects for their future. However, climate change information in school handbooks only covers 22-44% from UNICEF recommendation. The difference between student's CAI and handbook's CAI may shows the probability that students use another book besides the main reference handbooks from Education and Cultural Ministry with more climate change information or school handbooks are not the only source of information. Students can get more information apart from school handbooks.

Student's understanding to climate change phenomena can be improved by the social-economic factors and the source of information. The improvement in social-economic factors can be done by increasing student's access to climate change information and fixing the uneven information distributions problem by doing improvement of facilities and infrastructures, improving the access of students to books and the internet, and increaseing the teacher's quality and equiping them with good knowledge in climate change. Besides, it may need to research the influence of student's prosperity. By the sources of information, especially in formal education, the quantity of climate change information must be augmented, specifically in reference books of most students, teachers, and the other publisher which published by Ministry of Education and Cultural, also in the curriculum. In addition, the teacher's knowledge of climate change also need to be improved to support them in delivering good information for students.

ACKNOWLEDGMENT

This paper and the research behind it would not have been possible without the exceptional support of my previous Institution, PI Area, which provide researchers with so many ideas and input in the process of making this paper in Climate Change and Education sector. Alia Harza as the researcher's colleagues at IPB University, have also looked over the transcriptions of this paper and answered with unfailing patience numerous questions about the language on this paper.

REFERENCES

Chen, X. (2011). Why do people misunderstand climate change? Heuristics, mental models and ontological assumptions. *Climatic Change*, *108*(1-2), 31–46. doi:10.100710584-010-0013-5

Crowley, T. J. (2000). Causes of climate change over the past 1000 years. *Science*, *289*(5477), 270–276. doi:10.1126cience.289.5477.270 PMID:10894770

Currie, J., & Deschenes, O. (2016). Children and climate change: Introducing the issue. *Children and Climate Change*, *26*(1), 3–10. doi:10.1353/foc.2016.0000

Direktorat Jenderal Pembangunan Daerah Tertinggal. (2016). *Kabupaten Lembata*. https://ditjenpdt.kemendesa.go.id/potensi/district/44-kabupaten-lembata

Direktorat Jenderal Pembangunan Daerah Tertinggal. (2016). *Kabupaten Timor Tengah Utara*. https://ditjenpdt.kemendesa.go.id/potensi/district/41-kabupaten-timor-tengah-utara

Ghazvini, S. D., & Khajehpour, M. (2011). Gender differences in factors affecting academic performance of high school students. [). Elsevier.]. *Procedia: Social and Behavioral Sciences*, *15*, 1040–1045.

Gibbons, E. D. (2014). Climate change, children's rights, and the pursuit of intergenerational climate justice. *Health and Human Rights Journal*, *16*(1), 19–31.

Givin, J. G., & Shrader, J. (2016). Temperature extremes, health, and human capital. *Children and Climate Change*, *26*(1), 31–50.

Leiserowitz, A., Smith, N., & Marlon, J. R. (2011). *American Teens' Knowledge of Climate Change. Yale Project on Climate Change*. Yale University.

Marbun, D., Bachtiar, P. P., & Mawardi, S. (2015). *Iklim Usaha di Kabupaten Timor Tengah Utara (TTU): Kajian Kondisi Perekonomian dan Regulasi Usaha*. SMERU Research Institute.

National Oceanic and Atmospheric Administration. (2016). *Talking the Children About Climate Change*. https://oceanservice.noaa.gov/education/climate-stewards/talking-about.html

Ochieng, M. A., & Koske, J. (2013). The level of climate change awareness and perception among primary school teachers in Kisumu Municipality, Kenya. *International Journal of Humanities and Social Science*, *3*(21), 174–179.

Oppenheimer, M., & Anttila-Hughes, J. K. (2016). The science of climate change. *Children and Climate Change*, *26*(1), 11–30.

Perdinan, Santikayasa, I. P., Sugiarto Y., Amri, A., Atmaja, T., Adi, R. F., Basit, R. A., & Arini, E. Y. (2016). *Studi Baseline Adaptasi Perubahan Iklim Fokus Anak (4CA) Wilayah Kajian Timur Tengah Utara dan Lembata Provinsi Nusa tenggara Timur*. Generasi Hijau Indonesia.

Pitpitunge, A. D. (2013). Students' perceptions about climate change. *Asian Journal of Biology Education*, *7*, 2–11.

Rahman, S. M. A., Tasmin, S., Uddin, M. K., Islam, M. T., & Sujauddin, M. (2014). Climate change awareness among the high school students: Case study from a climate vulnerable country. *International Journal of Built Environment and Sustainability*, *1*(1), 18–26.

Stanis, S., Supriharyono, & Bambang, A. N. (2007). Pengelolaan sumberdaya pesisir dan laut melalui pemberdayaan kearifan lokal di Kabupaten Lembata. *Jurnal Pasir Laut*, *2*(2), 67–82.

Sugiyono. (2007). *Statistika untuk Penelitian*. ALFABETA.

United Nations Children's Fund. (2011). *Education Kit on Climate Change and Child Rights*. United Nations Children's Fund.

Chapter 12
Teaching Climate Change at Primary and Secondary Levels for Sustainable Development:
Young Indians Becoming Climate-Friendly

Sobha D. Namboodhiri
https://orcid.org/0000-0001-5905-8950
SRM University, Sonipat, India

Raghavendra G. Rao
SRM University, Sonipat, India

ABSTRACT

The young minds are provided with climate change and sustainable development education through various means. Environmental education is being taught at the primary level in Central Board of Secondary Education Schools while it is being taught at secondary level in the Social Science and Science Subject in C.B.S.E. Schools of India with various tools and methods. In the present study, the authors have incorporated some methods which help in teaching about climate change at primary and secondary levels. The study shows how the adopted methods are effective in disseminating knowledge about climate change. The investigator decided to adopt the survey method of research to know the method of teaching being adopted for teaching of climate change in environmental science at primary level and in science or social science, eco club at secondary level. Findings were that students should be taught in the school curriculum itself about climate change and its effects. The teaching strategies should be adopted based on the level of the child, his level of cognitive ability, accessibility, and ease of use.

DOI: 10.4018/978-1-7998-7512-3.ch012

INTRODUCTION

Earth is warming at a rate unprecedented and the last decade was the warmest and contained 9 of the 10 warmest years ever recorded. Climate change is a very broad term that tells us about the change of climate in the world. It is something that is continuously changing and is not the same as normal change of seasons occur at regular intervals on earth. Changes are generally welcome for all but all changes need not be good. Some are not good for us as they can result in some serious consequences.

Climate change is a term that has now become quite serious and world leaders are meeting every year to stop this climate change. Climate change results in fluctuations in weather patterns, blurring, and rise of sea levels, melting of glaciers, and depletion in ozone level. Climate change also results in a fall in agricultural productivity, an increase in the number of infectious diseases, instability in the water supply, and the dying out of many species of plants, animals, and birds. The effects of climate change are predominantly negative and impact most heavily on low-income countries and developing countries.

Mainlay and Tan (2012) have discussed in their research that adaptation to climate change capacity varies from country to country and the least developed countries have a lower adaptive capacity. The ability of a country depends on a nation's resources which are factors like knowledge, technology, support institutions as these can only help them to meet the Sustainable Development Goals.

The best resources for a country are the children who have to be developed in such a manner that they save natural resources and do not contribute to increasing climate change and this can be only brought about by providing proper education from the primary level.

Figure 1. Students performing a Dance drama depicting climate change, its consequences, methods to adapt to climate change.

Plutzer E. and Hannah A.L. in their article 'Teaching Climate Change in middle schools and high schools: investigating STEM education's deficit model', Climate Change, Vol.149, Pg.305-317(2018) have written that teachers are not well equipped to teach about climate change in the USA due to less content knowledge and their political inclinations. Also, teachers encourage students to debate established findings like global warming and its effects.

Siegner A. and Stapert N. (2019): 'Climate change education in the humanities classroom': a case study of the Lowell school curriculum pilot, Environmental Education Research analysed and evaluated a climate change curriculum implemented in a private school, a non-profit organization, and a government agency (NOAA). It was found that students developed a high scale climate literacy and it is a good study for developing the climate change curriculum with a socio-scientific approach to climate change education.

According to Dr Saminathan B. (2015), greenhouse gases have increased which has resulted in changes in the ecosystem and are affecting human health. Around 54 percent of total emissions of GHG's are done by the developed world while the South Asian region contributed only about 4 percent in the total emissions in the year 2000.

School students should be made aware of the concept of climate change, its causes and effects like-decrease in yield of plants due to excess of CO_2 emission by 2100 which would be about 500 ppm resulting in ruining of plant growth, the decline in the absorption of nutrients which are essential for plant growth and human wellbeing resulting in poor nutrition of mother and child. Climate change would also result in an increase in mortality due to an increase in the frequency and intensity of heatwaves. Natural disasters like floods and droughts increase due to an increase in the temperature of the earth. Vector-Borne diseases, infectious diseases, and respiratory diseases, and water borne diseases increase due to floods. Increased incidences of flood, Tsunami, and droughts causes displacement which can result in stress, trauma, and other mental illness. Biodiversity loss occurs as there is a decline in genetic, phenotypic materials and diversity of organisms like bleaching of coral reefs, reduction in several fishes resulting in affecting the human beings through the food chains and food webs. There is the threat of rising sea levels which would result in submergence of low-lying coastal regions resulting in displacement and environmental refugees would be seen all over the world.

Adaptation strategies will pose difficulties of implementation in low-income countries. Early planning can help reduce future adverse health impacts and mitigation strategies can help to solve or at least to bring to halt some of these problems.

In this study, strategies are discussed which are used by various educators for teaching about climate change on earth. This study is primarily done to know how the teaching at Primary and Secondary level about climate change is done by various teachers. It also studies how in teaching about climate change the strategy varies for each teacher. Topic-wise it has been calculated that how climate change can be explained and how much it has influenced the children. Hence the means employed for the teaching of climate change is studied and the aim is to understand how climate change can be effectively taught to students.

Figure 2. School exhibition project by a student depicting rainwater harvesting to save 'water- a precious natural resource.'

BACKGROUND

Environment Education in the Indian Scenario

According to the National Council of Educational Research and Training (NCERT),1975, the document "The curriculum for the ten-year school-A framework. "Environment education was one of the essential components of school education and in the primary classes, the science should be taught as environmental studies, in classes I and II as a composite course including both natural and social environment and later on as two subjects viz, environmental studies I (natural sciences) and environmental studies II (social science). The purpose should not be to stuff the minds of the children with facts and information, but to sharpen their senses, to enable them to observe their environment and to enrich their experience."

According to The National Policy on Education,1986 "The protection of the environment is a value which, along with certain values, must form an integral part of the curriculum at all stages of education." Environmental education was made a part of the curriculum and it was taught by exposure to entities of the surroundings of a child. Students are encouraged to participate in projects, exhibitions, and discussions related to saving natural resources like rainwater harvesting, utilization of solar power and wind energy.

According to The National Education Policy 2020

School education will promote community engagement, value-based education, and environmental education. The students would be prepared from the primary level of education to think and adopt an eco-friendly lifestyle. Their education would not be limited to some subjects but education would include environmental awareness, values, and climate education. The 17 sustainable development goals of the UN include education and this has been primarily incorporated with social development and children are provided Global Citizenship Education (GCED), to understand global challenges and solve them.

Figure 3.A school project exhibited in school depicting the earth of 2050 showing all types of pollution

Materials and Methods

Objectives

1. To study the teaching methods adopted for teaching the Topic- 'Climate Change' as a part of the Curriculum of EVS at the Primary Level in two schools of Panipat.
2. To study teaching methods adopted for teaching Climate Change in Social Science and Science at the Secondary Level in two schools of Panipat.
3. To study the difference in teaching the topic- 'Climate Change' between the Primary and Secondary School children of the two schools of Panipat.

4. To study the teaching methods adopted to teach climate change in Eco-club at the Secondary Level in Delhi Public School and Shiv Deep Vidya Mandir of Panipat.
5. To study the difference in teaching climate change in curriculum and eco-club in the two schools of Panipat.

Hypotheses

The following hypotheses are taken into account when we study this chapter-

1. There is a significant difference in teaching about Climate Change at the Primary and Secondary levels.
2. There is a significant difference in teaching about Climate Change in Curriculum and Eco-club.

Delimitations

1. The study is delimited to the Panipat District in Haryana.
2. The study is delimited to two schools of Panipat District which are Delhi Public School and Shiv Deep Vidya Mandir.
3. The study is delimited to teachers of Primary Level teaching EVS.
4. The study is delimited to teachers at Secondary Level teaching Science and Social Science.
5. The study is delimited to teachers teaching Eco-Club in school.
6. The study is delimited to 50 teachers.

The reasons for delimitations are-

a. This study is conducted during the early phase of lockdown so a wide area could not be covered as the schools got closed by March-end in India.
b. Environmental studies is taught in detail in some schools which are willing to impart holistic education to children.
c. Most of the schools declined participation in the study as they are not willing to disclose their methodologies of teaching.
d. Rural schools mainly of Government are having the least infrastructure for providing proper education and they are stressing upon imparting 'Sarva Siksha Abhiyan' A scheme of GOI for providing literacy to all.
e. Only Primary and Secondary Level teachers have been included in the survey as in Senior Secondary level the teachers are busy imparting knowledge for competitive exams.
f. Interest of students and their involvement in exhibitions and conservation issues are seen more in junior-level classes as in Primary and Secondary Level due to less pressure of education at these levels.
g. Rural area schools are more inclined to provide Basic education due to the financial constraints of the schools as well as the learners.

Research Area

The study is held in two different schools of Panipat which are Delhi Public School, Panipat, and Shiv Deep Vidya Mandir.

These two schools are representatives of most of the schools in this area as the Central Board of Education is being followed in both of these schools. The students in these schools belong to both Urban and Rural areas. These schools attract most of the students of the district of Panipat being famous for imparting a good level of education to children.

Sample Size

The total teachers taken in the sample are 50 which includes 25 from each school by the method of systematic sampling. Teachers teaching Environmental Science, Science, and Social Studies are included in the sample.

Methodology

The investigator decided to adopt the descriptive method involving survey method of research to know the method of teaching being adopted for teaching of Climate Change in environmental science at primary level and in Science or Social Science, Eco club at Secondary Level.

Teachers were chosen from the two different schools based on subjects taught and the level of teaching. Teachers teaching Science, Social Studies, and Environmental Science are selected which includes teachers from both the Primary and the Secondary levels.

These selected 50 teachers are seated and then they are provided with the questionnaire which they have to answer in a period of half an hour. They are not at all forced to participate as participation is done voluntarily by each teacher.

After the teachers answer the questions the analysis of questions is done to determine whether hypotheses are true or not.

Tools Used

For the present study Self Prepared Questionnaire is used. The questions are made clear, unambiguous, and are made in a manner that could be easily answered by the 50 teachers selected from both the schools. Then an analysis of the results is done by proper tabulation of the data. The percentage of respondents answering is plotted graphically.

DESCRIPTION OF THE QUESTIONNAIRE

Self-prepared Questionnaire

The questionnaire for studying the methods being adopted for teaching Climate Change is prepared by the investigator herself. The questionnaire includes some general questions like- Name, Age, Teaching Subject, Years of Experience.

Then a Topic Specific Questionnaire is made which has 50 questions that include questions on Teaching Strategies, Climate Change, Global Warming, Activity-based EVS Teaching.

Each item consisted of a topic and five options for teaching methods. There was a fifth option also "Any other". The teachers involved in the study were asked to select one of the teaching strategies or methods used to teach a topic related to climate change. If they used some other method of teaching then they were given a fifth option to write their method of teaching. These methods of teaching include- Role Play, Story writing, Poem recitation, Activity method, any other method in which the teachers would mention their method of teaching. This questionnaire was prepared by the investigator and methodologies used are based on the experience of teaching of the investigator.

COLLECTION OF DATA

In the present study, the investigator collected the data personally. For data collection, the investigator first took permission from the head of the institute of the sample concerned. The investigator distributed the questionnaire to the teachers and assured all the teachers that their responses would be kept strictly confidential and would be utilized for research purposes only. The purpose of the questionnaire was also explained to motivate them. Though there was a time limit of 30 minutes for the questionnaire yet the respondents took only 15-20 minutes in completing the questionnaire.

STATISTICAL TECHNIQUE:

SANDLER'S A test technique is used for analysing the data. SANDLER'S 'A' test technique is used for measuring the validity of analysis of the difference between mid-values of two matched groups.

Formula: The formula is as follows:

Where D_i = Difference in variables

Using the following steps, A-test is employed-

1. Subtract the hypothesized mean of the obtained answer from each teacher to obtain the difference in the variables. (Here variables are obtained answers).
2. Square each difference and then obtain the sum of such squares.
3. Then using the above-given formula 'A' statistic is obtained.
4. The critical value of A is obtained by reading the table of A-statistic for n-1 degrees of freedom at a given level of significance which is taken as 5% here.
5. Finally inference is drawn as - When the calculated value of A is equal to or less than the table value then the hypothesis is rejected while if it is greater than its table value the hypothesis is accepted.

Degree Of Freedom

The formula is as follows:

$$Degree\ of\ Freedom = No.of\ pairs - 1 = n - 1$$

Note-The lesser the value of the 'A' test more is the level of significance.

DATA ANALYSIS AND INTERPRETATION

Different methods are used for teaching different topics by the teachers and there is a significant difference in the teaching methods being adopted in various schools of the same district regarding the teaching of Environmental Science as a subject. Of the total 50 questions asked based on Environmental issues, few of the interpretations are shown here-

About 75% of teachers are using Audio Visual Aids which helps the child at Primary Level to understand the topic properly. A.V. aids are used by maximum teachers for the teaching of "Our earth, its weather, and climate".

The field trip method is used by most of the teachers (69%) in the Eco club for the teaching of "natural regions of the earth" while in primary schools discussion method (38%) is used by a large number of teachers.

Illustrations with the example method are most popular among the teachers for the teaching of "Climate of various regions" in Secondary schools. In this field, Audio-Visual aids and discussion methods are used by most of the teachers of Primary schools.

Roleplay and activity method is most popular among the teachers of Secondary (40%) and primary schools (31%) for the teaching of "Water Pollution & its effects".

Storytelling method is most popular among the teachers of primary (80%) secondary schools (31.25%). for teaching "Effects of Air Pollution".

The field trips method is most popular among the teachers of Eco Club (75%), Secondary Schools (43%), primary school (20%). After that, the most popular method is observed demonstration cum discussion method and discussion method for the teaching of "Rain Water Harvesting and its methods."

For teaching "Our health, good habits and diseases" the most popular method is an illustration with an example as 54% of teachers in primary schools 50% teachers in eco-club, 21% in secondary schools utilize illustration method. After it the Discussion method is used widely for this unit.

For teaching "Air and its uses" illustration with example and activity is considered the most popular method. 50% of teachers in primary, 30% of teachers in secondary,69% of teachers in eco-club 10% of teachers use the illustration method. After it, discussion method is a widely used method for this unit.

For teaching 'Global Environmental Crisis and Sustainable Development' Audio-visual aids and demonstration cum discussion is used by 50% of teachers in Primary level,78% in Eco-club while at Secondary Level more discussion method is used.

RESULTS

1. Audio-visual Aids, Storytelling, and Illustration with examples are being used more often by teachers at the Primary Level to teach topics related to Global Warming and Climate Change.
2. Topics in the Secondary Level are being taught more by using Discussion or Demonstration cum Discussion and Field Trips.
3. Environment Clubs like Eco-Clubs use Field Trips or Activities for teaching environmental issues like Climate Change.

Figure 4. Field trip of students to Rainwater harvesting pit

Hence the study tells that there is no significant difference in teaching climate change at the Primary and Secondary Levels as the educational strategy depends upon the topic to be covered. But there is the little level of significance in Eco-club and Educational Strategy used in Secondary Level as Eco-Club uses more activity-based learning with field trips while Secondary Education uses more Demonstration, discussion and illustration method.

Figure 5. Science Exhibition and participation in National Science Congress by students, explaining about Drip irrigation technique to judges.

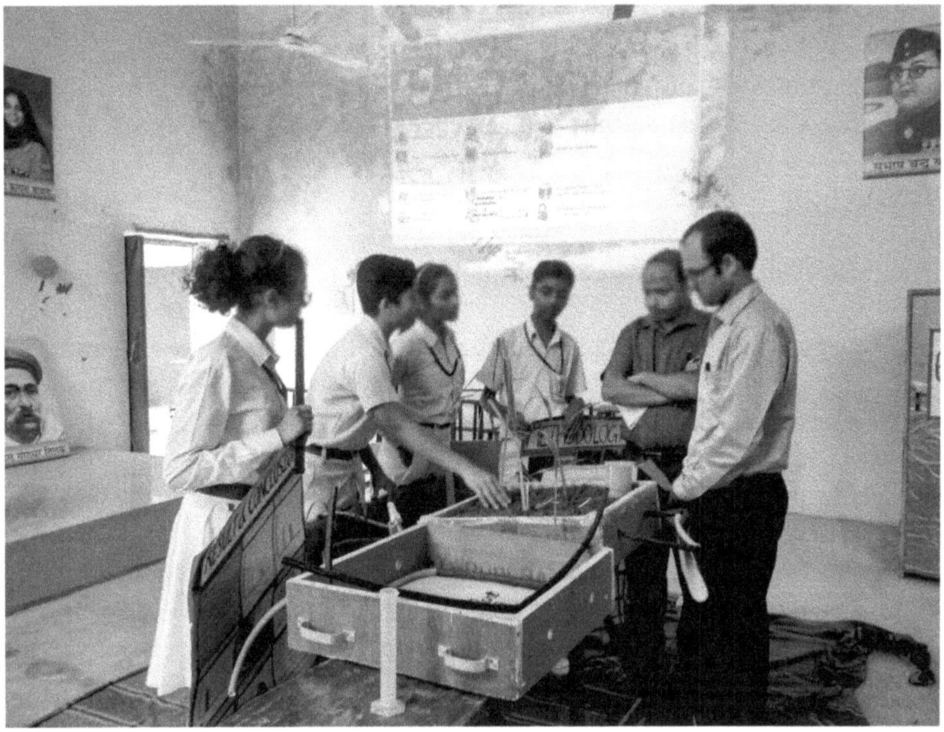

CONCLUSION

1. Proper teaching methods help the students to achieve the objectives not only related to the lower level of cognitive ability but higher abilities related to meta-cognitive, psycho-motor as well as affective domain.
2. Proper teaching methods help the students in self-learning, independent thinking, and arousing curiosity.
3. With the help of proper and effective teaching methods, the desired improvement in the quality of the teaching process and understanding can be achieved.
4. Teacher's efforts and proper teaching methods can increase the student's participation in the Classroom and can make them aware of the environment in which they are developing their personality and being useful to the Environment.
5. Different teaching methods help the students particularly of the secondary level in cognitive understanding and dissemination of the message in the society.

FUTURE ACTION PLANS

Educational Strategies

Environmental Education Strategies should be used to solve the problems associated with climate change not only in EVS, Eco-Club, Social Sciences, and Science subjects but also in other subjects as-

Primary Education

Environmental Education should be part of the child's education from the elementary level.

A. Teaching strategy should make the child understand the various organisms and then their importance on earth.
B. He should be ready to understand that man has affected the natural resources, wildlife, and plants in such a way that now it is becoming tough to go back to the original level of natural diversity.
C. Child should be made to understand the Food Chain and Food Web to know how organisms are dependent on each other.
D. At the primary level children should be made to understand the variety of organisms through activities like-
    1. Drawing, Painting and Activities of pasting pictures, Clay modeling, Block Games, and short plays.

2. Curricular activities like -

Advantages

1. Children develop sensitivity to the various types of organisms.
2. They understand the importance of plants and animals in their life.
3. Children can understand that all organisms should be taken care of.
4. Children know which activities can be harmful to organisms and should be stopped.
5. Natural resources their uses and importance are understood by them.
6. Field visits help them to know that natural habitat is essential for animals and plants. Hence, they help in the preservation and protection of biodiversity.
7. They influence their parents in the protection of the environment.
8. They influence society by their drawings and paintings.
9. They understand about co-habitation.
10. Children can share with friends; care and nurture-Plants, domestic animals.

Secondary Education

Teaching Strategy at the secondary level can help the adolescents by relating their body developments with nature. They can learn the following in various subjects-

Figure 6. A play on Environment Day about the consequences of pollution by students.

Other than the curriculum the children can participate in Science and Environment Clubs and Exhibition to show about effects and results of Climate Change.

Advantages

1. Children can understand the reasons for climate change.

Table 1.Teaching strategies for teaching Climate Change at Primary Level

Subjects	Activities
English	Rhymes, Stories, Short Plays, Small Comprehensions
Hindi	Recitation, Short stories, Short paragraph writing, Worksheets
Maths	Addition, subtraction, multiplication, and division of organisms, activities to show their number and distribution
EVS	Environmental Sensitive Activities, Field visits, Nature Walks

Table 2. Teaching Strategies for Climate Change at Secondary Level

Subject	Activities
English	Essay writing, Speechwriting, and Songs, Poetry
Hindi	Stories, Poems, and Writings
Science	Field trip, Illustrative with Discussion Method
Geography	Audio-Visual Aids, Discussion with example

2. Students can make Slides and PPT to show others about the effects of climate change and spread messages to everyone.
3. Climate change can be stopped by the active participation of these children.

Figure 7. Song by students on Environment Day to spread the message to save the environment

4. Children take part or organize exhibitions on this topic to spread message to parents.
5. Children draw posters or write slogans about climate change.
6. Science exhibition is held on this topic showcasing all the ill-effects and how to stop it.
7. Climate change is depicted through plays, mono acts, or Nukkad Nataks which can be shown in villages too.
8. Children do surveys in hospitals to know the bad effects and find out new solutions to the problems at hand.
9. Students collectively take an oath to stop climate change.
10. They stop using sprays which can result in ozone depletion and tell elders to practice carpooling, using public transports and cycles.
11. They tell industries to reduce pollution.
12. Through social media they tell everyone to reduce climate change.

Eco-Club

It is Environment Club in the Schools which is generally held once a week under the Socially Useful and Productive Work involving children in activities and Audio-visual or Field Trips which helps them to learn the concepts by doing. They also spread the message to society.

Advantages

1. Children can learn by doing.
2. They remember the concepts.
3. Children spread the message to society.

Educational Implications

1. Environment Education should be a part of the Curriculum from the Primary till Senior Secondary Level.
2. Students should be provided with proper facilities to exhibit in shows or science fairs about Climate Change.
3. Environment Clubs should be set up in schools where children can become Eco-warriors and spread the message in society to stop Pollution.
4. Environment Awareness Camps or Assemblies can be organized in Schools or Communities where children arrange stalls to make society aware about Ill effects of Climate Change.
5. Students should be taught to minimise use of scooters and cars, they should travel by public transport like school buses to reduce air pollution.
6. Students can go to houses of the local community and give them placards explaining about Climate Change.
7. Students should be encouraged to show some plays or nukkad nataks(Street Plays) in their community to make the people aware of climate change.

Figure 8. An Eco-warrior spreading the message to keep our area green and clean on Environment Day(5th June)

8. Students can work with medical officers in spreading the message about diseases like malaria whose incidence increases due to climate change.
9. Schools should provide students teaching about various techniques of saving the planet and promote them so that they are followed by students.
10. Schools should organize study tours to science or environment museums where children learn more and see the practical applications of various methods of saving our planet earth.

SUMMARY

Climate change has to be stopped and this can be achieved if the new generation of children are made aware of its dangers and how to stop it.

As Dr Pachauri (2015) has stated "Many Adaptation and Mitigation options can help address climate change. These include investment in environmentally sound technologies and behavioural and lifestyle choices."

According to Stevenson R.B., Nicholls J. and Whitehouse H. 'What is Climate Change Education?' Springer Link (2017) In Australian Schools when climate change mitigation measures are studied it is found that the educators have to practise inquiry and co-learning with students to understand the complex nature of education for climate change and adaptation. Co-curricular activities along with community initiatives for student's participation is must. Teachers must develop the capacity of students to think critically and respond meaningfully to any climate change adaptations and mitigation measures.

Anthony L. UNICEF Executive Director,2015 Nov.24, New York/Geneva, Press Release has stated "Today's children are the least responsible for climate change, but they, and their children, are the ones who will live with its consequences. And, as is so often the case, disadvantaged communities face the gravest threat."

Students should be taught in the school curriculum itself about Climate Change and its effects. The teaching strategies should be adopted based on the level of the child, his level of cognitive ability, availability, and ease of use.

The study shows that Primary and Secondary School Children can be made aware of the solutions and can be motivated to follow them. They can spread the message in the society and show the elders that how climate change can be stopped. School students are provided proper education about climate change and its causes using various methods by teachers. This has tremendous effect on their perspective towards environment and they are able to spread the message in the society. Various methods and teaching styles have their own benefits in making the students understand about climate change and its ill effects.

Appeal done by children is heard by all. So like Greta Thunberg from Sweden children can call the world and tell them to stop the harm-causing pollutants.

Children being the harbingers of the future should be taught the guidelines of reducing pollution of air, water, and soil and they should know how these things are going to affect them and the coming generations. They should spread the message in the society as the future is only going to be bright for them if they follow sustainable methods of using natural resources.

Students should discipline themselves and society so that earth is saved and it remains as a blue planet for them and their future generations.

REFERENCES

Bharti, K. S. (2003, November 1). Teaching strategies in a pre-school program in the states of Haryana and Rajasthan. *Journal of Indian Education, 29*, 63.

Brophy, J. (2004). *Motivating Students to learn* (2nd ed.). Erlbaum. doi:10.4324/9781410610218

Haines, A., Kovats, R. S., Campbell, D., & Corvalan, C. (n.d.). Climate Change and Human Health- Impacts, vulnerability, and public health. *The Journal of the Royal Institute of Public Health and Hygiene.*

Jamir & Amba. (2004). *Environmental education in school system-overview and challenges.* UNESCO/JAPAN Asia pacific environmental education research seminar.

Jony & Fei. (2012). *Climate change, Mainstreaming gender and climate change in Nepal.* International Institute for Environment and Development. https://www.jstor.org/stable/resrep01234.6

National Council of Educational Research and Training (NCERT). (1975). *The curriculum for the ten-year school-A framework.* Author.

Pachauri, R.K. (2015 Feb 1). Climate Change Adaptation and Mitigation, The SDG Nexus. *Terra Green, 7*(11).

Plutzer, E., & Hannah, A. L. (2018). Teaching Climate Change in middle schools and high schools: Investigating STEM education's deficit model. *Climatic Change*, *149*(3-4), 305–317. doi:10.100710584-018-2253-8

Raj, C. (2009, Dec. 21). Enemies of the Earth, Copenhagen Summit. *India Today*.

Rajagopalan, R. (2018). *Environmental Studies, From Crisis to Cure* (3rd ed.). Oxford University Press.

Rees & Anthony. (2015, Nov 6). *Division of Data, Research and Policy, Strategy and Networks Section, Unless we act now, The impact of climate change on children*. Academic Press.

Saminathan, B. (2015). The Syllabus adopted from 2015-16 onwards for B. Ed(II)Year, Environmental Education.

Santosh, K., Swamy, S., Prasad, A., & Singh, D. B. (2011). Teaching Learning Strategies or methods being adopted for teaching EVS at Primary Level. *International Journal of Transformation in Business Management*. https://www.ijtbm.com/

Siegner, A., & Stapert, N. (2019). Climate change education in the humanities classroom: A case study of the Lowell school curriculum pilot. *Environmental Education Research*.

Stevenson, R. B., Nicholls, J., & Whitehouse, H. (2017). *What is Climate Change Education?* Springer Link.

The National Policy on Education. (1986). Government of India.

The National Policy on Education. (2020). Ministry of Human Resource Development, Government of India.

KEY TERMS AND DEFINITIONS

Climate: It is the long-term pattern of weather conditions for a given area.

Eco-Club: Environment Club run by schools as a part of Socially Useful Productive Work.

Global Warming and Climate Change: Increase in temperature of the earth due to increased levels of Carbon Dioxide, Methane, CFCs, Water Vapour have made the twenty first century warmest on record, this has resulted in drastic changes in weather patterns causing climate change (Rajagopalan, 2018).

Primary Education: Formal education from Kindergarten to 5th standard.

Secondary Education: Formal Education from 6th to 10th Standard.

Weather: The condition of the atmosphere at a particular place and time such as humidity, temperature, rain, etc.

APPENDIX

Questionnaire

The teachers were asked the following general questions first-

1.Name of the teacher
2.Place of work
3.Age of teacher
4.Teaching Subjects
5. Years of experience
6.What is the class taught and do they teach environment as a topic or subject?

It was followed by the following topic related questions -

1. Air and it uses
2. Weather and Seasons
3. Natural regions of the earth
4. Climate of various regions
5. Ganga water pollution
6.Industrial development
7.Yellowing of Taj Mahal
8. Droughts in Vidarbha
9.Effects of Air Pollution
10.Natural Disasters mitigation
11.Reduction in forests
12.Waterborne diseases
13.Good habits of students
14. Health and hygiene
15. Rise in asthma cases
16.Poaching of wildlife
17.Smog in Delhi in winters
18.Rising loss of biodiversity
19.Water availability issues
20. Global Environmental Crisis
21. Sustainable Development
22. Changing patterns of weather
23. Global warming and its cause
24. Diseases due to global warming
25.Sea level rise as a threat of future

26. Malnutrition and deficiency diseases
27. Cloud bursts as in Leh, Ladakh
28. Earth- structure, and composition
29. Rise in particulate matter in the air
30. Health-related absenteeism in school
31. Bhuj and other major earthquakes in India
32. Nuclear plants and nuclear disasters
33. Dam constructions and related problems
34. Extinction of wildlife and plants
35. Water Pollution & its effects
36. Eutrophication of water bodies
37. Gangotri's of Himalaya and risk of glacial melting
38. Rain Water Harvesting and its methods
39. Stubble burning and associated air pollution
40. Increased incidences of water problems
41. Dal lake of Kashmir and associated problems
42. Mid-day meal and its need
43. Population and associated problems
44. Lack of proper amenities in cities
45. Problem of waste management
46. Rising incidences of crime for basic amenities
47. Malaria and other vector-borne diseases
48. Unprecedented flood of states like Kerala
49. Some major movements like-Narmada Bachao Andolan(Movement to save river Narmada)
50. Incidence of Coronavirus and associated problems

Chapter 13
Climate Change Adaptation and Sustainability in the Bangladeshi School Curriculum

M. Mahruf C. Shohel
https://orcid.org/0000-0002-4048-0577
University of Sussex, UK

Goutam Roy
https://orcid.org/0000-0001-7355-044X
University of Rajshahi, Bangladesh

Tanjeeba Chowdhury
https://orcid.org/0000-0003-4524-1046
Dan Church Aid, Bangladesh

Atm Shafiul Alam
https://orcid.org/0000-0002-8299-754X
Queen Mary University of London, UK

Shahriar Shams
https://orcid.org/0000-0001-6233-5365
Universiti Teknologi Brunei, Brunei

ABSTRACT

Bangladesh is one of the most affected countries by climate change and extremely vulnerable to environmental degradation. Young people, as the new generation, need to understand the consequences of climate change and its impact on their lives and the lives of their fellow citizens. This chapter presents the findings from a systematic evaluation of Bangladeshi school curriculum, and how it prepares the future generation for climate change adaptation and ensures a sustainable future for them. This chapter highlights the recommendations from the findings and suggests how the national curriculum could address

DOI: 10.4018/978-1-7998-7512-3.ch013

climate change mitigation and adaptation for sustainability by providing contents to offer grounding for the young people as global citizens to act in the local context. Although there is some content included in the school textbooks related to climate change and sustainable development issues, there is a need for inclusion of more relevant contents in the national school curriculum so that school textbooks provide enough contents on climate change, its mitigation and adaptation.

1. INTRODUCTION

Climate change is one of the most complex issues humanity is facing at the present time. It has many dimensions including scientific, economic, social, political and environmental (UNESCO, 2002, UNESCO, 2010), which have raised many moral, ethical and legal questions. While it is a global problem, every nation must come forward to tackle climate change on local scales. Bangladesh is an extremely vulnerable and affected country concerning climate change. It is one of the densely populated countries in the world. Its population massively suffers from global warming and becomes victims of environmental degradation, which destroys many people's livelihood as a result of climate change.

The most threatening global environmental problem is climate change while being a considerable political and scientific controversy in the present time (Singer & Seitz, 1998). However, it affects the economy, culture, and society (UNESCO, 2002; UNESCO, 2010). The Article 6 of the United Nations Framework Convention on Climate Change (UNFCCC) mainly concerns about climate change education (CCE), training and public awareness to mitigate the impacts of climate change and ensure a sustainable future. These aspects of learning have great importance to building capacity within society, especially among the young people, and for the activists working on climate justice (YOUNGO, 2011).

Climate change refers to ''the alterations in the atmosphere that are over and above natural climate variation and that are a result of human activity'' (Makrakis, Larios and Kaliantzi, 2012, p.55). In response to climate change, there are the two main strategies that are followed by policy-makers and practitioners: i) *climate change mitigation* (CCM) that addresses the root causes of climate change by reducing greenhouse gas emissions, and ii) *climate change adaptation* (CCA) which seeks to lower the risks posed by the consequences of climatic changes. However, the concept of CCA generally refers to human responses to global warming. The Intergovernmental Panel on Climate Change (IPCC) defines 'climate change adaptation' as ''the process of adjustment to actual or expected climate and its effects. In human systems, adaptation seeks to moderate or avoid harm or exploit beneficial opportunities. In some natural systems, human intervention may facilitate adjustment to expected climate and its effects'' (IPCC, 2014). That means in reality through appropriate actions or by taking the appropriate measures that may arise from the process, the effects of climate change could be prevented or mitigated. It has been shown that well planned early adaptation action saves the environment, lives and money in the longer term (Global Commission on Adaptation, 2019). Both strategies are necessary (and inter-linked) because even if emissions are dramatically decreased in the next decade, adaptation still needs to deal with the global changes that have already been set in motion.

According to the IPCC, there are three main factors which determine vulnerability to climate change. These are 'exposure to hazards (such as reduced rainfall), sensitivity to those hazards (such as an economy dominated by rain-fed agriculture), and the capacity to adapt to those hazards (for example, whether farmers have the money or skills to grow more drought-resistant crops)' (Loria, 2018, Online). However, adaptation measures can help reduce vulnerability, for instance, by lowering sensitivity or building adaptive capacity and allowing populations to benefit from opportunities of climatic changes.

Climate change impacts many arenas such as ecosystem and species survival, human health, water, energy and food supplies, including their production and usage, and social, economic, political and cultural stability. Thus, it is significantly affecting the vulnerable populations of the world, mostly from developing countries. In those countries, people suffer from inequalities and exclusions due to their age, gender, location, ethnicity and economic conditions (Agostino, 2010). Although there are disagreements about the consequences of unchecked threats, scientific evidence suggests that global warming is happening and climate change became evident in everyday life (Houghton et al., 1996).

Climate change adaptation measures may be planned in advance or put in place spontaneously in response to different local pressures. One adaptation approach would be large-scale infrastructure changes, such as building coastal defenses (Storm surge barriers, dykes, groynes) to protect against a sea-level rise and erosion from waves or diverting water during flooding (embankments, canals, retention basins) to protect lives and properties. Other approaches could be behavioral shifts such as individuals using less water by practicing water conservation, rainwater harvesting, farmers planting different crops with fewer water footprints and more households and businesses buying flood insurance to find a financial solution during emergencies.

Raising awareness and capacity building through early education is crucial to address the impact of climate change. For preparing future generations to deal with the impact of climate change, the emphasis has been given by educators, researchers and education planners on educating young people to take a positive attitude toward social, economic and environmental dimensions of sustainability issues by providing climate literacy (Shafer, 2008; Stephens & Graham, 2008; Dupigny-Giroux, 2010). According to the National Oceanic and Atmospheric Administration (NOAA, 2011), people with adequate knowledge of climate should understand the mandatory principles of all aspects of the Earth system that governs different climate patterns. This knowledge includes assessing scientific information, collecting evidence, exchanging meaningful dialogues with different stakeholders, and making scientifically informed, verified and ethically responsible decisions related to climate change and associated issues. Therefore, a climate literate citizen should be able to assess the importance of preserving climate and the impacts of climate change on oneself and society. A climate literate citizen should also be able to contribute to the development of a climate-resilient society and environment.

2. THE CONTEXT OF BANGLADESH

Bangladesh is a fertile, densely-populated, low-lying, mainly riverine delta country of South Asia, which is located in the delta of the Padma (Ganges) and Jamuna (Brahmaputra) rivers. It is bordered by the Bay of Bengal, India and Myanmar (formerly Burma). It extends between latitudes 20°34' and 26°38' North and longitudes 88°01' and 92°41' East (GoB, 2017).

According to the Bangladesh Bureau of Statistics 2018, the total population is 162.7 million (BBS, 2018), making Bangladesh the eighth-most populous country in the world. The capital of Bangladesh (Dhaka) is the largest city, with over 15 million people. According to the 2011 census (GoB, 2019), Muslims constituted the majority with 90.39%, Hindus came second at 8.54%, followed by others (1.07%).

The national language of Bangladesh is Bengali, which makes up 98% of Bangladesh's total population, making it one of the most ethnically homogeneous states in the world. The World Food Programme's Country Brief 2020 report says that Bangladesh is among the five fastest-growing economies globally, with an average growth rate of 6.5 percent (WFP, 2020). A significant portion of contributions comes from exports of garments and shrimp as well as remittances from the Bangladeshis working abroad, and they provide the country's three primary sources of foreign exchange income. However, 31.5% of the total population still live in poverty (WFP, 2020).

With full of natural beauty, many archaeological and historical sites, the longest uninterrupted natural beach in the world and many more, the country covers an area of 1,47,570 square kilometres, criss-crossed by innumerable rivers and streams (GoB, 2017). The landscape of Bangladesh provides the major natural resources of water, land, fisheries, forests, and wildlife. There are several UNESCO World Heritage Sites in Bangladesh including the Sundarbans - one of the largest mangrove forests in the world and the home of the Royal Bengal Tigers.

Due to its geographic location, Bangladesh has a sub-tropical monsoon climate characterized by wide seasonal variations in rainfall, high temperatures and humidity. During the summer season (from March to September), the climate is very hot (21° C - 38° C) and humid while the climate is cold (11° C - 20° C) and dry in winter (from November to March) (GoB, 2017). April is the warmest month in most parts of the country, and January is the coldest month. In the rainy season (from June to August), the maximum rainfall occurs, and Bangladesh receives a very high amount of precipitation that amounts in the range between 1,100 mm to 3,400 mm (GoB, 2017). During this season, most rivers overflow their banks and inundate the countryside, rising most heavily in September or October and receding quickly in November. Moreover, the country is vulnerable to cyclones, which can be devastating. For example, a cyclone killed 138,000 people in April 1991, and Cyclone Sidr hit the southern coastal strip of Bangladesh in November 2007, affecting thousands of people.

The most significant issues in Bangladesh are severe overpopulation, natural calamities such as floods, tropical cyclones, tornadoes, tidal bores, soil degradation and river erosion, saltwater intrusion in coastal areas, contamination of groundwater by naturally occurring arsenic. Additionally, as a low-lying country, Bangladesh stands to be adversely affected by rising sea levels because of environmental degradation and climate change. As a result, ''Bangladesh is an example of how a rise in sea level and global temperature could lead to the displacement of millions of people. Improving the country's biodiversity is essential to prevent the depopulation and to sustain the development; resilience and adaptation are mandatory to avoid and fight next and future climate calamities'' (Loria, 2018, Online).

Because of its geographic position, Bangladesh is suffering from different natural calamities, including floods, droughts, cyclones, and so on, resulting in the loss of human life and damage to property and communication systems. The aftermath of natural calamities is often followed by a shortage of drinking water, and unhygienic condition that leads to the outbreak of diseases (Diarrhea, Typhoid). It has been observed that there can be differential risks to the people and the risk distribution is also differential in terms of the community classes (Islam, Rahman & Mallick, 2010). Bangladesh is one of the most vulnerable nations to climate change, and the vulnerability is accelerating due to increased temperature (Islam, Rahman & Mallick, 2010). According to Climate Action by the European Commission (EC, 2018), humans are increasingly influencing the climate and the earth's temperature by burning fossil fuels, cutting down rainforests and converting agricultural lands for urbanization. Besides, the emission of greenhouse gases in the atmosphere due to increased industrialization has an enormous effect on global warming.

While there are several policy responses related to climate change made by the Government of Bangladesh including the Bangladesh Climate Change Strategy and Action Plan (BCCSAP) 2009 (Ministry of Environment and Forests, 2009), institutional responses for CCA and CCM can indirectly reduce the severity of the vulnerability (Islam, Rahman & Mallick, 2010; UNESCO, 2015). Therefore, it is of utmost importance to provide a state-of-the-art curriculum from an early age to focus on climate change awareness as precautionary measures for both CCA and CCM. In other words, educating our next generation is a way forward to combat the battle against climate change, and how we incorporate the know-how and consequences of climate change in the current curriculum must be prioritized.

3. NATIONAL CURRICULUM AND TEXTBOOK BOARD

The National Curriculum and Textbook Board, Bangladesh (NCTB), is the organization responsible for developing school curriculum, production and distribution of the school textbooks in Bangladesh. It is the largest autonomous organization under the Ministry of Education (MoE) that develops and reviews the national curriculum for Pre-Primary, Primary, Secondary and Higher Secondary level. It also produces relevant teaching and learning materials based on the National Curriculum. In Bangladesh, all the public and private schools follow the NCTB curriculum apart from international and English medium schools. Every year NCTB distributes textbooks of all subjects to all students studying from Grade 1 to Grade 10 free of cost. Two versions of these school textbooks are usually produced - one version in Bengali and another version in English (NCTB, 2020).

Based on the number of publishing textbooks, the NCTB is the largest public publishing house in the world. In 2020, the NCTB distributed a total of 3,53,144,554 copies of textbooks among more than 42 million students of pre-primary, primary, *ebtedayee* (equivalent to primary level education provided by the faith-based *Madrasa Education* subsystem), secondary, *dakhil* (equivalent to secondary level education provided by the faith-based *Madrasa Education* subsystem) and vocational institutions (The Daily Star, 2020).

The NCTB envisions developing quality curriculum and textbooks for the students of Bangladesh. Preparation of quality curriculum and textbooks to build students as creative, efficient and ethical citizens is their primary mission. According to the National Curriculum and Textbook Board Ordinance, 1983, the major functions of the NCTB include examining school curriculum and syllabus, evaluating its effectiveness and suggesting revision, publishing and distributing the books. In addition, it provides a list of approved textbooks, library and reference books. They also encourage scientific, literary and cultural works by providing grants and donations and other activities assigned by the Government from time to time.

4. CLIMATE CHANGE EDUCATION, SUSTAINABLE DEVELOPMENT AND YOUNG PEOPLE

According to UNESCO, ''Education is an essential element of the global response to climate change. It helps young people understand and address the impact of global warming, encourages changes in their attitudes and behaviors, and helps them adapt to climate change-related trends'' (UNESCO, n.d., Online). The crisis of climate change, and the breakdown of ecosystems caused by excessive use of natural resources, are not only questions of science, but also behind economic disasters, plunging hundreds of millions of people trapped into poverty and social injustice. As a result, these engender social harmony and create social unrest, leading to other crises, including conflict and migration (Brown, 2008). By proper integration in curriculum and teaching and learning provisions, climate change issues could fit into subject areas across the curriculum, not only physics, chemistry, biology and geography but also economics, history, social studies, media, arts and food technology even religious studies.

Historically, human beings have been adapting to their surrounding environment in their behaviors, cultural practices, and innovative ways and tools to adjust to the changing environment. They always adapt to their local conditions for their livelihood from the Mediterranean siesta to the Vietnamese practice of building homes on stilts to protect against monsoonal rains. However, climate change raises the high possibility of climatic shifts (in temperature, storm frequency, flooding and other factors) over the time that had been experienced by existing societies in the past. Due to climate change, the changing of the seasons is unreliable and has an impact on nature, biodiversity, and the livelihood of the people in Bangladesh. For instance, increased salinity forced people from coastal areas of Bangladesh to change their profession and even migrated to urban areas (Shams & Shohel, 2016; Shams, Shohel & Ahsan, 2014).

Sustainability, which is a multifaceted concept, is the crucial issue for both CCM and CCA. In reality, the current level of greenhouse gas emissions defines what the future world will look like for the new generation. Therefore, we need to change our behavior and develop moral and ethical responsibilities for all human beings living on the earth so that our future generation can survive in a healthy environment and they should not face any excurcity for resources they need. Considering the current situation related to climate change across the globe, the burning question is: Are we doing enough to avoid the worst? or Will it be a world locked into catastrophic warming and frequent disaster? We can live our own lives and our lives can end at any moment. Nonetheless, the knowledge needs to be transferred, and skills need to be developed to overcome the consequences of natural calamities resulting from climate change and global warming. Being human, we should think about how we are going to spend our time on earth and what we are doing to preserve the earth.

While many organizations and world leaders have been working towards the issue of climate change, the International Alliance of Leading Educational Institutions (IALEI, 2009) provided eight specific recommendations related to climate change and sustainable development. Those could be summarized as below:

1. Sustainable Development became an urgent priority due to climate change. Therefore, the policies and practices related to the education for sustainable development (ESD) can play an important role in introducing better global, national and local climate change policies.
2. As the priority, there must be radical changes in behavioral patterns including production, distribution and in societies to meet the needs and the challenges of climate change the global citizens face.
3. While ESD will create society demands, the educational institutions will implement the issues by providing effective teaching and by ensuring sustainable practices.
4. Introducing the ‘whole school’ approach would be a better strategy for reorienting schooling in societies to emphasize the importance of interdisciplinary themes in school curriculum. This approach will enhance participation in addressing authentic sustainability challenges within the context and help students interact with others from the community outside the school contexts.

5. Universities can offer post-graduate diplomas and certificates in ESD to the teachers. Besides, specific ESD modules could be integrated with the pre-service and in-service training for the teachers.
6. Sustainability must be integrated into the school curriculum. Schools should be given enough resources, and time should be allocated to students and teachers to experiment with sustainability issues.
7. It is essential to establish a nexus of multi-stakeholders (e.g., researchers, educators, NGO professionals, public officers) and others working in the field of ESD to make a collective effort to make a difference in practice. This kind of initiative already exists, which could be used as the model in other places as examples of good practices.
8. Research on ESD should be enhanced in values by focusing on: i) the documentation and identification of existing practice and future practice roles of the state, ii) the educational outputs and evaluation process concerning ESD, and iii) the demands, scope of needs and challenges of general relevance.

The purpose of the national curriculum is to 'introduce young minds to the great minds of the past' with the essential knowledge to become good and efficient citizens. Therefore, the aim of it should be to provide young people what is best for their development and future to appreciate human creativity and achievement. The national curriculum provides a clear guideline of what the Government expects schools should teach their students. However, the current national curriculum has offered very little on human, physical and scientific mechanisms of climate change, its mitigation or adaptation. Teaching about climate change is by no means easy, but it is a challenge worth tackling for saving the only planet we have to live (Ledley, Rooney-Varga, & Niepold, 2017).

As an interdisciplinary topic, climate change can open young peoples' minds to deeper avenues of thought and reinforce learning in social science. Moreover, climate change education opens the door for young peoples to work actively in a real situation. The effect of this work might be not visible shortly. However, young people could get the self-satisfaction that they are doing something for the planet and the future's betterment. However, they may learn the effects of climate change in great detail throughout their time in school, but for the sake of hope, we need to take them through ways of mitigating those effects and teach them how to adapt the changes.

5. EXPLORING SCHOOL TEXTBOOKS AT PRIMARY AND SECONDARY LEVELS

5.1 Research Questions

Based on the discussion in the above sections, the main research question is: *What do the current Bangladeshi school textbooks present following the national curriculum related to climate change and its mitigation, and adaptation?*

5.2 Methodology

This qualitative interpretative study was based on document analysis (Bowen, 2009). Relevant information about the curriculum and contents was collected from the government curriculum and textbooks. A checklist was developed to explore the content from the textbooks provided by the NCTB for the primary and secondary levels. Two core subjects, *Bangla* and *Bangladesh and Global Studies* from Grade 1 to Grade 10, as well as *Science* and *Geography* textbooks of Grade 9 and 10, were explored by using the checklist and then thematic analysis was carried out to find out topics and what they covered concerning climate change, its mitigation and sustainability.

5.3 Findings

This study found that there are limited topics related to climate change and sustainable development in the current textbooks. All these textbooks are developed following the national school textbooks by the National Curriculum and Textbooks Board (NCTB).

5.3.1 Textbook Analysis: Pre-Primary and Primary level

5.3.1.1 Bangla

We have reviewed the textbooks of both Pre-Primary and Primary levels. The main finding is that Bangla textbooks do not emphasize climate change or any environmental issues. Though in lower grades such as Pre-Primary, Grade 1 and Grade 2, there are some chapters in which some components of environments such as rainy seasons, the importance of trees, and weather are described but in other grades, no such description is found. For instance, in the introductory book for the Grade 1 titled My Bangla Book, although there is a brief story on tree plantation and, however, the importance of the tree plantation can be unnoticed by many children (Tree plantation, p. 67). This is only the chapter in the textbooks of all grades, particularly covered the environmental issues. In some chapters of other grades, the description of some seasons, mainly the rainy season, are described. However, these do not present the importance and effect of the environment and climate change.

5.3.1.2 Bangladesh and Global Studies

We have also reviewed the textbooks of Primary level (from Grade 3 to Grade 5) named Bangladesh and Global Studies. The textbook of Grade 3, the first chapter gives an introductory idea and definition of the natural environment along with the social environment. Focus is given on the social environment rather than the natural one (The natural and social environment, p. 2). While in chapter 6, tree plantation at home is encouraged as a household work (Improving our social environment, p. 34) and planting trees and taking care of it is encouraged as a daily school activity (p. 36); chapter 7 deals with protecting the environment. This chapter titled 'Protecting our environment against pollution' gives an initial idea about the types of environmental pollution that humans can do (p. 38). This chapter also discusses the consequences of different types of pollution in human life and nature. In addition, how these pollutants are caused is also mentioned. For example, how animals, plants and humans suffer from pollution. What diseases are triggered by pollution is mentioned and how humans pollute the environment is also given (p. 40). An introductory idea about how students can prevent pollution in the school, community, and home is also provided. However, while chapter 9 titled 'Our Bangladesh' discusses Bangladesh's existing forests and encourages planting trees and forest reservation (p. 56), the link between overpopulation and environment pollution is mentioned in chapter 12 titled 'Population of Bangladesh'.

In the textbook of Grade 4, there are a number of discussions on environmental issues. In the chapter 'Our Environment and Society', the environment is discussed a little more in detail (elements are mentioned with pictures) than the previous grade. The term 'climate' is introduced here for the first time. The regional difference of the environment of Bangladesh (between north and south) is discussed here (p. 2). Besides, how the natural environment influences the social environment is discussed in detail here. However, the social environment leaving its impact on the natural environment is mentioned but not in details (p. 4).

In chapter 9, namely, 'Developing Our Locality', the facilities that people need for the development of a social environment of both rural and urban areas are discussed. However, the importance of trees/natural environment is not mentioned properly (p. 44, 46). Chapter 11, Geography of Bangladesh, explained the climate of Bangladesh with the main seasons and the characteristics of these seasons (p. 54). Three disasters were discussed in chapter 12, Disaster Management. Among them, two were natural disasters- flood and cyclone (p. 60). The overall discussion revolves around causes, effects and precautions for disasters. It is mentioned that these two disasters are caused by natural causes and the effects signify drastically because of man-made changes. In the cyclone part, it is mentioned that cyclones have a link with the gradual increase of the earth's temperature. However, the climate term was not directly mentioned anywhere (p. 62). Chapter 12, named Population of Bangladesh, discusses the negative effects of population density environment pollution as mentioned along with other five major points.

In previous Grade 4, it is mentioned that overpopulation has an adverse effect on the environment. In the textbook of Grade 5, in chapter 5, it is explained with examples and linked with climate change (p. 42). in this topic as the solution of overpopulation the following strategic areas are mentioned - food, housing, health, education, skills development, the balance of trades and environment.

The definition of climate, the difference between weather and climate is clearly mentioned in chapter 6 (Climate and Disaster) for the first time. The climate of Bangladesh, how climate is changing all over the world, and the risks of climate change in Bangladesh are mentioned here (p. 48). Followed by climate change discussion, several natural disasters are discussed in detail here, the first one was riverbank erosion. The causes and effects are discussed here, and causes include natural and human-made points (p. 50). The part of Bangladesh that suffers most from drought is also discussed. The man-made causes behind the drought and how it impacts nature are explained along with the adverse effects (p. 52). It is also mentioned that Bangladesh is at risk of earthquakes because of its geographical location and what measures can be taken to mitigate the loss after an earthquake occurs (p. 54).

5.3.2 Textbook Analysis: Secondary Level

5.3.2.1 Bangla

We have reviewed all the Bangla textbooks of Grade 6 to 10 but did not find any chapter that discusses climate change and environmental issues. In some chapters of all grades, the descriptions of nature and seasons, particularly rainy season are presented, but these are not aligned with the climate change, disaster or environment.

5.3.2.2 Bangladesh and Global Studies

In chapter 3 (Bangladesh in the Global Geographical Environment) of the textbook of Grade 6, an introduction about the climate of the Asian continent is described. However, while explaining causes and effects of dissimilarity in birth rate climate was mentioned in chapter 4 (Introduction to Population of Bangladesh) as a cross-cutting issue along with lack of education, early marriage and polygamy, low living standard and infant mortality rate (p. 36). Similarly, while explaining causes of discrepancy in mortality rate geographical reason and natural disaster and accidents were mentioned as a cross-cutting issue along with various diseases, age, poverty, and wars and battles (p. 37).

Chapter 5 shows how human life is influenced by the natural and geographical environment and explained in detail with examples from Bangladesh. The influences include growth of civilization, lifestyle, occupation, industries, dress, and dwelling etc. (pp. 42-43). Chapter 9, Environment of Bangladesh, explains human relation with the environment, the causes of environmental problems, and the action plan to prevent and control environmental issues in Bangladesh. It is expected that learners will become conscious of the importance of the environment from this lesson. This chapter also explains how people are causing damage to the environment in details mentioning relatable examples. It also explains how the environment is changing through time and is becoming harmful in many ways. It focused on the causes of pollution resulting from over consumption, unplanned growth and industrialization. Lastly, this chapter provided some key points to reduce pollution and take care of the planet as future citizens. It emphasized on tree plantation, preserving nature, managing waste, reducing sound pollution, making responsible development, raising awareness and keeping the natural balance at any cost (pp. 76-79). Sustainable development is introduced in chapter 13 for the first time. The definition is explained with examples in lesson 2, the 17 goals fixed by the UN are mentioned (pp. 101-104).

In chapter 7 of Grade 7, namely, 'Climate of Bangladesh', the climate of Bangladesh the causes of climate change in Bangladesh are discussed. Disasters related to climate change like cyclones and tidal waves, floods, river erosion, drought, chill, tornado, nor wester etc. are mentioned. People's role in mitigating disasters happening due to climate change and the impacts of climate change on the social-economic condition of Bangladesh are also described (pp. 63-71). Like previous, chapter 13 includes a more detailed description of SDGs than the previous grade while giving a background on it (pp. 121-126).

The chapter 8 of Grade 8 titled 'Disasters in Bangladesh' explains the idea of global warming, analyzes the causes and effects of global warming, explains the disaster, mentions the types of disaster, describes different types of natural disasters like cyclone, flood, drought, earthquake, tsunami, fire accident etc. It narrates the causes of the natural disaster of Bangladesh, analyzes the impacts of these disasters on the life and economy of Bangladesh, gives advice and plans to prevent natural disasters, undertake necessary steps to save lives and assets from disasters and be aware of the environment (pp. 97-111).

The textbook of Grade 9 and 10 also discussed the topic of climate change and environment-related issues. Chapter 3 titled 'The solar system and the earth' is linked to the changes of the seasons from a universal perspective which can help to grasp the overall idea of environment and climate. However, this chapter describes the geographical location and periphery of Bangladesh and encourages preserving the environment. This chapter also compares the features of the climate of Bangladesh to that of some countries of South Asia like India, Myanmar and Nepal which also explain the influence of climate over the people's lives and their livelihood in Bangladesh. Besides, this chapter explains the surmise of earthquakes and what are the reasons behind them. It also explains why Bangladesh is called an earthquake-prone region, preparedness and necessary steps to confront the risk of earthquakes (pp. 75-93).

Chapter 5 titled 'The Rivers of Bangladesh and the Natural Resources', explains the interdependence of rivers and dwellings, analyze the cause of water crisis in certain parts of Bangladesh, realize the necessity of preserving the water resources to overcome water crisis, explain the concept of water resources management and encourages to be aware of preserving the natural resources (p. 105).

The textbook title Science for Grade 9 and 10 is also reviewed. Chapter 2 'Water for Life' explains the reasons for water pollution at sources in Bangladesh and the effects of water pollution. Besides, this chapter explains the effects of global warming on freshwater in Bangladesh and describes the strategy to prevent water pollution and the responsibility of civil society. On the other hand, chapter 10 titled 'Sustainable Development Goals' analyzes the effects of climate change in Bangladesh and other parts of the world and explains the causes of environmental problems. This chapter also describes the causes and prevention strategies and immediate measures to be taken and explains the importance of Conservation of Nature.

A number of issues are explained in the book Geography and Environment. Chapter 1 titled 'Geography and Environment' explains the concept of geography and environment, the interrelationship among different elements of geography and environment and the importance of studying geography and environment. On the other hand, the chapter 2 titled 'The Universe and Our Earth' provide the details of earth's rotation and revolution, the causes and duration of day and night, its impact as well as causes of season change and its impact.

The title of chapter 5 is 'Atmosphere'. This chapter explains the elements of air, the layers of the atmosphere, the difference between weather and climate and factors of climate. Besides, the interrelationship of global warming and climate change, as well as global climatic change and its impact on different regions of the world, are also discussed. This also includes the impact of climatic change on the people of Bangladesh, their living conditions and economic activities and the environment.

Chapter 13, Development activities of Bangladesh and environmental balance, explains the balance and imbalance of the environment and some mentionable development activities of Bangladesh. Besides, how the environment was polluted due to the development activities resulting from environmental imbalance are also discussed. On the other hand, how the environmental balance influences the development activities and the measures of protecting the ecological balance during the implementation of the development activities are explained clearly. Besides, raising awareness of environmental imbalance got an emphasis in this chapter.

Chapter 14 titled as 'Natural disaster of Bangladesh' describes the disaster and hazard, different kinds of natural disasters of Bangladesh, disaster management cycle, disaster management of coastal areas of Bangladesh and making people aware about how to confront natural disasters and what the measures are to mitigate it. Besides, the last chapter titled 'Sustainable Development Goals (SDG)', explains the importance of partnership in achieving SDG, results and challenges of achieving SDG, and actions needed to address challenges of achieving SDG. This chapter is very similar to the chapter which is included in Bangladesh and Global Studies textbook of Grade 9-10.

6. DISCUSSION

6.1 Pedagogical Aspects

The banking concept of Freire's pedagogy (1968) critique the 'act of depositing, in which the students are the depositories and the teacher is the depositor' (Freire, 2005, p.72) as the teacher and learner are the *co-investigators* who are engaged in 'acts of cognition, not transferral of information' (Freire, 2005, p.79). According to Freire (2008, pp. 61-62), ''The teacher is no longer the-one-who-teaches, but one who is himself taught in dialogue with the students, who in turn while being taught also teach. They become jointly responsible for a process in which all grow''. Through the dialogues, the role of today's students is to educate the world on the urgent need for climate action.

Education towards sustainable development (SD) should clearly address climate change adaptation and mitigation (Shohel & Howes, 2011). While there was a clear indication of limited professional education and training for climate change adaptation, it is urgently needed to be introduced as suggested by different professional bodies. The approaches to integrating climate change mitigation and SD with education and training are needed to address the problems of the climate change the world is facing (UNCC Learn, 2013). The study suggests that learning about CCA and CCM should be integrated with accredited education and training courses (Lyth, Nichols & Tilbury, 2007).

In the learning paradigm of education towards sustainable development (Shohel & Howes, 2011), three theoretical approaches of learning must be accommodated (Makrakis & Kostoulas-Makrakis, 2012). These theoretical approaches are experiential learning, constructivist learning and transformative learning. In combining these approaches, teachers should give a practical and valid activity related to any sustainable development issue so that students can explore and be able to identify the problem, its causes, its impacts and solutions for sustainable practices. In doing so, students should be able to show some mastery on collecting the necessary information, analyze and interpret the collected information using different inquiry-based methods, tools and techniques. For this, the teaching-learning procedure involves both inquiry-based and problem-based learning competences. Therefore, educating about climate change, its mitigation and adaptation become a relevant and essential need for every nation. The encouraging fact that an increasing number of schools in different countries have already started to embed the principles of education towards sustainable development in their curriculum planning, designing and delivery (Henderson, & Holman, 1993; Johnson et al., 2008; Shepardson et al., 2009).

Climate change education should be evidence-based, experienced-oriented and practice-centered so that it can assist in reducing the causes and impacts of climate change (Boakye, 2015). It should also focus on preparing societies for adaptability to the risk and physical environmental changes and the issues of the production of greenhouse gasses. Additionally, it is essential for students to change their unsustainable practices through learning-centered actions. These actions will provide them with the opportunities to experience and reflexively review their values, attitudes and practices for providing better solutions to the effects and impacts of climate change. In practice, for example, students could be taught how to implement energy-saving measures and strategies in their classrooms, in their schools and in their homes. By reflecting on personal and collective issues, students could be engaged in further reading and observing what is happening by collecting data, exploring multiple perspectives for gaining new experience to find the meaning and understanding of sustainability in every aspect of life.

For the students to make sense, construct knowledge, develop life-skills, individual and joint efforts are needed to reflect on their personal learning experiences. Their learning experiences help them to develop more abstract understandings by conceptualizing. This will also help them to co-create their individual and shared meaning by constructing and reconstructing. That is why involvement through a shared inquiry (problem-based group project) helps young people enrich their understanding by active experimentation, continuous reflection, and reconceptualization of issues surrounding them. Through a problem-based inquiry, learning becomes an active and contextualized process for them to construct meaningful knowledge based on their experiences and everyday practice. Constructed knowledge and meaning would be meaningful when it creates opportunities for them to act, which is also the key to develop active citizenship. Through combining knowledge and their meaning with action (*acting*) which indicates a change agency and active citizenship, young children would be able to act as change agents, become empowered to transform their experience through participation in actions including active experimentation, critical thinking and critical reflection. On the other hand, praxis, a process of transformation from critical reflection into action, makes the students able to transform oneself and society (*transforming*).

New values, positive attitudes, creative thinking, problem-solving and innovation for the best practices are the fundamental driving forces for mitigating, adapting and transforming the process of climate change to move toward sustainability. Based on these fundamental driving forces, critical analysis of causes and consequences helps students construct knowledge that subsequently leads to action. However, for integrating education for sustainable development and global citizenship in practice, teachers need to adopt appropriate pedagogies following the principles and values of the experiential, constructive and transformative learning. To become the co-learners, co-creator of knowledge, and facilitators of students' learning journey, teachers must move out from instructors as the sole source of information and knowledge. For doing so, teachers have to use multiple sources of information and knowledge and motivate their students and support them in the process of self-directed learning and enquiry. In fact, there are active and passive learners in the same cohort of the students. But for transformative learning, teachers must encourage the process of learning, whereby students participate actively and engage wholeheartedly in learning activities. Regarding climate change education, students need to act as active learners and activists through their actions. Therefore, they have to do various hands-on activities by collecting, analyzing and interpreting data to build new knowledge and understanding of climate change issues and good sustainable practices.

6.2 Curriculum Aspects

Including new contents on the causes and consequences of climate change and solutions to these causes and consequences within the curriculum is not enough to provide climate change education and prepare young people (Stevenson, Nicholls, & Whitehouse, 2017). Along with the vertical approach, both horizontal and spiral approaches are essential in designing the curriculum. The curriculum planning approach must be clearly described how to cluster knowledge and skills related to climate change adaptation and mitigation around core areas or sub-themes of learning. Within the integrated horizontal approach, a holistic view of the issues related should be encouraged through cross-fertilization of knowledge, values and skills within and between subjects taught in school; opportunities should be created for students to locate learning experience in the relevant contexts (Shohel & Howes, 2011); spaces must be provided for them to participate and explore interdisciplinary and cross-disciplinary questions and issues by promoting higher-level critical thinking skills (Shohel, Cann & Atherton, 2020); and teachers should be encouraged to collaborate with each other including team teaching within different subjects and active participation and collaboration among students must be encouraged (Shohel & Howes, 2008).

The teachers must be involved in organizing the core curriculum (both process skills and content knowledge) by accommodating climate change issues and sub-themes of sustainable development as openly and flexibly as possible so that students could be involved in reorganizing the course contents during the delivery of the subject contents. During the curriculum design plan, it is also essential to focus on the process so that the learning system could be a measure for understanding the outcome. To do so, both inter-disciplinary and cross-disciplinary approaches need to be adopted. However, diverse choices of participations and interactions could also be provided through web-based learning opportunities, including interactions between teacher and student, student and student, student and resources as well as contents. Moreover, all kinds of activities ranging from individual to group must be present in a decent curriculum.

An integrated curriculum is needed to address the issues of climate change and sustainability for Bangladeshi students at the primary and secondary levels. In the curriculum delivery process, an integrated thematic approach should be in place to teach climate change education effectively in conjunction with other school subjects. In this arrangement, connections between different themes and sub-themes could be easily identified as students get into a topic and explore more in-depth, they will be able to understand using various approaches. Again, cross-curricular projects might enable them to identify and understand how knowledge, values and skills are related to different school subjects. They will also help them understand how the knowledge is constructed in a particular context. The acquired skills could be used in other situations, especially in real-life contexts. Therefore, project-based learning could be used as a model to implement an integrated thematic approach for engaging students in different action-oriented activities.

6.3 Technology for Teaching and Learning about Climate Change and Sustainability

Information and Communication Technology (ICT) could be used in curriculum delivery for educating climate change in different ways. In the first place, different types of technologies or combinations of various technologies can be used in the teaching and learning process. Furthermore, technological aspects should be the core contents of the curriculum as technologies help to measure, collect, analyze and retrieve information. To show the interrelations between concepts and attributes, ICT tools could be used to help the students get useful learning experience. Besides, contemporary learning techniques such as peer learning or collaborative learning can be initiated to implement the integrated curriculum in open and distance learning (ODL) settings. In these settings, different digital tools and resources could be used for exploring and retrieving contents. The new ICT-based instructional practices can also help use the online and offline digital tools and facilitate interactions during synchronous and asynchronous learning activities among teachers and students.

A holistic framework for assessment should be in place to assist students to be aware of their personal thinking processes by using tools like multiple-choice questions (MCQs) with extended responses. Other assessment tools and techniques (such as concept mapping, scenario building, simulation, Venn diagrams, observation, checklists, portfolios, reflection, reflexivity, journalizing, case-study analysis) could be used depending on learning outcomes and contents. However, formative assessment, as the most critical process, provides effective and practical feedback opportunities. It also allows the active participation of students in designing and developing an integrated curriculum. For developing critical thinking, different tools (i.e., concept maps) and different structural knowledge representation techniques (i.e., symbolic, or hierarchical representation) could be extremely useful (Trumpowe & Sarwar, 2010). Through formative assessment, students should be given regular feedback and allowed to monitor their learning progress. However, peer learning support for assessment should be provided to students so that they can develop their skills to design self-assessment practice (Liang & Kim, 2004). Based on new knowledge, it will help them construct their own meaningful realities for processing abstract concepts or creating mental images through different visualization tools, including conceptual maps that could be used to design materials (Jonassen & Reeves, 1996; Jonassen, Peck & Wilson, 1998).

A well-planned climate change education program helps students understand across and within the curriculum how to address the causes and consequences of climate change and what could be better options for climate change mitigation. However, this kind of education program will also highlight the consequences which are related to climate change adaptation by adopting a transformative learning approach for making behavioral changes in practice. The causes and consequences of climate change, which are basically anthropogenic, must be identified and dealt with through human actions. However, there is a great deal of importance for identifying relevant solutions to adaptation practices at the local level. This is also equally significant to share good practices and transfer knowledge about different social, political and cultural strategies, economic and financial models, and technological innovation and integrations of new solutions to climate change adaptation and mitigation across the world.

Climate change education programs should be based on practically focused learning-oriented processes and methods that transform individuals of any age to reduce consumption of food, water and energy, change consumption patterns, use renewable forms of energy and depend on the circular economy. That means structural changes in social, cultural and economic systems would be part of the practice where the problems of sustainability lie. For doing so, there is a need for changing the curriculum and pedagogical approaches to transform values and practices that hinder sustainability. Only then climate change education will truly fulfil the 21st Century's needs and ensure the healthy environment of the planet earth.

Climate change education is a process of learning which enables the students to understand the causes of climate change and their consequences so that they know to respond to nature. It also highlights challenges of global warming in general to the earth and in particular to human life. In this effort of educating the new generations about climate change, curriculum development should include the ESD aspects by outlining the clear distinctions between different scientific concepts and processes related to climate change. It should also emphasize knowledge and abilities to understand climate change and how climate change is related to uncertainties, projections and risks. It should incorporate knowledge regarding the history and interrelated causes of climate change. It must also include the knowledge of climate change adaptation and mitigation practices that can contribute to the wider social, economic, cultural, political and ecological transformation towards sustainability.

Climate change education highlights the knowledge regarding the causes and consequences of global warming and climate change, and what is being learned about climate mitigation and adaptation. Therefore, it presents a good understanding of climate change dynamics and how to reduce the current greenhouse gas emissions. It also explores how to improve the quality of life for present generations and how to ensure security and development options for the generations to come. It also provides the understanding of different interest groups (for instance, business interests, consumer interests, farmers' interests, political interests, future generations' interests, etc.) and how their actions shape different responses to climate change. Concerning the public good, climate change education also gives the necessary skills to judge the validity of these interests critically. It also includes critical media literacy to understand the business motives and address the causes of over-consumption, exploitation of natural resources and importance of circular economy to develop capacity to make better lifestyle choices so that everyone can participate in the solutions of climate change.

The following stages can be set as part of young people's learning activities:

- Young people can investigate what the visible consequences of climate change are. They can give examples related to their school, family, nutrition, health, etc. They can also explore their responsibilities to mitigate the consequences given the examples from their everyday life (in school, home, neighborhood, local area) using different sources, including the media and internet.
- They can explore the impacts of climate change within the context of their local community, society, economy, employment, transportation, landscape, forestry and agriculture.

- Following the previous step, they can also analyze their local and national ecosystem and the repercussions climate change brings upon it. Then think about the deforestation and afforestation processes and then develop a champagne strategy for preventing deforestation and introducing reforestation.
- They can then learn about the physics of the atmospheric phenomenon of climate change. Then they can even go further to learn about the chronological impacts of climate change and predict the future.
- They can also formulate scenarios and prepare themselves for the future, having the knowledge from the past. Then they can analyze the consequences of using energy, fossil fuels and waste generation.

Young people are often very conscious and look for the ways which promote sustainable living e.g., using renewable sources of energy and new ways to handle the waste. Therefore, they could be engaged in different activities related to dealing with climate change issues into their doorsteps and their local environment. These activities could show them the ways how children can act to protect their environment. Alongside the activities conducted in the classroom, many other outdoor activities could be suggested, which will have positive impacts on behavioral changes to sustainability. During these in school activities, teachers could help young people understand the magnitude of climate change by gradually referring different examples from the local to the national and global levels.

Online-based learning environments could be created for providing education for sustainable development and global citizenship. However, the impacts of climate change could be explored through open and flexible digital contents, app and online-based e-learning opportunities and affordable ICT-based projects as these allow young people to learn anytime from anywhere. By providing opportunities to young people using emerging technologies to get CCE to develop their knowledge and understanding, engaging them in transformative learning seems easier and more manageable.

6.4 Moral and Ethical Emphasis on Changing Behaviors

Willingness and moral obligation of the politicians could play a vital role to tackle climate change, and in this regard, Greta Thunberg, an environmental activist has raised a deep and thought-provoking question, 'What is the point of learning facts when the most important facts given by the finest scientists are ignored by our politicians?' (The Guardian, 2018, Online). In addressing the question, the easiest way is to enhance moral and ethical understanding of young people to change their behaviors to learn and practice how to become a global citizen and act ethically and morally justified ways.

7. IMPLICATIONS, RECOMMENDATIONS AND FUTURE DIRECTION

7.1 Approaches of School, Education System and Public-Private Partnership

Climate change education needs interdisciplinary approaches, which require more work to equip students for the problems and discussions they might face outside of the classroom. Schools have the opportunity to help young people by providing climate change education to evaluate different evidence to come up with the conclusions by their own. Schools can help young people by providing climate change education to evaluate relevant evidence to come up with the conclusions on their own. However, without educating them about climate change, they will not be able to address the issues they face without getting trained to use the information and conclude based on different views and debates on global warming and its impacts (Graham, 2018). The economic, social and ecological crises faced by the current generation have its roots in the behaviors of previous and current generations to treat the environment over the years (Makrakis, Larios & Kaliantzi, 2012). At present, climate change mitigation and adaptation strategies are crying needs to save the only earth to be inhabited. Otherwise, the earth would be destroyed, and future generations' lives will be jeopardized.

Education systems across the globe need to be reoriented to understand the reasons, explore the consequences, and find out the solutions to climate change if expected changes to be in place so that people's thinking moves towards sustainability. To teach about climate change, a practical issue needs different pedagogical approaches to learn and bring changes in human behaviors. Therefore, it is now a significant necessity to understand the process of teaching and the system of learning related to the climate change issues.

In order to address the cause and consequence of climate change in our everyday practices, it requires a paradigm shift in pedagogical approaches, learning materials, content, and procedure in different levels of education. That is why climate change education puts emphasis on improving the capacity for mitigation, alternation and familiarization at all levels, particularly in personal and societal levels. Therefore, providing climate change education through formal education is vital for future generations. However, at the same time, informal and nonformal education channels should be used to win the heart of other people. As such, Public-Private Partnership is essential to face the challenges of climate change, environmental degradation as well as making people aware of the wellbeing of the children. Active citizenship and global learning are also equally crucial without which we cannot achieve the goals of sustainable practices. Additionally, indigenous and religious perspectives on environmental issues should be highlighted through religious education (RE), cultural practice and social mobilization campaigns.

7.2 Local Curriculum and Community-based Adaptation

Although a part of the current curriculum contains the basic information on climate change and sustainable development issues, it is of utmost importance to have inclusive changes in the national school curriculum on climate change, its mitigation and adaptation due to the recently predicted threat of climate change, especially for low-lying countries like Bangladesh.

It is vital to encourage young people to become change agents in their community and globally. Different mechanisms should be adopted to raise awareness about climate change through informal, nonformal and formal education programs. Therefore, the school curriculum should provide quality climate change education for young people so that they can understand the intensity of climate change and associate issues. The policy and practice also should be in place at local, national and global levels to encourage climate change education to be part of the school curriculum. In this climate change education process, different innovative teaching and learning approaches should be encouraged to deliver different contents related to climate change issues. Different media, networking and partnerships channels should be used in this educating effort along with formal side education (Global Center on Adaptation, 2020).

During national curriculum planning, pedagogical approaches must be adapted, especially in the content, to understand the impact of global warming and to increase 'climate literacy' among young people. The incorporation of the climate change topics into school curriculum should prioritize bringing the classroom teaching and learning to the real life setting of the communities. That would design and deliver in a way which will assist the fight against climate change in practice. Moreover, bringing schools and communities together is also vital so that climate change education could be multiplied through the collective responses. Along with learning the science behind climate change, students also need to learn how businesses, communities, and individuals deal with problems together, which could impact national, regional, and global levels. By doing that, they understand the larger picture of global changes in policy and practice.

7.3 Taking Responsibility: Individually and Collectively

Climate change education provides a learning experience about the responsibilities at the individual and societal level and how to live an eco-friendly life. Educators have a responsibility to teach children and young people to be mindful citizens. It is important to teach students about their responsibilities towards climate change and assist them to understand the importance of preserving the planet earth for themselves and future generations. They also need to learn how to reverse the current impact of climate change collectively so that the health of the earth could be retrieved. More than that, recently visible consequences of climate change remind us that we must act accordingly in response to climate change and consider the impacts of our actions beyond the present. Therefore, in our practice, we need to emphasize the circular economy by reducing, reusing, and recycling through our efforts. That should be the core of consumer behaviors and everyday practice in every sphere of our lives.

7.4 Hope in the Dark: Action Needed

The future of humankind is deeply uncertain, but the present is full of hope to create a better future for the next generations. The changes in the ways of human living on the earth that are both sustainable and eco-friendly can change global warming and environmental degradation (Agostino, 2010). Renewable energies are now cheaper than fossil fuels, energy storage is advancing, and electric cars will replace petrol sooner than we think. These developments illustrate the advancement of technologies, but they are not good enough to prevent climate change threats. There are many reasons to believe that we human beings can save our planet if we contribute our fair share of responsibilities for a sustainable environment. However, the truth is that more and more people are aware of the impact of human activity on the climate and actively engaged in tackling it in their personal, professional and political lives.

7.5 Making Climate Change Education Integral Part of National Curriculum

Some countries have already made compulsory classes on climate change in their primary and secondary schools. Climate change education should be an integral part of the national school curriculum in Bangladesh. As in Bangladesh, there is a centralized school curriculum and the same textbook across the country, and it would be easier to plan and implement the content and practice throughout the country. However, curriculum developers need to understand, what to incorporate in the curriculum and at what levels? How to educate and train future generations from an early age for the preparedness of climate change adaptation and sustainability?

7.6 Technological Innovation and Adaptation

Through the school curriculum, the Government should emphasize the importance of science and technology in response to climate change mitigation, adaptation and sustainability. Science and technology can play a vital role to provide adequate climate education in schools. Based on the ICT-enabled integrated curriculum, learning about climate change, its mitigation and adaptation could be an enjoyable experience for students.

7.7 Technology Enhanced Teaching and Learning

Emerging technologies can help integrate reflection in the learning process and provide the opportunities of problem based learning and provide different tools that increase awareness of climate change to students. Technologies are already playing a vital role in improving curriculum delivery and assessments, however, how it can be integrated for climate change adaptation and sustainability should be a major concern while adapting the emerging technologies in teaching and learning. So, the key question here is: How can the integration of science and technology contents be in primary and secondary curriculums as reflective/problem-based learning plays a key role in tackling climate change?

To understand the actual situation for raising awareness, particularly at the grassroot level, data analysis and prediction are essential for which technology plays a significant role. Data-driven decisions also help gain practical knowledge that facilitates the poor and marginalized to be empowered and raise their voice. To realize the potential of information and communication technologies, geospatial technologies such as Geographic information system (GIS) can be used to teach and learn concerning climate change. However, GIS is a set of technologies which allow users to view, understand, question, interpret and visualize spatial data in many ways that reveal relationships, patterns and trends in various forms. It helps enhance spatial reasoning and support problem-solving in the classroom. Furthermore, the proper implementation of GIS instruction may promote climate change literacy and interdisciplinary or cross-disciplinary learning in the school. Apart from GIS, the internet offers multiple perspectives of outstanding and different resources which could be assessed and used to support ESD components in the school curriculum.

8. CONCLUSION

Climate change is a complex global issue affecting every living being on earth. It could be addressed through shared responsibility and requires collective efforts from humankind who is mostly responsible for. It is the hard truth that we all fail to act according to our individual or group interests and not be active and conscious in our sustainability behaviors. This chapter highlights the importance of integrating content related to climate change, its mitigation and adaptation in school textbooks in Bangladesh. As shown in the finding section of this chapter, more contents and activities need to be included in school textbooks to inform young people about the issues of climate change, its mitigation and adaptation more widely. The provided content in the textbooks should be based on problem-oriented real-life aspects that would provide specific and authentic learning situations related personal experiences. This should build on what they encountered in their daily life and what may happen to them and their family due to climate change and what can be done to reverse the climate change effect. Young people could be involved in real-life problems, studying stories and case studies, and they can explore websites, articles or videos (real or animated) and different digital learning resources.

Besides the three main factors (identified by IPCC) that determine vulnerability to climate change, adaptation measures can help reduce vulnerability, e.g., by lowering sensitivity or building adaptive capacity and allowing populations to benefit from opportunities of climatic changes, such as growing new crops in previously unsuitable areas. The crisis of climate change must be addressed, and one of the effective and collective actions right now is to integrate climate change education in the school curriculum to teach next-generation citizens about climate change, its mitigation and adaptation. Our findings suggest that while the national curriculum provides a clear guideline of what the Government expects schools should teach their students, the current national curriculum, however, offers little on human, physical and scientific mechanisms of climate change, its mitigation or adaptation. Nevertheless, teaching about climate change - an interdisciplinary topic, is by no means easy, and it is a challenge worth tackling for saving the only planet we have to live on.

An ambitious yet achievable climate change education embedded in the curriculum of primary and secondary schools could empower young people to understand the natural world. It will provide them with an opportunity to see themselves as part of it, before giving them the chance to engage with it. It is important that school must nurture their passions and help them shape the future they stand to inherit. In this way, the future generation will be prepared to address the issues of climate change and sustainability.

REFERENCES

Agostino, A. (2010). Gender equality, climate change and education for sustainability. *Newsletter for beyond access: Gender, education and development*, 24. Retrieved on April 12, 2020, from: http://www.e4conference.org/wp-content/uploads/2010/02/Equals24.pdf

Bangay, C., & Blum, N. (2010). Education Responses to Climate Change and Quality: Two Parts of the Same Agenda? *International Journal of Educational Development*, *30*(4), 335–450. doi:10.1016/j.ijedudev.2009.11.011

Bangladesh Bureau of Statistics (BBS). (2018). *Bangladesh statistics 2018*. Statistics and Informatics Division (SID) Ministry of Planning. Retrieved on June 28, 2020, from http://bbs.portal.gov.bd/sites/default/files/files/bbs.portal.gov.bd/page/a1d32f13_8553_44f1_92e6_8ff80a4ff82e/Bangladesh%20%20Statistics-2018.pdf

Boakye, C. (2015). Climate Change Education: The Role of Pre-Tertiary Science Curricula in Ghana. *SAGE Open*, *2015*(4), 1–10. doi:10.1177/2158244015614611

Bowen, G. A. (2009). Document analysis as a qualitative research method. *Qualitative Research Journal*, *9*(2), 27–40. doi:10.3316/QRJ0902027

Brown, O. (2008). Migration and Climate Change. Geneva, Switzerland: International Organisation for Migration (IOM). doi:10.18356/26de4416-en

Dupigny-Giroux, L. A. (2010). Exploring the challenges of climate science literacy: Lessons from students, teachers and lifelong learners. *Geography Compass*, *4*(9), 1203–1217. doi:10.1111/j.1749-8198.2010.00368.x

European Commission (EC). (2018). *Climate Action: Our planet, our future - fighting climate change together*. Luxembourg: Publications Office of the European Union. Retrieved June 28, 2020, from https://ec.europa.eu/clima/sites/clima/files/youth/docs/youth_magazine_en.pdf

Freire, P. (2008). *Education for critical consciousness* (3rd ed.). Continuum.

Freire, P. (2005). *Pedagogy of the oppressed* (30th anniversary ed.). The Continuum Publishing Company.

Freire, P. (1968). *Pedagogy of the oppressed*. The Continuum Publishing Company.

Global Commission on Adaptation. (2019). Adapt Now: A Call for Leadership on Climate Resilience. Rotterdam: Global Center on Adaptation and Washington: World Resources Institute.

Global Center on Adaptation. (2020). *Innovation in Education and training as a tool to enhance youth participation in adaptation action: Integration of Bottom-up and Top-down Approaches*. Rotterdam: Global Center on Adaptation and Washington: World Resources Institute. Retrieved on December 23, 2020, from http://tep-a.org/policy-briefs/

Government of Bangladesh (GoB). (2017). *An introduction to Bangladesh*. Ministry of Foreign Affairs, Bangladesh. Retrieved June 28, 2020, from https://mofa.gov.bd/site/page/6dde350b-1ca6-4c69-becd-a3f12cf14ac1/Bangladesh--An-Introduction

Government of Bangladesh (GoB). (2019). *Statistical yearbook of Bangladesh 2019*. Bangladesh Bureau of Statistics, Ministry of Planning, Government of Bangladesh.

Graham, A. (2018). *Why Should Schools Teach Climate Education?* UN CC: Learn. Retrieved on December 23, 2020, from: https://medium.com/uncclearn/why-should-schools-teach-climate-education-f1e101ebc56e

Henderson, S., & Holman, S. R. (1993). Global climate change education: Technology transfer to schools. *Climate Research*, *3*(1-2), 137–140. doi:10.3354/cr003137

Houghton, J. T., Meira Filho, L. G., Callander, B. A., Harris, N., Kattenberg, A., & Maskel, K. (1996). *Climate change 1995: The science of climate change*. Cambridge University Press.

Intergovernmental Panel on Climate Change (IPCC). (2014). *Glossary. Intergovernmental Panel on Climate Change*. Retrieved on May 2, 2020, from https://www.ipcc.ch/site/assets/uploads/2018/02/WGIIAR5-AnnexII_FINAL.pdf

International Alliance of Leading Education Institutes (IALEI). (2009). *Climate change and sustainable development: The response from education*. Available at: https://www.intlalliance.org/fileadmin/user_upload/documents/DPU_recommendations.pdf

Islam, T., Rahman, M. A., & Mallick, F. (2010). Impacts of climate change induced hazards and adaptation processes in Bangladesh: An overview. In *Climate Change Adaptation and Disaster Risk Reduction: An Asian Perspective*. Emerald Group Publishing Limited. doi:10.1108/S2040-7262(2010)0000005008

Johnson, R. M., Henderson, S., Gardiner, L., Russell, R., Ward, D., Foster, S., Meymaris, K., Hatheway, B., Carbone, L., & Eastburn, T. (2008). Lessons learned through our climate change professional development program for middle and high school teachers. *Physical Geography*, *29*(6), 500–511. doi:10.2747/0272-3646.29.6.500

Jonassen, D. H., & Reeves, T. C. (1996). Learning with technology: Using computers as cognitive tools. In D. H. Jonassen (Ed.), *Handbook of research for educational communications and technology* (pp. 693–719). Macmillan.

Jonassen, D. H., Peck, K. L., & Wilson, B. G. (1998). *Learning with technology: A constructivist perspective*. Prentice-Hall.

Ledley, T. S., Rooney-Varga, J., & Niepold, F. (2017). Addressing Climate Change Through Education. In *Oxford Encyclopaedia of Environmental Science*. Oxford University Press. doi:10.1093/acrefore/9780199389414.013.56

Liang, X., & Kim, C. (2004). Classroom assessment in a web-based institutional environment: Instructors' experiences. *Practical Assessment, Research & Evaluation*, *9*(7). Retrieved June 12, 2020, from https://scholarworks.umass.edu/pare/vol9/iss1/7/

Loria, E. (2018). The Great Inequity of Climate Change. *One Natural Energy*. Retrieved on December 19, 2020, from https://www.onlynaturalenergy.com/the-great-inequity-of-climate-change/

Lyth, A., Nichols, S., & Tilbury, D. (2007). *Shifting towards sustainability: Education for climate change adaptation in the built environment sector.* A report prepared by the Australian Research Institute in Education for Sustainability. Retrieved on April 15, 2020, from http://aries.mq.edu.au/projects/ClimateChange/files/climateChange.pdf

Makrakis, V., & Kostoulas-Makrakis, N. (2012). The challenges of ICTs to online climate change education for sustainable development: The ExConTra learning paradigm. In S. A. Anwar (Ed.), *Proceedings of the 5th Conference on eLearning Excellence in the Middle East – Sustainable Innovation in Education* (pp. 594–605). Dubai, UAE: Hamdan Bin Mohammed e-University.

Makrakis, V., Larios, N., & Kaliantzi, G. (2012). ICT-enabled climate change education for sustainable development across the school curriculum. *Journal of Teacher Education for Sustainability*, *14*(2), 54–72. doi:10.2478/v10099-012-0009-5

Ministry of Environment and Forests. (2009). *Bangladesh climate change strategy and action plan 2009*. Author.

National Curriculum and Textbook Board (NCTB). (2020, May). *Mission and Vision of NCTB*. Retrieved on May 18, 2020, from http://www.nctb.gov.bd/

National Oceanic and Atmospheric Administration (NOAA). (2011). *Increasing public demand for NOAA'S climate-related data and services.* United States Department of Commerce: NOAA. Retrieved on May 28, 2020, from https://www.noaa.gov/climate.html

Reid, A. (2019). Climate change education and research: Possibilities and potentials versus problems and perils? *Environmental Education Research*, *25*(6), 767–790. doi:10.1080/13504622.2019.1664075

Shafer, M. A. (2008). Climate literacy and a national climate service. *Physical Geography*, *29*(6), 561–574. doi:10.2747/0272-3646.29.6.561

Shams, S., & Shohel, M. M. C. (2016). Food security and livelihood in coastal area under increased salinity and frequent tidal surge. *Environment & Urbanization Asia*, *7*(1), 22–37. doi:10.1177/0975425315619046

Shams, S., Mahruf, M., Shohel, C., & Ahsan, A. (2014). Housing problems for middle and low income people in Bangladesh: Challenges of Dhaka megacity. *Environment & Urbanization Asia*, *5*(1), 175–184. doi:10.1177/0975425314521538

Shepardson, D. P., Choi, S., Niyogi, D., & Charusombat, U. (2009). Seventh grade students' mental models of the greenhouse effect. *Environmental Education Research*, *15*(5), 549–570. doi:10.1080/13504620903114592

Shohel, M. M. C., Cann, R., & Atherton, S. (2020). Enhancing students' engagement using a blended learning approach: Case studies of first-year undergraduate students. *International Journal of Mobile and Blended Learning*, *12*(4), 51–68. doi:10.4018/IJMBL.2020100104

Shohel, M. M. C. & Howes, A. J. (2008). Informality of teaching and learning in nonformal schools: Sociocultural process as meso-systems of student development. *Education 3-13*, *36*(3), 293-309.

Shohel, M. M. C., & Howes, A. (2011). Models of education for sustainable development and nonformal primary education in Bangladesh. *Journal of Education for Sustainable Development*, *5*(1), 129–139. doi:10.1177/097340821000500115

Singer, F. S., & Seitz, F. (1998). *Hot talk, cold science: Global warming's unfinished debate*. Washington, DC: The Independent Institute.

Stephens, J. C., & Graham, A. C. (2008). Climate science to citizen action: Energizing nonformal climate science education. *Eos (Washington, D.C.)*, *89*(22), 204–205. doi:10.1029/2008EO220010

Stevenson, R. B., Nicholls, J., & Whitehouse, H. (2017). What is Climate Change Education? *Curriculum Perspectives*, *37*(3), 67–71. doi:10.100741297-017-0015-9

Tatenda, M. (2016). *Curriculum Reforms: Essential Tool for Effective Implementation of Climate Change Adaptation Strategies*. Retrieved on May 3, 2020, from https://climatestrategies.org/wp-content/uploads/2016/04/Tatenda-Curriculum-Reforms-for-Climate-Adaptation-Strategies.pdf

The Daily Star. (2020). *Prepare yourselves as worthy citizens: Zakir Hossen to primary students*. Retrieved on May 18, 2020, from https://www.thedailystar.net/country/textbook-distribution-festival-2020-held-amid-enthusiasm-1848082

The Guardian. (2018). *I'm striking from school to protest inaction on climate change – you should too.* Retrieved on December 19, 2020, from https://www.theguardian.com/commentisfree/2018/nov/26/im-striking-from-school-for-climate-change-too-save-the-world-australians-students-should-too

Trumpowe, D., & Sarwar, G. S. (2010). Concept maps: Making learning meaningful. In A. J. Sánchez, A. J. Cañas & J. D. Novak (Eds.), *Formative structural assessment: Using concept maps as assessment for learning. Proceedings of the 4th International Conference on Concept Mapping* (pp. 132–136). Lom Ediciones SA Publisher.

UNCC Learn. (2013). *Resource Guide for Advanced Learning on Integrating Climate Change in Education at Primary and Secondary Level.* United Nations Institute for Training and Research (UNITAR).

United Nations Educational, Scientific and Cultural Organisation (UNESCO). (2002). *UNESCO universal declaration on cultural diversity.* UNESCO.

United Nations Educational, Scientific and Cultural Organisation (UNESCO). (n.d.). *Climate change education.* Retrieved on January 2, 2020, from https://en.unesco.org/themes/education-sustainable-development/cce

United Nations Educational, Scientific and Cultural Organisation (UNESCO). (2015). *Climate change education for sustainable development in Bangladesh. UNESCO programme and meeting document.* Retrieved on June 28, 2020, from https://unesdoc.unesco.org/ark:/48223/pf0000246779

United Nations Educational, Scientific and Cultural Organisation (UNESCO). (2010). *The UNESCO climate change initiative: Climate change education for sustainable development.* Retrieved May 14, 2020, from https://unesdoc.unesco.org/images/0019/001901/190101e.pdf

World Food Programme (WFP). (2020). *Country Brief: Bangladesh 2020.* Retrieved June 26, 2020, from https://docs.wfp.org/api/documents/WFP-0000116570/download/?_ga=2.131938381.319260259.1593310272-46580275.1593310272

Youth Non-Governmental Organisation (YOUNGO). (2011). *Enhanced youth participation and education. In climate change: The Article 6 implementation toolkit. A collaborative document created by international youth.* Retrieved September 14, 2012, from the YOUNGO Article 6 Working Group. Available at: https://unfccc.int/cc_inet/files/cc_inet/information_pool/application/pdf/enhanced_youth_participation.pdf

KEY TERMS AND DEFINITIONS

Climate Change: Changes happen in nature and physical environment due to carbon dioxide increase at atmospheric because of using fossil fuels and cause global warming. As a result, there is a change in climate patterns globally and regionally. The change in climate patterns is called climate change.

Climate Change Adaptation: Climate change adaptation means adapting to lifestyles and livelihood due to climate change. It requires adjusting to present ongoing changes in climate and preparing for expected future climate. The main reason for climate change adaptation is to reduce vulnerability due to the harmful effects of climate change.

Climate Change Education: Climate change education is a process through which different issues related to climate change incorporated in the learning process using diverse pedagogical approaches in informal, nonformal and formal educational programs. It increases awareness of global warming and the importance of living a sustainable life.

Climate Change Mitigation: Climate change mitigation is the efforts that reduce or prevent greenhouse gas emissions. Actions for mitigation include using renewable energies and energy-efficient technologies, following the circular economy, and changing consumer behavior.

Climate Justice: Climate justice is the notion which emphasizes action against global warming as an ethical and political issue, rather than an environmental or physical issue related to nature. The central part of climate justice involves legal action by the growing global body and supporting the activist against climate change issues.

Education for Sustainable Development: Education for sustainable development (ESD) is defined as the educating process that encourages changes in knowledge, values, attitudes and skills to enable a more sustainable and just society for all. ESD encourages and empowers individuals for informed decision making and acts considerately for social justice by taking care of environmental integrity, economic viability, for present and future generations.

Global Citizenship: Global citizenship is a notion that encourages the idea of citizenship beyond political or geographical identity. It is based on moral and ethical responsibilities related to the universal rights of humankind to live in peace and harmony in the global village where interdependence and interconnectedness are intertwined, and sustainable and fairer living must be practiced.

School Curriculum: School curriculum is the curriculum which dictates the academic activities at the school level. In general, the curriculum means the courses offered by a school or any educational institution. However, in the broader sense, the totality of experiences an individual has throughout the schooling process including a planned sequence of instruction or a view of the student's experiences in terms of the educator's or school's instructional goals.

Sustainability: Sustainability means meeting the needs of current demand without compromising the needs of future generations. It is the ability to exist continually without harming the ecological harmony of the earth. The biosphere and humankind can co-exist and which has three dimensions, namely social, economic and environmental.

Sustainable Development Goals: The Sustainable Development Goals (SDGs) are a set of goals adopted by all United Nations Member States in 2015. The purposes of these goals are to end poverty, protect the planet and ensure that all people enjoy peace and prosperity by 2030. These 17 interlinked goals aim for achieving a better and more sustainable future for all.

Young People: The term 'young people' refers to the older children who better understand their worlds or their surroundings. They are more experienced and able to make the decisions for themselves.

Section 4
Climate Change and Environment

Chapter 14
Determinants of Household Pro-Environmental Practices:
An Application of Australian Survey Data

Scott Baum
https://orcid.org/0000-0003-1711-2087
Griffith University, Australia

ABSTRACT

The decision by individuals and households to undertake recycling and to adopt green technology is driven by a large array of factors. Understanding the role of these factors is in turn important for developing appropriate green policies and in enacting sustainable and efficient education and information campaigns. The main aim of this chapter is to outline the factors impacting on pro-environmental behavior of a sample of Australian households. Using data drawn from a survey undertaken across the State of Queensland, the chapter will consider how a factors such as socio-economic background, beliefs around climate change, previous exposure to environmental hazards, and beliefs about individual and government abilities to determine outcomes help explain patterns of pro-environmental behavior in various settings.

INTRODUCTION

The decision by individuals and households to undertake recycling and to adopt green technology is driven by a large array of factors including the level of environmental awareness, political views and socioeconomic status and background (Clark, Kotchen, & Moore, 2003; Li, Zhao, Ma, Shao, & Zhang, 2019; Trotta, 2018). Understanding the role of these factors is in turn important for developing appropriate green policies or programs and in enacting sustainable and efficient education and information campaigns (Goldman et al., 2020; Pooley & O'Connor, 2000; White & Wall, 2008). In developing educational materials especially, understanding the motivations and barriers to pro-environmental behavior is important in developing effective pedagogy (Pooley & O'Connor, 2000; Skamp, Boyes, & Stanisstreet, 2013). Within this context, the aim of this chapter is to analyze the factors impacting on pro-environmental practices and choices of households using a sample from the state of Queensland in Australia.

DOI: 10.4018/978-1-7998-7512-3.ch014

BACKGROUND

In Australia, as elsewhere, there has been increasing concern around the impact of human activity on climate change and with this, increasing interest in understanding how individuals and households can change their behavior or practices to become more environmentally friendly. To effect these changes, there has been an ever increasing body of literature around the issue of environmental education (Braun, Cottrell, & Dierkes, 2018; Gifford & Sussman, 2012; Pooley & O'Connor, 2000). While the link between environmentally friendly practices and climate concerns is often seen as being straight-forward, the factors that influence such behaviors are in fact complex and driven by a large array of factors (Giefer, Peterson, & Chen, 2019; Gifford & Nilsson, 2014; Li et al., 2019), an issue that has direct impact on the content and role of environmental education programs (Pooley & O'Connor, 2000). That is, if the role of environmental education is partly to motivate people to work towards solutions, then the importance of understanding the drivers of particular environmental practices is clear.

Within the literature, environmentally friendly behavior is often referred to as 'pro-environmental' behavior or behavior that can act to reduce a negative impact on the environment (Li et al., 2019). As such, pro-environmental behavior includes recycling and improving waste management, changing transport use, consuming 'green' energy purchasing energy efficient products or devices and undertaking political activism (Blankenberg & Alhusen, 2019), each being viewed as part of a smaller subset of behaviors including consumer behavior, environmental citizenship and policy support (Stern, 2000).

Developing a comprehensive understanding of pro-environmental behavior requires, in turn, an examination of the multiple factors thought to induce, determine and change the way behavior is undertaken (Steg & Vlek, 2009). Often pro-environmental behaviors can be seen as habitual such as setting the air conditioning to a given temperature and are therefore not necessarily a conscious behavior. At other times a given behavior is income constrained or context specific as in the case of installing photovoltaic cells for electricity generation or insulating homes (Stern, 2000). Given the potential complexity of the factors impacting on pro-environmental behavior, researchers including Stern (2000) have turned to theories and broad conceptualizations to help develop understanding. Stern (2000) for example utilizes a conceptualization built around a value-belief-norm theory that explains behavior in terms of: attitudinal factors such as norms and beliefs; contextual factors such as monetary incentives and costs, community expectations or government regulations; personal capabilities such as knowledge, skills and resources; and habit or routine which requires changing or breaking. Similar conceptual approaches are also present in the work of researchers including (Kollmuss & Agyeman, 2002) who developed a conceptual framework focusing on the role of demographic, external and internal factors in shaping behavior. External factors include institutional constraints, lack of financial control and means cultural norms and political support. Internal factors include an individual's motivation, environmental knowledge and awareness, their values, the extent to which they have control to undertake action (locus of control), as well as their priorities. With reference to understanding potential educational solutions, it is.

The bulk of the empirical work aiming to understand pro-environmental behavior has focused on identifying these conceptual factors and understanding how various factors differentiate between behavior outcomes. Several studies have aimed to provide insights into the differential uptake of various practices across households and individuals (Fielding & Head, 2012; Fielding, Louis, Warren, & Thompson, 2011; Newton & Meyer, 2013; Pearce, Willis, Mamerow, Jorgensen, & Martin, 2014). Fielding et al. (2011) find a significant variation in household's attitudes towards the up-take of pro-environmental practices such as the use energy efficient appliances and taking steps to conserve water, with gender and age important in differentiating outcomes. Trotta (2018) using British survey data was able to profile households to identify differences in willingness to undertake conservation and efficiency measures. These differences prompted the author to point out that 'policy makers and energy saving organisations need to devote more attention to the household profiles' when they develop policy and educational products (Trotta, 2018, p. 537). In a study of residents in Melbourne Australia, Newton and Meyer (2013) used a lifestyle segmentation approach to group households according to their level of environmental commitment. Comparing the 'committed greens' to the 'enviro-sceptics' showed significant differences in household structure, with committed greens being more educated, younger with higher incomes compared to the 'enviro-sceptics' who were more likely to be male and older. Looking at the up-take of solar energy production technologies Welsch and Kühling (2009) find that demographic factors such as years of education, types of employment and household size all significantly impact on the likelihood a household will install solar panels in their home.

Over and above the potential impact of demographic factors on behavior, researchers have also considered factors consistent with Kollmuss and Agyeman's (2002) category of external drivers. The potential impact that housing tenure may play in pro-environmental behavior has been picked up in several studies. Fielding et al. (2011) illustrate that home owners were more likely to have highly positive attitudes and a greater level of engagement in environmentally sustainable practices when compared to other tenure types. The constraint on renters of undertaking major renovation activities is illustrated by the research conducted by Welsch and Kühling (2009) who found that people who lived in a rented dwelling were significantly less likely to have solar equipment than those who own their house. A household's disposable income may also be an important influence of pro-environmental behavior with decisions being based on questions of costs and benefits and economic priorities. (Clark et al., 2003) looking at the decision to participate in a green electricity program identifies income as a significant factor with the willingness to participate increasing with as income rises. In contrast to these findings (Fielding et al., 2011) find that in general it is lower income households who tend to have more positive attitudes towards environmental sustainability and also it is these households who are more likely to engage in certain sustainability practices. This was especially the case in terms of energy and water conservation practices. The exception had to do with the installation of energy efficient appliances where higher income households were more likely to participate.

The internal factors discussed by Kollmuss and Agyeman (2002) cover a range of possibilities that include the extent to which an individual feels responsibly for environmental change and futures, the perceived level of control to make change and environmental concern, knowledge and experience. A range of existing studies has identified that responsibility and a sense of empowerment to act are important drivers in decisions to undertake positive environmental actions. In a study of young Australians (Fielding & Head, 2012) find that both a sense of community responsibility to act and also the ability to act (locus of control) were important in explaining both pro-environmental intentions and pro-environmental behavior. There is evidence that environmental knowledge and concerns are also an important driver of pro-environmental behavior. Studies have generally found that in samples where the level of environmental knowledge is high there is a commensurate high level of pro-environmental practices. In the study of 14 to 18 year olds on the west coast of the USA, Meinhold and Malkus (2005) identified that among other variables, attitudes and behavior were impacted by the knowledge respondents held regarding the environment. Similarly, in the study of young Australians Fielding and Head (2012) find that environmental knowledge is positively associated with pro-environmental intentions and behavior and negatively associated with environmental harmful behavior. These types of findings are also reflected in the meta-studies conducted by Bamberg and Möser (2007). Finally evidence from some studies suggest that exposure or experiences with environmental risk or hazards may act as a driver of pro-environmental action. Lorenzoni, Nicholson-Cole, and Whitmarsh (2007) in a study of UK residents in flood prone locations identified that experience of flood had a positive impact on engagement with issues about climate change. Also in the UK (Whitmarsh, 2009) found that experience in terms of air pollution whilst it did not impact on a respondents views regarding the cause of climate change it did impact on their pro-environmental values.

MAIN FOCUS OF THE CHAPTER

The main focus of this chapter is to consider the factors that impact on pro-environmental practices and consider how this can inform education around appropriate behaviour. Within this chapter the focus is on practices that include reducing the use of resources, installing energy efficient infrastructure and being engaged in political activism. Data on these practices, together with a range of socio-demographic factors, beliefs and knowledge were obtained from a sample of 1500 households across the state of Queensland Australia.

Respondents were asked to what extent they had considered or had undertaken the following practices:

- Reduce motor vehicle use
- Install energy efficient light bulbs
- Reduce amount of water used around the house and garden
- Use energy efficient appliances
- Vote for particular political candidates
- Sign a petition for (local or global) action
- Reduce gas and/or electricity used around the house
- Install rainwater tank(s)
- Install solar power panels
- Increase home insulation

Responses were recorded across three possible categories; not considered taking action, considered taking action or taken action. In order to develop an analysis of pro-environmental practices the separate questions were reduced to distinct categories similar to those identified by Stern (2000). These categories focused on the use of resources, the use or uptake of energy efficient infrastructure and the extent to which individuals engaged in political activism. While causal interpretation of the 10 types of practices suggest that each can be placed into one of the broader categories, Categorical Principle Components Analysis (CATPCA) was undertaken to confirm membership. The category action to reduce resource use comprised responses about reducing motor vehicle use, installing energy efficient light bulbs, reducing the amount of water used around the house and garden, using energy efficient appliances and reducing gas or electricity used around the home. The second category, uptake of energy efficient infrastructure, comprised responses to the questions regarding installing rainwater tanks, installing solar power systems and increasing or installing home insulation. The final category, engaged in political activism, comprised responses to questions regarding if household members had signed a petition or had voted for a particular political party. The final indicator for each of the three pro-environmental practice categories was developed using weights obtained from the principle components analysis to compute a simple weighted index.

To understand pro-environmental practices in the context of the chapters aims, it is necessary to consider how each of the categories of behavior outlined above is associated with a range of demographic, internal and external drivers (Kollmuss & Agyeman, 2002). The choice of socio-economic/ demographic indicators presented in the paper is driven by suggestions from existing research (Fielding & Head, 2012; Fielding et al., 2011; Kollmuss & Agyeman, 2002; Welsch & Kühling, 2009) and the data that was availability from the survey instrument. The variables included are outlined in Table 1 and include the sex of the respondent, the age of the respondent (differentiated along generational lines) and the level of education. The external drivers (Kollmuss & Agyeman, 2002) of pro-environmental practices are accounted for by 6 variables. Three account for the ability of a household to engage in pro-environmental practices (renters, high income households and proactive planning), two represent dummy variables indicating the way geographical location may impact on the ability to undertake certain practices and a final variable accounting for if the household had been impacted by an extreme weather event. The final set of variables account for drivers that are considered internal to the respondent or their household (Kollmuss & Agyeman, 2002). As beliefs have been shown to be important in understanding pro-environmental practices an indicator of the extent to which the respondent believes climate change is real is included. The extent to which respondents considered that individuals or governments are capable of preventing harm in relation to environmental hazards was assessed using responses to the question 'in your opinion, to what extent may the following people or organizations be capable of preventing harm in relation to the environment'. A variable accounting for *locus of control* was included that measured the extent to which respondents agreed that household members had appropriate opportunities to influence decision on important issues. The extent to which respondents trust in the actions of political leaders and others in the neighborhood was measured using dichotomous variables associated with the statements 'I am confident that political leaders will act in the best interests of my household' and 'I trust the people in my neighborhood to act in ways that will not cause me harm'.

Understanding Pro-environmental Practices

Figure 1 contains information regarding the different types of pro-environmental practices measured in the survey. Across the ten categories of actions, while the large percentage of households had taken no action on many of the items, significant variation is evident. For actions associated with daily consumption (reducing water use, using efficient appliances etc.) large proportions of households had taken action. The majority of respondents (over 60%) stated that their household used energy efficient light bulbs, saved water and reduced power use. Given the spread out nature of Australian cities, generally poor public transport links and the need to travel for daily activities (Currie & Stanley, 2007; Mees, 2009), it is not surprising that only 1/3 of respondents said they had reduced motor vehicle use. Solar energy production and rain water harvesting was undertaken by approximately one third of all households in the sample, while almost 40 per cent installed or improved home insulation. Of all the actions considered the lowest level of take-up was associated with signing a petition (17.3 per cent) and voting for a particular candidate or political party (24.3per cent).

While the individual actions described are in themselves interesting, in order to analyze the drivers of pro-environmental practices in a more meaningful way the 10 separate actions were grouped into three meaningful classifications using the approach described above (Figure 2). The individual actions

Table 1. Independent variables used to assess pro-environmental behavior

Variable name	Variable detail
Demographic drivers	
Male	Coded 1 if respondent is male, 0 if the respondent is female
Baby boomers	Coded 1 if respondent was born between 1944 and 1964, 0 otherwise
Generation X	Coded 1 if respondent was born between 1965 and 1979, 0 otherwise
Higher education	coded 1 if the respondent held a university degree, 0 if otherwise
External drivers	
Renters	Coded 1 if the respondent lived in a rented house, 0 if otherwise
High income households	Coded 1 if the respondent lived in a household with annual income greater than $120,000, 0 if otherwise
Major city	Coded 1 if respondent lives in a major city, 0 otherwise
Inner regional area	Coded 1 if respondent lives in an inner regional area, 0 otherwise
Household has been impacted by natural event	Coded 1 if the household had been impacted by an environmental or climate hazard, 0 otherwise
Internal drivers	
Doesn't think climate change is influencing environmental hazards in their area	Coded 1 if the respondent didn't think climate change was influencing environmental hazards in their area, 0 otherwise
Individuals were capable of preventing harm in relation to the environment	Coded 1 if the respondent believed that individuals were capable of preventing harm in relation to the environment, 0 otherwise
Governments were capable of preventing harm in relation to the environment	Coded 1 if the respondent believed that governments were capable of preventing harm in relation to the environment, 0 otherwise
Locus of control	Coded 1 if the respondent believed that household members had appropriate opportunities to influence decision on important issues, 0 otherwise
Trust in the actions of political leaders	Coded 1 if the respondent had trust in the actions of political leaders, 0 otherwise
Had trust in the actions of neighbors	Coded 1 if the respondent had trust in the actions of neighbors, 0 otherwise
Household is proactive in planning for the future	Coded 1 if the respondent believed that their household is proactive in planning for the future, 0 otherwise

associated with reducing resource use (i.e. using efficient light bulbs etc.) were grouped into a single indicator action to reduce resource use. For this indicator 25.1 per cent of all respondents had taken action on most or all possible measures, with just under 30 per cent taking no action on any measures. The indicator uptake of energy efficient infrastructure combined responses to improving insulation, installing solar panels and installing rain water tanks. Just over 22 per cent of all respondents stated they had undertaken most or all of these measures, while almost one-third had taken no action. The final category, engaged in political activism, recorded the lowest aggregate action with just on one-tenth of all respondents saying they had undertaken both of these actions.

To consider the main research questions of this chapter, a regression analysis was applied to the three aggregate pro-environmental practices and the range of independent variables. The results are presented in table 2-4.

Figure 1.

Action to Reduce Resource use

The results for the aggregate variable action to reduce resource use are presented in table 2. While many of the variables identified as potential socio-economic drivers were of the expected direction of influence, only the variables accounting for age were significant. Compared to respondents categorised as millennials (those born between 1980 and 1994), if a respondent was categorised as either a Baby Boomers (1944 to 1964) or Generation X (1965 to 1979) it would be expected that they had undertaken more of the actions to reduce resource use.

Several of the variables accounting for beliefs and experience were significant. Respondents who had experienced an environment or climate relate event (Household has been impacted by natural event) were associated with a higher score on the dependent variable. Respondents who did not believe that that climate change was impacting on hazards in their area were associated with a lower score on the dependent variable. Respondents who agreed with the statement that ‘individuals were capable of preventing harm in relation to the environment’ were associated with a higher score on the dependent variable. he

Figure 2.

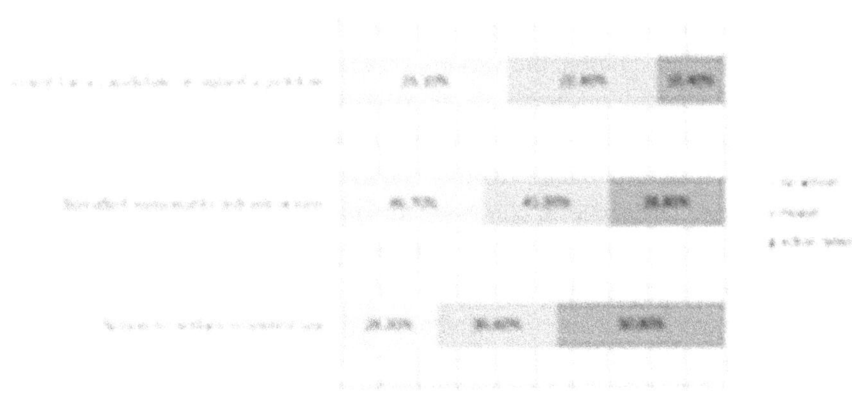

Table 2. Action to reduce resource use, regression results

	Coefficients	Significance
Constant	-0.264	0.02
Male	-0.072	0.17
Baby boomers	0.331	00.0*
Generation X	0.179	0.01*
Higher education	0.089	0.13
Renters	-0.009	0.87
High income households	-0.044	0.59
Major city	-0.114	0.14
Inner regional area	-0.067	0.44
Household has been impacted by natural event	0.318	0.00*
Doesn't think climate change is influencing environmental hazards in their area	-0.355	0.00*
Individuals were capable of preventing harm in relation to the environment	0.135	0.00*
Governments were capable of preventing harm in relation to the environment	-0.022	0.70
Locus of control	0.071	0.17
Trust in the actions of political leaders	0.04	0.47
Had trust in the actions of neighbors	-0.07	0.20
Household is proactive in planning for the future	0.166	0.00*

Source: authors calculations

final significant variable related to the extent to which households planned ahead. Respondents who agreed with the statement 'My household is proactive in planning for the future', were associated with a higher score on the dependent variable.

Uptake of Energy Efficient Infrastructure

The results for the aggregate variable uptake of energy efficient infrastructure are presented in table 3. As with households taking actions to reduce resource use, several of the independent variables showed significant associations in relation to the uptake of energy efficient infrastructure. Of the socio-demographic variables included in the equation, the variables for age, education and tenure were significant. Respondents categorized as Generation X were associated with a higher score on the dependent variable, suggesting this group of respondents installed energy efficient infrastructure such as solar panels or rain water harvesting tanks more than respondents in other generations. There was also a significant positive relationship between holding a tertiary level education and installing energy efficient infrastructure. In contrast to age and education level, being a renter, when compared to an owner, was associated in a lower score on the dependent variable, suggesting that renters installed less energy efficient infrastructure.

Respondents who had experienced an environment or climate relate event were associated with a higher score on the dependent variable, while those who did not believe that that climate change was impacting on hazards in their area were associated with a lower score on the dependent variable. Respondents who agreed with the statement that individuals were capable of preventing harm in relation

Table 3. Uptake of energy efficient infrastructure, regression results

	Coefficients	Significance
Constant	-0.223	0.06
Male	-0.072	0.18
Baby boomers	0.109	0.07
Generation X	-0.18	0.00*
Higher education	0.315	0.00*
Renters	-0.181	0.02*
High income households	-0.043	0.60
Major city	-0.075	0.34
Inner regional area	0.056	0.52
Household has been impacted by natural event	0.288	0.00*
Doesn't think climate change is influencing environmental hazards in their area	-0.326	0.00*
Individuals were capable of preventing harm in relation to the environment	0.119	0.02*
Governments were capable of preventing harm in relation to the environment	-0.026	0.65
Locus of control	0.07	0.18
Trust in the actions of political leaders	0.042	0.46
Had trust in the actions of neighbors	-0.068	0.21
Household is proactive in planning for the future	0.138	0.009*

Source: authors calculations

to the environment were associated with a higher score on the dependent variable. The final significant variable related to the extent to which households planned ahead. Respondents who agreed with the statement 'My household is proactive in planning for the future', were associated with a higher score on the dependent variable.

Voted for a Candidate or Signed a Petition

The results for the aggregate variable voted for a candidate or signed a petition are presented in table 4. Of the socio-economic variables, the ones accounting for sex of the respondent and age were significant. Respondents who were male had a lower score on the dependent variable. Compared to respondents characterized as Millennials, Baby Boomers and Generation Xers were associated with higher scores on the dependent variable.

Respondents who had experienced an environment or climate relate event were associated with a higher score on the dependent variable, while those who did not believe that that climate change was impacting on hazards in their area were associated with a lower score on the dependent variable. Respondents who agreed with the statement that individuals were capable of preventing harm in relation to the environment were associated with a higher score on the dependent variable. Respondents who agreed with the statement 'household members had appropriate opportunities to influence decision on important issues' (Locus of Control) were associated with a higher score on the dependent variable. Finally, respondents

Table 4. Voted for a candidate or signed a petition, regression results

	Coefficients	Significance
Constant	-0.359	0.00
Male	-0.108	0.03*
Baby boomers	0.314	0.00*
Generation X	0.202	0.00*
Higher education	0.094	0.10
Renters	0.031	0.56
High income households	-0.059	0.46
Major city	-0.002	0.97
Inner regional area	-0.056	0.50
Household has been impacted by natural event	0.338	0.00*
Doesn't think climate change is influencing environmental hazards in their area	-0.403	0.00*
Individuals were capable of preventing harm in relation to the environment	0.141	0.00*
Governments were capable of preventing harm in relation to the environment	0.034	0.54
Locus of control	0.105	0.03*
Trust in the actions of political leaders	0.02	0.72
Had trust in the actions of neighbors	-0.097	0.07
Household is proactive in planning for the future	0.17	0.00*

Source: authors calculations

who agreed with the statement 'My household is proactive in planning for the future', were associated with a higher score on the dependent variable.

SOLUTIONS AND RECOMMENDATIONS

The analysis in this chapter has suggested that over and above of socio-economic / demographic and external drivers such as age, income and education, drivers associated with Kollmuss and Agyeman's (2002) internal drivers are important in understanding pro-environmental practices of the households included in data sample used. As such this has important implications for environmental education as it appears that a range of drivers amenable to educational intervention are important in influencing household and individual practices.

The importance of knowledge in determining pro-environmental practices has been widely identified (López-Mosquera, Lera-López, & Sánchez, 2015; Saphores, Ogunseitan, & Shapiro, 2012). In the data considered in this chapter, variables indicating a particular level of knowledge— 'Household has been impacted by natural event' and 'Doesn't think climate change is influencing environmental hazards in their area'— were significant with regard to all three of the pro-environmental practices considered. Responses to the first statement were associated with an increased likelihood of positive pro-environmental practices, while responses on the second statement reduced the likelihood of the same practices. Assuming that these two statement might be taken as representing two opposite ends of a spectrum of knowledge, then educational programs that target knowledge around climate change and environmental outcomes, and in particular knowledge around climate change impacts and human induced drivers, may be important in increasing positive pro-environmental practices.

In line with previous research (Botetzagias, Dima, & Malesios, 2015; Lizin, Van Dael, & Van Passel, 2017) attitudes regarding the ability of individuals to make a difference, measured by the statement 'individuals are capable of preventing harm in relation to the environment' and the variable measuring locus of control, are important for understanding pro-environmental practices. In both cases positive responses on these variables were associated with an increasing likelihood that a household had undertaken some form of pro-environmental practice. A further attitudinal variable relates to a household ability in planning for the future. In situations where a household was future focused there was a higher likelihood of positive pro-environmental practices. From an education standpoint, programs that aim to impact on attitudes and values around individual responsibility and being future focused are likely to have an impact on pro-environmental outcomes.

FUTURE RESEARCH DIRECTIONS

This chapter has argued that understanding the drivers that impact on the willingness or ability for households to undertake pro-environmental practices is important to developing appropriate education and policy strategy. While the information provided in this chapter makes a contribution to our understanding, there is clear imperative for future research to maintain knowledge. While the data used in the chapter was collected only recently (2017) the landscape around individual and household behavior and practices is undergoing change. For example, the mass youth protests of 2019 regarding climate change maybe an indicator of shifting mindsets of younger generations, a finding that was not suggested in the data used here. Consequently, further up to date survey work may be appropriate.

CONCLUSION

This chapter has discussed the issue of individual and households willingness and ability to undertake pro-environmental practices in the context of developing appropriate educational materials. It is clear from the data presented and associated discussion, that there is a need for educational programs to focus on those factors that are likely to influence a household's decisions around undertaking pro-environmental practices. Important here is the need for programs to address both knowledge around climate change, causes and the impacts as well as the attitude of individuals and households towards the necessity of individuals to take action to address climate concerns.

REFERENCES

AdditionalREADING

Bamberg, S., & Möser, G. (2007). Twenty years after Hines, Hungerford, and Tomera: A new meta-analysis of psycho-social determinants of pro-environmental behaviour. *Journal of Environmental Psychology, 27*(1), 14–25. doi:10.1016/j.jenvp.2006.12.002

Blankenberg, A.-K., & Alhusen, H. (2019). On the determinants of pro-environmental behavior: A literature review and guide for the empirical economist. *Center for European, Governance, and Economic Development Research (CEGE),* (350).

Botetzagias, I., Dima, A.-F., & Malesios, C. (2015). Extending the theory of planned behavior in the context of recycling: The role of moral norms and of demographic predictors. *Resources, Conservation and Recycling, 95*, 58–67. doi:10.1016/j.resconrec.2014.12.004

Braun, T., Cottrell, R., & Dierkes, P. (2018). Fostering changes in attitude, knowledge and behavior: Demographic variation in environmental education effects. *Environmental Education Research, 24*(6), 899–920. doi:10.1080/13504622.2017.1343279

Clark, C. F., Kotchen, M. J., & Moore, M. R. (2003). Internal and external influences on pro-environmental behavior: Participation in a green electricity program. *Journal of Environmental Psychology, 23*(3), 237–246. doi:10.1016/S0272-4944(02)00105-6

Currie, G., & Stanley, J. (2007). *No way to go: Transport and social disadvantage in Australian communities*. Academic Press.

Diekmann, A., & Franzen, A. (2019). Environmental concern: A global perspective. In *Einstellungen und Verhalten in der empirischen Sozialforschung* (pp. 253–272). Springer. doi:10.1007/978-3-658-16348-8_11

Ertz, M., Karakas, F., & Sarigöllü, E. (2016). Exploring pro-environmental behaviors of consumers: An analysis of contextual factors, attitude, and behaviors. *Journal of Business Research, 69*(10), 3971–3980. doi:10.1016/j.jbusres.2016.06.010

Fielding, K. S., & Head, B. W. (2012). Determinants of young Australians' environmental actions: The role of responsibility attributions, locus of control, knowledge and attitudes. *Environmental Education Research, 18*(2), 171–186. doi:10.1080/13504622.2011.592936

Fielding, K. S., Louis, W. R., Warren, C., & Thompson, A. (2011). Understanding household attitudes and behaviours toward waste, water and energy conservation. *Urban Consumption*, 199-214.

Garay, L., Font, X., & Pereira-Moliner, J. (2017). Understanding sustainability behaviour: The relationship between information acquisition, proactivity and performance. *Tourism Management, 60*, 418–429. doi:10.1016/j.tourman.2016.12.017

Gatersleben, B., Murtagh, N., Cherry, M., & Watkins, M. (2019). Moral, wasteful, frugal, or thrifty? Identifying consumer identities to understand and manage pro-environmental behavior. *Environment and Behavior, 51*(1), 24–49. doi:10.1177/0013916517733782

Giefer, M. M., Peterson, M. N., & Chen, X. (2019). Interactions among Locus of Control, Environmental Attitudes and Pro-Environmental Behaviour in China. *Environmental Conservation, 46*(3), 234–240. doi:10.1017/S0376892919000043

Gifford, R., & Nilsson, A. (2014). Personal and social factors that influence pro-environmental concern and behaviour: A review. *International Journal of Psychology*, *49*(3), 141–157. doi:10.1002/ijop.12034 PMID:24821503

Gifford, R., & Sussman, R. (2012). *Environmental attitudes*. Academic Press.

Goldman, D., Hansmann, R., Činčera, J., Radović, V., Telešienė, A., Balžekienė, A., & Vávra, J. (2020). Education for Environmental Citizenship and Responsible Environmental Behaviour. In *Conceptualizing Environmental Citizenship for 21st Century Education* (pp. 115–137). Springer. doi:10.1007/978-3-030-20249-1_8

Kollmuss, A., & Agyeman, J. (2002). Mind the gap: Why do people act environmentally and what are the barriers to pro-environmental behavior? *Environmental Education Research*, *8*(3), 239–260. doi:10.1080/13504620220145401

Kollmuss, A., & Agyeman, J. (2002). Mind the gap: Why do people act environmentally and what are the barriers to pro-environmental behavior? *Environmental Education Research*, *8*(3), 239–260. doi:10.1080/13504620220145401

Li, D., Zhao, L., Ma, S., Shao, S., & Zhang, L. (2019). What influences an individual's pro-environmental behavior? A literature review. *Resources, Conservation and Recycling*, *146*, 28–34. doi:10.1016/j.resconrec.2019.03.024

Lizin, S., Van Dael, M., & Van Passel, S. (2017). Battery pack recycling: Behaviour change interventions derived from an integrative theory of planned behaviour study. *Resources, Conservation and Recycling*, *122*, 66–82. doi:10.1016/j.resconrec.2017.02.003

López-Mosquera, N., Lera-López, F., & Sánchez, M. (2015). Key factors to explain recycling, car use and environmentally responsible purchase behaviors: A comparative perspective. *Resources, Conservation and Recycling*, *99*, 29–39. doi:10.1016/j.resconrec.2015.03.007

Lorenzoni, I., Nicholson-Cole, S., & Whitmarsh, L. (2007). Barriers perceived to engaging with climate change among the UK public and their policy implications. *Global Environmental Change*, *17*(3-4), 445–459. doi:10.1016/j.gloenvcha.2007.01.004

Mees, P. (2009). How dense are we? Another look at urban density and transport patterns in Australia, Canada and the USA. *Road & Transport Research: A Journal of Australian and New Zealand Research and Practice, 18*(4), 58.

Meinhold, J. L., & Malkus, A. J. (2005). Adolescent environmental behaviors: Can knowledge, attitudes, and self-efficacy make a difference? *Environment and Behavior*, *37*(4), 511–532. doi:10.1177/0013916504269665

Newton, P., & Meyer, D. (2013). Exploring the attitudes-action gap in household resource consumption: Does "environmental lifestyle" segmentation align with consumer behaviour? *Sustainability*, *5*(3), 1211–1233. doi:10.3390u5031211

Pearce, M., Willis, E., Mamerow, L., Jorgensen, B., & Martin, J. (2014). The prestige of sustainable living: Implications for water use in A ustralia. *The Geographical Journal*, *180*(2), 161–174. doi:10.1111/geoj.12016

Pooley, J. A., & O'Connor, M. (2000). Environmental education and attitudes: Emotions and beliefs are what is needed. *Environment and Behavior*, *32*(5), 711–723. doi:10.1177/0013916500325007

Pothitou, M., Hanna, R. F., & Chalvatzis, K. J. (2016). Environmental knowledge, pro-environmental behaviour and energy savings in households: An empirical study. *Applied Energy*, *184*, 1217–1229. doi:10.1016/j.apenergy.2016.06.017

Saphores, J.-D. M., Ogunseitan, O. A., & Shapiro, A. A. (2012). Willingness to engage in a pro-environmental behavior: An analysis of e-waste recycling based on a national survey of US households. *Resources, Conservation and Recycling*, *60*, 49–63. doi:10.1016/j.resconrec.2011.12.003

Skamp, K., Boyes, E., & Stanisstreet, M. (2013). Beliefs and willingness to act about global warming: Where to focus science pedagogy? *Science Education*, *97*(2), 191–217. doi:10.1002ce.21050

Steg, L., & Vlek, C. (2009). Encouraging pro-environmental behaviour: An integrative review and research agenda. *Journal of Environmental Psychology*, *29*(3), 309–317. doi:10.1016/j.jenvp.2008.10.004

Stern, P. C. (2000). New environmental theories: Toward a coherent theory of environmentally significant behavior. *The Journal of Social Issues*, *56*(3), 407–424. doi:10.1111/0022-4537.00175

Trotta, G. (2018). Factors affecting energy-saving behaviours and energy efficiency investments in British households. *Energy Policy*, *114*, 529–539. doi:10.1016/j.enpol.2017.12.042

Welsch, H., & Kühling, J. (2009). Determinants of pro-environmental consumption: The role of reference groups and routine behavior. *Ecological Economics*, *69*(1), 166–176. doi:10.1016/j.ecolecon.2009.08.009

White, K., Habib, R., & Hardisty, D. J. (2019). How to SHIFT consumer behaviors to be more sustainable: A literature review and guiding framework. *Journal of Marketing*, *83*(3), 22–49. doi:10.1177/0022242919825649

White, T., & Wall, R. (2008). National, regional and local attitudes towards climate change: Identifying appropriate target audiences for communications. *Local Environment*, *13*(7), 589–607. doi:10.1080/13549830802260126

Whitmarsh, L. (2009). Behavioural responses to climate change: Asymmetry of intentions and impacts. *Journal of Environmental Psychology*, *29*(1), 13–23. doi:10.1016/j.jenvp.2008.05.003

Whitmarsh, L., & O'Neill, S. (2010). Green identity, green living? The role of pro-environmental self-identity in determining consistency across diverse pro-environmental behaviours. *Journal of Environmental Psychology*, *30*(3), 305–314. doi:10.1016/j.jenvp.2010.01.003

KEY TERMS AND DEFINITIONS

Green Consumer Behaviour: Consumer willingness to purchase green products.

Green Policy: A documented strategy of the organization in terms of its actions relating to the environment.

Green Technology: Technology whose use is intended to mitigate or reverse the effects of human activity on the environment.

Locus of Control: Is the degree to which people believe that they, as opposed to external forces (beyond their influence), have *control* over the outcome of events in their lives.

Pro-Environmental Practice: Practices that consciously seeks to minimize the negative impact of one's actions on the environment.

Chapter 15
Role of Cities in Native Woody Species Conservation and Carbon Storage Insight From Niger

Soulé Moussa
Kwame Nkrumah University of Science and Technology, Ghana

Saley Karim
Université Dan Dicko Dankoulodo Maradi, Niger

Abasse Amadou Tougiani
Institut National de la Recherche Agronomique, Niger

Boateng Kyereh
Kwame Nkrumah University of Science and Technology, Ghana

Mahamane Saadou
Université Dan Dicko Dankoulodo Maradi, Niger

ABSTRACT

Loss of native tree species constitutes a major socio-economic and ecological problem. Nevertheless, there is scanty data on the role of Sahelian cities in native woody species conservation and climate change mitigation. This study tried to determine the native woody diversity, structure, and carbon stocks in Niamey, and Maradi, Niger. The study used the plot method to inventory dendrometric variables based on a stratified sampling approach from peri-urban areas to the built environment. Biodiversity indices and allometric models were used for the data analysis. A total of 63 native woody species representing 47 genera and 24 families were recorded in the overall urban forests in two cities. In Niamey, 46 species belonging to 35 genera and 19 families were recorded while in Maradi, 50 species, 41 genera, and 19 families were inventoried. Fabaceae was the dominant families in both cities with equal richness (13 species).

DOI: 10.4018/978-1-7998-7512-3.ch015

1. INTRODUCTION

Native tree species are under threat due to human activities and climate change. For instance, 1.8% of all biodiversity hotspot areas will be taken up by spatial urban expansion by 2030 (Seto et al., 2012). In addition to that, 90% of the protected areas are likely to be modified by urbanization mostly in developing countries (Mcdonald et al., 2008). Urbanization destroys the natural environment (Elmqvist et al., 2013a) with the introduction of exotic species which also leads to biotic homogenisation (Mckinney, 2006). According to Dunn and Heneghan, (2011) native species richness decreases and exotic species increases with urbanization gradient. Furthermore, Pysek, (1998) reported that the native biodiversity decreases from villages to the urban areas and cities have a greater number of exotic species with respect to the total floristic list. Urbanization influences negatively vegetation cover (Seto et al., 2012) leading to a great reduction in biomass (Arroyo-Rodríguez et al., 2017).The depletion of native species due to urbanization leads to the reduction of biodiversity in urban ecosystems (Aronson et al., 2014; McKinney, 2002; McPherson and Simpson, 2003), and ecosystems services such as carbon biomass (Arroyo-Rodríguez et al., 2017) therefore it has biogeochemical cycles (Wang et al., 2017) and regional and global climate (Seto and Shepherd, 2009).

Although urban areas expansion continue towards the natural vegetation, urban areas play a crucial role in the biodiversity conservation (Elmqvist et al., 2013b; McKinney, 2002). In addition that urban areas uptake considerable amount of atmospheric carbon dioxide through urban green areas (Chen, 2015; Nowak et al., 2013a; Reynolds et al., 2017) such as indigenous woody species in urban areas (Stoffberg et al., 2010).

Empirical studies across the world pointed out cities can serve as native woody biodiversity refugia. In urban areas in Asia (Yan and Yang, 2017) found an important proportion of native woody species in Beijing. Similar results found that cities are home of native woody plant such as in Central European cities as reported by (Pysek, 1998), in Hannover (Matthies et al., 2017). Cities in African also have been reported to be a habitat of native woody species such as in cities of the Eastern Cape, South Africa (Kuruneri-Chitepo and Shackleton, 2011), in Lomé in Togo (Raoufou et al., 2011), Abuja in Nigeria (Agbelade et al., 2017), Kumasi in Ghana (Nero et al., 2017) and Nairobi in Kenya (Nyambane et al., 2016). These studies reported on woody plant species (native and exotic) used in urban green spaces, their systematic composition and growth characteristics. These studies do not however, provide information on the extent to which urban areas can be key especially to native woody species conservation and the role of native woody flora to climate change mitigation. In addition, lacking in the literature is a comparison between peri-urban native woody vegetation and different land use and land cover (LULC) types found in built areas. Besides, there are no study in the international literature on the City Biodiversity Index (CBI) such as native biodiversity in the city and ecosystems services provided by biodiversity in the city from West Africa Sahel cities. This impedes effort to generalise native urban woody vegetation characteristics across different climatic zones as well as regions with contrasting socio-economic, cultural and environmental characteristics like West Africa Sahel.

Niamey and Maradi are two cities in Niger, a country that lies right in the centre of the Sahel. With an annual population growth rate of 3.9% and mass migration of human population from rural to urban centres (INS, 2016), about 19% of the total population of Niger are urbanites (INS, 2016). Niger is facing serious urbanization issues (Issaka, 2013) such as depletion of public urban green areas and local biodiversity depletion. For instance green belt of Niamey has initial land mass of 2500 hectares which is reduced to 2336 hectares due to the construction in the green belt of Niamey (Ministère de l'Environnement du

Niger, 2010). In addition to that, 90% of urban areas in Niger used wood as source of domestic energy (Ministère de l'Environnement du Niger, 2010). The spatial expansion of Niamey and Maradi cities has led to the depletion of natural vegetation and loss of emblematic local woody species (Communication Personal, Saadou, Mahamane, 2018) to Niamey and Maradi regions. The urbanization model of such cities is posited to modify native woody vegetation composition, resulting in either fragmentation or loss of biodiversity (Blair, 1999; Mckinney, 2006) or conservation of indigenous woody biodiversity in urban areas (Yan and Yang, 2017).

Assessment of the native woody flora in these cities can provide City Biodiversity Index (CBI) such as the number of native woody species and carbon storage that close this information gap. It will also provide a baseline information for native tree urban planting programme such as school planting trees, which will serve for educational purpose. Focusing on native woody species, this study aimed to assess differences in the diversity and structure of native woody vegetation in Niamey and Maradi across urban land use and land cover gradient. The study also assessed the effect of native woody species richness on structure (basal area, tree cover and stems density), effect of stem density on tree cover and effect of size of land use and land cover richness. We hypothesised that (i) built areas have high native woody species diversity that peri-urban areas and (ii) native woody specie richness influences structure (basal area, tree cover and stems density) from peri-urban areas to the built environment.

2. METHODS AND MATERIALS

2.1 Study Site

The study was conducted in Niamey and Maradi, which are two Sahelian cities in Niger (Figure 1). The two cities belong to Sahel agroclimatic zone with the yearly average rainfall that varies from 150 mm to 350 mm (Conseil National de l'Environnement pour un Développement Durable, 2016). Niamey is the capital and biggest town in Niger with an extended area of over 552.27 km^2, an urbanized area of 297.46 km^2 and a population of 1,026,848 people (INS, 2017). Niamey is located at latitude 13°20'-13°35'N and longitude 2°00'-2°15'E in western Niger. While Maradi is the economic capital of Niger and the third largest city with 326,804 inhabitants and covers 86 km^2 (INS, 2016) (Fig. 1). Maradi city is located at latitude 13°29′- 13°49'N and longitude 7°5′ -7°09'E in south-central Niger with 27.3 °C as average yearly temperature and 29.2°C in Niamey.

The two cities belong to different indigenous vegetation. The native vegetation of Niamey belongs to the South Western Sahelian compartment (B1) with *Combretum* thicket as dominant woody formation found on the plateau lateritic soils and steppe found on sandy terraces, in dry valleys and on the dunes (Saadou, 1990). Niamey natural vegetation consists also of riparian forest along the Niger River that traverses the city. There are also some wetlands vegetation within the city. The peri-urban native forest consists of agroforestry parklands. However, some native species are planted in public urban green spaces such as green belt of Niamey. Meanwhile Maradi belongs to the Southern Central Sahelian (B2) with some vegetation types such as savannah on the southern sandy terraces, steppes on dunes and dry valleys and *Combretum* thicket on lateritic plateaus (Saadou, 1990). Maradi has also some gallery forest along the existing water body that traverses the city.

Figure 1. Study area

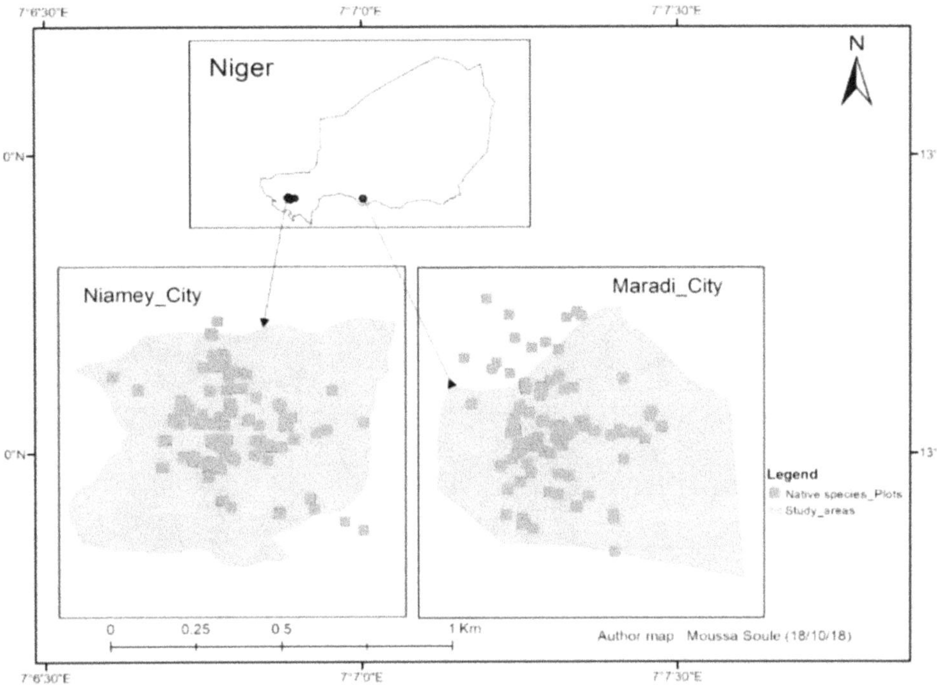

2.2 Pictorial View of Native Species Planted or Kept in two Cities

Our study population was any native woody species found in built areas and peri-urban area in two cities weather natural stand or planted. Native species are used in urban forestry in both cities by the private or public for several purpose as illustrated in the photos 1. Some of the native species are kept during construction as illustrated in the figure 2.

2.2 Data Collection

2.2.1 Native Urban Forest Tree Species Inventory

A stratified random sampling approach was used for native woody species inventory as used in (Nowak et al., 2016a) to study urban forest structure and ecosystem services in Syracuse, USA . As urban forest is the collection of woody species found in peri-urban areas and urban settings (FAO, 2016), based on this definition we considered seven urban forests types or land use and land cover types (LULC) that constitute our strata. Our land use and land cover types are seven strata that are:

1. **Commercial:** market, shops, restaurant and garage;
2. **Roads:** main streets and boulevards;
3. **Residential:** Mosque, church and houses;
4. **Schools:** primary schools, secondary schools, universities, polytechnics, training colleges etc. The schools were private as well as public;

Figure 2. Planted and natural stem of native species kept in Niamey and Maradi. a, Planted Acacia senegal in a house in Maradi; b, A natural stand of Piliostigma reticulatum kept in a house in Maradi; c, Natural stand of Borassus aethiopum kept along a road in Maradi; d, Faidherbia albida kept on the road in Niamey; e, A planted Ficus platyphylla in the front of house in Niamey; f, A planted Tamarindus indica in a house in Niamey. Source: Moussa Soulé

5. **Administrative areas:** public and private services;
6. **Urban-forested areas:** urban green spaces (green belt, bosquet), agroforestry systems, and wetlands, irrigated farmlands and botanical gardens found within urban areas. The first six LULC constitute the built areas in both cities.
7. **Peri-urban forests**: consist of peri-urban vegetation such as croplands and pastoral fields.

The LULC types were randomly selected from five districts in Niamey and thee in Maradi. A random list for sampling was selected from an inventory of schools, administrative posts, urban green spaces, roads, markets and residential compounds obtained from directorates of education, environment, urban equipment and habitat and the communes in Niamey and Maradi.

We inventoried native species through a survey in plots of 50 m x 50 m, consistent with guidelines for inventories in the agroforestry systems in West Africa Sahel (Thiombiano et al., 2016). The plot size varied in some cases and was less than 0.25 ha if a randomly selected point could not allow a plot of a 50 m x 50 m to be demarcated without getting into another LULC type. 2.5 cm was considered as minimum diameter at breast height (DBH) as recommended for drylands areas (Pearson et al., 2007). But for the above ground biomass estimation, the minimum DBH of 5 cm due to the model DBH restriction (Chave et al., 2014). One of limitations was the inaccessibility to some urban land use and land cover types such as embassies, barracks and presidential residence. The main limitation of the study is that we did not get permit to assess the tree vegetation in barracks, embassies and some private urban land use.

2.2.2 Dendrometric Measurement

All native woody plants within the plots were noted, DBH, total height and crown diameter were measured. DBH was measured at 1.30 m from the ground using a diameter tape. Woody plants with forks before 1.3 m were considered as multi-stemmed; their individual stems were measured separately. The total height (H) was measured using graduated ringing poles. The crown diameter (largest and perpendicular) was measured using tape meter. In field, we used also the book of (Arbonnier, 2002) for the specie identification. The photos of native woody species were collected. However, for unknown specie, specie identification was done with the assistance of the botanists of the Departments of Biology of Abdou Moumouni University of Niamey and University Dan Dicko Dankoulodo of Maradi. During the fieldwork, we used a questionnaire to ask questions whether the native specie was planted or it was natural stand kept during Niamey and Maradi construction.

2.3 Data Analysis

2.3.1 Native Woody Species Diversity in Niamey and Maradi

To appreciate the native woody species diversity in both cities, we used Shannon-Wiener diversity (H'), richness (S), evenness (J), number of families and genera. Native specie richness was calculated as the total number of native woody species found in each LULC type. Shannon diversity index was calculated as the proportion of the species relative to the total number of species using the formula: H' $= -\sum[(ni/N)*Ln(ni/N)]$, where *ni* is the number of individuals of species *i*, *N* is the total number of individuals per LULC types and *Ln* is the natural logarithm. Pielou's evenness (J') was used as a measure of relative abundance of tree and shrub species, calculated using the formula: $J = H'/\text{Ln}(S)$, where *H'* is the Shannon diversity index and *S* is the number of native woody species. The formulas of Shannon diversity and Pielou's evenness were described in (Thiombiano et al., 2016).

2.3.1 Structure of Native Woody Plantations in Both Cities

The multi-stem diameters were calculated as the square root of the sum of squares of the individual stems as described in (Thiombiano et al., 2016) in order to have the overall DBH. Native woody plants DBH data were grouped into five DBH classes [5-10[, [10-20[, [20-30[, [30-40] and > 40 cm to provide the number of stems in each diameter class LULC in two cities. The diameter and height were calculated across land use and land cover types of the urban forests. To characterize the native woody species stand in Niamey and Maradi, the following structural parameters were used, the formulas were described in (Thiombiano et al., 2016): **Basal area** (m^2): $D^2 \times \frac{\pi}{4}$, with $\pi = 3.14$. The total basal (m^2/ha).

Stem density (ha): Native woody stems density is the number of individuals of each per unit area. For the stem density, the DBH was ≥ 2.5 cm.

Tree cover (%)**:** crown cover (m^2)/(area (m^2)*100, with crown cover (m^2) = $R_1*R_2*3.14$ where R_1(m) = radius 1 = (perpendicular crown diameter)/2 and R_2 (m) = radius 2 = (largest crown diameter)/2 as described in (Jennings et al., 1999).

We calculated the total number of native woody stems by LULC and the percentage of planted native woody species and natural woody species kept during the buildings of Niamey and Maradi.

2.3.2 Biomass and Carbon Estimation

Estimation of the woody plant biomass of each specie was not feasible due to the lack of species-specific model or suitable model for a given region or vegetation types in the literature (Henry et al., 2013). For instance, estimating woody species biomass of urban forest may not be possible due to urban areas are areas of high specie richness (Nero et al., 2017a; Yan and Yang, 2017). However, (Chave et al., 2014) developed generalized above ground biomass (AGB) equation for different vegetation types and, which are used by authors when destructive methods are not allowed (Henry et al., 2013) mostly in the urban areas where destructive methods cannot be possible. Allometric equation has been widely used to estimate the biomass woody species in urban settings (Nero et al., 2017b; Nowak et al., 2013b; Stoffberg et al., 2010; Ning et al., 2015). Therefore, via a critical review of the literature allometric model potentially suitable for estimating native woody plants biomass in Niamey and Maradi cities was identified. The pantropical equation developed by (Chave et al., 2014) for tropical forest, subtropical forests and woodland savannah were identified useful, because the model holds across tropical woody vegetation types, with no detectable effect of region or environmental factors (Chave et al., 2014) (Table1). This model was used to quantify AGB of semi-natural woody vegetation such as West Africa Sahel agroforestry systems (Dayamba et al., 2016) and urban forests in Kumasi, Ghana (Nero et al., 2017b). Allometric equations for woody species below ground biomass (BGB) are also rare in the literature (Cairns et al., 1997). Thus, BGB was estimated as a proportion of AGB using root shoot ratio developed by (Cairns et al., 1997) for tropical vegetation (Table 1).

Table 1. Allometric models used for estimating native woody plant biomass

Source	Allometric equations
(Chave et al., 2014)	AGB (kg) = $0.0673(\rho D^2 H)^{0.976}$
(Cairns et al., 1997)	BGB (kg) = AGB0.25

Where: ρ = wood density (gcm^3), D = diameter at breast height (1.3 m) (cm), H = total tree height (m), AGB = aboveground biomass in kg. The wood density was obtained from the (Intergovernmental Panel on Climate Change (IPCC), 2006) database. The total native woody biomass were obtained by summing up the AGB and BGB, which were converted to carbon stocks using the default IPCC carbon fraction value of 0.47 (IPCC, 2006).

2.3.3 Statistical Analysis

Prior to the statistical analysis, the Ryan-Joiner test and Levene 'test were used to check the normality and homogeneity of the data. An ANOVA test was used to test the differences in the mean of biodiversity indices (Shannon, Evenness, specie richness) and structure (tree cover, basal area and stem density) across the LULC types at alpha 0.05 level of significance. We calculated the effect size with ANOVA between groups; we used η^2 (Eta squared) which is the sum of squares between groups over total sum of squares as described by (Vacha-haase and Thompson, 2004). Linear regression was used to assess the effects of

Figure 3. Number of woody native species relationship and 7 LULC areas (ha).

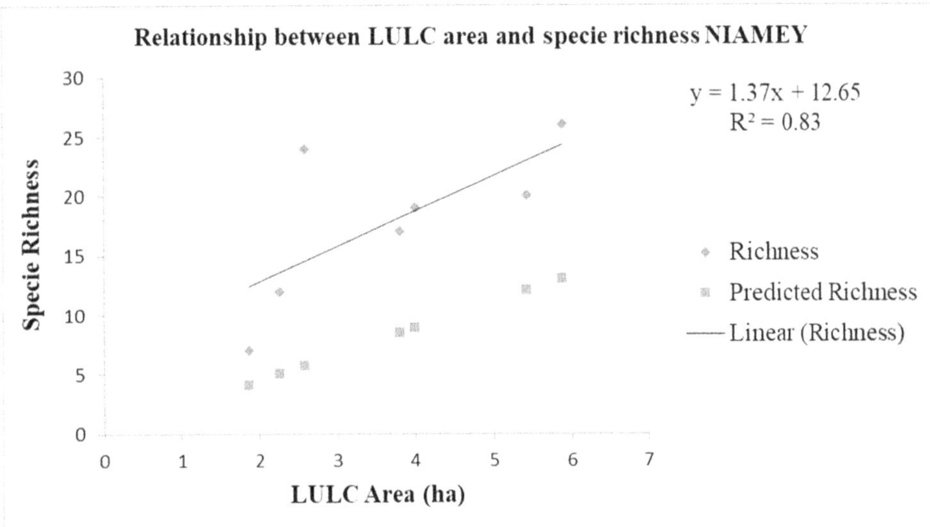

LULC area size on specie richness, the effect of native species richness on structure (tree cover, basal area and density) and the effect of stem density on tree cover. Excel were used for the statistical analysis.

3. RESULTS

3.1 Native Species Diversity in Niamey and Maradi

A total of 63 native woody species were identified within Niamey and Maradi cities (Table 2). The native woody plants belonged to 24 families and 47 genera (Table 7). Fabaceae was the most dominant families with equal number of species (13 species) in both cities (Table 7). The number of families (19) was also equal in both cities. We recorded 46 native woody species in Niamey belonging to 35 genera. Thirteen native woody species were common to Niamey. The overall Shannon index was 2.72 and evenness was 0.72 in Niamey. The three most common species (referring to the highest stem numbers), *Balanites aegyptiaca*, *Faidherbia albida* and *Acacia senegal* in Niamey (Table 7).

While in Maradi, we recorded 50 native woody plants belonging to 41 genera. Seventeen native woody species were common to Maradi (Table 7). The overall Shannon index was 3.28 and evenness was 0.85 in Maradi. The three most common species (referring to the highest stem numbers) were *Vachellia nilotica* subsp. *nilotica, Balanites aegyptiaca* and *Prosopis africana* in Maradi (Table 7).

The native woody species biodiversity indices varied significantly from peri-urban forests to urban built areas ($F(2, 12) = 49.09$, $P < 0.001$) in Niamey with effect size $\eta^2 = 0.84$. The forested areas had the highest native woody specie richness (26 species) in Niamey followed by residential areas (24 species). The lowest was observed in commercial areas (7 species) and road (12 species). The size of the land area per LULC has significant effect on native woody species richness ($R^2 = 0.83$, $F = 49.09$, Significance $F = 0.0016$, $df = 1$) in Niamey. Figure 2 shows the relationship between the seven LULC area (ha) and number of specie per LULC.

Similar trend was observed in Maradi city, where there was significant difference in the biodiversity indices from peri-urban areas to the built areas ($F (2, 12) = 59.05$, $P < 0.001$) with effect size $\eta^2 = 0.87$. The forested areas had the highest native woody specie richness (31 species) in Maradi followed by administrative areas (26 species). The lowest was observed in commercial areas (12 species). The size of the land area per LULC has significant effect on native woody species richness ($R^2 = 0.85$, $F = 33.43$, Significance $F = 0.0022$) in Maradi. Figure 3 shows the relationship between the seven LULC area (ha) and number of specie per LULC in Maradi.

Figure 4. Number of woody native species relationship and 7 LULC areas (ha).

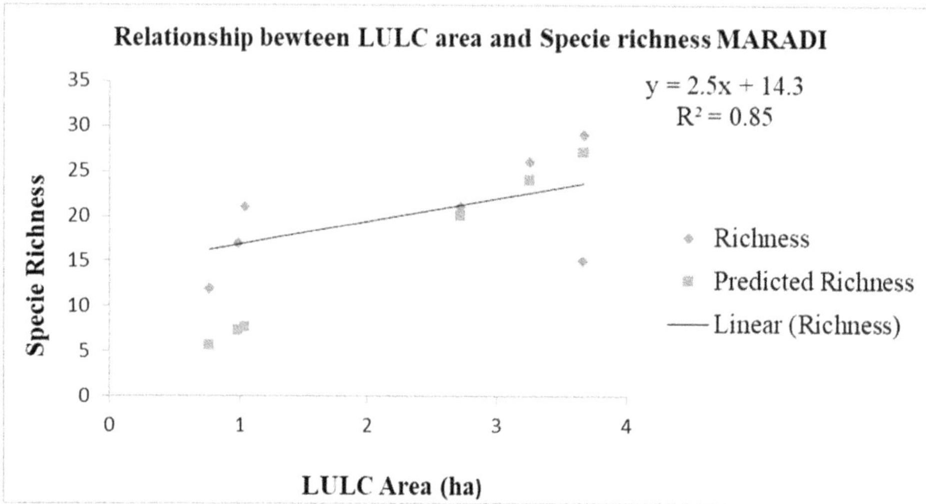

3.2 Structure of Native Woody Plantations in Both Cities

We measured 1407 stems (DBH ≥ 2.5 cm) of native woody species in Niamey (818 stems) and in Maradi (589 stems) (Table 3 and 4). About 58% of the native woody stems inventoried in Niamey were natural stands kept during construction and 42% were planted. A reserve trend was observed in Maradi, where 62% of the total native woody stems were planted and only 38% of the total stems were natural stand kept during construction. The number of natural and planted stems varied from peri-urban forest to the built areas (Table 4) though the difference was not significant ($F (1, 13) = 0.003$, $P = 0.95$) with effect size $\eta^2 = 0.0001$ in the two cities.

The structural characteristics (basal area, tree cover, stem density and number of stems per LULC) varied significantly from peri-urban areas to the built areas in Niamey ($F (3, 18) = 20.54$, $P < 0.001$) with effect size $\eta^2 = 0.87$ and in Maradi ($F (3, 18) = 10.20$, $P < 0.001$) with effect size $\eta^2 = 0.50$. The overall structural characteristics and its standard deviation were for stem density (30.66±12.11 stem/ha), basal area (4.60±3.15 m^2/ha) and tree cover (17.37±9.60%) (Table 3) in Niamey. Mean stem density (37.56±13.41 stem/ha), basal area (3.74±2.65 m^2/ha) and tree cover (14.28±9.77%) (Table 3) were found in Maradi.

Table 2. Native woody species diversity and number of plots over the inventoried area by LULC

Cities	LULC	Number of plots	Area (ha)	Richness	Shannon	Evenness
Niamey	Administrative areas	19	3.81	17	2.06	0.73
Maradi		21	3.25	26	2.71	0.83
Niamey	Commercial areas	6	1.86	7	1.57	0.81
Maradi		6	0.76	12	2.27	0.92
Niamey	Forested areas	27	5.87	26	2.50	0.78
Maradi		15	3.67	31	2.74	0.74
Niamey	Residential areas	32	2.58	24	2.76	0.88
Maradi		15	1.04	21	2.92	0.96
Niamey	Road	30	2.26	12	1.91	0.77
Maradi		21	0.99	18	2.30	0.81
Niamey	School	20	5.43	20	2.14	0.71
Maradi		17	3.66	15	2.39	0.88
Niamey	peri-urban forest	16	4.00	19	2.26	0.77
Maradi		12	2.72	21	2.64	0.87
Niamey	Overall	150	25.81	46	2.72	0.72
Maradi		107	16.09	50	3.28	0.85

Native woody species show high structural values in Niamey (Table 3). Road had the highest basal area (11.50 m^2/ha) followed by peri-urban forest (4.61 m^2/ha) in Niamey. The lowest basal area (< 3 m^2/ha) was observed commercial areas, forested areas and school in Niamey. Pei-urban had the highest stem density (45 stem/ha) followed by road (39.85 stem/ha) and administrative areas (38.85 stem/ha). The lowest native woody species stem density in Niamey was observed in commercial areas (8.60 stem/ha). The highest native woody plants cover was in road (37.04%) followed by administrative areas (21.40%). While native woody plants show, high structural values also in Maradi (Table 3). Road had the highest basal area (9.4m^2/ha) followed by administrative areas (4.30 m^2/ha). The lowest basal area (< 2 m^2/ha) was observed in commercial areas (1.85 m^2/ha) and school (1.8 m^2/ha) (Table 3). Forested areas had the highest stem density (52.86 stem/ha) in Maradi followed by residential areas (48.08 stem/ha) and road (46.46 stem/ha). The lowest stem density was observed in school (15.30 stem/ha) and in commercial areas (25 stem/ha). The highest native woody species cover was observed in road (32.92%) followed by residential areas (17.50%). The lowest tree cover was observed in School (2.65%).

Native woody plants density has effect on native woody plants cover ($F = 32.69$, $R^2 = 0.84$, $P = 0.0023$,$df = 1$) in Niamey. The native woody specie richness has effect on stem density ($F= 34.77$, $R^2 = 0.85$, $P = 0.0019$, $df =1$) in Niamey. Specie richness has also effect on tree cover ($F = 1.69E+32$, $R^2 = 1$, $P = 0.001$, $df =1$) in Niamey. Specie richness has effect on basal area even though the effect was not significant ($F = 7.54$, $R^2 = 0.56$, $P = 0.04$, $df =1$) in Maradi. While in Maradi stem density has effect on tree cover even though the difference was not significant ($F = 5.17$, $R^2 = 0.51$, $P = 0.07$, $df =1$) in Maradi. Specie richness has also weak effect on stem density though the difference was not significant ($F = 4.40$, $R^2 =0.47$, $P= 0.09$, $df= 1$). Richness does not have effect on tree cover ($F = 0.19$, $R^2 = 0.03$,

Table 3. Native woody species structural characteristics

Cities	LULC	Stem density (ha)	Basal area (m^2/ha)	Tree cover (%)
Niamey	Administrative areas	38.85	4.07	21.4
Maradi		36.62	4.30	15.0
Niamey	Commercial areas	8.60	2.33	11.94
Maradi		25.00	1.85	5.63
Niamey	Forested areas	29.47	2.75	10.18
Maradi		52.86	2.43	14.01
Niamey	Residential areas	26.74	3.98	12.40
Maradi		48.08	3.29	17.50
Niamey	Road	39.83	11.50	37.40
Maradi		46.46	9.40	32.92
Niamey	School	26.15	2.94	12.18
Maradi		15.30	1.80	2.65
Niamey	Peri-urban forest	45.00	4.61	16.07
Maradi		38.6	3.10	12.25
Niamey	Mean ± standard deviation	30.66±12.11	4.60±3.15	17.37±9.60
Maradi		37.56±13.41	3.74±2.65	14.28±9.77

Table 4. Distribution of the planted and natural stems

Cities	LULC	Planted stems	Natural stems	Total stems
Niamey	Administrative areas	53	95	148
Maradi		70	49	119
Niamey	Commercial areas	6	10	16
Maradi		18	1	19
Niamey	Forested areas	86	87	173
Maradi		127	67	194
Niamey	Residential areas	26	43	69
Maradi		39	11	50
Niamey	Road	58	32	90
Maradi		26	20	46
Niamey	School	98	44	142
Maradi		44	12	56
Niamey	Peri-urban forests	15	165	180
Maradi		43	62	105
Niamey	Total	342	476	818
Maradi		367	222	589

Table 5. DBH classes and its contribution to AGB

Cities	DBH classes (cm)	AGB (kg)	Number of stems	Contribution (%) to AGB
Niamey	[5-10[	595.39	80	0.00
Maradi		418.07	89	0.00
Niamey	[10-20[	6025.20	166	0.01
Maradi		5229.03	123	0.02
Niamey	[20-30[	22728.94	130	0.04
Maradi		16882.32	100	0.08
Niamey	[30-40]	40142.94	116	0.08
Maradi		27538.14	118	0.13
Niamey	> 40	439067.57	293	0.86
Maradi		165245.10	135	0.77
Niamey	Total	508560.04	785	1.00
Maradi		215312.67	565	1.00

P =0.68, df =1) in Maradi. Richness also does not have effect on the basal area (F = 0.008, R^2 = 0.001, P = 0.9, df = 1).

3.3 Diameter Classes and AGB in the Both Cities

We sampled a total of 1350 stems (Niamey = 785 stems, Maradi 565 stems) with DBH > 5 cm in both cities (Table 5). The large diameter class (DBH > 40 cm) was the leading class in terms of the AGB and number of stems in the two cities (Table 5). The large DBH > 40 cm contributed about 86% of AGB in Niamey and 77% in Maradi. While the small trees [5 10 [class represented small amount of AGB and number of stems in both cities (Table 5) with zero contribution to AGB.

3.4 Native Woody Species Carbon Density in Both Cities

Mean carbon stock density per hectares of native woody species varies in the two cities across LULC (Table 6). The mean native woody species carbon density was 25.71±7.71 t/ha in Niamey and in Maradi 20.18±8.20t/ha (Table 6). The native specie mean carbon density varied from peri-urban areas to the built areas though the differences were not significant (F = 7.84, P = 0.02, df = 1) with effect size η^2 = 0.40 in Niamey and in Maradi (F = 3.85, P= 0.07, df =1) with effect size η^2 = 0.24.

Among the LULC, school had the highest mean carbon density (62.40±18.80t/ha) followed by commercial areas (42±12.70t/ha) and road (30.51±7.90t/ha) in Niamey (Table 6). The lowest mean carbon density was observed in forested areas (7.90±2.53t/ha) in Niamey. While road (67.40±31.30t/ha) followed by residential areas (21.08±7.15t/ha) in Maradi. The lowest native woody specie mean carbon density was observed in commercial areas (2.50±1.23t/ha) in Maradi.

Table 6. Mean carbon density of native specie by LULC in Niamey and Maradi

LULC	Mean carbon density± standard error (t/ha)	
	Niamey	Maradi
Administrative areas	11.91±3.06	17.41±9.40
Commercial areas	42±12.70	2.50±1.23
Forested areas	7.90±2.53	7.30±3.03
Residential areas	10.16±2.8	21.08±7.15
Road	30.51±7.90	67.40±31.30
School	62.40±18.80	12.90±5.80
peri-urban forest	15.09±3.70	12.70±3.30
Overall	25.71±7.71	20.18±8.20

4. DISCUSSION

Providing the City Biodiversity Index such as the number native biodiversity and climate regulation service such as carbon storage in urban areas represents an important part of understanding whether an urban area is growing sustainably or not as urban green areas are key component to good urban life (Borelli and Conigliaro, 2018). Our study has revealed several distinctive patterns of native woody species diversity in Niamey and Maradi from the peri-urban to the built areas. The results has shown a high overall values of the Shannon diversity index for the two cities was greater than 2.0, which indicates medium to high diversity (Magurran, 2004). The specie richness in the different land use and land cover types varied significantly, this reflected the wide use of native woody species in urban greening by the city authority in two Sahel cities in Niger as reported also by (Ministère de l'Environnement du Niger, 2010). It may reflect also the use of native woody flora by the private for their urban forestry. It indicated also that native woody flora is integrated during construction in both cities as our study reported that native woody species are kept and planted in Niamey and Maradi cities. This could be explained that the city residents in both cities may use native woody flora for numerous and varied benefits such as for educational purpose, medicinal, food, etc. It could also due to the human activities such as irrigation, which may explain the presence of high demanding water species the two Sahelian cities. The built areas had higher native woody species richness than the peri-urban areas in the two cities, which confirmed the findings of (Pickett et al., 2001) who reported that urban areas are having a high level of number of species when compared to the surrounding environment. Built areas tend to have high number of species because of urbanite's planting efforts such as the public green spaces, management such as protection and irrigation and diversity such as the diversity of habitat, socio-economic characteristics of the urban people (Matthew McConnachie et al., 2008) in Niamey and Maradi cities. For instance, public urban lands in Niger such as schools, administrative areas, roads etc. often receive planting materials during national arbour day. The greening efforts in built areas more intense than in peri-urban due to may be proximity to water. On the other hand, peri-urban forest in the two cities consist of agroforestry systems which are under anthropogenic pressure such as pastoral and agricultural activities and collection of wood for energy (Larwanou and Saadou, 2005). High richness in the built areas could serve for native woody species environmental education for the urban youth generation for biodiversity conservation. The study has shown that the urban land use and land cover areas has effect on specie richness in the two cities. This may indicate that in the two cities richness is a function of planting spaces indicating that space availability is key to urban forestry (Elmqvist et al., 2013). The forested areas had the highest number of native species in both cities, which due to the urban-forested areas in the two consists of botanical garden, public mixed species green spaces, and urban wetlands. Biodiversity is the foundation for all ecosystem services. Therefore, the native woody species found in two cities may due the benefits the urban people derive from the native woody flora (Bolund and Hunhammar, 1999). Added to that native plants are key for urban biodiversity conservation such as birds as reported by (Burghardt et al., 2009). This study provided City Biodiversity Index related to the native species as they are poorly studied in urban areas mostly from rapid urbanization perspective (McKinney, 2002) which can be used as reference level in urban land use planning and urban forest management in both cities.

Fabaceae was the most dominant families with equal number of species (13 species) in both cities. This could be due the Fabaceae species are multipurpose tree species. For example, *Faidherbia albida* is a fodder and shade tree species while *Acacia senegal* used for gum production. The dominance of Fabaceae species in the two cities is of great importance to the vitality of urban ecosystems as Fabaceae species are nitrogen provider.

Our results have shown that native woody species richness had effect on structure (tree cover, basal area and stem density) in Niamey although the difference was not significant in Maradi. Specie richness has effect on basal areas, this confined the finding of (Bohn and Huth, 2017) who revealed that forest specie diversity is the dominant driver of basal area. Urban vegetation structure is important to the urban ecosystem service provision (Nowak et al., 2016b).The vegetation structure is also important for biodiversity conservation such as avian biodiversity (Burghardt et al., 2009). Our findings suggest that specie richness is an important predictor for native woody vegetation in Niamey and Maradi. More specifically for exemple, the more diverse in a given areas, the high is the tree cove, the basal area and density. As in urban areas, one of possible reason of tree plantation is for shading purpose. Therefore, the combination of many native woody species is a predictor of tree shade in both cities according to our results.

Native woody vegetation within urban settings provides numerous benefit to the society but relatively little is known about their services such as carbon sequestration (Weissert et al., 2014). Our study provided the carbon storage density in the two cities from peri-urban to the built areas. Our results showed that urban vegetation in Sahel cities of West Africa stores a considerable amount of carbon. The role of urban trees to store carbon has been widely demonstrated in the world cities (Chen, 2015; Nero et al., 2017a; Nowak et al., 2013c; Tang et al., 2016) mostly indigenous trees (Stoffberg et al., 2010). Our findings showed clearly that Niamey and Maradi act as carbon sink through the native woody vegetation. Nevertheless, some areas such as commercial areas showed small amount of carbon density in Maradi that must be increased by the urban greening policy, as commercial areas are traffic zones thus zone of carbon dioxide emission. Forested areas showed also the low carbon density in two cities that may be because those areas are not zone of human activities mostly related to the carbon dioxide emission. On the other hand, the roads in the both cities showed high carbon density, which may be due they, are traffic zones. As about 75% of the carbon dioxide emission comes from urban areas (United Nations, 2017) which is the master contributor to climate change (Intergovernmental Panel on Climate Change (IPCC), 2014). Although Niger is not a contributor to the global greenhouse gases (Conseil National de l'Environnement pour un Développement Durable, 2016). However, climate change is worldwide issue

Table 7. List of native species and their stems number inventoried

Species	Families	Number of stems_Niamey	Number of stems_Maradi
Acacia senegal (L.) Willd.	*Fabaceae*	81	5
Adansonia digitata L.	*Bombacaceae*	16	29
Albizia chevalieri Harms	*Fabaceae*		7
Annona senegalensis Pers.	*Annonaceae*	8	10
Anogeissus leiocarpus (DC.) Guill. & Perr.	*Combretaceae*	1	1
Balanites aegyptiaca (L.) Del.	*Balanitaceae*	217	69
Bauhinia rufescens Lam.	*Fabaceae*	7	20
Bombax costatum Pellegr. & Vuillet	*Bombacaceae*		1
Borassus aethiopum Mart.	*Arecaceae*	3	3
Boscia angustifolia A. Rich.	*Capparaceae*	2	
Boscia salicifolia Oliv.	*Capparaceae*		1
Boscia senegalensis (Pers.) Lam. ex Poir.	*Capparaceae*	1	
Boswellia dalzielii Hutch.	*Burseraceae*		2
Boswellia papyrifera (Del.) A. Rich.	*Burseraceae*		3
Calotropis procera (Ait.) Ait. f.	*Asclepiadaceae*	5	12
Cassia sieberiana DC.	*Fabaceae*	3	
Ceiba pentandra (L.) Gaertn.	*Bombacaceae*	5	4
Combretum aculeatum Vent.	*Combretaceae*	2	
Combretum glutinosum Perr. ex DC.	*Combretaceae*	24	7
Combretum nigricans Lepr. ex Guill. et Perr.	*Combretaceae*	4	
Commiphora africana (A. Rich.) Engl.	*Burseraceae*		1
Daniellia oliveri (Rolfe) Hutch. & Dalz.	*Fabaceae*		1
Diospyros mespiliformis Hochst. ex A. Rich.	*Ebenaceae*	7	13
Euphorbia balsamifera Ait.	*Euphorbiaceae*	3	4
Faidherbia albida (Del.) Chev.	*Fabaceae*	106	61
Ficus platyphylla Del.	*Moraceae*	10	8
Ficus sycomorus ssp. *gnaphalocarpa* (Miq.) C.C. Berg	*Moraceae*	4	9
Ficus thonningii Blume	*Moraceae*	2	4
Gardenia erubescens Stapf & Hutch.	*Rubiaceae*		2
Grewia bicolor Juss.	*Tilliaceae*	2	
Guiera senegalensis J.F. Gmel.	*Combretaceae*	2	8
Hyphaene thebaica (L.) Mart.	*Arecaceae*	24	11
Khaya senegalensis (Desr.) A. Juss.	*Meliaceae*	75	14
Kigelia africana (Lam.) Benth.	*Bignoniaceae*	7	8
Lannea microcarpa Engl. & K. Krause	*Anacardiaceae*	4	8
Maerua angolensis DC.	*Capparaceae*	1	
Maerua crassifolia Forssk.	*Capparaceae*	5	6
Mitragyna inermis (Willd.) Kuntze	*Rubiaceae*		1
Mimosa pigra L.	*Fabaceae*	2	
Neocarya macrophylla (Sabine) Prance	*Chrysobalanaceae*	22	
Parkia biglobosa (Jacq.) R. Br. ex G. Don	*Fabaceae*	1	12
Phoenix dactylifera L.	*Arecaceae*		5
Phoenix reclinata Jacq.	*Arecaceae*		1
Phyllanthus reticulatus Poir.	*Euphorbiaceae*		2
Piliostigma reticulatum (DC.) Hochst.	*Fabaceae*	5	31
Prosopis africana (Guill. & Perr.) Taub.	*Fabaceae*		30

continued on following page

Table 7. Continued

Species	Families	Number of stems_Niamey	Number of stems_Maradi
Sclerocarya birrea (A. Rich.) Hochst.	*Anacardiaceae*	7	17
Senna singueana (Del.) Lock	*Fabaceae*		2
Sesbania sesban (L.) Merrill	*Fabaceae*	1	
Sterculia setigera Del.	*Sterculiaceae*		5
Stereospermum kunthianum Cham.	*Bignoniaceae*		2
Tamarindus indica L.	*Fabaceae*	5	13
Tapinanthus dodoneifolius (DC.)	*Loranthaceae*	2	
Vachellia nilotica subsp. *nilotica*	*Fabaceae*	18	70
Vachellia seyal Del.	*Fabaceae*	50	16
Vachellia sieberiana DC.	*Fabaceae*	1	
Vachellia tortilis (Forssk.) Galasso & Banfi	*Fabaceae*	6	1
Vitellaria paradoxa Gaertn. f.	*Sapotaceae*	1	1
Vitex doniana Sweet	*Verbenaceae*	19	5
Vitex simplicifolia Oliv.	*Verbenaceae*	1	14
Ziziphus mauritiana Lam.	*Rhamnaceae*	39	16
Ziziphus mucronata Willd.	*Rhamnaceae*	7	
Ziziphus spina-christi (L.) Desf.	*Rhamnaceae*		13
Total		818	589

therefore there is for protection native woody species and further use of native woody species for urban afforestation in Niamey and Maradi. The names of the recorded species are in the table 7.

5. CONCLUSION

The study highlighted the native woody species diversity and structure in Niamey and Maradi along land use gradient. Niamey and Maradi are the native woody biodiversity refuge. A total of 63 native woody species representing 47 genera and 24 families were recorded in the overall urban forests in two cities. Native woody species are used for urban greening the two cities which uptake considerable amount of carbon. The study recommends that the expansion of the two cities should increase the native woody diversity peri-urban areas. Native woody biodiversity conservation should be the core of city land use planning in Niamey and Maradi for carbon conservation. The city authority of the two cities should make a law that encourages use of native tree species for the urban public and private urban forestry development.

6. ACKNOWLEDGEMENT

Moussa Soulé acknowledges the Federal Ministry of Education and Research (BMBF) and West African Science Centre on Climate Change and Adapted Land Use (WASCAL, www.wascal.org) for providing the scholarship and financial support for this programme. Moussa Soulé acknowledges the support of Regional Direction of the Environment of Niamey and Maradi cities.

REFERENCES

Arbonnier, M. (2002). *Arbres, arbustes et lianes des zones sèches d'Afrique de l'Ouest* (C. Isabelle, Ed.). CIRAD-MNHN-UICN.

Aronson, M. F. J., La Sorte, F. A., Nilon, C. H., Katti, M., Goddard, M. A., Lepczyk, C. A., Warren, P. S., Williams, S. G., Cilliers, S., Clarkson, B., Dobbs, C., Dolan, R., Hedblom, M., Klotz, S., Kooijmans, J. L., Kuhn, I., MacGregor-Fors, L., McDonnell, M., Mortberg, U., . . . Winter, M. (2014). A global analysis of the impacts of urbanization on bird and plant diversity reveals key anthropogenic drivers. *Proc. R. Soc., 281*, 20133330. 10.1098/rspb.2013.3330

Arroyo-Rodríguez, V., Melo, F. P. L., Martínez-Ramos, M., Bongers, F., Chazdon, R. L., Meave, J. A., Norden, N., Santos, B. A., Leal, I. R., & Tabarelli, M. (2017). Multiple successional pathways in human-modified tropical landscapes: New insights from forest succession, forest fragmentation and landscape ecology research. *Biological Reviews of the Cambridge Philosophical Society*, *92*(1), 326–340. doi:10.1111/brv.12231 PMID:26537849

Blair, R. B. (1999). Birds and butterflies along an urban gradient : Surrogate taxa for assessing biodiversity? *Ecological Applications*, *9*(1), 164–170. doi:10.1890/1051-0761(1999)009[0164:BABAAU]2.0.CO;2

Bohn, F. J., & Huth, A. (2017). The importance of forest structure to biodiversity – productivity relationships. *Royal Society Open Science*, *4*(1), 12. doi:10.1098/rsos.160521 PMID:28280550

Bolund, P., & Hunhammar, S. (1999). Ecosystem services in urban areas. *Ecological Economics*, *29*(2), 293–301. doi:10.1016/S0921-8009(99)00013-0

Borelli, S., & Conigliaro, M. (2018). Urban forests in the global context. *Unasylva*, *69*, 3–79.

Burghardt, K. T., Tallamy, D. W., & Gregory Shriver, W. (2009). Impact of native plants on bird and butterfly biodiversity in suburban landscapes. *Conservation Biology*, *23*(1), 219–224. doi:10.1111/j.1523-1739.2008.01076.x PMID:18950471

Cairns, M.A., Helmer, E.H., & Baumgardner, G.A. (1997). Root biomass allocation in the world ' s upland forests. *Oecologie*, 1–11.

Chave, J., Mechain, M. R., Burquez, A., Chidumayo, E., Colgan, M. S., Delitti, W. B., Alvaro, D., & Tron, E. (2014). Improved allometric models to estimate the aboveground biomass of tropical trees. *Global*, *20*(10), 3177–3190. doi:10.1111/gcb.12629 PMID:24817483

Chen, W. Y. (2015). The role of urban green infrastructure in offsetting carbon emissions in 35 major Chinese cities: A nationwide estimate. *Cities (London, England)*, *44*, 112–120. doi:10.1016/j.cities.2015.01.005

Conseil National de l'Environnement pour un Développement Durable. (2016). *Troisieme Communication Nationale a la Conference des Parties de la Convention Cadre des Nations Unies sur les Changements Climatiques*. Niamey.

Dayamba, S. D., Djoudi, H., Zida, M., Sawadogo, L., & Verchot, L. (2016). Agriculture, Ecosystems and Environment Biodiversity and carbon stocks in different land use types in the Sudanian Zone of Burkina Faso, West Africa. *Agriculture, Ecosystems & Environment*, *216*, 61–72. doi:10.1016/j.agee.2015.09.023

de l'Environnement du Niger, M. (2010). Foresterie Urbaine, version revue Niger. Academic Press.

Dunn, C. P., & Heneghan, L. (2011). Composition and diversity of urban vegetation. Oxford University Press.

Elmqvist, T., Fragkias, M., Goodness, J., Marcotullio, P. J., McDonald, R. I., Parnell, S., Schewenius, M., Sendstad, M., Seto, K. C., & Wilkinson, C. (2013a). *Urbanization, Biodiversity and Ecosystem Services: Challenges and Opportunities*. Springer. doi:10.1007/978-94-007-7088-1

Elmqvist, T., Zipperer, W.C., & Gii, B. (2013b). *Urbanization, habitat loss and biodiversity decline Solution pathways to break the cycle*. Academic Press.

FAO. (2016). *Guidelines on urban and peri-urban forestry (No. 178)*. FAO Forestry Paper. FAO.

Henry, M., Bombelli, A., Trotta, C., Alessandrini, A., Birigazzi, L., Sola, G., Vieilledent, G., Santenoise, P., Longuetaud, F., Valentini, R., Picard, N., & Saint-André, L. (2013). GlobAllomeTree: International platform for tree allometric equations to support volume, biomass and carbon assessment. *IForest (Viterbo)*, *6*(6), 326–330. doi:10.3832/ifor0901-006

INS. (2016). Le Niger en Chiffres 2016. Niamey.

INS. (2017). Annuaire Statistique Regional 2012 - 2016. Niamey.

Intergovernmental Panel on Climate Change (IPCC). (2006). *2006 IPCC Guidelines for National Greenhouse Inventories – A primer*. Prepared by the National Greenhouse Gas Inventories Programme, Intergovernmental Panel on Climate Change National Greenhouse Gas Inventories Programme.

Intergovernmental Panel on Climate Change (IPCC). (2014). *Climate Change 2014: Synthesis Report. Contribution of Working Groups I, II and III to the Fifth Assessment Report of the Intergovernmental Panel on Climate Change, Core Writing Team, R.K. Pachauri and L.A. Meyer. Geneva, Switzerland.* doi:10.1017/CBO9781107415324.004

Issaka, H. (2013). Privilège urbain:conditions de vie et santé au Niger. *Revue-Urbanites*, 42–52.

Jennings, S. B., Brown, N. D., & Sheil, D. (1999). Assessing forest canopies and understorey illumination : Canopy closure, canopy cover and other measures. *Forestry*, *72*(1), 72. doi:10.1093/forestry/72.1.59

Kuruneri-Chitepo, C., & Shackleton, C. M. (2011). The distribution, abundance and composition of street trees in selected towns of the Eastern Cape, South Africa. *Urban Forestry & Urban Greening*, *10*(3), 247–254. doi:10.1016/j.ufug.2011.06.001

Larwanou, M., & Saadou, M. (2005). Biodiversity of ligneous species in semi-arid to arid zones of southwestern Niger according to anthropogenic and natural factors. *Agriculture, Ecosystems & Environment*, *105*(1-2), 267–271. doi:10.1016/j.agee.2004.03.004

Magurran, A.E. (2004). *Measuring biological diversity: Rejoinder, Environmental and Ecological Statistics*. Blackwell Science Ltd. doi:10.2989/16085910409503825

Matthew McConnachie, M., Shackleton, C. M., & McGregor, G. K. (2008). The extent of public green space and alien plant species in 10 small towns of the Sub-Tropical Thicket Biome, South Africa. *Urban Forestry & Urban Greening*, *7*(1), 1–13. doi:10.1016/j.ufug.2007.12.003

Matthies, S. A., Rüter, S., Schaarschmidt, F., & Prasse, R. (2017). Determinants of species richness within and across taxonomic groups in urban green spaces. *Urban Ecosystems*, *20*(4), 897–909. doi:10.100711252-017-0642-9

Mcdonald, R. I., Kareiva, P., & Forman, R. T. T. (2008). The implications of current and future urbanization for global protected areas and biodiversity conservation. *Biological Conservation*, *141*(6), 1695–1703. doi:10.1016/j.biocon.2008.04.025

McKinney, M. L. (2002). Urbanization, Biodiversity, and Conservation. *Bioscience*, *52*(10), 883–890. doi:10.1641/0006-3568(2002)052[0883:UBAC]2.0.CO;2

Mckinney, M. L. (2006). Urbanization as a major cause of biotic homogenization. *Biological Conservation*, *3*(3), 247–260. doi:10.1016/j.biocon.2005.09.005

McPherson, E.G., & Simpson, J.R. (2003). Potential energy savings in buildings by an urban tree planting programme in California. *Urban For. Urban Green, 2*, 73–86. doi:10.1078/1618-8667-00025

Nero, B. F., Callo-Concha, D., Anning, A., & Denich, M. (2017a). Urban Green Spaces Enhance Climate Change Mitigation in Cities of the Global South: The Case of Kumasi, Ghana. *Procedia Engineering*, *198*, 69–83. Advance online publication. doi:10.1016/j.proeng.2017.07.074

Nero, B. F., Campion, B. B., Agbo, N., Callo-concha, D., & Denich, M. (2017b). Tree and trait diversity, species coexistence, and diversity- functional relations of green spaces in Kumasi, Ghana. *Procedia Engineering*, *198*, 99–115. doi:10.1016/j.proeng.2017.07.164

Nowak, D. J., Green, E. J., Hoehn, R. E., & Lapoint, E. (2013a). Carbon storage and sequestration by trees in urban and community areas of the United States. *Environ. Pollut. J.*, *178*, 229–236. doi:10.1016/j.envpol.2013.03.019 PMID:23583943

Nowak, D.J., Green, E.J., Hoehn, R.E., & Lapoint, E. (2013b). *Carbon storage and sequestration by trees in urban and community areas of the United States*. Academic Press.

Nowak, D. J., Green, E. J., Hoehn, R. E., & Lapoint, E. (2013c). Carbon storage and sequestration by trees in urban and community areas of the United States. *Environ. Pollut.*, *178*, 229–236. doi:10.1016/j.envpol.2013.03.019 PMID:23583943

Nowak, D. J., Hoehn, R. E., Bodine, A. R., Greenfield, E. J., & Neil-dunne, J. O. (2016a). Urban forest structure, ecosystem services and change in Syracuse, NY. *Urban Ecosystems*, *19*(4), 1455–1477. doi:10.100711252-013-0326-z

Nowak, D. J., Hoehn, R. E., Bodine, A. R., Greenfield, E. J., & Neil-dunne, J. O. (2016b). Urban forest structure, ecosystem services and change in Syracuse, NY. *Urban Ecosystems*, *19*(4), 1455–1477. doi:10.100711252-013-0326-z

Pearson, T.R.H., Brown, S.L., & Birdsey, R.A. (2007). Measurement Guidelines for the Sequestration of Forest Carbon. *North. Res. Stn.*, 42.

Pickett, S. T. A., Cadenasso, M. L., Grove, J. M., Nilon, C. H., Pouyat, R. V., Zipperer, W. C., & Costanza, R. (2001). Urban Ecological Systems: Linking Terrestrial Ecological, Physical,and Socioeconomic Components of Metropolitan Areas. *Annual Review of Ecology and Systematics*, *32*(1), 127–157. doi:10.1146/annurev.ecolsys.32.081501.114012

Pysek, P. (1998). Alien and native species in Central European urban floras: A quantitative comparison. *Journal of Biogeography*, *25*(1), 155–163. doi:10.1046/j.1365-2699.1998.251177.x

Reynolds, C. C., Escobedo, F. J., Clerici, N., & Zea-Camaño, J. (2017). Does "greening" of neotropical cities considerably mitigate carbon dioxide emissions? The case of Medellin, Colombia. *Sustain.*, *9*(5), 785. Advance online publication. doi:10.3390u9050785

Saadou, M. (1990). *La végétation des milieux drainés nigériens à l'Est du fleuve Niger*. Université de Niamey.

Seto, K. C., Guneralp, B., & Hutyra, L. R. (2012). Global forecasts of urban expansion to 2030 and direct impacts on biodiversity and carbon pools. *Proceedings of the National Academy of Sciences of the United States of America*, *109*(40), 16083–16088. doi:10.1073/pnas.1211658109 PMID:22988086

Seto, K. C., & Shepherd, J. (2009). Global urban land-use trends and climate impacts. *Current Opinion in Environmental Sustainability*, *1*(1), 89–95. doi:10.1016/j.cosust.2009.07.012

Stoffberg, G. H., Van Rooyen, M. W., Van Der Linde, M. J., & Groeneveld, H. T. (2010). Carbon sequestration estimates of indigenous street trees in the City of Tshwane, South Africa. *Urban Forestry & Urban Greening*, *9*(1), 9–14. doi:10.1016/j.ufug.2009.09.004

Tang, Y., Chen, A., & Zhao, S. (2016). Carbon Storage and Sequestration of Urban Street Trees in Beijing, China Carbon Storage and Sequestration of Urban Street Trees in Beijing, China. *Frontiers in Ecology and Evolution*, *4*, 1–11. doi:10.3389/fevo.2016.00053

Thiombiano, A.R., Glele KaKai, P., Bayen, J.I.B.A., & Mahamane, A. (2016). Méthodes de collecte et d'analyse des données de terrain pour l'évaluation et le suivi de la végétation en Afrique Methods for sampling and analysis of field data to evaluate and monitor vegetation in Africa. *Annales des sciences agronomiques, FSA/UAC.*

UN. (2017). *Goal 11-Sustaimnable Cities and Communities*, https://www.unenvironment.org/explore-topics/sustainable-development-goals/why-do sustainable-development-goals-matter/goal-11

Wang, H., Marshall, C. W., Cheng, M., Xu, H., Li, H., Yang, X., & Zheng, T. (2017). Changes in land use driven by urbanization impact nitrogen cycling and the microbial community composition in soils. *Scientific Reports*, *7*(1), 1–12. doi:10.1038rep44049 PMID:28281565

Weissert, L. F., Salmond, J. A., & Schwendenmann, L. (2014). A review of the current progress in quantifying the potential of urban forests to mitigate urban CO2emissions. *Urban Climate*, *8*, 100–125. doi:10.1016/j.uclim.2014.01.002

Yan, P., & Yang, J. (2017). Species diversity of urban forests in China. *Urban Forestry & Urban Greening*, *28*, 160–166. doi:10.1016/j.ufug.2017.09.005

Zhu, H. N., Robert, C., & Kamran, A. (2015). Modeling air pollutant removal, carbon storage, and CO2 sequestration potential of urban forests in Scotlandville, Louisiana, USA. *Biogeosciences For.*, *009*, 1–8. doi:10.3832/ifor1845-009

KEY TERMS AND DEFINITIONS

Carbon Sequestration Potential: Ability of a green plant to remove carbon dioxide from the atmosphere via photosynthesis process.

City Biodiversity Index: Number of native woody species and the carbon stock are examples of City biodiversity Index.

Native Woody Species: Local or indigenous woody plants.

Native Woody Species Conservation: Any act or practice to protect and preserve a variety of indigenous woody species in a given area of the planet.

Urban Forest Structure: Biophysical aspects of the urban forests such as the name of species, stem density, tree cover, etc.

Urban Forests: A collection of woody species found in urban and peri-urban areas.

Urban Woody Species Values: different benefits or services such as carbon dioxide reduction, fodder, oxygen, fruit and leaves for human consumption, etc. provided by urban woody flora.

Chapter 16
Effects of Urbanization on the Degradation of the Palm Grove Around the City of Marrakech

Mohamed Radi
Faculty of Sciences Semlalia, Cadi Ayyad University, Morocco

Abdelilah Meddich
Cadi Ayyad University, Morocco

Elmehdi Ouatiki
Cadi Ayyad University, Morocco

Lahcen Ouahmane
Cadi Ayyad University, Morocco

Mohamed Hafidi
Cadi Ayyad University, Morocco

Abdelkrim Ben Salem
https://orcid.org/0000-0002-2283-5928
Research Center BioBio, FS, Mohammed V University in Rabat, Morocco

ABSTRACT

The city of Marrakech is experiencing a very significant development in terms of urbanization which leads to a significant reduction of the agricultural activity and increase of the rate of metallic pollution in soils. The impact of this metallic pollution on the microbial activity in these soils can negatively affect the growth and the development of the date palm (Phoenix dactylifera L). In addition, the urbanization has direct effects on the degradation of palm groves in the Marrakech region, following a decrease in agricultural and forestry productivity after anthropogenic activities which are at the origin of high metallic trace element (MTE) contents in soils. In order to assess the impacts of this pollution, the authors carried out a study in the rhizospheric soils of the date palm (Phoenix dactylifera L) monitoring the levels of heavy metals (Pb, Zn). Eight sites were chosen: five urban sites (three gardens, two boulevards) and two peri-urban sites (palm groves).

DOI: 10.4018/978-1-7998-7512-3.ch016

INTRODUCTION

In urban and pre-urban areas, arid and semi-arid zones, drought and soil pollution are major constraints considerably limiting plant production. Anthropogenic activities are the source of many pollutants disseminated in the environment (Goyal et al., 2020), case of diesel (diesel) and gasoline of vehicles and industrial activities (construction, mining, steel, chemical pollution) (Magali Sammut 2007) (Torkshavand et al., 2014). Heavy metals such as lead, cadmium, copper and zinc are continuously added to the soil and cannot be biodegraded and therefore persist in the environment for long periods of time(Singh et al., 2020). Those which constitute the major constraint considerably limiting plant production. The concentration and type of heavy metals influence phytotoxicity, and in some cases plant growth may be reduced (Shah et al., 2007; Salt et al., 1998; Singh et al., 2003).

On the other hand, a great heterogeneity of the forms of drought is encountered in these regions (Annerose; 1991). The scarcity of rains, the drop in the level of groundwater and dams, very high temperatures, and the soils are generally poor in nutrients such as nitrogen and phosphorus (Diem et al., 1981; Mikola, 1987). These environmental conditions lead to a deterioration of the physico-chemical characteristics of the soil, and a gradual loss of soil fertility. These forms constitute major constraints considerably limiting plant production(Verma et al., 2020).

In Morocco, palm groves know several problems of degradation under the action of drought and also of anthropogenic activities which was the principal origin of high levels of metallic trace elements (MTEs) in the soils. The date palm occupied a large area with an effective of 250 000 productive feet. Currently the area occupied by the species is limited to 44000 ha with about 4.4 million feet. It covers several regions, including Tafilalet and Drâa, which contain more than 70% of the total national effective in date palms (Haddouchi 1996). In the Marrakesh region, as in the other oases of southern Morocco, several sites have been contaminated by MTEs that have caused a loss of the multifunctionality of the soil and the biodiversity of edaphic microorganisms. Resulting in the degradation of numerous palm groves, thus increasing the mortality rate of date palm.

The objective of this work is to study the effects of water stress and heavy metals (Pb, Zn, Cd) pollution in the rhizosphere soils of the date palm (*Phoenix dactylifera* L) on the degradation of date palm groves around Marrakesh city.

MATERIAL AND METHODS

1. Study Sites

We conducted our study in the palm grove of Marrakesh (Morocco). The choice of sites took into account the condition of the palm trees and human activity. We thus selected eight study sites, including six in urban areas (three gardens and three boulevards) and two in peri-urban areas (two palm groves) for the sampling of the rhizospheric soil from the feet of date palms (Table 1). For each site, 15 soil and rhizospheric roots samples from the date palm were collected. Indeed, for each tree and 1 m from the trunk, four samples of soil and roots (about 2 kg) were taken in the four cardinal directions to a depth of 10 to 40 cm and subsequently mixed to have homogeneous sample of the medium. The physicochemical composition of the soils at the eight sites studied was analyzed.

Table 1. Presentation of the studied sites and their characteristics.

Study sites		Locations	Characteristics
Peri-urban environment	Palm groves	P1: Palm groves Azzouzia	• Presence of agricultural activity • Presence of irrigation system (wastewater) of associated cultures
		P2: Circuit Palm groves	• Palm trees in their natural state • No agricultural activity • No irrigation or maintenance system;
urban environment	Gardens	J1: Arset Moulay Abdesslam	• Presence of agricultural activity • Presence of irrigation system and maintenance works;
		J2: Menara	
		J3: Rwidate	
	Boulevards	BV1: Abdelkrim Khattabi	• Palm trees in their natural state • No agricultural activity • No irrigation or maintenance system • A wide variety of anthropogenic actions (road traffic, waste disposal, etc.).
		BV2: Prince Moulay Abdallah	
		BV3: Mouzdalifa	

2. Cultivation Process

After sampling, we used plastic pots (17 × 8 cm) containing 1 kg of soil from each of the eight sites studied for the development of the root system of young plants. The date palm seeds of the "Boufeggous" variety from the Marrakesh palm grove were disinfected by immersion in a 15% hydrogen peroxide solution for 20 minutes followed by three washes with sterile distilled water. They were germinated in the prepared pots at the rate of one seed / pot. The pot experience was conducted in the nursery of the urban commune of Marrakesh under the following conditions: T° average maximum temperature: 33.26 °C; T° average min temperature: 22.31 °C; The average maximum relative humidity: 48.2%; The average minimum relative humidity: 14.33%. The palm seedlings are grown under normal water conditions for ten days, then watered to saturation and left to drain for 48 hours to weigh the soil to determine the maximum water retention capacity (field capacity of the soil). The method used for the application of water stress is that described by El achouri [12] and Tobar et al [13]. W1 was the weight of the pot full of dry soil. The soil was then watered until saturation, then it was allowed to drip under the effect of gravity, to constant weight. The soil was thus at field capacity (FC). W2 was the weight of the pot after the excess water has flowed out. The difference (W2 - W1) will correspond to the volume of water necessary to obtain the FC of the soil used (100% FC). To have equivalent humidity of soil, at 75%, 50% and 25% of the FC, we added to the series of pots containing dry soil, the water volumes corresponding to 0.75 × (W2 - W1) ; 0.50 × (W2 - W1) and 0.25 × (W2 - W1) respectively. We then note the weights W3, W4 and W5 of the pots which successively correspond to 75%, 50% and 25% of the FC. During the experiment, in the three water treatments used, the pots were weighed and then distilled water was added until the weight corresponding to the imposed water stress is obtained (75%, 50% or 25% of the FC). During the whole experiment, the pots were weighed twice a day (Ohaus DS4 capacity 44 × 02 Ib type balance) and the lost water is replaced [13]. We performed 30 repetitions for each treatment used.

3. Measured Parameters

During the experiment, the morphological parameters of date palm growth were carried out and monitored periodically during the eight months of cultivation under the various water constraints imposed (75%, 50% or 25% of the FC). After eight months of cultivation, the response of the date palm seedlings to the water constraint condition was estimated by determining the number of leaves formed, the leaf area, the extension of the aerial part of the plant and the production of biomass fresh matter and dry matter (obtained after drying in an oven at 105 °C. for 24 hours). The relative water content (TRE) was determined by measuring the mass of fresh matter (PF), that of dry matter (PS), and that of relative full turgor (PPT) which is done according to the method described by Barrs [17]. The relative water content is deduced by the following formula:

Other physico-chemical analyzes of the different soil samples studied were carried out. The pH was measured in a soil paste saturated with distilled water in a ratio 1/2 (w/v). 10 g of soil are dissolved in 20 ml of distilled water. After 30 minutes of stirring, the measurements were carried out using a pH meter (CRISON). The principle of the Total Limestone method consists in dissolving calcium carbonates ($CaCO_3$) by acid attack (HCl) and measuring the volume of CO_2 released, using a Bernard Calcimeter. The organic carbon content present in the soil was determined by cold oxidation, using a mixture of potassium dichromate in an acid medium (Anne, 1945). The measurement of total nitrogen was based on the transformation of organic nitrogen into ammoniacal nitrogen. The sample undergoes mineralization with concentrated sulfuric acid in the presence of the Kjeldahl catalyst (copper sulfate, potassium sulfate and selenium), then the ammonia formed was displaced by NaOH (40%). Next, the ammonia entrained by water vapor was fixed by boric acid and titrated with sulfuric acid (0.02N) (Rodier, 1984). For labile phosphorus, the Olsen method consists in extracting the orthophosphates with a sodium bicarbonate solution. The phosphorus assay was done by colorimetry based on the formation and reduction of a phosphomolybdic complex which was accompanied by a blue coloration measured with the spectrophotometer at 820 nm. The determination of total phosphorus in the soil was carried out by the same method used for phosphorus in the plant. After calcination at 600 ° C for 6 hours, the samples were cooled (overnight) and the ashes were taken up in 0.5 ml of H_2O_2 (110 volumes) with dry evaporation in a sandbox, then with 2 ml of concentrated HCl and dry evaporation in the sandbox. Finally, the residue was taken up in 5 ml of HCl diluted 1/10, the suspension obtained was filtered on filter paper. The filtrate was adjusted to 100 ml. Then we proceed as Olsen and the DO was read in a spectrophotometer at 820 nm.

The total concentration of metals was determined after mineralization of the soils according to AFNOR NF X 31-151 standard. One g of each sample was calcined at 550 ° C for 4 h and was attacked by hydrofluorodric acid (5 ml). After evaporation of the acid, the residue obtained was dissolved with a mixture of HNO_3 / HCl (1: 1). Then acidified to 1% with nitric acid and analyzed by the Atomic Absorption Spectrometry with flame (SAA).

4. Statistical Data Analysis

Statistical analysis was performed by one-way ANOVA, followed by Student–Newman–Keuls grouping test. The statistical significance was declared when $p \leq 0.05$. All statistical evaluations were performed using the XLSTAT software.

RESULTS AND DISCUSSION

1. Physico-chemical Properties of Soil Samples

The results of the analysis of the physico-chemical parameters of the collected soil samples are presented in Figures (2-8). Statistical analysis has shown that all the physicochemical parameters of the soil (pH, electrical conductivity, CEC,% $CaCO_3$, contents in Nitrogen and Phosphorus…) vary significantly depending on the sites studied. The results obtained show that the pH (Figure 2) was relatively alkaline at all sites, but it reaches very important values at the level of garden sites and lot of palm grove 2 (8.6) because this zone is located next to mine tailings rich in basic mineral elements.

Total nitrogen and organic (Figure 6) and mineral nitrogen compounds exist in the soil. Indeed, the concentration of this element in the soils tested differs from one site to another. Thus, it was higher at the site level J3 (0.16%) in accordance to the values noted in the other soils of the various sites.

The content for total phosphorus (Figure 5) and labile phosphorus (Figure 3) obtained vary from one site to another. Thus, the maximum values recorded for total phosphorus were noted at the soil level of site J2 (123.31 mg / g). Whereas for labile phosphorus, the maximum values were noted at the level of the boulevard 2 site (0.39 mg / g). Indeed, for the two types of phosphorus, the levels recorded in the soils of the other sites generally remain less important compared to the previous values.

The organic carbon content (Figure 4) present in the soil was very high in the soils of the boulevards sites, notably the soil of the BV1 site (2.29%). While for the other soils tested, the contents were generally lower, especially the value obtained at site level J3 (0.39%).

Regarding the composition of the soils tested in total limestone (Figure 7), analyzes show that the lowest values were recorded at garden sites 3 and at lot 2 of the palm grove. While the other soils tested take very significant values at the site of lot 1 of the palm grove which reaches 21.32%.

The mean contents calculated in total MTE in the soil samples, from 3 samples per station, are presented in Table 2.

Statistical analysis showed significant differences in the distribution of concentrations between heavy metals. Regarding the rhizospheric levels of lead, cadmium, copper, zinc and iron, the rhizospheric soils of the boulevards were more contaminated compared to other sites. With the exception of lead, which was a little concentrated in garden soils, the concentrations of other forms were equal between palm groves and gardens.

The results of the pH measurement show that the soils of all the studied sites were alkaline with a pH varying between 7.7 and 8.6. Whereas, relatively high total limestone ($CaCO_3$) contents (greater than 18.8%) have been recorded in anthropized sites P1, BV1, BV2 and J1. In addition, the mineral contents of the various soils (N and P) as well as in total carbon, were generally low. The labile phosphorus varies significantly from one site to another; the soils of sites BV1, BV2 and J3 have the highest levels of labile phosphorus. On the other hand, Total soil phosphorus shows significantly variable levels between the sites studied. Generally, the soils of boulevards sites have high levels of mineral elements compared to garden and palm grove sites, which increases their electrical conductivity (figure 8). The mobility of ETM in soils was strongly influenced by soil parameters, in particular the pH, the contents of elements and organic matter and the mineral composition.

According to Anne Tremel-Schaub and Isabelle Feix (2005). The metallic trace elements enter into the constitution of the solid phase of the soil. They can be feebly linked to the soil by electrostatic connections (exchangeable form). They can also be more or less strongly linked to the soil by ionic or

Figure 1. Pictures of the sites studied

Site 1: Palm grove 1 North East (P1)

Site 2: Palm grove 2 North Ouest (P2)

Site 3: Boulevard 1 (BV1)

Site 4: Boulevard 2 (BV2)

Site 5: Boulevard 3 (BV3)

Site 6 : garden 1 (J1)

Site 7: garden 2 (J2)

Site 8: garden3 (J3)

Figure 2. Variation of the pH soil samples from the studied sites

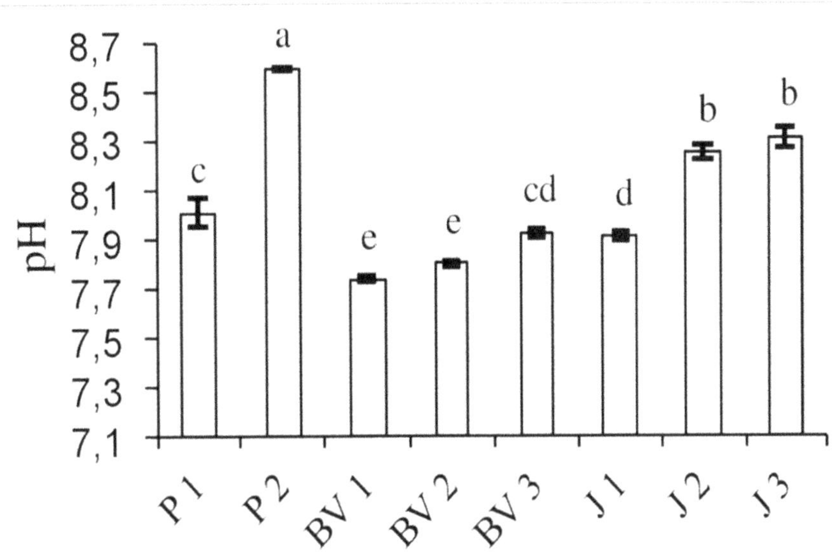

molecular liaison (organic complex) or also form organo-metallic complexes (inorganic speciations). While some were more mobile than others (Cd and Zn), although others were immobile (Cu and Pb) and easily form organic complexes with fulvic acids (Prasad & Hagemeyer, 1999). The increase the content of a form of the metal induced in the increase of the intensity of soil contamination.

Figure 3. Variation of the labile phosphorus content of soils from the studied sites

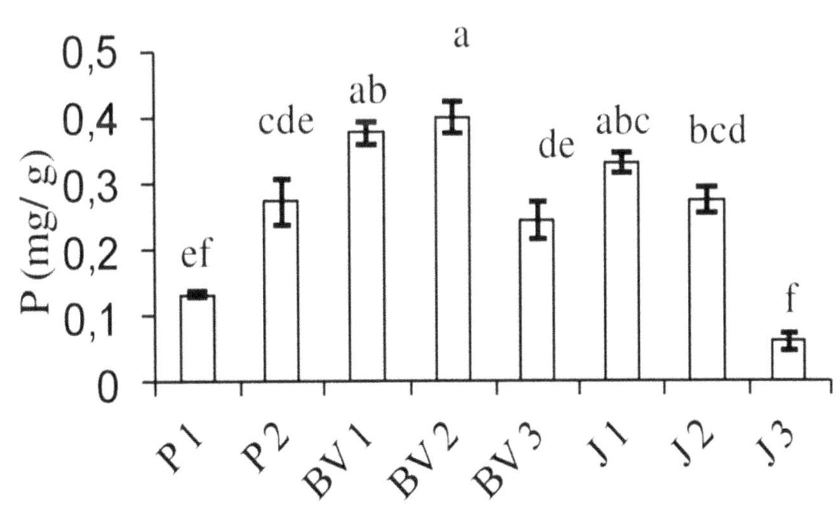

Figure 4. Variation of total organic carbon content in soil samples from the studied sites

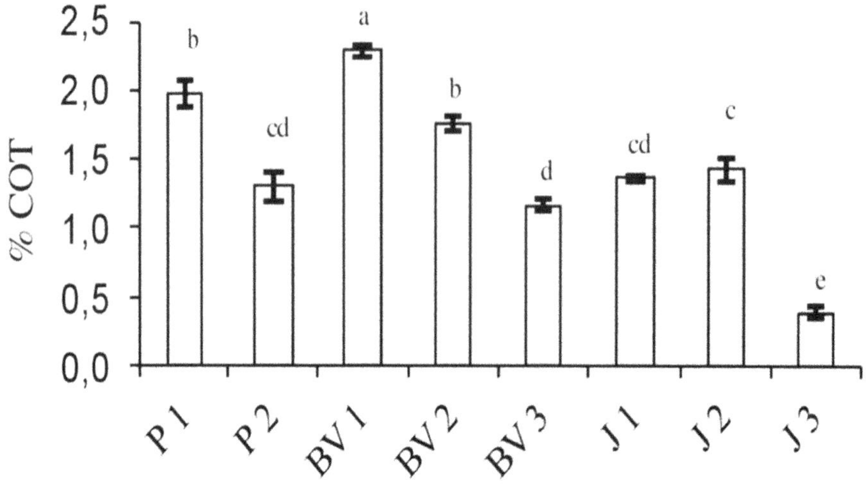

2. Effect of Water Deficit on the Growth of the Date Palm

The results obtained have shown that water stress has a negative effect on most of the studied morphological parameters of the date palm. The palm leaf area was significantly increased by the application of the water regime by 75% of the FC and this in the eight study sites (Table 3). Similarly, the leaf area (LA) of the palm seedling planted in the rhizospheric soils of gardens and palm groves was generally greater in comparison with the other boulevards sites (Table 3). The LA was (34.15, 26.61) cm^2 in the seedling of palm planted in the soil of the Rwidate Garden (J3) and in the boulevard (BV3) respectively

Figure 5. Variation of the total phosphorus content in soil samples from the studied sites

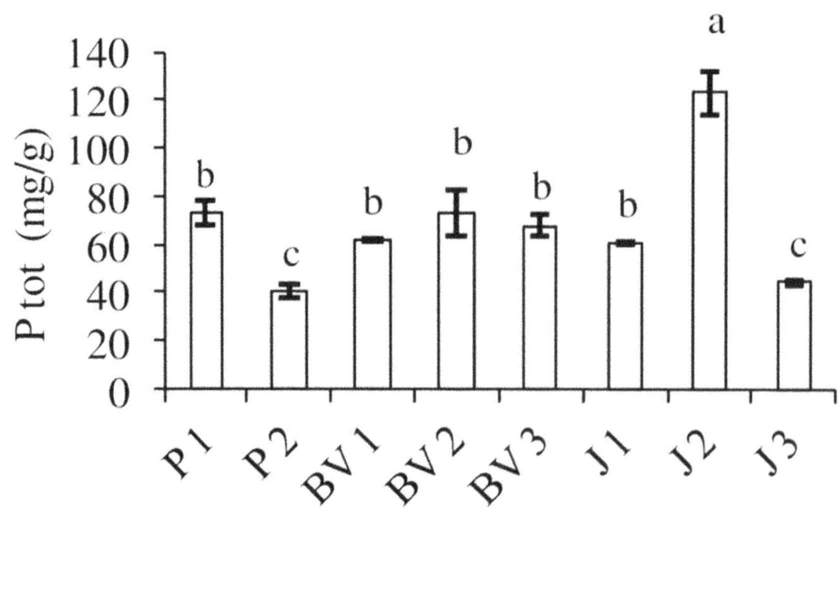

Figure 6. Variation of the total nitrogen content in the soil samples of the studied sites

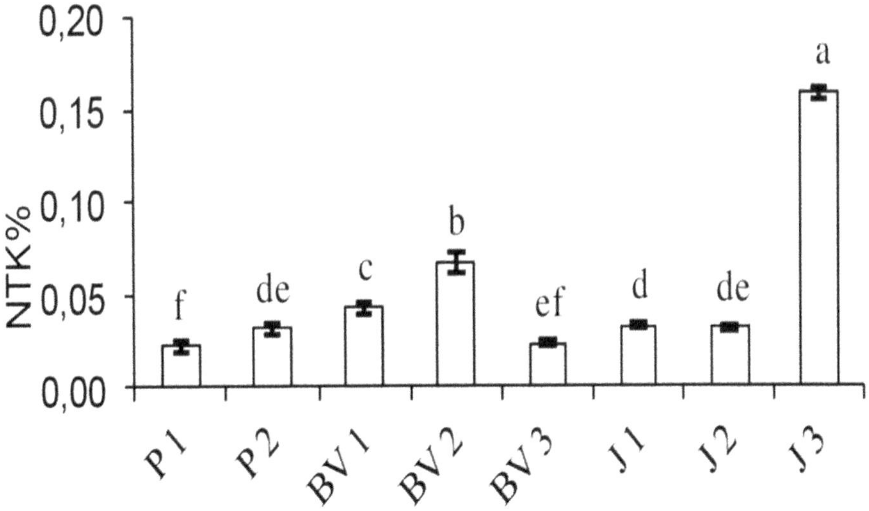

for a water treatment of 75% of the FC. This LA parameter was measured only on the first best developed leaf of the same rank 1 for all of the plantations tested and this at the level of the different water regimes applied and for the different repetitions used. For water stress corresponding to 25 and 50% of the FC, there was a non-significant decrease in the LA. Similarly, the number of leaves formed decreases when the available water reserve of the soil decreases. The study of the rapidity of appearance of the leaves according to the water treatments has shown that the application of the water regime of 75% of the

Figure 7. Variation of the total limestone content in the soil samples of the studied sites

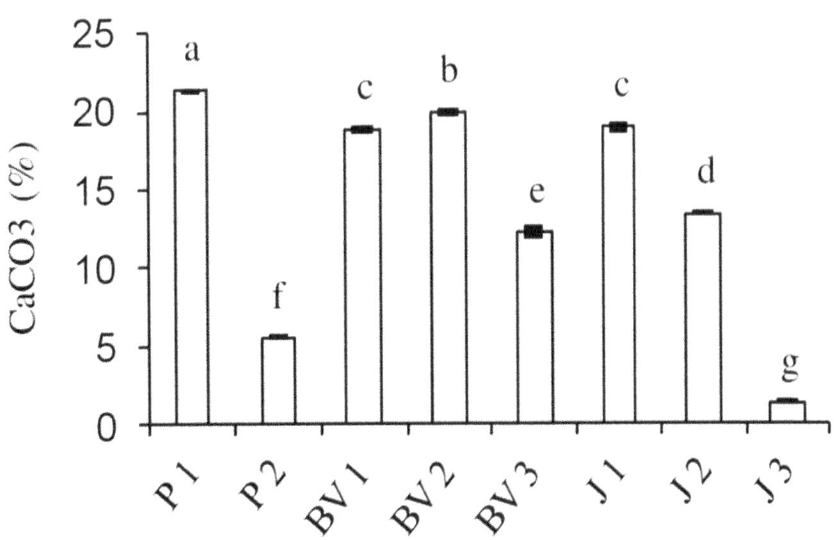

Table 2. Average contents of MTE in the studied soil samples

Sites	Palm groves	Boulevards	Gardens
Lead	174,3[a]	137,53[a]	77,33[b]
Cadmium	5.23[b]	9,153[a]	4,88[b]
Copper	72 [b]	100,33 [a]	78,53 [a]
Zinc	77 [b]	123,33[a]	82,133 [b]
Iron	6050 [a]	6597,8 [a]	6203,133 [a]

Note: The values followed by the same letter(.(a and b) are not significantly different P <0.05 (Newman and Keuls test).

FC permits the formation of a number of leaves significantly greater than that produced by the plants cultivated under the other water regimes and this in the eight sites studied (Table 3). The application of water stress also had a negative effect on the height of the aerial part of the palm from the eighth week of growth. When the water stress increases, the elongation of the aerial part of the palm was reduced (Table 3). These decreases in elongation were significantly greater with the application of the regime of 50, 75 and 100% FC in the rhizospheric soil of the palm of the gardens sites and palm groves than in the sites of the boulevards. The application of the water constraint had no significant effect on the height of the aerial part of the plant of the urban site (Boulevard My Abdellah: BV2).

The production of aerial dry matter was increased more significantly in the palm of garden sites and palm groves by the application of water regimes of 75, 50 and 25% of FC. The application of water stress of 25% FC had a significant effect on the water content of the plant in the sites of gardens and palm groves. Among the sites of the urban environment studied, the water content was highest in the garden site (J3) which was 34.94% and it was only 20.16% in the boulevard site (BV3) (Table 3)

Generally, when the content of heavy metals increases the growth of the seedlings decreased, such as at the boulevards sites, the results show a weak growth, because of a high content of ETM. In contrast,

Figure 8. Variation of electrical conductivity in soil samples from the studied sites

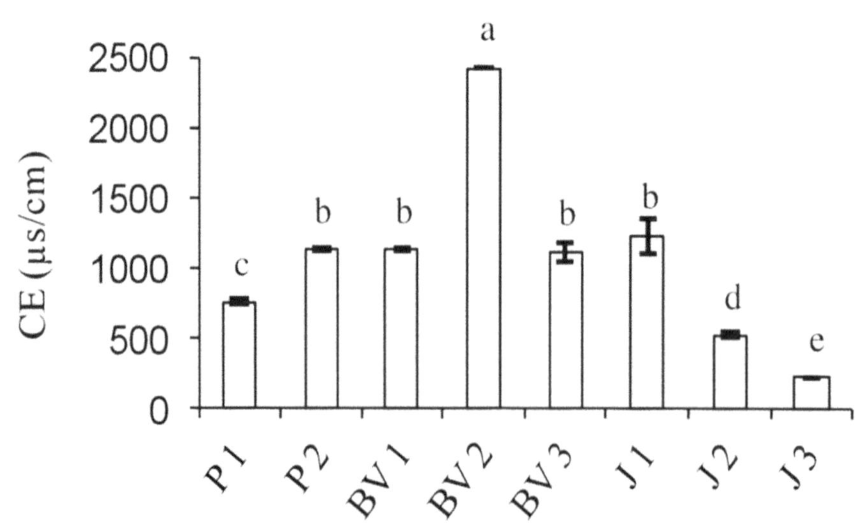

Table 3. Effect of water stress on the growth parameters of the date palm after eight months of culture

	Water stress imposed (% CC)	J 1	J 2	J 3	P 1	P 2	BV1	BV 2	BV 3
SF (cm^2)	25	26,72^{a}	25,79^{a}	57,65^{b}	18,03^{a}	20,43^{a}	12,89^{a}	18,46^{a}	17,92^{a}
	50	29,22^{a}	27,75^{a}	30,43^{a}	28,68^{a}	27,86^{a}	24,97^{a}	26,17^{a}	27,15^{a}
	75	31,36^{ab}	32,34^{ab}	34,15^{b}	32,45^{ab}	30,98^{ab}	30,38^{ab}	29,12^{ab}	26,61^{a}
	100	39,61^{a}	43,38^{a}	46,17^{a}	34,42^{a}	36,55^{a}	33^{a}	35,46^{a}	41,14^{a}
HT	25	37,33^{b}	37,33^{b}	34,83^{b}	35,66^{b}	35,83^{b}	33,16^{b}	29,66^{a}	28^{a}
	50	42,5^{b}	38,33^{c}	37^{bc}	36,16^{bc}	37,66^{bc}	33,16^{ab}	29,83^{a}	31,5^{a}
	75	44,83^{d}	40,83^{bcd}	38,5^{bc}	38^{bc}	41,83^{cd}	37,83^{bc}	30,5^{a}	35,16^{b}
	100	45,83^{d}	40,5^{bc}	39,33^{bc}	38,5^{bc}	42,83^{cd}	57,5^{bc}	31,66^{a}	36,5^{b}
NF	25	4,66^{a}	4,66^{a}	4,66^{a}	4,66^{a}	4,66^{a}	4,66^{a}	4^{a}	4^{a}
	50	4,66^{a}	4,66^{a}	4,66^{a}	5^{a}	4,66^{a}	5^{a}	4,66^{a}	4^{a}
	75	5,66^{b}	5,33^{ab}	5,33^{ab}	5^{ab}	5,33^{ab}	5^{ab}	5^{ab}	4^{a}
	100	5,66^{a}	5,66^{a}	5,33^{ab}	5,66^{a}	5,66^{a}	5,66^{a}	5^{a}	5,66^{a}
MSF (g)	25	6,22^{b}	5,43^{ab}	4,79^{ab}	4,76^{ab}	4,52^{ab}	4,41^{ab}	4,83^{ab}	3,58^{a}
	50	6,84^{b}	5,49^{ab}	4,98^{ab}	5,53^{ab}	4,75^{ab}	4,95^{ab}	4,56^{ab}	3,95^{a}
	75	7,82^{b}	6,44^{ab}	6,32^{ab}	5,9^{ab}	5,32^{ab}	5,55^{ab}	4,95^{ab}	4,70^{a}
	100	8,12^{a}	6,98^{a}	6,53^{a}	6,21^{a}	6,30^{a}	5,743^{a}	5,68^{a}	5,15^{a}
TE (%)	25	21,11^{ab}	22,87^{ab}	34,94^{b}	32,7^{3b}	27,52^{ab}	4,43^{a}	21,75^{ab}	20,16^{ab}
	50	19,83^{a}	42,85^{a}	43,93^{a}	21,24^{a}	39,94^{a}	19,56^{a}	33,53^{a}	25,54^{a}
	75	20,92^{a}	42,77^{a}	34,98^{a}	26,45^{a}	35,61^{a}	37^{a}	35,62^{a}	26,03^{a}
	100	21,67^{a}	35,07^{a}	37,34^{a}	26,29^{a}	27,3^{a}	183,4^{a}	29,81^{a}	20,38^{a}

SF: leaf area; NF: number of sheets; HT: height of the aerial part; MS: leaf dry matter; TE: water content; For each parameter, the values followed by the same letter(.(a;b;c and d) are not significantly different $P < 0.05$ (Newman and Keuls test).

in the soils of the garden and palm grove sites, which were characterized by a low ETM content, was noted the increase of the growth of the seedlings. This can be explained by the great pollution existing at the level of the boulevards sites which are diffused in heavy metals, in mineral elements and in micropollutants, this pollution acts directly on the production and the density of the spores and on the specific richness, which decreases with time (Ortega-Larrocea et al., 2007).

Nemmar, (1983) and Moursi et al. (1979) showed that drought reduces the height and diameter of the stem, the length of the internodes, the number of leaves, the water content of the leaves and the weight of the dry matter as well as the leaf area in plants. In our study, the results obtained also showed that water stress has a negative effect on plant growth, and the more intense the stress, the lower the plant's morphological parameters.

The results obtained have shown that the combination between pollution (by heavy metals) and drought (by the application of water stress) have harmful and negative effects on the growth and development of date palm seedlings those who observed in the boulevards sites in particular in the water regimes 25% and 50% FC, on the other hand the increase in the growth of these seedlings was well noted in the soils of the garden and palm grove sites, this can be explained by the poverty of these heavy metal soils, and their richness in organic matter.

Indeed, soil pollution by heavy metals, mineral elements and micropollutants, acts directly on the production and density of spores and on the specific richness, which decreases over time (Ortega-Larrocea et al., 2007). In recent years, the city of Marrakesh has experienced a very significant development in terms of urbanization, which leads to a significant reduction in all agricultural activity and increased the rate of metal pollution of the soil, in addition to the drought and salinity of the soil. Accentuating the problem which tends towards a negative and alarming situation. The impact of this metal pollution, on the microbial activity of these soils, can negatively affect the growth and development of host plants (Phoenix dactylifera L). Degradation of many palm groves thus increasing the mortality rate of the feet of date palm (*Phoenix dactylifera* L).

3. Role of the Mohammed VI Foundation for the Protection of the Environment in the Development and Protection of the Palm Grove of Marrakesh

The palm grove of Marrakesh dating from the Almoravid period was an emblematic heritage of the ocher city, it covers 12 000 hectares. The main constraints of the date palm groves were: Aging palm trees; lack of rejuvenation operations; drought and salinity; lack of site maintenance and security and Urban Extension.

Conscious of the need to take immediate action involving all components of society, His Majesty King Mohammed VI entrusted the Mohammed VI Foundation for the Protection of the environment under the Presidency of His Royal Highness Princess Lalla Hasnaa to mobilize all the intervenors around a program capable of safeguarding this centuries-old heritage while developing it.

In this context, and on the foundation of exhaustive proposals presented by the Wilaya of Marrakesh, a substantial program has been initiated and validated by the Foundation and its partners, with the main objective of reviving this space, replanting nearly 450 000 palm trees in 10 years on an area of 258 ha and develop education and awareness in order to perpetuate the actions to be engage.

4. Launch of the Palm-groves Safeguard and Development Plan

The Mohammed VI Foundation for the Protection of the Environment in partnership with the Wilaya, the City and Economic Operators launched an ambitious program aiming at:

The restoration of public green spaces and in particular by the maintenance of existing palm trees and the planting of new palm trees at large scale; the creation of environmental conditions for the protection of the palm grove by the classification of it as site of biological interest and its irrigation; the development of the culture of the protection of the Palm grove among the Marrakesh people then also the visitors by the creation of the ecomuseum; and the development and equipment of a walking circuit with the necessary infrastructure.

The Mohammed VI Foundation for Environmental Protection had organized several conferences on the environmental education. For example, the 7th World Environmental Education Congress, which was organized in Marrakesh in June 2013 also marked by the presence of renowned international institutions such as UNESCO, UNEP and ISESCO. The different actors in the field of the sustainable development of date palm groves had highlighted the following actions.

Increase support from public authorities, international organizations and donors for environmental education projects; Recognize the action of civil society in environmental education and strengthen its role in this area.; Strengthen concertation and coordination between all stakeholders with a view to better synergy in the definition of strategies and in the implementation of environmental education projects; Establish networks of actors in environmental education in order to facilitate the sharing of knowledge, know-how, experiences and good practices; Encourage the consideration of territorial specificities and the mobilization of local actors in the development and implementation of environmental education projects emphasizing respect for nature, the values of eco-citizenship and solidarity; Develop innovative and adapted teaching tools and materials making more use of new information and communication technologies while continuing to develop conventional tools and finally promote research and development in different areas of environmental education: learning methods, assessment tools, and development of attitudes, values and behaviors more respectful of the environment.

CONCLUSION

This present work has made it possible to study the effect of drought and metal pollution from anthropogenic activities on the growth and development of the date palm. (*Phoenix dactylifera* L.) This study shows that the date palm, originating from eight study sites including six in urban areas (three gardens and three boulevards) and two in peri-urban areas (two palm groves), has different aptitudes to develop and resist various applied water constraints. Our results indicate that at the level of water stress 25% and 50% the growth of seedlings of date palm is lower.

The study of the heavy metals (Cd, Cu, Fe, Pb and Zn) in soil samples shows that their contents vary depending on the sampling site. Thus, the contents of heavy metals generally increase in boulevards comparing to the other sites studied. The low growth recorded on the boulevards can be explained by the diffuse pollution of heavy metals, mineral elements and micropollutants. These have a detrimental effect on the growth of date palm plants (*Phoenix dactylifera* L).The intense drought and urbanization occuring in the city of Marrakesh during the year and the diffuse pollution by heavy metals recorded in most boulevards and gardens are considered as serious problems facing the development of many palm groves. The mortality rate in (*Phoenix dactylifera* L) trees is alarming and consequently sustainable development strategies must be identified to preserve this international heritage.

REFERENCES

Anne, P. (1945). Carbone organique (total) du sol et de l'humus. *Annales Agronomiques*, *15*, 161–172.

Diem, H.G., Gueye, L., Gianinazzi-Pearson, V., Fortin, J.A., & Dommergues, Y.R. (1981). Ecology of VA mycorrhizae in the tropics: The semi-arid zone of Senegal. *Acta Oecologica/Oecologia Plantarum*, *16*, 53-62.

Elachouri, M. (1993). Diploma of in-depth study. *Faculty of Sciences of Tunis*, *II*, 89.

Goyal, T., Mitra, P., Singh, P., Sharma, S., & Sharma, P. (2020). Assessement of Blood Lead and Cadmium Levels in Occupationally Exposed Workers of Jodhpur, Rajasthan. *Indian Journal of Clinical Biochemistry*. Advance online publication. doi:10.100712291-020-00878-6

Haddouchi, M. (1996). Situation actuelle et perspectives de développement du palmier dattier au Maroc. *Option méditerranéenne A, 28*, 63–79.

Mikola, P. (1987). Mycorrhizae under tropical stresses. *Journal of Applied Botany and Food Quality, 61*, 15–23.

Prasad, M. N. V., & Hagemeyer, J. (Eds.). (1999). *Heavy metal stress in plants. From molecules to ecosystems*. Springer.

Rodier, K. R. (1984). *L'analyse de l'eau, eaux naturelles, eaux résiduaires, eau de mer* (7th ed.). Dunod.

Salt, D. E., Smith, R. D., & Raskin, I. (1998). Phytoremediation. *Plant Physiol. Plant Mol. Biol., 49*, 643–668.

Sammut. (2007). *Speciation du cadmium, du plomb et du zinc dans les poussieres d'emissions atmospheriques d'origine siderurgique Approche de l'impact toxicologique des poussières* (Thesis).

Shah, K., & Nongkynrih, J. M. (2007). Metal hyperaccumulation and bioremediation. *Biologia Plantarum, 51*, 618–634.

Singh, O. V., Labana, S., Pandey, G., Budhiraja, R., & Jain, R. K. (2003). Phytoremediation: An overview of metallic ion decontamination from soil. *Applied Microbiology and Biotechnology, 61*, 405–412.

Singh, P., Kumar, A., & Borthakur, A. (2020). *Abatement of environmental pollutants*. Academic Press.

The Mohammed VI Foundation for Environmental Protection site. (n.d.). https://www.fm6e.org/

Torkshavand, Z., Gholami, M., Farzadkia, M., & Esrafili, A. (2014). Adsorption of Cu^{2+} from aqueous solution onto modified glass beads with 3aminopropyltriethoxysilane. *Iranian Journal of Health, Safety and Environment, 1*(3), 101–110.

Tremel-Schaub & Feix. (2005). *Delivered Contamination des sols transferts des sols vers les plantes*. Academic Press.

Verma, R. K., Shivay, Y. S., Prasanna, R., Choudhary, M., & Ghasal, P. C. (2020). Nutrient management options modulating soil physico-chemical and biological properties under direct-seeded rice-based cropping systems. *Archives of Agronomy and Soil Science, 0*(0), 1–16. Advance online publication. doi:10.1080/03650340.2020.1808627

KEY TERMS AND DEFINITIONS

Anthropogenic Activities: All the phenomena which can be consequent of the presence or the action of the human being.

Date Palm (*Phoenix dactylifera L.*)**:** Monocotyledonous plant of the Arecaceae family, perennial of significant socio-economic importance. It is very suitable for arid and semi-arid regions and is a vital component of oasis ecosystems.

Heavy Metals: Toxic metallic chemical elements in low concentrations. Heavy metals include mercury (Hg), cadmium (Cd), arsenic (As), chromium (Cr), thallium (Tl) and lead (Pb).

Palm Groves: Agricultural plot intended for the cultivation of palm trees.

Phytotoxicity: A toxic effect of a compound on plant growth including trace metals, salinity, pesticides.

Pollution: The degradation of an ecosystem or the biosphere by the introduction, generally human, of entities (physical, chemical, or biological).

Urbanization: A process of development of cities and of concentration of human populations.

Chapter 17
Heavy Metals Pollution and Its Studies in Coimbatore District, Tamil Nadu

Rajendrakumar S.
https://orcid.org/0000-0003-0594-3174
Centre for Sustainable Future, Department of Chemical Engineering & Materials Science, Amrita Vishwa Vidyapeetham, Coimbatore, India

Ramachandran V. S.
Centre for Environmental Studies, Amrita Vishwa Vidyapeetham, Coimbatore, India

ABSTRACT

This chapter is a review of published research papers, books, book chapters, unpublished scientific reports prepared by research organisations, and expert talk about heavy metal pollution. This chapter also summarises the research studies conducted in Coimbatore and a summary of the studies says that very tiny information is available on heavy metals in soil, water, and air components. Till now heavy metals in plants, plant products, animals, animal products, bioaccumulation in humans, and its impacts are not accounted in this region. Further, these chapters identify the research gaps, and research interventions are required to assess the heavy metals contaminations in different ecosystems. It also highlights the need for long-term monitoring of heavy metals in the ground, surface water, food materials, agricultural products produced in the contaminated regions. The chapter concludes with a need for study on human health impacts, restoration, and remediation measures required in Coimbatore regions.

HEAVY METALS IN ENVIRONMENT

Naturally, heavy metals are present in the Earth's crust, some metals are very useful to human biological system and few metals are damage humans' system. Heavy metals like Copper (Cu), Selenium (Se), Zinc (Zn) are vital in very small quantity for the proper functioning of the biological systems. Lack of these vital heavy metals could cause various disorders in the human body (Azeh Engwa et al., 2019). While

DOI: 10.4018/978-1-7998-7512-3.ch017

others like Mercury (Hg), Cadmium (Cd), Arsenic (As), Chromium (Cr), Thallium (Tl) and Lead (Pb) toxic to living organisms and can cause severe health disorders if exposed for long term even in minimal quantities. Presence of toxic heavy metals in water, soil and air is an important issue as they can undergo geoaccumulation in due course of time (Ali, Khan and Ilahi, 2019). They can enter biological (plant and animal) tissue through food chain and undergo bioaccumulation and biomagnification in long term, causing severe health issues in human beings (Jaishankar et al., 2014). The unplanned, rapid urbanization and industrialization carried out without giving due consideration for the environment is the root cause for the presence of heavy metals in quantities larger than what natural ecosystem can absorb and clean.

The bioaccumulation and geoaccumulation of heavy metals are significantly alter the quality of abiotic components and impacted the biotic components of an ecosystems (Jaiwan and Ajay, 2011). The numerous researchers are stated the impacts of heavy metals (Tchounwou et al., 2014; Masindi and Muedi, 2018; Sall et al., 2020). Well documented evidence on bio-toxic effects to human on long-term exposure of heavy metals causes gastrointestinal disorders, vomiting, depression, diarrhea, stomatitis, tremor and paralysis (Verma and Dwivedi, 2013). The long-term exposure of Cd dysfunction the kidneys. Lung disease by Pb in dust and fumes, chest pain, cough with foamy and bloody sputum is an effects of chronic exposure of Pb. Metal like Cd associated with bone defects, increased blood pressure and mycocardic dysfunctions. Zn causes symptoms like Pb, but impair the growth, reduce the function of reproductive organs of humans. The clinical signs of Zn are bloody urine, icterus, liver, kidney failure and anemia (Verma and Dwivedi, 2013). The inorganic forms mercury is highly toxic to human, causes spontaneous abortion, congenital malformation and gastrointestinal disorders. Information on heavy metals reduces the quality of water, air and soil components is required study in in-depth, but through abiotic component it accumulates into plants, animals causes damages, at higher level it affects the human (Jaiwan and Ajay, 2011).

Studies on Heavy Metals Assessment in the Environmental Components

Many researchers have recorded heavy metals in groundwater, surface water, air and soil components (Boukhalfa, 2007; Clemente, Dickinson, and Lepp 2008). Heavy metals in the atmosphere of rural areas of the United Kingdom (UK) was studied by Centre for Ecology and Hydrology (2012). Fifteen sites were continuously monitored to record heavy metal deposition in soils of industrial areas along the Jinxi River and surrounding Lakes of Qingshan, China (Abdelhafez and Li, 2015). These studies focused on the release of heavy metals by unorganised industrial area and their impact on the health of residents of the adjoining area. Deposition of heavy metals like Pb, Cd, Zn, Cu and Mn in the kahikatea trees in Christchurch and west coast of South Island, New Zealand, was studied to find out the impacts of urbanization and industrialization on the ecosystem (Stewart, Norton and Fergusson, 1991). Soil and dust particles were monitored to record the concentration of metals in Industrial Areas of Misurata, Libya (Elbagermi, Edwards and Alajtal, 2013). Water samples of Sion and MNPS lakes of Mumbai (Shenai and Quadros, 2014) and Batiano River of Philippines (Sarinas et al., 2014), vegetables and other biological materials of Kathmandu Valley (Sharma and Chhetri, 2005) were monitored to study the heavy metals in food chains.

The impacts of toxic metals on the agricultural lands adjacent to textile industries in Sanganer (Jaipur, Rajasthan) and industrial area in the Manipal (Mangalrore, Karnataka) was studied recently by Jaishree and Khan (2014) and Poornima, Vasudevan, and Venkateshwarlu (2011) respectively. Both the study showed higher concentration of Cr, Nickel (Ni), Zn and Pb in the agricultural lands. Composition of trace metals in ground water was carried out in Dhemaji District of Assam (Sharma and Sharma, 2013). Selvam et al. (2017) assessed heavy metals contamination in ground water of Gulf of Mannar Seashore using Geographical Information System (GIS) tools. The study said groundwater along the shore lines is not fit for human consumption because of high concentration of heavy metals in groundwater. Similar result was obtained by a GIS based study conducted in the waste disposal site converted into sports ground in Galway City of Ireland (Carr et al., 2008). Lin et al. (2010) identified underground channels beneath the irrigation canals that were carrying pollutants along with groundwater in Changhua of Central Taiwan. Li et al. (2004), developed urban environmental planning based on the regional variations of heavy metals accumulation in Kowloon area of Hong and Kong.

Studies are available for the bioaccumulation of heavy metals in plants (Pandey and Pandey, 2009), (Stobrawa and Lorenc-Plucińska, 2008), animals and human beings (de Vries, Römkens, and Schütze, 2007; Lagisz and Laskowski, 2008; Korashy and El-Kadi, 2008). Blakbern (2003), estimated heavy metal accumulation in soil of Chaktal Biosphere Reserve, Uzbekistan and assessed potential health risk to plants, carnivores and humans. Animals raised by feeding crop plants grown on farmlands where sewage sludge has been disposed may pose potential risk to food chains (Winder, Merrington and Green, 1999). Zhuang, Zou, and Shu (2009) studied bio-transfer of heavy metals through food chains of contaminated sites and recorded steady increase in the concentration of heavy metals like Pb, Zn, Cu and Cd in each trophic level. Metal accumulation has been reported in snails (Devkota and Schmidt, 2000), insects (Merrington, Winder and Green, 1997) and invertebrates (Milton, Cooke, and Johnson, 2003; Gasparik et al., 2004; Vandecasteele et al., 2004). Khan et al. (2008) reported health risk of heavy metals contaminated food crops that were cultivated with industrial wastewater in Beijing. Lokeshwari and Chandrappa (2006) studied the transfer of metals from soil to crops in the plants grown near Bellandur Lake (Bengaluru) where plants were irrigated with water mixed with sewage.

Studies on Restoration of Heavy Metal Polluted Land

In recent years' researchers are focusing on restoration and remediation of contaminated ecosystems with an integrated approach, that uses any of the combinations of physical, chemical and biological approaches in order to recover the heavy metals effectively. Economically viable integrated methods can effectively remove heavy metals from environmental components (Cameselle et al., 2013; Mao et al., 2016; Selvi and Aruliah, 2018; Selvi et al., 2019; Abbas et al., 2018; Oyekanmi et al., 2019). One of the successful approach is the use of Chemical and Biological methods where one method follows the other and this way gives a polishing effect on ecosystems (Selvi et al., 2019). A study conducted by Ahmed et al. (2016) reported the removal of 99.3% of total Cr and 98.4% of Cr (VI) by chemical precipitation followed by biological treatment. In biological treatment fungi called *Fusarium chlamydosporum* removed Turbidity, Chemical Oxygen Demand and 62.33% of Cr (Sharma and Malaviya, 2014).

In Electro-Kinetic (EK) method, electrical current and acidic and alkalophilic microbes were used to precipitate heavy metals from the soil. Rosestolato Bagatin, and Ferro (2015) recovered 400kg of Hg from contaminated land using bio-electrokinetic method. With the help of electric current and *Lysinibacillus fusiformis* bacteria 78% of Hg was removed from contaminated land within seven days of time (Azhar et al., 2016). Eighty-nine percentage of Zn was removed by EK assisted bioremediation using *Pseudomonas putida*. Gold (Au), Cobalt (Co), Iron (Fe) (Varia et al., 2013), Ar, Cu (Kim et al., 2012; Cameselle et al., 2013) Zn, Cr and Pb (Cameselle et al., 2013) were recovered from polluted sites using EK method. Depending on the uptake mechanism of plants, different strategies like, phytoextraction, phyto-evaporation, phytostabilization, rhizo-degradation, and rhizo-filtration were employed in phytoremediation (Mao et al., 2016; Lotfy and Mostafa, 2014). Metal like Cr and Cd were removed using *Brassica juncea* (Nirmal, Bhargavi and Sudha, 2015), Pb (Lim, Salido and Butcher, 2004) Cu, Zn (Cang et al., 2012) by mustard plant. *Lemna minor* coupled with EK was used to remove the heavy metals from water (Selvi et al., 2019). The fungi and bacteria living within the plants, especially in the roots - rhizomicrobe (Díaz de Villegas, Villa and Frías, 2002), free living algae (Al-Homaidan, Al-Ghanayem and Alkhalifa, 2011) and fungi (Wu, Ye, and Wong 2009; Orłowska, Godzik, and Turnau 2012) were used in the recovery of heavy metals. In order to speed up the phyto-remediatation process genetically engineered plants, microbes were employed (Selvi et al., 2019). Other methods like EK-geosynthetic (Jones, 2010) and EK-membrane filtration (Peng et al., 2015) were also used to remove the heavy metals.

STUDIES ON IMPACT OF INDUSTRIALISATION ON THE ENVIRONMENT OF COIMBATORE DISTRICT

Srinivasan et al. (2014) studied the current status of Noyyal River basin where they looked into ground water level, water usage, agricultural practices, soil, water contamination and impact of contaminated water on the health of people in the regions. Appasamy and Nelliyat (2007) examined water, soil and biological factors and they concluded that loss of ecosystem services of Noyyal river basin are due to industrial pollution. The study also estimated the cost of ecosystem services, economic and ecological loss of river system. The Noyyal river was monitored by dividing into five major zones like Chadivayal, Mangalam, Anakuli, Orathupalayam dam and Noyyal village. The Water quality and heavy metals contamination were studied, the concentration of pollution and heavy metals was high industrial regions it resulted the Noyyal river lost its self-purification capacity (Babunath and John, 2017).

Mathew et al. (2002) examined, the physico-chemical characteristics of water in six wetlands (Map 1: Selvachinthamani kulam, Singanallur kulam, Periyakulam, Kovaipudur Eri, Valan kulam and Amman kulam). The presence of heavy metals in the lakes studied were due to the mixing of effluents from textile and electroplating industries. Chandra, Nishadh, and Azeez (2010) monitored the water quality of Ukkadam, Perur, Kuruchi and Chinna kulam wetlands (Map 1) of Coimbatore and reported that, rapid degradation of environmental parameters in cityscape is deteriorating water quality of wetlands; resulting in increased level of pH and Total Dissolved Solids (TDS) when compared with that of previous decades. Study by Priya et al. (2010), on seven major wetlands of Coimbatore (Kurichi, Perur, Selvapuram, Kumaraswami, Ukkadam, Narasampahi and Singanallur) showed that concentration of pollutants was high in pre-monsoon time. They examined the quality of water for irrigating crop plants based on Sodium Adsorption Ratio (SAR), Exchangeable Sodium Ratio, Magnesium Hazard, Residual Sodium Chloride, Electrical Conductivity (EC) and concluded that the water in the lakes is not suitable for irrigation. Concentration of pollutants and heavy metals are very high in eight lakes (Selvachinthamani, Singanallur, Ukkadam, Perur, Valakulam, Ammankulam, Selampatti and Kumarasamy) of Coimbatore (Mohanraj *et al.,* 2000) due to release of untreated industrial effluent and domestic sewage. Waste dumping, encroachment, mixing of industrial effluents, weed growth were the major reasons for the environmental degradation in the 24 wetlands (Ukkulam, Pudukulam, Kolarampathy, Narasampathy, Krishnampathy, Selvampathy, Kumaraswamy, Selvachinthamani, Sottaiyandi kuttai, Gangarayaananasamudharam, Puttuvikki, Senkulam, Kuniyamuthur, Periyakulam, Valankulam, Kurichi, Vellalore, Singanallur, Pallapalayam, Kannampalayam, Irugur, Achankulam, Sulur Big and Sulur Small Lake) of Coimbatore connected with

Figure 1. Map 1: Location, Taluks and Distributions of Wetlands in Coimbatore District of Tamil Nadu.

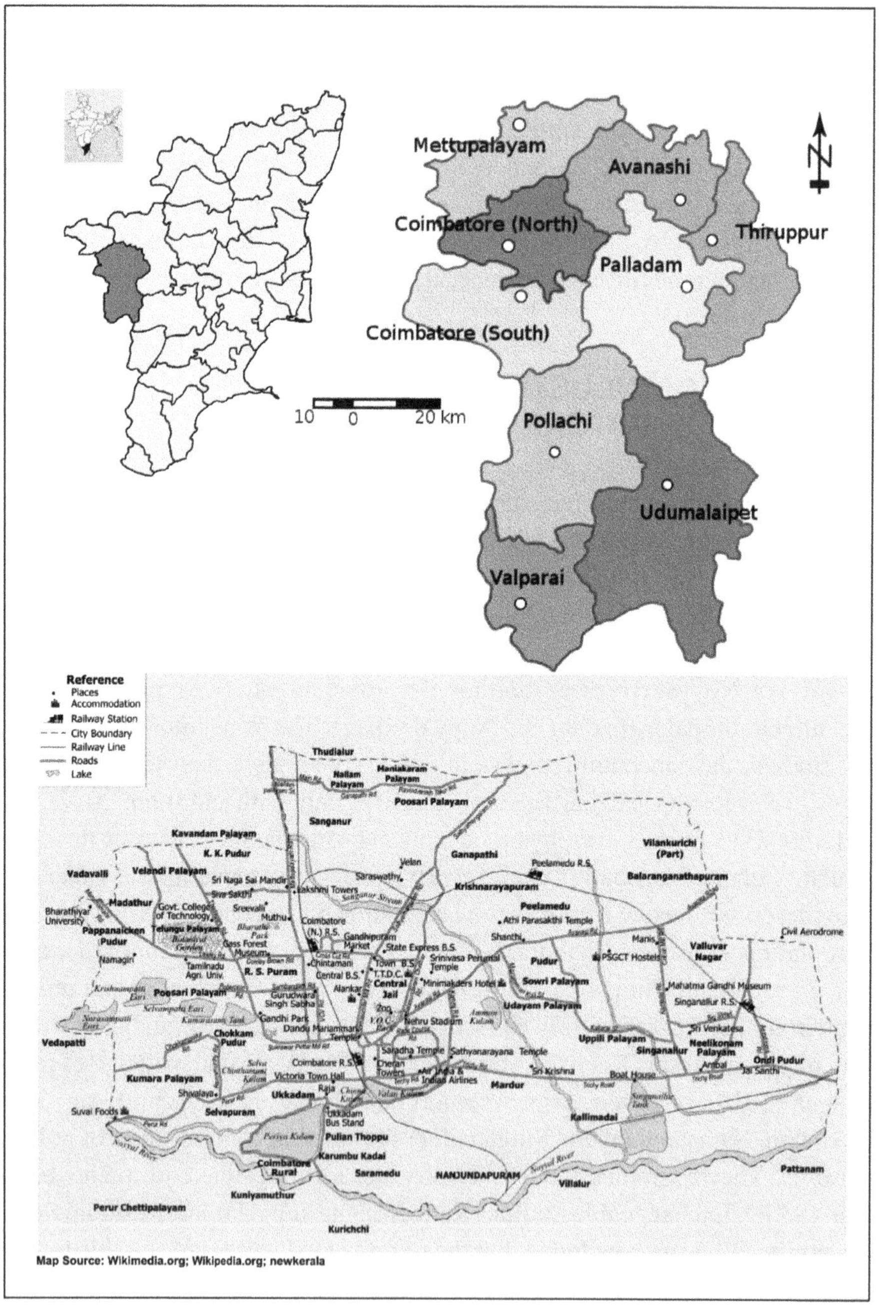

the Noyyal river basin (Anonymous[1], 2013). A study conducted by Pavithrapriya, Mahimairaja, and Sivasubramanian (2015) reported high concentration of heavy metals in 12 wetlands of Coimbatore region. The water samples collected from Periyakulam, Singanallur, Valankulam, Muthanankulam, Perur, Sulur, Kurichkulam, Narasampathi, Krishnampathi and Selvapuram wetlands (Map 1) recorded Cd, Cr, Cu, Ni

Table 1. Studies on Impacts of Environmental Components in Coimbatore District

Authors	Study Type	Location	Results
Srinivasan et al. (2014)	Water Usage and Ground Water pollution	Noyyal River	Ground Water, Surface Water, Soil are polluted due to ongoing industrial activities.
Appasamy and Nelliyat (2007)	Industrial pollution and Ecosystem Loss	Noyyal River	Loss of ecosystem services, Economic and Ecological loss of Noyyal river system was discussed.
*Mathew et al. (2002)	Wetlands water quality and Heavy Metals concentration	Selvachinthamani kulam, Singanallur kulam, Periyakulam, Kovaipudur Eri, Valan kulam and Amman kulam wetlands	Heavy metals in the lakes are due to the mixing of effluents from textile and electroplating industries.
Chandra, Nishadh, and Azeez (2010)	Wetlands water quality	Ukkadam, Perur, Kuruchi and Chinna kulam wetlands	Water quality deteriorated due to increased level of pH and Total Dissolved Solids (TDS) compared with that of previous decades.
Priya et al. (2010)	Wetlands water quality	Kurichi, Perur, Selvapuram, Kumaraswami, Ukkadam, Narasampahi and Singanallur Wetlands	The water not suitable for irrigation reference to high SAR, ESR, Magnesium Hazard, Residual Sodium Chloride and EC.
*Mohanraj et al. (2000)	Wetlands water quality and Heavy Metals concentration	Selvachinthamani, Singanallur, Ukkadam, Perur, Valakulam, Ammankulam, Selampatti and Kumarasamy	Pollution and heavy metals level was high due to discharge of untreated industrial effluent and domestic sewage in wetlands.
Anonymous[1], (2013)	Wetland degradation	Ukkulam, Pudukulam, Kolarampathy, Narasampathy, Krishnampathy, Selvampathy, Kumaraswamy, Selvachinthamani, Sottaiyandi kuttai, Gangarayaananasamudharam, Puttuvikki, Senkulam, Kuniyamuthur, Periyakulam, Valankulam, Kurichi, Vellalore, Singanallur, Pallapalayam, Kannampalayam, Irugur, Achankulam, Sulur Big and Sulur Small Lake	Waste dumping, encroachment, mixing of industrial effluents, weed growth were the major reasons for the environmental degradation of wetlands in Coimbatore regions.
*Pavithrapriya, Mahimairaja, and Sivasubramanian (2015)	Heavy metals pollution	Periyakulam, Singanallur, Valankulam, Muthanankulam, Perur, Sulur, Kurichkulam, Narasampathi, Krishnampathi and Selvapuram wetlands	Cd, Cr, Cu, Ni and Pb concentration was higher than the limits set by the WHO for drinking and FAO for irrigation.
Manish kumar et al. (2016)	Microbial Contamination	Sanganur canal	Disposal of human and animals waste increases a microbial contamination in Sanganur canal.
*Jenelapriscy, Jesvinshobini, and Reyaissac (2017)	Heavy Metals pollution (Review Study)	Wetlands of Coimbatore	Industrial activities released heavy metals to wetlands
Karthiyayini, Ponnammal, and Joseph (2005)	Pollution Remediation	Coimbatore – Ooty Highways	Recommenced suitable plants to be planted along roadsides to control dust particles in atmosphere.
Kumar, Elangovan and Srimathi (2019)	Air pollution	Coimbatore	Public transport increases the air pollution, it is comparatively higher than the industrial activities causes air pollution in this region.
Janaranjani et al. (2017)	Wetland degradation	Wetlands of Coimbatore	Reduction of wetlands areas due to encroachment, construction and waste disposal.
Balakrishnan, Ramu and Murugesan (2014)	Assessment on Pollution and Biodiversity	Marudhamalai, Coimbatore	Environmental conditions of Marudhamalai temple were deteriorated due to tourism related activities and recommendations were proposed to improve the landscapes.
CSE (2013)	Ground Water Pollution	Coimbatore	Irrigation water showed high level of EC. The water used of irrigation it pollute the agricultural lands, either partially or fully.
Sundar and Saseetharan (2008)	Ground Water Pollution	Ground Water along the Noyyal River Basin	EC in pre monsoon is high due to excess amount of salt content in groundwater, making groundwater not suitable for irrigation and industrial purposes
Usharani et al. (2010)	River Water, Ground Water Pollution	Ground Water, Noyyal River Basin of Perur region, Coimbatore	The excess amount of phosphate in water, increases the salinity and Water quality parameters exceeds the permissible level of WHO.
*Gandhimathi and Meenambal (2012)	Heavy metals in Soil	Coimbatore	Higher concentration of Heavy Metals in Soil of Industrial regions.
*Mohanraj, Azeez and Priscilla (2004)	Heavy Metals in Air	Coimbatore	In Respirable and Non-Respirable particles, the concentration of Pb, Cu increased due to vehicle transportation and Zn, Ni, Cr, Cd were increased by industrial activities.
Mohanraj and Azeez (2005)	Air Pollution	Coimbatore	Increase of Respirable Suspended Particulate Matter (RPSM) in the city, which is well above the prescribed Air quality standards.
Vedamadavan and Sarithabanuraman (2012)	Air Pollution	Coimbatore	Ambient Air Quality is poor in industrial zone.

continued on following page

Table 1. Continued

Authors	Study Type	Location	Results
Eswaran, Bharathiraj, and Jayanthi (2019)	Air Pollution	Coimbatore	Industrial zone recorded with higher concentration of air pollutants than the residential and commercial zones.
Saravanakumar, Sivalingam, and Elangovan (2016)	Air Pollution	Coimbatore	Industrial activities contributed to more atmospheric pollution than the vehicular emission in commercial regions.
Govindarajalu (2003)	Water Pollution impacts on Health	Noyyal River Basin	Pollution impacts on health like skin allergy, respiratory infections, allergy, gastritis and ulcers.
*Malarkodi et al. (2007)	Heavy Metals in Soil	Coimbatore	Soils irrigated with sewage and industrial effluents showed that the concentration of Pb, Ni, Cd, Cr, Cu and Zn and the levels exceeded the critical limits.
*Antisha and Priyadharsini (2015)	Bioaccumulation in Plants	Cultivated land of Coimbatore	Concentration of Cr and Zi metals were high and Pb was below deductible in crops cultivated near contaminated waterbed.
Rajamanickam and Nagan (2018)	Environmental Quality	Kuruchi Industrial cluster of Coimbatore	Industrial units implemented pollution control measures that improves the environmental quality of Coimbatore.
*Babunath and John (2017)	River water quality and Heavy metal contaminations	Noyyal River	Heavy Metal concentration, pollutants were high industrial regions. The Noyyal river lost its self-purification capacity due to contaminations of industrial wastewater.
Sarahamani et al. (2014)	Water Pollution, Social and Economic Impacts	Pollachi, Coimbatore district	The lower income group are affected due lack on access of good quality water in their regions
Manju et al. (2018)	Ambient Air Quality and Air Pollution	Coimbatore	The higher concentration of O^3 and CO in summer and PM_{10}, $PM_{2.5}$, SO_2 and NO_2 high during southwest monsoon.
*Sherene (2009)	Heavy metals in Soil	Coimbatore	Higher concentration of Cd, Cu, Mn, Fe, Pb, Ni, Cr, Zn metals in electroplating, textile, casting industrial sites than the other industrial operation sites.
*Radha et al. (2014)	Soil heavy metals	Coimbatore	The concentration of Cd is high in soils of Sewage disposal sites.

* studies on heavy metals

and Pb metals higher than the limits set by the WHO for drinking and FAO for irrigation. Apart from heavy metals, the improper disposal of human and animals waste increased the microbial contamination in Sanganur canal of Coimbatore District (Manish kumar et al., 2016). Jenelapriscy, Jesvinshobini, and Reyaissac (2017) reviewed research articles on the accumulation of metals in the wetlands of Coimbatore, which said industrial activities released heavy metals to wetlands of Coimbatore Districts. Karthiyayini, Ponnammal, and Joseph (2005) studied tolerance levels in plants towards air pollution and based on the outcome they recommended species of plants to be planted along roadsides to control dust particles in air. A GIS based approach was done to understand the air pollution caused by public transport in major roads of the city (Kumar, Elangovan and Srimathi, 2019). They categorized bus routes bus stops based on air pollution. Land use, pollution status, encroachment of wetlands of Coimbatore was studied by Janaranjani et al. (2017), the study results were indicated decline of wetlands size due to waste dumping, construction, silting and canal encroachment. Pollution Assessment was carried out in the hilly regions of Subramaniyaswami Temple of Marudhamalai. Air, Soil, Water, Biodiversity and Socio-economic factors was assessed to study the environmental conditions of the region. Environmental conditions were deteriorated in this region due to tourism and related activities and recommendations were proposed to improve the conditions (Balakrishnan, Ramu and Murugesan, 2014).

The study by the CSE (2013), reported that the effluents of textile and dyeing industries were discharged directly into the land and water ecosystem of Coimbatore and Tiruppur district. Resulted irrigation water showed high level of EC. It polluted 60% of agricultural lands, either partially or fully. Physico-chemical characteristic analysis of groundwater samples collected were collected along the Noyyal River were basin was analysed by the Sundar and Saseetharan (2008). A comparative study on pre and post monsoon showed increased value of EC in pre monsoon due to excess amount of salt content in groundwater, making groundwater not suitable for irrigation and industrial purposes. In addition, Usharani et al. (2010) examined physico-chemical and biological characteristics of water of Noyyal River basin and nearby ground water sources of Perur town panchayat. The quality parameters are exceeding the permissible level of WHO due to disposal of sewage from domestic and commercial activities. The excess amount of phosphate in water, increases the salinity, if water used agricultural activity.

Gandhimathi and Meenambal (2012) measured the concentration of heavy metals in the soil of industrial region of Coimbatore District. The assessment on heavy metals contaminations in soils irrigated with sewage and industrial effluents showed that the concentration of Pb, Ni, Cd, Cr, Cu and Zn were exceeded the critical limits (Malarkodi et. al., 2007). The soils industrial regions of electroplating, textile, casting, found higher concentration of Cd, Cu, Mn, Fe, Pb, Ni, Cr, Zn. Specially the electroplating industrial regions were higher concentration metals than other industrial operation regions (Sherene, 2009). The record of higher concentration of Cd in soils of Sewage disposal areas of Coimbatore was reported by Radha, Kumutha, and Marimuthu (2014). Heavy metal association with Respirable and Non-Respirable particulate matter was examined (Mohanraj, Azeez and Priscilla, 2004). The concentration of Pb and Cu increased due to vehicle transportation and Zn, Ni, Cr, and Cd were increased by industrial activities. Mohanraj and Azeez (2005) worked on urban development and particulate matter in the Coimbatore city. They recorded increase in Respirable Suspended Particulate Matter (RPSM) in the city, which is well above the prescribed standards. Additionally, they reported that wind speed is positively correlated while temperature is negatively correlated with particulate matter dispersion in the city. Ambient Air Quality was measured in residential, commercial and industrial of zone of Coimbatore (Vedamadavan and Sarithabanuraman, 2012; Eswaran, Bharathiraj, and Jayanthi 2019; Saravanakumar, Sivalingam, and Elangovan 2016). Based on Ambient Air Quality Index, the industrial zone showed high pollution level than two other zones. The study highlighted that industrial activities contributed to more atmospheric pollution than the vehicular emission in commercial area. To support these results a study on spatial variations on air pollutant in relation with meteorological factors high concentration of air pollutant in industrial zone followed by traffic and commercial areas (Manju et. al., 2018). It results in higher concentration of O_3 and CO in summer and PM_{10}, $PM_{2.5}$, SO_2 and NO_2 high during southwest monsoon. While the study was conducted by Rajamanickam and Nagan (2018), using Comprehensive Environmental Pollution Index (CEPI) to assess the quality of Kurichi Industrial Cluster of Coimbatore region. The study illustrate that the Industrial units were implemented various pollution control measures and that bring down the CEPI score than other critically polluted regions. Govindarajalu (2003), conducted a survey on water pollution and its impact on the health of villagers along the Noyyal river basin of Coimbatore, Erode and Karur districts. His study was based on the survey of 31 villages along the river which documented health problems like skin allergy, respiratory infections, allergy, gastritis and ulcers. Also study on economic impact of water pollution on agriculture and rural households in Makinampatti, Chinnampalayam, chandrapuram, Unjavelampatti and Solapalayam of Pollachi regions of Coimbatore district indicates that the lower income group are affected due lack on access of good quality water in their regions (Sarathamani, Govindarajalu, Divya, 2014). A study on heavy metals accumulation in *Amranthus dubius* crops cultivated in near contaminated waterbeds regions and organic farms. The study revealed that the concentration of Cr and Zi metals were high and Pb was below deductible in crops cultivated near contaminated water bed regions (Antisha and Priyadharsini, 2015).

Table 2. Studies on Heavy Metals in Environmental Components in Coimbatore District

Types of Studies	Monitoring of Heavy Metals					Heavy Metals Impacts			Treatment	
Environmental Components/ Authors	**Air**	**Soil**	**Water**	**Plants and Crops**	**Animals**	**Plants and Crops**	**Animals and Foods**	**Human Impacts**	**Remediation of Ecosystem**	**Restoration of Ecosystem**
Mathew et al. (2002)			✓							
Mohanraj et al. (2000)			✓							
Pavithrapriya, Mahimairaja, and Sivasubramanian (2015)			✓							
Jenelapriscy, Jesvinshobini, and Reyaissac (2017)			✓							
Gandhimathi and Meenambal (2012)		✓								
Mohanraj, Azeez and Priscilla (2004)	✓									
Malarkodi et. al. (2007)		✓								
Antisha and Priyadharsini, (2015)				✓						
Babunath and John (2017)			✓							
Sherene (2009)		✓								
Radha et al. (2014)		✓								

WHY COIMBATORE NEED TO BE STUDIED FOR HEAVY METAL POLLUTION?

- Coimbatore is the third biggest city of Tamil Nadu and one of the fastest growing cities in South India (Map 1). It is surrounded by mountains on the west, Reserve Forests (RF) and the Nilgiri Biosphere Reserve (NBR) on the northern side. It has borders with the State of Kerala on its western side.
- In National Population Census Report, Coimbatore district has a population of 34,58,045 with density of 731 people sq/km, 27.27% people are living in Rural area and 75.73% people dwelling in urban limit (Anonymous, 2011).
- Coimbatore supports 110 types of industrial activities; it is home for 50,000 small and large scale industries (Anonymous[2], 2014). This includes textile, machinery, cement, rubber, transport, equipment, foundry, engineering and a number of other associated industrial establishments.

- The city known as the "Manchester of South India". The first textile mill in Coimbatore was started in 1888. Now there are 2175 units of Cotton and Textile mills and 5567 units of Ready-made Garments and Embroidery industries. Further it supports 702 chemical industries, 465 metals casting industries and 915 Electrical and Electronic industries (Anonymous[2], 2014).
- Many Industries uses dyes based on benzidine structures with a combination of heavy metals (CSE, 2013) like, Pb, Cr, Cd and Cu (Halimoon, Putra and Halimoon, 2010).
- The waste from electroplating, lead smelting, painting industries contribute Hg, Ni, Pb and Cr to the environment (Santhosh and Sridevi, 2013).
- The city is situated on the banks of River Noyyal (Map 1). There were 30 wetlands constructed on both sides of the River Noyyal for irrigation and flood mitigation (Anonymous[1], 2013).
- The treated, untreated and partially treated effluents from industries are mostly discharged into the terrestrial (land) and aquatic ecosystem (Noyyal River; Anonymous[1], 2013).
- Wetlands serve as a home to many native and migratory birds, fishes and other life forms in pristine glory. Being natural filters, they help to purify water by trapping many pollutants and disease causing organisms.

SCOPE OF STUDIES ON THE IMPACT OF HEAVY METAL IN THE ENVIRONMENT OF COIMBATORE

Past 130 years of industrial activities, associated dumping of chemical and hazardous waste in different landscapes of the city there are very high possibilities of heavy metals accumulation in biological components, especially food crops as well food materials. The previous studied conducted in the Coimbatore district were concentrated on aquatic ecosystem (rivers and wetlands) (Table 1) and very few studies reported heavy metals in soil and air (Table 2). Studies are not available for the accumulation of the heavy metals in biological tissues (plants, animals or humans) (Table 2). Further, quantification of heavy metals released by organised and unorganised sectors in city and in the rural regions did not get attention from the researchers.

The long term monitoring of contaminated lands, water ecosystems, agricultural lands, harvested products, contamination of ground water, movement in underground water resources, air components, depositions on biological components has to be initiated for Coimbatore. The need of the hour is a holistic approach to find the link between the concentration of heavy metals in the abiotic components of the ecosystem, their accumulation in the biological system and its impact on the biological system. Work has to be initiated to remediate heavy metals impacted water bodies and land. All these activities require intervention at policy level by the Government – local and the State.

ACKNOWLEDGMENT

The authors wish to acknowledge Amrita Vishwa Vidyapeetham for support and facilities provided during writing stage of this paper.

REFERENCES

Abbas, S. Z., Rafatullah, M., Hossain, K., Ismail, N., Tajarudin, H. A., & Khalil, H. P. (2017). A review on mechanism and future perspectives of cadmium-resistant bacteria. *International Journal of Environmental Science and Technology*, *15*(1), 243–262. doi:10.100713762-017-1400-5

Abdelhafez, A. A., & Li, J. (2014). Environmental Monitoring of Heavy Metal Status and Human Health Risk Assessment in the Agricultural Soils of the Jinxi River Area, China. *Human and Ecological Risk Assessment*, *21*(4), 952–971. doi:10.1080/10807039.2014.947851

Ahmed, E., Abdulla, H. M., Mohamed, A. H., & El-Bassuony, A. D. (2016). Remediation and recycling of chromium from tannery wastewater using combined chemical–biological treatment system. *Process Safety and Environmental Protection*, *104*, 1–10. doi:10.1016/j.psep.2016.08.004

Al-Homaidan, A. A., Al-Ghanayem, A. A., Alkhalifa, A., & Arabia, S. (2011). Green Algae as Bioindicators of Heavy Metal Pollution in Wadi Hanifah Stream, Riyadh, Saudi Arabia. *International Journal of Water Resources and Arid Environments*, *1*(1), 10–15.

Ali, H., Khan, E., & Ilahi, I. (2019). Environmental Chemistry and Ecotoxicology of Hazardous Heavy Metals: Environmental Persistence, Toxicity, and Bioaccumulation. *Journal of Chemistry*, *2019*, 1–14. doi:10.1155/2019/6730305

Anonymous. (2011). *Coimbatore District Population Censucs 2011-2020, Tamil Nadu literacy sex ratio and density*. Directorate of Census Operations, Tamil Nadu, Ministry of Home Affairs. https://www.census2011.co.in/census/district/32-coimbatore.html

Anonymous. (2013). *Environmental degradation of the Coimbatore Wetlands in the Noyyal River Basin.* India Environment Portal | News, reports, documents, blogs, data, analysis on environment & development | India, South Asia, EIA Resource and Response Centre (ERC)Kotagiri, Nilgiri, India. http://www.indiaenvironmentportal.org.in/content/384872/environmental-degradation-of-the-coimbatore-wetlands-in-the-noyyal-river-basin/

Anonymous. (2014). *Brief Industrial Profile ofCoimbatore District2012-13*. http://dcmsme.gov.in/dips/IPS coimbatore 2012_t.pdf

Antisha & Priyadharsini, D.V. (2017). A Study on the Contamination of Heavy Metals in Arakeerai (*Amaranthus dubius*) Cultivated in Different Sites of Coimbatore in India. *International Journal of Science and Research, 6*(8), 120-125.

Appasamy, P. P., & Nelliyat, P. (2007). *Compensating the Loss of Ecosystem Services Due to Pollution in Noyyal River Basin, Tamil Nadu.* Development Economics Working Papers 22493. East Asian Bureau of Economic Research. https://ideas.repec.org/p/eab/develo/22493.html

Azhar, A. T., Nabila, A. T., Nurshuhaila, M. S., Zaidi, E., Azim, M. A., & Farhana, S. M. (2016). Assessment and Comparison of Electrokinetic and Electrokinetic-bioremediation Techniques for Mercury Contaminated Soil. *IOP Conference Series. Materials Science and Engineering*, *160*, 012077. doi:10.1088/1757-899X/160/1/012077

Babunath, R., & John, G. (2017). A Study on Physico Chemical and Heavy Metals Characterstics of River Noyyal, Tamil Nadu, India. *Environemntal Science: An Indian Journal*, *13*(1), 1–10.

Balakrishnan, A., Ramu, A., & Murugesan, A. (2014). Assessment of Heavy Metal Distribution in Groundwater in and around Gulf of Mannar Seashore Area Using GIS Technique. *Research Journal of Chemical Sciences*, *4*(1), 9–16.

Blakbern, A. A. (2003). Accumulation and migration of trace elements along trophic chains in ecosystems of the chatkal biosphere reserve (the Western Tien Shan, Uzbekistan). *Russian Journal of Ecology*, *34*(1), 68–72. doi:10.1023/A:1021827408038

Boukhalfa, C. (2007). Heavy metals in the water and sediments of Oued Es-Souk, Algeria, a river receiving acid effluents from an abandoned mine. *African Journal of Aquatic Science*, *32*(3), 245–249. doi:10.2989/AJAS.2007.32.3.3.302

Cameselle, C., Gouveia, S., Eddine, D., & Belhadj, B. (2013). *Advances in Electrokinetic Remediation for the Removal of Organic Contaminants in Soils*. Organic Pollutants - Monitoring, Risk and Treatment. doi:10.5772/54334

Cang, L., Zhou, D., Wang, Q., & Fan, G. (2012). Impact of electrokinetic-assisted phytoremediation of heavy metal contaminated soil on its physicochemical properties, enzymatic and microbial activities. *Electrochimica Acta*, *86*, 41–48. doi:10.1016/j.electacta.2012.04.112

Carr, R., Zhang, C., Moles, N., & Harder, M. (2007). Identification and mapping of heavy metal pollution in soils of a sports ground in Galway City, Ireland, using a portable XRF analyser and GIS. *Environmental Geochemistry and Health*, *30*(1), 45–52. doi:10.100710653-007-9106-0 PMID:17610027

CEH. (2012). *Heavy Metal Deposition Mapping: Concentrations and Deposition of Heavy Metals in Rural Areas of the UK*. Centre for Ecology and Hydrology. Available at: https://uk-air.defra.gov.uk/assets/documents/reports/cat13/1511251439_AQ0716_Heavy_Metals_Final_Report2012-15NOV12.pdf

Chandra, R., Nishadh, K. A., & Azeez, P. A. (2009). Monitoring water quality of Coimbatore wetlands, Tamil Nadu, India. *Environmental Monitoring and Assessment*, *169*(1-4), 671–676. doi:10.100710661-009-1206-0 PMID:19902369

Clemente, R., Dickinson, N. M., & Lepp, N. W. (2008). Mobility of metals and metalloids in a multi-element contaminated soil 20years after cessation of the pollution source activity. *Environmental Pollution*, *155*(2), 254–261. doi:10.1016/j.envpol.2007.11.024 PMID:18249071

CSE. (2013). *Noyyal River Pollution excreta happens*. Centre for Sciences and Environment. Available at: https://www.yumpu.com/en/document/view/12073282/noyyal-river-pollution-excreta-happens-centre-for-science-and-

de Vries, W., Römkens, P. F., & Schütze, G. (2007). Critical soil concentrations of cadmium, lead, and mercury in view of health effects on humans and animals. *Reviews of Environmental Contamination and Toxicology*, *191*, 91–130. doi:10.1007/978-0-387-69163-3_4 PMID:17708073

Devkota, B., & Schmidt, G. (2000). Accumulation of heavy metals in food plants and grasshoppers from the Taigetos Mountains, Greece. *Agriculture, Ecosystems & Environment, 78*(1), 85–91. doi:10.1016/S0167-8809(99)00110-3

Díaz de Villegas, M. E., Villa, P., & Frías, A. (2002). Evaluation of the siderophores production by *Pseudomonas aeruginosa* PSS'. *Revista Latinoamericana de Microbiologia, 44*(3–4), 112–117. PMID:17061484

Elbagermi, M. A., Edwards, H. G., & Alajtal, A. I. (2013). Monitoring of Heavy Metals Content in Soil Collected from City Centre and Industrial Areas of Misurata, Libya. *International Journal of Analytical Chemistry, 1-5*, 1–5. Advance online publication. doi:10.1155/2013/312581 PMID:23762061

Engwa, Ferdinand, Nwalo, & Unachukwu. (2019). Mechanism and Health Effects of Heavy Metal Toxicity in Humans, In *Poisoning in the Modern World - New Tricks for an Old Dog?* IntechOpen. Available from: https://www.intechopen.com/books/poisoning-in-the-modern-world-new-tricks-for-an-old-dog-/mechanism-and-health-effects-of-heavy-metal-toxicity-in-humans doi:10.5772/intechopen.82511

Eswaran, S., Bharathiraj, L. T., & Jayanthi, S. (2019). Modelling of ambient air quality, Coimbatore, India. E3S Web of Conferences, 117, 2. doi:10.1051/e3sconf/201911700002

Gandhimathi, A., & Meenambal, T. (2012). Spatial Prediction of Heavy Metal Pollution for Soils based on ANN and Kriging Model. *International Journal of Computers and Applications, 29*(10), 52–63.

Gasparik, J., Massányi, P., Slamecka, J., Fabis, M., & Jurcik, R. (2004). Concentration of Selected Metals in Liver, Kidney, and Muscle of the Red Deer (*Cervus elaphus*). *Journal of Environmental Science and Health. Part A, Toxic/Hazardous Substances & Environmental Engineering, 39*(8), 2105–2111. doi:10.1081/ESE-120039378 PMID:15332672

Govindarajalu, K. (2003). Industrial Effluent And Health Status - A Case Study Of Noyyal River Basin. In *Proceedings of the Third International Conference on Environment and Health*. Department of Geography, University of Madras and Faculty of Environmental Studies, York University.

Halimoon, N., Putra, U., & Halimoon, N. (2010). Removal of Heavy Metals from Textile Wastewater using Zeolite. *Environment Asia, 3*(Special issues), 124–130. doi:10.14456/ea.2010.51

Jaishankar, M., Tseten, T., Anbalagan, N., Mathew, B. B., & Beeregowda, K. N. (2014). Toxicity, mechanism and health effects of some heavy metals. *Interdisciplinary Toxicology, 7*(2), 60–72. doi:10.2478/intox-2014-0009 PMID:26109881

Jaishree, & Khan, T. I. (2014). Monitoring of Heavy Metal in Textile Waste Water of Sanganer, Jaipur (Rajasthan). *International Journal of Scientific and Research Publications, 4*(3), 1–4.

Jaiwan, S., & Ajay, K. S. (2011). Effects of Heavy Metals on Soil, Plants Human Health and Aquatic Life. *International Journal of Research in Chemistry and Environment, 1*(2), 15–21.

Janaranjani, M., Varunprasath, K., Priya, R., Sutharsan, L., & Lakeshmanaswamy, M. (2017). Status of Wetlands in the Coimbatore District, Tamilnadu, India. *Asian Journal of Environment & Ecology, 4*(4), 1–12. doi:10.9734/AJEE/2017/36421

Jenelapriscy, J., & Jesvinshobini, S., & ReyaIssac. (2017). Heavy Metal Pollution in Wetlands around-Coimbatore. *International Journal of Chemtech Research, 10*(10), 119–128.

Jones, C. J. (2010). *Earth reinforcement and soil structures*. T. Telford.

Karthiyayini, R., Ponnammal, N. R., & Joseph, R. (2005). Air Pollution Tolerance Index of Certain Plants of Coimbatore-Ooty Highways, Near I.T.I Area, Coimbatore, Tamilnadu. *Pollution Research*, *24*(2), 363–365.

Khan, S., Cao, Q., Zheng, Y., Huang, Y., & Zhu, Y. (2008). Health risks of heavy metals in contaminated soils and food crops irrigated with wastewater in Beijing, China. *Environmental Pollution*, *152*(3), 686–692. doi:10.1016/j.envpol.2007.06.056 PMID:17720286

Kim, H., Lee, K., Lee, B., Kim, S., & Kim, K. (2012). Comparative study of simultaneous removal of As, Cu, and Pb using different combinations of electrokinetics with bioleaching by Acidithiobacillus ferrooxidans. *Water Research*, *46*(17), 5591–5599. doi:10.1016/j.watres.2012.07.044 PMID:22921395

Korashy, H. M., & El-Kadi, A. O. (2008). Modulation of TCDD-mediated induction of cytochrome P450 1A1 by mercury, lead, and copper in human HepG2 cell line. *Toxicology In Vitro*, *22*(1), 154–158. doi:10.1016/j.tiv.2007.08.003 PMID:17889500

Kumar, V. S., Elangovan, K., & Srimathi, N. (2019). GIS-based air pollution distribution mapping of emissions from public buses for Coimbatore Corporation, Tamil Nadu, India. *Indian Journal of Geo-Marine Sciences*, *48*(6), 947–951.

Lagisz, M., & Laskowski, R. (2007). Evidence for between-generation effects in carabids exposed to heavy metals pollution. *Ecotoxicology (London, England)*, *17*(1), 59–66. doi:10.100710646-007-0176-7 PMID:17940866

Li, X., Lee, S., Wong, S., Shi, W., & Thornton, I. (2004). The study of metal contamination in urban soils of Hong Kong using a GIS-based approach. *Environmental Pollution*, *129*(1), 113–124. doi:10.1016/j.envpol.2003.09.030 PMID:14749075

Lim, J. M., Salido, A. L., & Butcher, D. J. (2004). Phytoremediation of lead using Indian mustard (*Brassica juncea*) with EDTA and electrodics. *Microchemical Journal*, *76*(1–2), 3–9. doi:10.1016/j.microc.2003.10.002

Lin, Y., Chu, H., Wu, C., Chang, T., & Chen, C. (2010). Hotspot Analysis of Spatial Environmental Pollutants Using Kernel Density Estimation and Geostatistical Techniques. *International Journal of Environmental Research and Public Health*, *8*(1), 75–88. doi:10.3390/ijerph8010075 PMID:21318015

Lokeshwari, H., & Chandrappa, G. (2006). Impact of heavy metal contamination of Bellandur Lake on soil and cultivated vegetation. *Current Science*, *91*(5), 622–627.

Lotfy, S., & Mostafa, A. (2014). Phytoremediation of contaminated soil with cobalt and chromium. *Journal of Geochemical Exploration*, *144*, 367–373. doi:10.1016/j.gexplo.2013.07.003

Malarkodi, M., Krishnasamy, R., Kumaraperumal, R., & Chitdeshwari, T. (2007). Characterization of Heavy Metal Contaminated Soils of Coimbatore District of Tamil Nadu. *Journal of Agronomy*, *6*(1), 147–151. doi:10.3923/ja.2007.147.151

Manish kumar, B., Anusiya dev, K., Lekeshmanaswamy, M., & Vasuki, C.A. (2016). Studies on the bacterial contamination of Sanganur canal from Nanjundapuram origin to Noyyal end point, Coimbatore District, Tamilnadu, India. *Kongunadu Research Journal*, *3*(1), 70–72. doi:10.26524/krj132

Manju, A., Kalaiselvi, K., Dhananjayan, V., Palanivel, M., Banupriya, G. S., Vidhya, M. H., Panjakumar, K., & Ravichandran, B. (2018). Spatio-seasonal variation in ambient air pollutants and influence of meteorological factors in Coimbatore, Southern India. *Air Quality, Atmosphere & Health*, *11*(10), 1179–1189. doi:10.100711869-018-0617-x

Mao, X., Han, F. X., Shao, X., Guo, K., Mccomb, J., Arslan, Z., & Zhang, Z. (2016). Electro-kinetic remediation coupled with phytoremediation to remove lead, arsenic and cesium from contaminated paddy soil. *Ecotoxicology and Environmental Safety*, *125*, 16–24. doi:10.1016/j.ecoenv.2015.11.021 PMID:26650421

Masindi & Muedi. (2018). Environmental Contamination by Heavy Metals. *Heavy Metals*. Available from: https://www.intechopen.com/books/heavy-metals/environmental-contamination-by-heavy-metals doi:10.5772/intechopen.76082

Mathew, M., Kumar, M. S., Azeez, P. A., Sivakumar, R., & Pattabi, S. (2002). Sediment Quality of Wetlands in Coimbatore, Tamilnadu, India. *Bulletin of Environmental Contamination and Toxicology*, *68*(3), 389–393. doi:10.1007001280266 PMID:11993814

Merrington, G., Winder, L., & Green, I. (1997). The uptake of cadmium and zinc by the bird-cherry oat aphid *Rhopalosiphum padi* (Homoptera:Aphididae) feeding on wheat grown on sewage sludge amended agricultural soil. *Environmental Pollution*, *96*(1), 111–114. doi:10.1016/S0269-7491(96)00149-2 PMID:15093439

Milton, A., Cooke, J. A., & Johnson, M. S. (2003). Accumulation of lead, zinc, and cadmium in a wild population of *Clethrionomys glareolus* from an abandoned lead mine. *Archives of Environmental Contamination and Toxicology*, *44*(3), 405–411. doi:10.100700244-002-2014-5 PMID:12712302

Mohanraj, R. (2000). Pollution Status of Wetlands in Urban Coimbatore, Tamilnadu, India. *Bulletin of Environmental Contamination and Toxicology*, *64*(5), 638–643. doi:10.1007001280000051 PMID:10857445

Mohanraj, R., & Azeez, P. A. (2005). Urban development and particulate air pollution in Coimbatore city, India. *The International Journal of Environmental Studies*, *62*(1), 69–78. doi:10.1080/0020723042000261713

Mohanraj, R., Azeez, P. A., & Priscilla, T. (2004). Heavy metals in airborne particulate matter of urban Coimbatore. *Archives of Environmental Contamination and Toxicology*, *47*(2), 162–167. doi:10.100700244-004-3054-9 PMID:15386140

Nirmal Bhargavi, V. L., & Sudha, P. N. (2015). Removal of heavy metal ions from soil by electrokinetic assisted phytoremediation method. *International Journal of Chemtech Research*, *8*(5), 192–202.

Orłowska, E., Godzik, B., & Turnau, K. (2012). Effect of different arbuscular mycorrhizal fungal isolates on growth and arsenic accumulation in *Plantago lanceolata* L. *Environmental Pollution*, *168*, 121–130. doi:10.1016/j.envpol.2012.04.026 PMID:22609863

Oyekanmi, A. A., Latiff, A. A., Daud, Z., Mohamed, R. M., Ismail, N., Aziz, A. A., ... Abiodun, A. K. (2019). Adsorption of cadmium and lead from palm oil mill effluent using bone-composite: Optimisation and isotherm studies. *International Journal of Environmental Analytical Chemistry, 99*(8), 707–725. do i:10.1080/03067319.2019.1607318

Pandey, J., & Pandey, U. (2009). Accumulation of heavy metals in dietary vegetables and cultivated soil horizon in organic farming system in relation to atmospheric deposition in a seasonally dry tropical region of India. *Environmental Monitoring and Assessment, 148*(1–4), 61–74. doi:10.100710661-007-0139-8 PMID:18202901

Pavithrapriya, S., Mahimairaja, S., & Sivasubramanian, K. (2015). Pollution due to Heavy Metals in Coimbatore Wetlands, India. *Research Journal of Agriculture and Forestry Sciences, 3*(6), 1–5.

Peng, L., Chen, X., Zhang, Y., Du, Y., Huang, M., & Wang, J. (2015). Remediation of metal contamination by electrokinetics coupled with electrospun polyacrylonitrile nanofiber membrane. *Process Safety and Environmental Protection, 98*, 1–10. doi:10.1016/j.psep.2015.06.003

Poornima, K., Vasudevan, K., & Venkateshwarlu, M. (2011). Evaluation of Heavy metals in Textile effluents in relation to Soil and Pond water. *Electronic Journal of Environmental Sciences, 4*, 79–84.

Priya, & Gabriela Jennifer, Thankam, Thankam, & Mathew. (2010). Monitoring the Pollution Intensity of Wetlands of Coimbatore, Tamil Nadu, India. *Nature Environmentand Pollution Technology, 10*(3), 447–454.

Radha, R. V., Kumutha, K., & Marimuthu, P. (2014). Assessment of Cadmimum Contamination of Soils in Sewage Disposal Areas of Coimbaotre District, Tamil Nadu, India. *Current World Environment, 9*(2), 379–386. doi:10.12944/CWE.9.2.18

Rajmanickam, R., & Nagan, S. (2018). Assessment of Comprehensive Environmental Pollution Index of Kurichi Industrial Cluster, Coimbatore District, Tamil Nadu, India - a Case Study. *Journal of Ecological Engineering, 19*(1), 191–199. doi:10.12911/22998993/78747

Rosestolato, D., Bagatin, R., & Ferro, S. (2015). Electrokinetic remediation of soils polluted by heavy metals (mercury in particular). *Chemical Engineering Journal, 264*, 16–23. doi:10.1016/j.cej.2014.11.074

Sall, M. L., Diaw, A. K., Gningue-Sall, D., Aaron, S. E., & Aaron, J. (2020). Toxic heavy metals: Impact on the environment and human health, and treatment with conducting organic polymers, a review. *Environmental Science and Pollution Research International, 27*(24), 29927–29942. doi:10.100711356-020-09354-3 PMID:32506411

Santhosh, P., & Sridevi, A. (2013). A lab-scale study on reduction of heavy metals from electroplating effluent using conventional chemical precipitation. *Journal of Environmental Research and Development, 8*(1), 102–108.

Sarathamani, T., Govindarajalu, K., & Divya, K. R. (2014). Economic Impact of Water Pollution on Agricultural and Rural Households. *IOSR Journal of Economics and Finance, 3*(5), 54–57. doi:10.9790/5933-0355457

Saravanakumar, R., Sivalingam, S., & Elangovan, S. (2016). Assessment of air quality index of Coimbatore City in Tamil Nadu. *Indian Journal of Science and Technology*, *9*(41). Advance online publication. doi:10.17485/ijst/2016/v9i41/99185

Sarinas, B. G., Gellada, L. D., Jamolangue, E. B., Teruñez, M. R., & Flores, J. R. (2014). Assessment of Heavy Metals in Sediments of Iloilo Batiano River, Philippines. *International Journal of Environmental Sciences and Development*, *5*(6), 543–546. doi:10.7763/IJESD.2014.V5.542

Selvam, S., Ravindran, A. A., Venkatramanan, S., & Singaraja, C. (2015). Assessment of heavy metal and bacterial pollution in coastal aquifers from SIPCOT industrial zones, Gulf of Mannar, South Coast of Tamil Nadu, India. *Applied Water Science*, *7*(2), 897–913. doi:10.100713201-015-0301-3

Selvi, A., & Aruliah, R. (2018). A statistical approach of zinc remediation using acidophilic bacterium via an integrated approach of bioleaching enhanced electrokinetic remediation (BEER) technology. *Chemosphere*, *207*, 753–763. doi:10.1016/j.chemosphere.2018.05.144 PMID:29859487

Selvi, A., Rajasekar, A., Theerthagiri, J., Ananthaselvam, A., Sathishkumar, K., Madhavan, J., & Rahman, P. K. (2019). Integrated Remediation Processes Toward Heavy Metal Removal/Recovery From Various Environments-A Review. *Frontiers in Environmental Science*, *7*, 66. Advance online publication. doi:10.3389/fenvs.2019.00066

Sharma, B., & Chhetri, M. K. (1970). Monitoring of heavy metals in vegetables and soil of agricultural fields of Kathmandu Valley. *Ecoprint: An International Journal of Ecology*, *12*, 1–9. doi:10.3126/eco.v12i0.3189

Sharma, M., & Sharma, R. G. (2013). Identification, purification and quantification of Andrographolide from *Andrographis paniculata* (Burm. F.) nees by HPTLC at different stages of life cycle of crop. *Journal of Current Chemical & Pharmaceutical Sciences*, *3*(1), 23–32.

Sharma, S., & Malaviya, P. (2014). Bioremediation of tannery wastewater by chromium resistant fungal isolate fusarium chlamydosporium SPFS2-g. *Current World Environment*, *9*(3), 721–727. doi:10.12944/CWE.9.3.21

Shenai, D., & Quadros, G. (2014). Comparison between a manmade lake and a natural lake in mumbai using select water quality parameters. *Ecology. Environmental Conservation*, *20*(2), 777–782.

Sherene, T. (2009). Heavy Metals of Soils in Industrial Belts of Coimbatore District, Tamil Nadu. *Nature Environment and Pollution Technology*, *8*(3), 613–618.

Srinivasan, V., Suresh Kumar, D., Chinnasamy, P., Sulagna, S., Sakthivel, D., Paramasivam, P., & Lele, S. (2014, February 1). *Water Management in the Noyyal River Basin A Situation Analysis*. Retrieved from http://eprints.atree.org/52/

Stewart, C., Norton, D., & Fergusson, J. (1991). Historical monitoring of heavy metals in kahikatea ring wood in Christchurch, New Zealand. *The Science of the Total Environment*, *105*, 171–190. doi:10.1016/0048-9697(91)90340-K

Stobrawa, K., & Lorenc-Plucińska, G. (2008). Thresholds of heavy-metal toxicity in cuttings of European black poplar (*Populus nigra* L.) determined according to antioxidant status of fine roots and morphometrical disorders. *The Science of the Total Environment, 390*(1), 86–96. doi:10.1016/j.scitotenv.2007.09.024 PMID:17963819

Sundar, M. L., & Saseetharan, M. K. (2008). Ground water quality in Coimbatore, Tamil Nadu along Noyyal River. *Journal of Environmental Science & Engineering, 50*(3), 187–190. PMID:19552071

Tchounwou, P. B., Yedjou, C. G., Patlolla, A. K., & Sutton, D. J. (2012). Heavy Metal Toxicity and the Environment. *Experientia Supplementum Molecular, Clinical and Environmental Toxicology*, 133-164. doi:10.1007/978-3-7643-8340-4_6

Usharani, K., Umarani, K., Ayyasamy, P., Shanthi, K., & Lakshmanaperumalsamy, P. (2010). Physico-Chemical and Bacteriological Characteristics of Noyyal River and Ground Water Quality of Perur, India. *Journal of Applied Science & Environmental Management, 14*(2). Advance online publication. doi:10.4314/jasem.v14i2.57830

Vandecasteele, B., Samyn, J., Quataert, P., Muys, B., & Tack, F. M. (2004). Earthworm biomass as additional information for risk assessment of heavy metal biomagnification: A case study for dredged sediment-derived soils and polluted floodplain soils. *Environmental Pollution, 129*(3), 363–375. doi:10.1016/j.envpol.2003.12.007 PMID:15016458

Varia, J., Martínez, S. S., Orta, S. V., Bull, S., & Roy, S. (2013). Bioelectrochemical metal remediation and recovery of Au3, Co2 and Fe3 metal ions. *Electrochimica Acta, 95*, 125–131. doi:10.1016/j.electacta.2013.02.051

Vedamadavan & Sarithabanuraman. (2012). Assessment of Ambient Air Quality in Coimbatore City. *Civil and Environmental Research, 2*(1), 1–6.

Verma, R., & Dwivedi, P. (2013). Heavy Metals Water Pollution - A case study. *Recent Research in Science and Technology, 5*(5), 98–99.

Winder, L., Merrington, G., & Green, I. (1999). The tri-trophic transfer of Zn from the agricultural use of sewage sludge. *The Science of the Total Environment, 229*(1–2), 73–81. doi:10.1016/S0048-9697(99)00070-4

Wu, F. Y., Ye, Z. H., & Wong, M. H. (2009). Intraspecific differences of arbuscular mycorrhizal fungi in their impacts on arsenic accumulation by *Pteris vittata* L. *Chemosphere, 76*(9), 1258–1264. doi:10.1016/j.chemosphere.2009.05.020 PMID:19535126

Zhuang, P., Zou, H., & Shu, W. (2009). Biotransfer of heavy metals along a soil-plant-insect-chicken food chain: Field study. *Journal of Environmental Sciences (China), 21*(6), 849–853. doi:10.1016/S1001-0742(08)62351-7 PMID:19803093

KEY TERMS AND DEFINITIONS

Bioaccumulation: Accumulation of heavy metals or any other chemicals in the trophic levels.

Electro-Kinetic (EK) Method: An approach by using electrical current and acidic and alkalophilic microbes to precipitate heavy metals from the soil.

Heavy Metals: Naturally occurring chemical substances found in the earth crust. It exposed or distributed in environment due to anthropogenic activities like industrial operations, mining, smelting operations and other technological applications.

Long-Term Monitoring (LTM): A regular systematic data collection on amount of heavy metals in air, water and soil and other biological components in contaminated site over an extended period of time.

Phyto-Remediation: The process that uses the different types of plants to control, remove, detoxify pollutant or contaminants from the environmental components.

Remediation: An action on controlling heavy metal contaminations or distributions by adopting physical, chemical, and biological methods to reduce harm to human and the environment.

Restoration: The efforts to create the healthy habitats by creating physical, chemical, and biological structures to safeguard the environment, people and improving ecosystem functions.

Compilation of References

Abadi, S., Huda, M., Basiron, B., Ihwani, S.S., Jasmi, K.A., Hehsan, A., Safar, J., Mohamed, A.K., Embong, W.H.W., Mohamad, A.M., Noor, S.S.M., Novita, D., Maseleno, A., Irviani, R., Idris, M., & Muslihudin, M. (2018a). Implementation of fuzzy analytical hierarchy process on notebook selection. *International Journal of Engineering and Technology*, *7*(2.27), 238-243.

Abadi, S., Huda, M., Hehsan, A., Mohamad, A.M., Basiron, B., Ihwani, S.S., Jasmi, K.A., Safar, J., Mohamed, A.K., Embong, W.H.W., Noor, S.S.M., Brahmono, B., Maseleno, A., Fauzi, A.N., Aminudin, N., & Gumanti, M. (2018b). Design of online transaction model on traditional industry in order to increase turnover and benefits. *International Journal of Engineering and Technology*, *7*(2.27), 231-237.

Abadi, S., Huda, M., Jasmi, K.A., Noor, S.S.M., Safar, J., Mohamed, A.K., Embong, W.H.W., Mohamad, A.M., Hehsan, A., Basiron, B., Ihwani, S.S., Maseleno, A., Muslihudin, M., Satria, F., Irawan, D., & Hartati, S. (2018c). Determination of the best quail eggs using simple additive weighting. *International Journal of Engineering and Technology*, *7*(2.27), 225-230.

Abadi, S., Teh, K.S.M., Huda, M., Hehsan, A., Ripin, M.N., Haron, Z., Muhamad, N.H.N., Rianto, R., Maseleno, A., Renaldo, R., & Syarifudin, A. (2018e). Design of student score application for assessing the most outstanding student at vocational high school. *International Journal of Engineering and Technology*, *7*(2.27), 172-177.

Abadi, S., Teh, K.S.M., Nasir, B.M., Huda, M., Ivanova, N.L., Sari, T.I., Maseleno, A., Satria, F., & Muslihudin, M. (2018f). Application model of k-means clustering: insights into promotion strategy of vocational high school. *International Journal of Engineering and Technology*, *7*(2.27), 182-187.

Abadi, S., Hawi, A., Dacholfany, I., Huda, M., Teh, K. S. M., Walidi, J., ... Maseleno, A. (2019). Identification of Sundep, Leafhopper and Fungus of Paddy by Using Fuzzy SAW Method. *International Journal of Pharmaceutical Research*, *11*(1), 695–699.

Abadi, S., Huda, M., Teh, K. S. M., Haron, Z., Ripin, M. N., Hehsan, A., Sarip, S., Hehsan, M. R., Amrullah, M., & Maseleno, A. (2018d). Hazard Level of Vehicle Smoke by Fuzzy Multiple Attribute Decision Making with Simple Additive Weighting Method. *International Journal of Pharmaceutical Research*, *10*(4), 58–71.

Abbas, S. Z., Rafatullah, M., Hossain, K., Ismail, N., Tajarudin, H. A., & Khalil, H. P. (2017). A review on mechanism and future perspectives of cadmium-resistant bacteria. *International Journal of Environmental Science and Technology*, *15*(1), 243–262. doi:10.100713762-017-1400-5

Abdelhafez, A. A., & Li, J. (2014). Environmental Monitoring of Heavy Metal Status and Human Health Risk Assessment in the Agricultural Soils of the Jinxi River Area, China. *Human and Ecological Risk Assessment*, *21*(4), 952–971. doi:10.1080/10807039.2014.947851

Adams, R., Jeanrenaud, S., Bessant, J., Denyer, D., & Overy, P. (2016). Sustainability-oriented innovation: A systematic review. *International Journal of Management Reviews, 18*(2), 180–205. doi:10.1111/ijmr.12068

AdditionalREADING

Adela, A., Jasmi, K. A., Basiron, B., Huda, M., & Maseleno, A. (2018). Selection of dancer member using simple additive weighting. *IACSIT International Journal of Engineering and Technology, 7*(3), 1096–1107. doi:10.14419/ijet.v7i3.11983

African Manual. (2020). *African Organic Agriculture Training Manual.* Available at: https://www.organic-africa.net/training-manual.html

Agostino, A. (2010). Gender equality, climate change and education for sustainability. *Newsletter for beyond access: Gender, education and development*, 24. Retrieved on April 12, 2020, from: http://www.e4conference.org/wp-content/uploads/2010/02/Equals24.pdf

Ahmed, E., Abdulla, H. M., Mohamed, A. H., & El-Bassuony, A. D. (2016). Remediation and recycling of chromium from tannery wastewater using combined chemical–biological treatment system. *Process Safety and Environmental Protection, 104*, 1–10. doi:10.1016/j.psep.2016.08.004

Aiyelaagbe, I., Harris, P., & Olowe, V. (2015). Skills gaps in organic agriculture and SWOT analysis in higher educational institutions (HEIs) in Anglophone West Africa. *Organic Agriculture, 6*(2), 109–118. Advance online publication. doi:10.100713165-015-0119-1

Ajiboye, J., & Silo, N. (2008). Enhancing Botswana children's environmental knowledge, attitudes and practices through the school civic clubs. *International Journal of Environmental and Science Education, 3*(3), 105–114.

Alcaraz, J. M., & Thiruvattal, E. (2010). An interview with Manuel Escudero. The United Nations' principles for responsible management education: A global call for sustainability. *Academy of Management Learning & Education, 9*(3), 542–550. doi:10.5465/amle.9.3.zqr542

Al-Homaidan, A. A., Al-Ghanayem, A. A., Alkhalifa, A., & Arabia, S. (2011). Green Algae as Bioindicators of Heavy Metal Pollution in Wadi Hanifah Stream, Riyadh, Saudi Arabia. *International Journal of Water Resources and Arid Environments, 1*(1), 10–15.

Ali, A. B., Tajuddin, S. N. A. A., Marzuki, M., & Huda, M. (2020). (in press). Digital Approach of Malay Literature through Interactive Computer Method (ICM): A Proposed Framework Model. *Solid State Technology.*

Aliaga, M., & Gunderson, B. (2002). *Interactive Statistics*. Sage Publications.

Ali, H., Khan, E., & Ilahi, I. (2019). Environmental Chemistry and Ecotoxicology of Hazardous Heavy Metals: Environmental Persistence, Toxicity, and Bioaccumulation. *Journal of Chemistry, 2019*, 1–14. doi:10.1155/2019/6730305

Altarawneh, M. (2016). Determine the barriers of organic agriculture implementation in Jordan. *Bulgarian Journal of Agricultural Science, 22*, 10–15.

Altinay, F., Dagli, G., & Altınay, Z. (2017). Role of technology and management in tolerance and reconciliation education. *Quality & Quantity, 51*(6), 2725–2736. doi:10.100711135-016-0419-x

Álvarez-García, O., Sureda-Negre, J., & Comas-Forgas, R. (2015). Environmental education in pre-service teacher training: A literature review of existing evidence. *Journal of Teacher Education for Sustainability, 17*(1), 72–85. doi:10.1515/jtes-2015-0006

Aminin, S., Huda, M., Ninsiana, W., & Dacholfany, M. I. (2018). Sustaining civic-based moral values: Insights from language learning and literature. *International Journal of Civil Engineering and Technology, 9*(4), 157–174.

Amin, M. M., Nugratama, M. A. A., Maseleno, A., Huda, M., & Jasmi, K. A. (2018). Design of cigarette disposal blower and automatic freshner using mq-5 sensor based on atmega 8535 microcontroller. *IACSIT International Journal of Engineering and Technology*, *7*(3), 1108–1113. doi:10.14419/ijet.v7i3.11917

Aminudin, N., Fauzi, Huda, M., Hehsan, A., Ripin, M.N., Haron, Z., Junaidi, J., Irviani, R., Muslihudin, M., Hidayat, S., Maseleno, A., Gumanti, M., & Fauzi, A.M. (2018a). Application program learning based on android for students experiences. *International Journal of Engineering and Technology*, *7*(2.27), 194-198.

Aminudin, N., Huda, M., Ihwani, S.S., Noor, S.S.M., Basiron, B., Jasmi, K.A., Safar, J., Mohamed, A.K., Embong, W.H.W., Mohamad, A.M., Maseleno, A., Masrur, M., Trisnawati, & Rohmadi, D. (2018b). The family hope program using AHP method. *International Journal of Engineering and Technology*, *7*(2.27), 188-193.

Aminudin, N., Huda, M., Mohamed, A.K., Embong, W.H.W., Mohamad, A.M., Basiron, B., Ihwani, S.S., Noor, S.S.M., Jasmi, K.A., Safar, J., Natalie, L., Ivanova, Maseleno, A., Triono, A., & Nungsiati. (2018c). Higher education selection using simple additive weighting. *International Journal of Engineering and Technology*, *7*(2.27), 211-217.

Amirshokoohi, A. (2010). Elementary pre-service teachers' environmental literacy and views toward Science, Technology, and Society (STS) Issues. *Science Educator*, *19*(1), 56–63.

Anggraeni, E.Y., Huda, M., Maseleno, A., Safar, J., Jasmi, K.A., Mohamed, A.K., Hehsan, A., Basiron, B., Ihwani, S.S., Embong, W.H.W., Mohamad, A.M., Noor, S..S.M., Fauzi, A.N., Wijaya, D.A., & Masrur, M. (2018a). Poverty level grouping using SAW method. *International Journal of Engineering and Technology*, *7*(2.27), 218-224.

Anggraeni, E.Y., Pardimin, Dacholfany, M.I., Akla, Huda, M., Teh, K.S.M., Hehsan, A., Junaidi, J., Yusof, F.M., Abas, H., Husin, M.F.A., Apriani, D., & Maseleno, A. (2018b). Modelling effectivenes of IS learning methodology with AHP method. *IACSIT International Journal of Engineering and Technology*, *7*(4), 4708–4714.

Anishchenko, M. (2016). *Ecotourists' motivation and its relation to the Ecotourism's principles* (Magister Thesis). Department of Social and Economic Geography, Umea University.

Anne, P. (1945). Carbone organique (total) du sol et de l'humus. *Annales Agronomiques*, *15*, 161–172.

Anonymous. (2011). *Coimbatore District Population Censucs 2011-2020, Tamil Nadu literacy sex ratio and density*. Directorate of Census Operations, Tamil Nadu, Ministry of Home Affairs. https://www.census2011.co.in/census/district/32-coimbatore.html

Anonymous. (2013). *Environmental degradation of the Coimbatore Wetlands in the Noyyal River Basin.* India Environment Portal | News, reports, documents, blogs, data, analysis on environment & development | India, South Asia, EIA Resource and Response Centre (ERC)Kotagiri, Nilgiri, India. http://www.indiaenvironmentportal.org.in/content/384872/environmental-degradation-of-the-coimbatore-wetlands-in-the-noyyal-river-basin/

Anonymous. (2014). *Brief Industrial Profile ofCoimbatore District2012-13.* http://dcmsme.gov.in/dips/IPS coimbatore 2012_t.pdf

Ansell, C., & Gash, A. (2017). Collaborative platforms as a governance strategy. *Journal of Public Administration: Research and Theory*, *28*(1), 16–32. doi:10.1093/jopart/mux030

Anshari, M., Almunawar, M. N., Shahrill, M., Wicaksono, D. K., & Huda, M. (2017). Smartphones usage in the classrooms: Learning aid or interference? *Education and Information Technologies*, *22*(6), 3063–3079. doi:10.100710639-017-9572-7

Antadze, N., & McGowan, K. A. (2017). Moral entrepreneurship: Thinking and acting at the landscape level to foster sustainability transitions. *Environmental Innovation and Societal Transitions*, *25*, 1–13. doi:10.1016/j.eist.2016.11.001

Antisha & Priyadharsini, D.V. (2017). A Study on the Contamination of Heavy Metals in Arakeerai (*Amaranthus dubius*) Cultivated in Different Sites of Coimbatore in India. *International Journal of Science and Research, 6*(8), 120-125.

Appasamy, P. P., & Nelliyat, P. (2007). *Compensating the Loss of Ecosystem Services Due to Pollution in Noyyal River Basin, Tamil Nadu.* Development Economics Working Papers 22493. East Asian Bureau of Economic Research. https://ideas.repec.org/p/eab/develo/22493.html

Arbonnier, M. (2002). *Arbres, arbustes et lianes des zones sèches d'Afrique de l'Ouest* (C. Isabelle, Ed.). CIRAD-MNHN-UICN.

Ardoin, N., Bowers, A., Roth, W., & Holthuis, N. (2018). Environmental education and K-12 student outcomes: A review and analysis of research. *The Journal of Environmental Education*, *49*(1), 1–17. doi:10.1080/00958964.2017.1366155

Aronson, M. F. J., La Sorte, F. A., Nilon, C. H., Katti, M., Goddard, M. A., Lepczyk, C. A., Warren, P. S., Williams, S. G., Cilliers, S., Clarkson, B., Dobbs, C., Dolan, R., Hedblom, M., Klotz, S., Kooijmans, J. L., Kuhn, I., MacGregor-Fors, L., McDonnell, M., Mortberg, U., . . . Winter, M. (2014). A global analysis of the impacts of urbanization on bird and plant diversity reveals key anthropogenic drivers. *Proc. R. Soc., 281*, 20133330. 10.1098/rspb.2013.3330

Arroyo-Rodríguez, V., Melo, F. P. L., Martínez-Ramos, M., Bongers, F., Chazdon, R. L., Meave, J. A., Norden, N., Santos, B. A., Leal, I. R., & Tabarelli, M. (2017). Multiple successional pathways in human-modified tropical landscapes: New insights from forest succession, forest fragmentation and landscape ecology research. *Biological Reviews of the Cambridge Philosophical Society*, *92*(1), 326–340. doi:10.1111/brv.12231 PMID:26537849

Arslan, H. O., Cigdemoglu, C., & Moseley, C. (2012). A three-tier diagnostic test to assess pre-service teachers' misconceptions about global warming, greenhouse effect, ozone layer depletion, and acid rain. *International Journal of Science Education*, *34*(11), 1667–1686. doi:10.1080/09500693.2012.680618

Assahubulkahfi, M., Sam, Y. M., Maseleno, A., & Huda, M. (2018). LQR Tuning by Particle Swarm Optimization of Full Car Suspension System. *International Journal of Engineering & Technology, 7*(2.13), 328-331.

Association Marocaine de Biotechnologie et de Protection des Ressources Naturelles (MICROBIONA). (2018, March). *Rapport Scientifique*. Author.

Asunta, T. (2004). Knowledge sources, attitudes and self-reported behaviour of secondary-level science students concerning environmental topics. In A. Laine, J. Lavonen, & V. Meisalo (Eds.), *Current Research on Mathematics and Science Education. University of Helsinki. Research Report 253.*

Atmotiyoso, P., & Huda, M. (2018). Investigating Factors Influencing Work Performance on Mathematics Teaching: A Case Study. *International Journal of Instruction*, *11*(3), 391–402. doi:10.12973/iji.2018.11327a

Azhar, A. T., Nabila, A. T., Nurshuhaila, M. S., Zaidi, E., Azim, M. A., & Farhana, S. M. (2016). Assessment and Comparison of Electrokinetic and Electrokinetic-bioremediation Techniques for Mercury Contaminated Soil. *IOP Conference Series. Materials Science and Engineering*, *160*, 012077. doi:10.1088/1757-899X/160/1/012077

Babunath, R., & John, G. (2017). A Study on Physico Chemical and Heavy Metals Characterstics of River Noyyal, Tamil Nadu, India. *Environemntal Science: An Indian Journal*, *13*(1), 1–10.

Bahk, C. M. (2010). Environmental education through narrative films: Impact of medicine man on attitudes toward forest preservation. *The Journal of Environmental Education*, *42*(1), 1–13. doi:10.1080/00958960903479811

Baker, S., & Mehmood, A. (2015). Social innovation and the governance of sustainable places. *Local Environment*, *20*(3), 321–334. doi:10.1080/13549839.2013.842964

Balakrishnan, A., Ramu, A., & Murugesan, A. (2014). Assessment of Heavy Metal Distribution in Groundwater in and around Gulf of Mannar Seashore Area Using GIS Technique. *Research Journal of Chemical Sciences*, *4*(1), 9–16.

Bamberg, S. & Möser, G. (2007). Twenty years after Hines, Hungerford, and Tomera: a new meta-analysis of psycho-social determinants of pro-environmental behaviour. *Journal of Environmental Psychology, 27*, 14-25.

Bamberg, S., & Möser, G. (2007). Twenty years after Hines, Hungerford, and Tomera: A new meta-analysis of psycho-social determinants of pro-environmental behaviour. *Journal of Environmental Psychology*, *27*(1), 14–25. doi:10.1016/j.jenvp.2006.12.002

Bangay, C., & Blum, N. (2010). Education Responses to Climate Change and Quality: Two Parts of the Same Agenda? *International Journal of Educational Development*, *30*(4), 335–450. doi:10.1016/j.ijedudev.2009.11.011

Bangladesh Bureau of Statistics (BBS). (2018). *Bangladesh statistics 2018*. Statistics and Informatics Division (SID) Ministry of Planning. Retrieved on June 28, 2020, from http://bbs.portal.gov.bd/sites/default/files/files/bbs.portal.gov.bd/page/a1d32f13_8553_44f1_92e6_8ff80a4ff82e/Bangladesh%20%20Statistics-2018.pdf

Barkauskiene, K., & Snieska, V. (2013). Ecotourism as an Integral part of sustainable tourism development. *Economics and Management*, *18*(3), 2029–2038. doi:10.5755/j01.em.18.3.4272

Barth, M., & Timm, J. (2011). Higher education for sustainable development: Students' perspectives on an innovative approach to educational change. *Journal of Social Science*, *7*(1), 13–23. https://ssrn.com/abstract=1978184

Bascopé, M., & Caniguan, N. I. (2016). Propuesta pedagógica para la incorporación de conocimientos tradicionales de Ciencias Naturales en Primaria. *Revista Electrónica de Investigación Educativa, 18*(3), 162–175. https://redie.uabc.mx/redie/article/view/1143

Bascopé, M., Perasso, P., & Reiss, K. (2019). Systematic review of education for sustainable development at an early stage: Cornerstones and pedagogical approaches for teacher professional development. *Sustainability*, *11*(3), 719. doi:10.3390u11030719

Bassano, C., Carotenuto, A., Ferretti, M., Pietronudo, M. C., & Coskun, H. E. (2018, July). Smart University for sustainable governance in smart local service systems. In *International Conference on Applied Human Factors and Ergonomics* (pp. 206-216). Springer.

Baugher, J. E., Osika, W., & Robert, K. H. (2016). Ecological Consciousness, Moral Imagination, and the Framework for Strategic Sustainable Development. In *Creative Social Change: Leadership for a Healthy World* (pp. 119–142). Emerald Group Publishing Limited. doi:10.1108/S2058-88012016037

Baumgartner, R. J., & Rauter, R. (2017). Strategic perspectives of corporate sustainability management to develop a sustainable organization. *Journal of Cleaner Production, 140*, 81–92. doi:10.1016/j.jclepro.2016.04.146

Bebbington, J., & Unerman, J. (2018). Achieving the United Nations Sustainable Development Goals: An enabling role for accounting research. *Accounting, Auditing & Accountability Journal*, *31*(1), 2–24. doi:10.1108/AAAJ-05-2017-2929

Beier, M. E., & Rittmayer, A. D. (2008). Literature Overview: Motivational factors in STEM. Interest and Self-Concept. In B. Bogue & E. Cady (Eds.), *Applying Research to Practice (ARP) Resources*. Retrieved from https://www.engr.psu.edu/AWE/ARPresources.aspx

Ben Salem, A., Messouli, M., & Yacoubi-Khebiza, M. (2011). Developing an oasis-based water management tool: Ecohydrologic approach and weap software for a large arid catchment in Morocco. *International Journal of Water Resources and Arid Environments*.

Benmohammadi, A. (1998). Ensablement et désertification dans la vallée moyenne de l'oued Drâa: Le mythe de la fluctuation de la limite nord du désert. *Africa Geoscience Review*, *5*, 11–22.

Bennur, S. (2014, May 29). Schools directed to grow vegetables, fruits in kitchen gardens. *The Hindu*.

Bentley, G., Pugalis, L., & Shutt, J. (2017). Leadership and systems of governance: The constraints on the scope for leadership of place-based development in sub-national territories. *Regional Studies*, *51*(2), 194–209. doi:10.1080/00343404.2016.1181261

Berkes, F. (2017). Environmental governance for the anthropocene? Social-ecological systems, resilience, and collaborative learning. *Sustainability*, *9*(7), 1232. doi:10.3390u9071232

Berkowitz, A., Ford, M., & Brewer, C. (2005). A framework for integrating ecological literacy, civics literacy, and environmental citizenship in environmental education. In E. A. Johnson & M. J. Mappin (Eds.), *Environmental education and advocacy* (pp. 227–266). Cambridge University Press.

Berleant, A. (2017, June 4). *Environmental Aesthetics*. Academic Press.

Bharti, K. S. (2003, November 1). Teaching strategies in a pre-school program in the states of Haryana and Rajasthan. *Journal of Indian Education*, *29*, 63.

Bhatt, V. S. (1992). Handbook of Agriculture. Indian Council of Agricultural Research (ICAR).

Blair, R. B. (1999). Birds and butterflies along an urban gradient : Surrogate taxa for assessing biodiversity? *Ecological Applications*, *9*(1), 164–170. doi:10.1890/1051-0761(1999)009[0164:BABAAU]2.0.CO;2

Blakbern, A. A. (2003). Accumulation and migration of trace elements along trophic chains in ecosystems of the chatkal biosphere reserve (the Western Tien Shan, Uzbekistan). *Russian Journal of Ecology*, *34*(1), 68–72. doi:10.1023/A:1021827408038

Blamey, R. K., & Braithwaite, V. A. (1997). A social values segmentation of the potential Ecotourism market. *Journal of Sustainable Tourism*, *5*(1), 29–45. doi:10.1080/09669589708667274

Blankenberg, A.-K., & Alhusen, H. (2019). On the determinants of pro-environmental behavior: A literature review and guide for the empirical economist. *Center for European, Governance, and Economic Development Research (CEGE)*, (350).

BMEL. (2019). Ökologischer Landbau in Deutschland. Bundesministerium für Ernährung und Landwirtschaft (BMEL).

Boakye, C. (2015). Climate Change Education: The Role of Pre-Tertiary Science Curricula in Ghana. *SAGE Open*, *2015*(4), 1–10. doi:10.1177/2158244015614611

Boca, G., & Saracli, S. (2019). Environmental education and student's perception for sustainability. *Sustainability*, *11*(6), 1553. doi:10.3390u11061553

Boeve-de Pauw, J., Gericke, N., Olsson, D., & Berglund, T. (2015). The Effectiveness of Education for Sustainable Development. *Sustainability*, *7*(11), 15693–15717. doi:10.3390u71115693

Bohn, F. J., & Huth, A. (2017). The importance of forest structure to biodiversity – productivity relationships. *Royal Society Open Science*, *4*(1), 12. doi:10.1098/rsos.160521 PMID:28280550

Bolton, R., & Hannon, M. (2016). Governing sustainability transitions through business model innovation: Towards a systems understanding. *Research Policy*, *45*(9), 1731–1742. doi:10.1016/j.respol.2016.05.003

Bolund, P., & Hunhammar, S. (1999). Ecosystem services in urban areas. *Ecological Economics*, *29*(2), 293–301. doi:10.1016/S0921-8009(99)00013-0

Bonnett, M. (1999). Education for sustainable development: A coherent philosophy for environmental education? *Cambridge Journal of Education, 29*(3), 313–324. doi:10.1080/0305764990290302

Borchers, C., Boesch, C., Riedel, J., Guilahoux, H., Ouattara, D., & Randler, C. (2014). Environmental education in Côte D'Ivoire/West Africa: Extra-curricular primary school teaching shows positive impact on environmental knowledge and attitudes. *International Journal of Science Education, 4*(3), 240–259.

Borelli, S., & Conigliaro, M. (2018). Urban forests in the global context. *Unasylva, 69*, 3–79.

Bosselmann, K. (2016). *The principle of sustainability: transforming law and governance*. Routledge.

Botetzagias, I., Dima, A.-F., & Malesios, C. (2015). Extending the theory of planned behavior in the context of recycling: The role of moral norms and of demographic predictors. *Resources, Conservation and Recycling, 95*, 58–67. doi:10.1016/j.resconrec.2014.12.004

Botswana Government. (1994). *The revised National Policy on Education*. Government Printers.

Boukhalfa, C. (2007). Heavy metals in the water and sediments of Oued Es-Souk, Algeria, a river receiving acid effluents from an abandoned mine. *African Journal of Aquatic Science, 32*(3), 245–249. doi:10.2989/AJAS.2007.32.3.3.302

Bowen, G. A. (2009). Document analysis as a qualitative research method. *Qualitative Research Journal, 9*(2), 27–40. doi:10.3316/QRJ0902027

Bowker, R., & Tearle, P. (2007). Gardening as a learning environment: A study of children's perceptions and understanding of school gardens as part of an international project. *Learning Environment Research*.

Boyd, S. W., & Butler, R. W. (1996). Managing Ecotourism: An opportunity spectrum approach. *Tourism Management, 17*(8), 557–566. doi:10.1016/S0261-5177(96)00076-3

Bradley, J. C., Waliczek, T. M., & Zajicek, J. M. (1999). Relationship between environmental knowledge and environmental attitude of high school students. *The Journal of Environmental Education, 30*(3), 17–21. doi:10.1080/00958969909601873

Brandi, C. A. (2017). Sustainability standards and sustainable development–synergies and trade-offs of transnational governance. *Sustainable Development, 25*(1), 25–34. doi:10.1002d.1639

Brasseur, G. P., Jacob, D., & Schuck-Zöller, S. (2017). *Klimawandel in Deutschland: Entwicklung, Folgen, Risiken und Perspektiven*. Springer. doi:10.1007/978-3-662-50397-3

Braun, T., Cottrell, R., & Dierkes, P. (2018). Fostering changes in attitude, knowledge and behavior: Demographic variation in environmental education effects. *Environmental Education Research, 24*(6), 899–920. doi:10.1080/13504622.2017.1343279

Breiting, S., & Wickenberg, P. (2010). The progressive development of environmental education in Sweden and Denmark. *Environmental Education Research, 16*(1), 9–37. doi:10.1080/13504620903533221

Bringezu, S., Potočnik, J., Schandl, H., Lu, Y., Ramaswami, A., Swilling, M., & Suh, S. (2016). Multi-scale governance of sustainable natural resource use—Challenges and opportunities for monitoring and institutional development at the national and global level. *Sustainability, 8*(8), 778. doi:10.3390u8080778

Brophy, J. (2004). *Motivating Students to learn* (2nd ed.). Erlbaum. doi:10.4324/9781410610218

Brown, O. (2008). Migration and Climate Change. Geneva, Switzerland: International Organisation for Migration (IOM). doi:10.18356/26de4416-en

Brundtland Commission. (1987). *Our common future: Report of the World Commission on Environment and Development*. Oxford University Press.

Budiyanto, G., Ipnuwati, S., Al Gifari, S. A., Huda, M., Jalal, B., Maseleno, A., & Hananto, A. L. (2018). Web based expert system for diagnosing disease pest on banana plant. *IACSIT International Journal of Engineering and Technology*, *7*(4), 4715–4721.

Burghardt, K. T., Tallamy, D. W., & Gregory Shriver, W. (2009). Impact of native plants on bird and butterfly biodiversity in suburban landscapes. *Conservation Biology*, *23*(1), 219–224. doi:10.1111/j.1523-1739.2008.01076.x PMID:18950471

Caiman, C., & Lundegård, I. (2014). Pre-school children's agency in learning for sustainable development. *Environmental Education Research*, *20*(4), 437–459. doi:10.1080/13504622.2013.812722

Cairns, M.A., Helmer, E.H., & Baumgardner, G.A. (1997). Root biomass allocation in the world ' s upland forests. *Oecologie*, 1–11.

Cameselle, C., Gouveia, S., Eddine, D., & Belhadj, B. (2013). *Advances in Electrokinetic Remediation for the Removal of Organic Contaminants in Soils*. Organic Pollutants - Monitoring, Risk and Treatment. doi:10.5772/54334

Campbell, C., Speldewinde, C., Howitt, C., & MacDonald, A. (2018). STEM practice in the early years. *Creative Education*, *09*(1), 11–25. doi:10.4236/ce.2018.91002

Cang, L., Zhou, D., Wang, Q., & Fan, G. (2012). Impact of electrokinetic-assisted phytoremediation of heavy metal contaminated soil on its physicochemical properties, enzymatic and microbial activities. *Electrochimica Acta*, *86*, 41–48. doi:10.1016/j.electacta.2012.04.112

Carr, R., Zhang, C., Moles, N., & Harder, M. (2007). Identification and mapping of heavy metal pollution in soils of a sports ground in Galway City, Ireland, using a portable XRF analyser and GIS. *Environmental Geochemistry and Health*, *30*(1), 45–52. doi:10.100710653-007-9106-0 PMID:17610027

Casado-Belmonte, M. P., Martínez-Victoria, M. C., Martínez-Romero, M. J., Ramos-Requena, J. P., & Martínez-Alonso, R. (2019). *Students' Perceptions of Ethical Development in Business Education Programs. In Proceedings of ICERI2019*. IATED.

Castells, M. (2000). A era da informação: economia, sociedade e cultura. In *A Sociedade em rede*. Paz e Terra.

Ceballos-Lascurain, H. (1996). *Tourism, Ecotourism and Protected Areas*. Reino Unido: IUCN/The World Conservation Union.

Ceballos-Lascurain, H. (1987). The future of Ecotourism. *Mexico Journal*, *1*, 13–19.

Cebesoy, Ü.B. & Karisan, D. (2020). Teaching the role of forests in mitigating the effects of climate change using outdoor educational workshop. *Research in Science and Technological Education*. doi:10.1080/02635143.2020.1799777

Cebesoy, Ü. B. (2019). Pre-service teachers' opinions about a two-day climate change education. *International Research in Geographical and Environmental Education*, *28*(3), 211–227. doi:10.1080/10382046.2019.1579982

CEH. (2012). *Heavy Metal Deposition Mapping: Concentrations and Deposition of Heavy Metals in Rural Areas of the UK*. Centre for Ecology and Hydrology. Available at: https://uk-air.defra.gov.uk/assets/documents/reports/cat13/1511251439_AQ0716_Heavy_Metals_Final_Report2012-15NOV12.pdf

Chandra, R., Nishadh, K. A., & Azeez, P. A. (2009). Monitoring water quality of Coimbatore wetlands, Tamil Nadu, India. *Environmental Monitoring and Assessment*, *169*(1-4), 671–676. doi:10.100710661-009-1206-0 PMID:19902369

Chave, J., Mechain, M. R., Burquez, A., Chidumayo, E., Colgan, M. S., Delitti, W. B., Alvaro, D., & Tron, E. (2014). Improved allometric models to estimate the aboveground biomass of tropical trees. *Global*, *20*(10), 3177–3190. doi:10.1111/gcb.12629 PMID:24817483

Chelleri, L., Minucci, G., Ruiz, A., & Karmaoui, A. (2014). Responses to drought and desertification in the Moroccan Drâa Valley Region: Resilience at the expense of sustainability? *International Journal of Climate Change: Impacts & Responses*, *5*(2), 17–33. doi:10.18848/1835-7156/CGP/v05i02/37199

Cheng, A., & Gursoy, D. (2016). The influence of materialism on ecotourism attitudes and behaviors. *Journal of Travel Research*, *55*(2), 176–189. doi:10.1177/0047287514541005

Chen, W. Y. (2015). The role of urban green infrastructure in offsetting carbon emissions in 35 major Chinese cities: A nationwide estimate. *Cities (London, England)*, *44*, 112–120. doi:10.1016/j.cities.2015.01.005

Chen, X. (2011). Why do people misunderstand climate change? Heuristics, mental models and ontological assumptions. *Climatic Change*, *108*(1-2), 31–46. doi:10.100710584-010-0013-5

Choudhary, S., Saha, A. R., & Tiwary, N. K. (2019). The role of compulsory environmental education in higher learning: A study in the University of Delhi. *Applied Environmental Education and Communication*, 1–13. doi:10.1080/1533015X.2019.1605946

Cizel, B., Ajanovic, E., & Cakar, K. (2016). Prerequisites for effective and sustainable destination governance. *Anatolia*, *27*(2), 155–166. doi:10.1080/13032917.2015.1058831

Clark, A.-M. (2006). Changing classroom practice to include the project approach. *Early Childhood Research & Practice*, *8*(2), n2.

Clark, C. F., Kotchen, M. J., & Moore, M. R. (2003). Internal and external influences on pro-environmental behavior: Participation in a green electricity program. *Journal of Environmental Psychology*, *23*(3), 237–246. doi:10.1016/S0272-4944(02)00105-6

Clarke, I. III, Flaherty, T. B., & Mottner, S. (2001). Student perceptions of educational technology tools. *Journal of Marketing Education*, *23*(3), 169–177. doi:10.1177/0273475301233002

Clemente, R., Dickinson, N. M., & Lepp, N. W. (2008). Mobility of metals and metalloids in a multi-element contaminated soil 20years after cessation of the pollution source activity. *Environmental Pollution*, *155*(2), 254–261. doi:10.1016/j.envpol.2007.11.024 PMID:18249071

Cobbinah, P. B. (2015). Contextualising the meaning of ecotourism. *Tourism Management Perspectives*, *16*, 179–189. doi:10.1016/j.tmp.2015.07.015

Coertjens, L., Boeve-de Pauw, J., De Maeyer, S., & Van Petegem, P. (2010). Do schools make a difference in their students' environmental attitudes and awareness? Evidence from Pisa 2006. *International Journal of Science and Mathematics Education*, *8*(3), 497–522. doi:10.100710763-010-9200-0

Colbert, B. A., Nicholson, J., & Kurucz, E. C. (2018). Humanistic Leadership for Sustainable Transformation. In Evolving Leadership for Collective Wellbeing: Lessons for Implementing the United Nations Sustainable Development Goals (pp. 33-47). Emerald Publishing Limited. doi:10.1108/S2058-880120180000007004

Cole, M. S., Feild, H. S., & Harris, S. G. (2004). Student learning motivation and psychological hardiness: Interactive effects on students' reactions to a management class. *Academy of Management Learning & Education*, *3*(1), 64–85. doi:10.5465/amle.2004.12436819

Cole, R., & Snider, B. (2019). Managing in turbulent times: The impact of sustainability in management education on current and future business leaders. *Journal of Cleaner Production*, *210*, 1622–1634. doi:10.1016/j.jclepro.2018.11.135

Conseil National de l'Environnement pour un Développement Durable. (2016). *Troisieme Communication Nationale a la Conference des Parties de la Convention Cadre des Nations Unies sur les Changements Climatiques*. Niamey.

Coria, J., & Calfucura, E. (2012). Ecotourism and the development of indigenous communities: The good, the bad, and the ugly. *Ecological Economics*, *73*, 47–55. doi:10.1016/j.ecolecon.2011.10.024

Coyle, K. (2005). *Environmental literacy in America: What ten years of NEETF/Roper research and related studies say about environmental literacy in the U.S.* The National Environmental Education and Training Foundation.

Creswell, J. W., & Plano Clark, V. L. (2011). *Designing and conducting mixed methods research* (2nd ed.). Sage Publications.

Crim, C., Moseley, C., & Desjean-Perrotta, B. (2017). Strategies toward the inclusion of environmental education in educator preparation programs: Results from a national survey. *School Science and Mathematics*, *117*(3-4), 104–114. doi:10.1111sm.12211

Crowder, D., & Reganold, J. (2015). Financial competitiveness of organic agriculture on a global scale. *Proceedings of the National Academy of Sciences of the United States of America*, *112*(24), 7611–7616. doi:10.1073/pnas.1423674112 PMID:26034271

Crowley, T. J. (2000). Causes of climate change over the past 1000 years. *Science*, *289*(5477), 270–276. doi:10.1126cience.289.5477.270 PMID:10894770

CSE. (2013). *Noyyal River Pollution excreta happens*. Centre for Sciences and Environment. Available at: https://www.yumpu.com/en/document/view/12073282/noyyal-river-pollution-excreta-happens-centre-for-science-and-

CTA. (2019). *The Digitalization of African Agriculture Report, 2018-2019*. CTA.

Cucuzzella, C. (2016). Creativity, sustainable design and risk management. *Journal of Cleaner Production*, *135*, 1548–1558. doi:10.1016/j.jclepro.2015.12.076

Currie, G., & Stanley, J. (2007). *No way to go: Transport and social disadvantage in Australian communities*. Academic Press.

Currie, J., & Deschenes, O. (2016). Children and climate change: Introducing the issue. *Children and Climate Change*, *26*(1), 3–10. doi:10.1353/foc.2016.0000

Cutter, A., & Smith, R. (2001). Gauging primary school teachers' environmental literacy: An issue of 'priority'. *Asia Pacific Education Review*, *2*(2), 45–60. doi:10.1007/BF03026290

Dacholfany, M.I., & Islahudin, A.N., Kuliyatun, Ilyas, U., Sudrajat, A., Siregar, M., Cholis, N., Jandra, M., Nurmartiani, E., Huda, M., Maseleno, A., Hashim, A., Leh, F.C., Rosman, A.S., Ab Rahim, N.M.Z., Razzak, M.A., Rashid, M.H., Darawi, A.B.S., & Nasir, B.M. (2020). Investigating Islamic-based Moral and Professional Principles of Stem Cell-Cloning Technolog. *International Journal of Pharmaceutical Research*, *12*(SI2), 4478–4488.

Dada, D., Eames, C., & Calder, N. (2017). Impact of environmental education on beginning preservice teachers' environmental literacy. *Australian Journal of Environmental Education*, *33*(3), 201–222. doi:10.1017/aee.2017.27

Dangi, T. B. (2018). Exploring the Intersections of Emotional Solidarity and Ethic of Care: An Analysis of Their Synergistic Contributions to Sustainable Community Tourism Development. *Sustainability*, *10*(8), 2713. doi:10.3390u10082713

Darnton, A. (2008). *Reference report: An overview of behaviour change models and their uses. A GSR behaviour change knowledge review*. Centre for Sustainable Development, University of Westminster.

Datta, P., Gopalakrishna-Remani, V., & Bozan, K. (2015). The impact of sustainable governance and practices on business performance: An empirical investigation of global firms. *International Journal of Sustainable Society, 7*(2), 97–120. doi:10.1504/IJSSOC.2015.069912

Davis, J. M. (2015). *What is early childhood education for sustainability and why does it matter?* (J. M. Davis, Ed.). Cambridge University Press.

Dawson, V., & Carson, K. (2017). Using climate change scenarios to assess high school students' argumentation skills. *Research in Science & Technological Education, 35*(1), 1–16. doi:10.1080/02635143.2016.1174932

Dayamba, S. D., Djoudi, H., Zida, M., Sawadogo, L., & Verchot, L. (2016). Agriculture, Ecosystems and Environment Biodiversity and carbon stocks in different land use types in the Sudanian Zone of Burkina Faso, West Africa. *Agriculture, Ecosystems & Environment, 216*, 61–72. doi:10.1016/j.agee.2015.09.023

De Kock, A., Sleegers, P., & Voeten, M. (2004). New Learning and the Classification of Learning. *Review of Educational Research, 74*(2), 141–170. doi:10.3102/00346543074002141

de l'Environnement du Niger, M. (2010). Foresterie Urbaine, version revue Niger. Academic Press.

De Lange, H. J., Sala, S., Vighi, M., & Faber, J. H. (2010). Ecological vulnerability in risk assessment—A review and perspectives. *The Science of the Total Environment, 408*(18), 3871–3879. doi:10.1016/j.scitotenv.2009.11.009 PMID:20004002

De Schaetzen, S. (2019). *Organic Agriculture and the Sustainable Development Goals – Part of the Solution*. EOSTA.

de Vries, W., Römkens, P. F., & Schütze, G. (2007). Critical soil concentrations of cadmium, lead, and mercury in view of health effects on humans and animals. *Reviews of Environmental Contamination and Toxicology, 191*, 91–130. doi:10.1007/978-0-387-69163-3_4 PMID:17708073

De Wit, J., & Verhoog, H. (2007). Organic values and the conventionalization of organic agriculture. *NJAS - Wageningen Journal of Life Sciences, 54*(4), 449-462. doi:10.1016/S1573-5214(07)80015-7

DeChano, L. M. (2006). A multi-country examination of the relationship between environmental knowledge and attitudes. *International Research in Geographical and Environmental Education, 15*(1), 15–28. doi:10.2167/irgee/184.0

Deng, J., & Li, J. (2015). Self-identification of ecotourists. *Journal of Sustainable Tourism, 23*(2), 255–279. doi:10.1080/09669582.2014.934374

Derman, A., Sahin, E., & Hacieminoglu, E. (2016). Does outdoor education make any difference in environmental literacy of pre-service classroom teachers? *International Journal of Environmental and Science Education, 11*(15), 8491–8506.

Desjean-Perrotta, B. (2013). Raising a generation of environmentally literate children: Are preservice teachers ready? *Childhood Education, 89*(4), 260–267. doi:10.1080/00094056.2013.815556

Desmond, Grieshop, & Subramaniam. (2004). *Revisiting Garden-Based learning in basic education*. ResearchGate.

Devkota, B., & Schmidt, G. (2000). Accumulation of heavy metals in food plants and grasshoppers from the Taigetos Mountains, Greece. *Agriculture, Ecosystems & Environment, 78*(1), 85–91. doi:10.1016/S0167-8809(99)00110-3

Diamantis, D. (2018). Stakeholder Ecotourism management: Exchanges, coordination's and adaptations. *Journal of Ecotourism, 17*(3), 203–205. doi:10.1080/14724049.2018.1502122

Dias, J., Pinto, R. N., Vieira, C. B., & de Abreu Corrêa, A. (2018). Detection and quantification of human adenovirus (HAdV), JC polyomavirus (JCPyV) and hepatitis A virus (HAV) in recreational waters of Niterói, Rio de Janeiro, Brazil. *Marine Pollution Bulletin*, *133*, 240–245. doi:10.1016/j.marpolbul.2018.05.031 PMID:30041311

Díaz de Villegas, M. E., Villa, P., & Frías, A. (2002). Evaluation of the siderophores production by *Pseudomonas aeruginosa* PSS'. *Revista Latinoamericana de Microbiologia*, *44*(3–4), 112–117. PMID:17061484

Dieguez, T. (2018). Growth or Development? A Sustainable Approach. *Journal of Business Management and Economic Research*, *2*(8), 38–46. doi:10.29226/TR1001.2018.56

Diekmann, A., & Franzen, A. (2019). Environmental concern: A global perspective. In *Einstellungen und Verhalten in der empirischen Sozialforschung* (pp. 253–272). Springer. doi:10.1007/978-3-658-16348-8_11

Diem, H.G., Gueye, L., Gianinazzi-Pearson, V., Fortin, J.A., & Dommergues, Y.R. (1981). Ecology of VA mycorrhizae in the tropics: The semi-arid zone of Senegal. *Acta Oecologica/Oecologia Plantarum, 16*, 53-62.

Din, G., Omar, A., Said, C. S., Hanif, M. H. M., Ali, A. B., Huda, M., & Maseleno, A. (2020). (in press). Developing Software Application of Sasteraya: An Empirical Research. *International Journal of Pharmaceutical Research.*

Direktorat Jenderal Pembangunan Daerah Tertinggal. (2016). *Kabupaten Lembata*. https://ditjenpdt.kemendesa.go.id/potensi/district/44-kabupaten-lembata

Direktorat Jenderal Pembangunan Daerah Tertinggal. (2016). *Kabupaten Timor Tengah Utara*. https://ditjenpdt.kemendesa.go.id/potensi/district/41-kabupaten-timor-tengah-utara

Disinger, J. (1981). Environmental education in the K-12 schools: A national survey. In A. Sacks & ... (Eds.), *Current issues VII: The yearbook of environmental education and environmental studies* (pp. 141–156). ERIC/SMEAC.

Disinger, J. F., & Roth, C. E. (1992). *Environment Literacy*. ERIC Clearinghouse for Science, Mathematics and Environmental Education. Retrieved from https://files.eric.ed.gov/fulltext/ED351201.pdf

Djuric, M., & Filipovic, J. (2015). Human and social capital management based on complexity paradigm: Implications for various stakeholders and sustainable development. *Sustainable Development*, *23*(6), 343–354. doi:10.1002d.1595

Dögl, C., & Behnam, M. (2015). Environmentally sustainable development through stakeholder engagement in developed and emerging countries. *Business Strategy and the Environment*, *24*(6), 583–600. doi:10.1002/bse.1839

Dolnicar, S., Crouch, G. I., & Long, P. (2008). Environment-friendly tourists: What do we really know about them? *Journal of Sustainable Tourism*, *16*(2), 197–210. doi:10.2167/jost738.0

Dredge, D. (2017). Responsibility and Care in the Collaborative economy. In Dredge & Gyimóthy (Eds.), Collaborative Economy and Tourism, Tourism on the Verge. Springer International Publishing. doi:10.1007/978-3-319-51799-5_4

Dredge, D., Benckendorff, P., Day, M., Gross, M. J., Walo, M., Weeks, P., & Whitelaw, P. (2012). The philosophic practitioner and the curriculum space. *Annals of Tourism Research*, *39*(4), 254–2176. doi:10.1016/j.annals.2012.07.017

Dubelling, M., Wiskerke, H., & Moschitz, H. (2015). *Nachhaltige Lebensmittelversorgung in der Stadt*. RUAF Fondation und SUPURBFOOD-Projekt.

Dubey, R., & Gunasekaran, A. (2015). Shortage of sustainable supply chain talent: An industrial training framework. *Industrial and Commercial Training*, *47*(2), 86–94. doi:10.1108/ICT-08-2014-0052

Duhn, I. (2012). Making "place" for ecological sustainability in early childhood education. *Environmental Education Research*, *18*(1), 19–29. doi:10.1080/13504622.2011.572162

Dunn, C. P., & Heneghan, L. (2011). Composition and diversity of urban vegetation. Oxford University Press.

Dupigny-Giroux, L. A. (2010). Exploring the challenges of climate science literacy: Lessons from students, teachers and lifelong learners. *Geography Compass*, *4*(9), 1203–1217. doi:10.1111/j.1749-8198.2010.00368.x

Dyllick, T., & Muff, K. (2016). What does sustainability for business really mean? And when is a business truly sustainable? In S. Jeanrenaud, J. Gosling, & J. P. Jeanrenaud (Eds.), *Sustainable Business: A One Planet Approach* (pp. 381–408). Wiley.

Eagles, P. F. (1992). The Travel Motivations of Canadian Ecotourists. *Journal of Travel Research*, *31*(2), 3–7. doi:10.1177/004728759203100201

Eastman, J. K., Iyer, R., & Eastman, K. L. (2011). Business students' perceptions, attitudes, and satisfaction with interactive technology: An exploratory study. *Journal of Education for Business*, *86*(1), 36-43. doi:10.1080/08832321003774756

Edsand, H., & Broich, T. (2020). The impact of environmental education on environmental and renewable energy technology awareness: Empirical evidence from Colombia. *International Journal of Science and Mathematics Education*, *18*(4), 611–634. doi:10.100710763-019-09988-x

EEFEct. (2018). *Study on Ecotourism Development*. Erasmus+ EEFEct project, reference number: 2017-1- IT02-KA204-036713.

Eernstman, N., & Wals, A. E. (2013). Locative meaning-making: An arts-based approach to learning for sustainable development. *Sustainability*, *5*(4), 1645–1660. doi:10.3390u5041645

EIP-AGRI. (2020). *New Skills for Digital Farming*. Final Report. Available at: https://ec.europa.eu/eip/agriculture/en/publications/eip-agri-seminar-new-skills-digital-farming-final

Elachouri, M. (1993). Diploma of in-depth study. *Faculty of Sciences of Tunis*, *II*, 89.

Elbagermi, M. A., Edwards, H. G., & Alajtal, A. I. (2013). Monitoring of Heavy Metals Content in Soil Collected from City Centre and Industrial Areas of Misurata, Libya. *International Journal of Analytical Chemistry*, *1-5*, 1–5. Advance online publication. doi:10.1155/2013/312581 PMID:23762061

El-Hage Scialabba, N. (2012). *Current situation of organic market access. In: Let the good products flow! Global organic market access in 2012 and beyond.* Available at: http://www.fao.org/3/an455e/an455e00.pdf

Elkington, J. (1994). Towards the Sustainable Corporation: Win-Win-Win Business Strategies for Sustainable Development. *California Management Review*, *36*(2), 90–100. doi:10.2307/41165746

Elkington, J. (1998). *Cannibals with forks: The triple bottom line of 21st Century business*. New Society.

Elliott, S., & Young, T. (2016). Nature by default in early childhood education for sustainability. *Australian Journal of Environmental Education*, *32*(1), 57–64. doi:10.1017/aee.2015.44

Elmqvist, T., Zipperer, W.C., & Gii, B. (2013b). *Urbanization, habitat loss and biodiversity decline Solution pathways to break the cycle*. Academic Press.

Elmqvist, T., Fragkias, M., Goodness, J., Marcotullio, P. J., McDonald, R. I., Parnell, S., Schewenius, M., Sendstad, M., Seto, K. C., & Wilkinson, C. (2013a). *Urbanization, Biodiversity and Ecosystem Services: Challenges and Opportunities*. Springer. doi:10.1007/978-94-007-7088-1

Emanuel, R., & Adams, J. N. (2011). College students' perceptions of campus sustainability. *International Journal of Sustainability in Higher Education*, *12*(1), 79–92. doi:10.1108/14676371111098320

Engwa, Ferdinand, Nwalo, & Unachukwu. (2019). Mechanism and Health Effects of Heavy Metal Toxicity in Humans, In *Poisoning in the Modern World - New Tricks for an Old Dog?* IntechOpen. Available from: https://www.intechopen.com/books/poisoning-in-the-modern-world-new-tricks-for-an-old-dog-/mechanism-and-health-effects-of-heavy-metal-toxicity-in-humans doi:10.5772/intechopen.82511

Entwistle, N. (1988). Motivational factors in students' approaches to learning. In *Learning strategies and learning styles* (pp. 21–51). Springer. doi:10.1007/978-1-4899-2118-5_2

Erhabor, N., & Don, J. (2016). Impact of environmental education on the knowledge and attitude of students towards the environment. *International Journal of Environmental and Science Education, 11*(12), 5367–5375.

Ernst, J., & Monroe, M. (2004). The effects of environment-based education on students' critical thinking skills and disposition toward critical thinking. *Environmental Education Research, 10*(4), 507–522. doi:10.1080/1350462042000291038

Erskine, L., & Johnson, S. D. (2012). Effective learning approaches for sustainability: A student perspective. *Journal of Education for Business, 87*(4), 198–205. doi:10.1080/08832323.2011.590162

Erturgut, R. (2011). Increasing demand for logistics technician in business world and rising trend of logistics programs in higher vocational schools: Turkey case. *Procedia: Social and Behavioral Sciences, 15*, 2776–2780. doi:10.1016/j.sbspro.2011.04.187

Ertz, M., Karakas, F., & Sarigöllü, E. (2016). Exploring pro-environmental behaviors of consumers: An analysis of contextual factors, attitude, and behaviors. *Journal of Business Research, 69*(10), 3971–3980. doi:10.1016/j.jbusres.2016.06.010

Erzurumlu, S. S., & Erzurumlu, Y. O. (2015). Sustainable mining development with community using design thinking and multi-criteria decision analysis. *Resources Policy, 46*, 6–14. doi:10.1016/j.resourpol.2014.10.001

Eswaran, S., Bharathiraj, L. T., & Jayanthi, S. (2019). Modelling of ambient air quality, Coimbatore, India. E3S Web of Conferences, 117, 2. doi:10.1051/e3sconf/201911700002

Etikan, L., Musa, S. A., & Alkassim, R. S. (2015). Comparison of convenience sampling and purpose sampling. *American Journal of Theoretical and Applied Statistics*, 1–4.

European Commission (2020). *Communication from the Commission to the European Parliament, the Council, the European Economic and Social Committee and the Committee of the Regions - A Farm to Fork Strategy for a fair, healthy and environmentally-friendly food system* (COM/2020/381).

European Commission (EC). (2018). *Climate Action: Our planet, our future - fighting climate change together.* Luxembourg: Publications Office of the European Union. Retrieved June 28, 2020, from https://ec.europa.eu/clima/sites/clima/files/youth/docs/youth_magazine_en.pdf

Evans, J., Jones, R., Karvonen, A., Millard, L., & Wendler, J. (2015). Living labs and co-production: University campuses as platforms for sustainability science. *Current Opinion in Environmental Sustainability, 16*, 1–6. doi:10.1016/j.cosust.2015.06.005

Faguet, J., & Sanchez, F. (2014). Decentralization and access to social services in Colombia. *Public Choice, 160*(1-2), 227–249. doi:10.100711127-013-0077-7

Fang, W., Lien, C., Huang, Y., Han, G., Shyu, G., Chou, J., & Ng, E. (2018). Environmental Literacy on Ecotourism: A Study on Student Knowledge, Attitude, and Behavioral Intentions in China and Taiwan. *Sustainability, 10*(6), 1886. doi:10.3390u10061886

FAO. (2016). *Guidelines on urban and peri-urban forestry (No. 178).* FAO Forestry Paper. FAO.

FAO. (2017). *Greenhouse gas emissions from aquaculture – A life cycle assessment of three Asian Systems.* FAO Fisheries and Aquaculture Technical Paper 609.

FAO. (2020). *Sustainable Food and Agriculture.* Available at: http://www.fao.org/sustainability/en/

Farias, G., Farias, C. M., & Fairfield, K. D. (2010). Teacher as judge or partner: The dilemma of grades versus learning. *Journal of Education for Business, 85*(6), 336–342. doi:10.1080/08832321003604961

Fauzi, Huda, M., Teh, K.S.M., Haron, Z., Ripin, M.N., Hehsan, A., Abas, H., Hehsan, M.R., Irawan, J., & Maseleno, A. (2018). The Design of Fuzzy Expert System Implementation for Analyzing Transmissible Disease of Human. *International Journal of Pharmaceutical Research, 10*(4), 68–78.

Fauzi, I., Irviani, R., Muslihudin, M., Satria, F., Huda, M., Kamenez, N. V., & Maseleno, A. (2019). Revolutionizing Education through Artificial Intelligence: Fuzzy Multiple Attribute Decision Making Approach for Determining the Best Vocational High School. *Applied Mechanics and Materials, 892*, 234–239. doi:10.4028/www.scientific.net/AMM.892.234

Feddema, J. J., Oleson, K. W., Bonan, G. B., Mearns, L. O., Buja, L. E., Meehl, G. A., & Washington, W. M. (2005). The importance of land-cover change in simulating future climates. *Science, 310*(5754), 1674–1678. doi:10.1126cience.1118160 PMID:16339443

Fennell, D. (2008). *Ecotourism* (3rd ed.). Routledge.

Fielding, K. S., Louis, W. R., Warren, C., & Thompson, A. (2011). Understanding household attitudes and behaviours toward waste, water and energy conservation. *Urban Consumption*, 199-214.

Fielding, K. S., & Head, B. W. (2012). Determinants of young Australians' environmental actions: The role of responsibility attributions, locus of control, knowledge and attitudes. *Environmental Education Research, 18*(2), 171–186. doi:10.1080/13504622.2011.592936

Fisman, L. (2005). The Effects of Local Learning on Environmental Awareness in Children: An Empirical Investigation. *The Journal of Environmental Education, 36*(3), 39–50. doi:10.3200/JOEE.36.3.39-50

Fitrian, Y., Huda, M., Muhtar, A., Arifin, A. Y., Musa, N., Teh, M., ... Maseleno, A. (2019). Application Design for Determining Suitable Cosmetics with the Facial Skin Type Using Fuzzy Logic Approach. *Journal of Computational and Theoretical Nanoscience, 16*(5-6), 2153–2158. doi:10.1166/jctn.2019.7866

Florini, A., & Pauli, M. (2018). Collaborative governance for the sustainable development goals. *Asia & the Pacific Policy Studies, 5*(3), 583–598. doi:10.1002/app5.252

Font, X., & Mccabe, S. (2017). Sustainability and Marketing in Tourism: Its Contexts, Paradoxes, Approaches, Challenges and Potential. *Journal of Sustainable Tourism, 25*(7), 869–883. doi:10.1080/09669582.2017.1301721

Food and Agricultural Organization of United Nations. (2010). *A New Deal for School Gardens.* UN Report on School Gardens. FAO.

France 24. (2015, October 13). *Morocco's desert oases now home to eco-friendly toilets.* The Observers France 24. Retrieved December 21, 2020, from https://observers.france24.com/en/20151013-morocco-oasis-toilet-waste-fertilizer

Francis, C. A., Jordan, N., Porter, P., Breland, T. A., Lieblein, G., Salomonsson, L., Sriskandarajah, N., Wiedenhoeft, M., DeHaan, R., Braden, I., & Langer, V. (2011). Innovative Education in Agroecology: Experiential Learning for a Sustainable Agriculture. *Critical Reviews in Plant Sciences, 30*(1-2), 226–237. doi:10.1080/07352689.2011.554497

Freibauer, A., Rounsevell, M. D. A., Smith, P., & Verhagen, J. (2004). Carbon sequestration in the agricultural soils of Europe. *Geoderma, 122*(1), 1–23. doi:10.1016/j.geoderma.2004.01.021

Freire, P. (2005). *Pedagogy of the oppressed* (30th anniversary ed.). The Continuum Publishing Company.

Freire, P. (2008). *Education for critical consciousness* (3rd ed.). Continuum.

Galaitsi, S., Russell, R., Bishara, A., Durant, J., Bogle, J., & Huber-Lee, A. (2016). Intermittent domestic water supply: A critical review and analysis of causal-consequential pathways. *Water (Basel)*, *8*(7), 274. doi:10.3390/w8070274

Gandhimathi, A., & Meenambal, T. (2012). Spatial Prediction of Heavy Metal Pollution for Soils based on ANN and Kriging Model. *International Journal of Computers and Applications*, *29*(10), 52–63.

Garay, L., Font, X., & Pereira-Moliner, J. (2017). Understanding sustainability behaviour: The relationship between information acquisition, proactivity and performance. *Tourism Management*, *60*, 418–429. doi:10.1016/j.tourman.2016.12.017

Gasparik, J., Massányi, P., Slamecka, J., Fabis, M., & Jurcik, R. (2004). Concentration of Selected Metals in Liver, Kidney, and Muscle of the Red Deer (*Cervus elaphus*). *Journal of Environmental Science and Health. Part A, Toxic/Hazardous Substances & Environmental Engineering*, *39*(8), 2105–2111. doi:10.1081/ESE-120039378 PMID:15332672

Gatersleben, B., Murtagh, N., Cherry, M., & Watkins, M. (2019). Moral, wasteful, frugal, or thrifty? Identifying consumer identities to understand and manage pro-environmental behavior. *Environment and Behavior*, *51*(1), 24–49. doi:10.1177/0013916517733782

Gattinger, A., Müller, A., Haeni, M., Skinner, C., Fliessbach, A., Buchmann, N., Mäder, P., Stolze, M., Smith, P., Scialabba, N. H., & Niggli, U. (2012). Enhanced top soil carbon stocks under organic farming. *Proceedings of the National Academy of Sciences of the United States of America*, *109*(44), 18226–18231. doi:10.1073/pnas.1209429109 PMID:23071312

Geist, H. (2017). *The causes and progression of desertification*. Routledge. doi:10.4324/9781315240855

Gethke, K., Herbst, H., & Pinnekamp, J. (2007). Human urine-decomposition processes and nutrient recovery. *Gewasserschutz Wasser Abwasser*, *206*, 36.

Ghazvini, S. D., & Khajehpour, M. (2011). Gender differences in factors affecting academic performance of high school students. [). Elsevier.]. *Procedia: Social and Behavioral Sciences*, *15*, 1040–1045.

Gibbons, E. D. (2014). Climate change, children's rights, and the pursuit of intergenerational climate justice. *Health and Human Rights Journal*, *16*(1), 19–31.

Giefer, M. M., Peterson, M. N., & Chen, X. (2019). Interactions among Locus of Control, Environmental Attitudes and Pro-Environmental Behaviour in China. *Environmental Conservation*, *46*(3), 234–240. doi:10.1017/S0376892919000043

Gifford, R., & Sussman, R. (2012). *Environmental attitudes*. Academic Press.

Gifford, R., & Nilsson, A. (2014). Personal and social factors that influence pro-environmental concern and behaviour: A review. *International Journal of Psychology*, *49*(3), 141–157. doi:10.1002/ijop.12034 PMID:24821503

Givin, J. G., & Shrader, J. (2016). Temperature extremes, health, and human capital. *Children and Climate Change*, *26*(1), 31–50.

GKoumas. (2019). Evaluating a standard for sustainable tourism through the lenses of local industry. *Heliyon*, *5*, e027072.

Global Center on Adaptation. (2020). *Innovation in Education and training as a tool to enhance youth participation in adaptation action: Integration of Bottom-up and Top-down Approaches*. Rotterdam: Global Center on Adaptation and Washington: World Resources Institute. Retrieved on December 23, 2020, from http://tep-a.org/policy-briefs/

Global Commission on Adaptation. (2019). Adapt Now: A Call for Leadership on Climate Resilience. Rotterdam: Global Center on Adaptation and Washington: World Resources Institute.

Godemann, J., Haertle, J., Herzig, C., & Moon, J. (2014). United Nations supported principles for responsible management education: Purpose, progress and prospects. *Journal of Cleaner Production*, *62*, 16–23. doi:10.1016/j.jclepro.2013.07.033

Goksen, F., Adaman, F., & Zenginobuz, E. Ü. (2002). On environmental concern, willingness to pay, and postmaterialist values: Evidence from Istanbul. *Environment and Behavior*, *34*(5), 616–633. doi:10.1177/0013916502034005003

Goldman, D., Assaraf, O., & Shaharabani, D. (2013). Influence of a non-formal environmental education programme on junior high-school students' environmental literacy. *International Journal of Science Education*, *35*(3), 515–545. doi:10.1080/09500693.2012.749545

Goldman, D., Hansmann, R., Činčera, J., Radović, V., Telešienė, A., Balžekienė, A., & Vávra, J. (2020). Education for Environmental Citizenship and Responsible Environmental Behaviour. In *Conceptualizing Environmental Citizenship for 21st Century Education* (pp. 115–137). Springer. doi:10.1007/978-3-030-20249-1_8

Government of Bangladesh (GoB). (2017). *An introduction to Bangladesh*. Ministry of Foreign Affairs, Bangladesh. Retrieved June 28, 2020, from https://mofa.gov.bd/site/page/6dde350b-1ca6-4c69-becd-a3f12cf14ac1/Bangladesh--An-Introduction

Government of Bangladesh (GoB). (2019). *Statistical yearbook of Bangladesh 2019*. Bangladesh Bureau of Statistics, Ministry of Planning, Government of Bangladesh.

Govindarajalu, K. (2003). Industrial Effluent And Health Status - A Case Study Of Noyyal River Basin. In *Proceedings of the Third International Conference on Environment and Health*. Department of Geography, University of Madras and Faculty of Environmental Studies, York University.

Goyal, T., Mitra, P., Singh, P., Sharma, S., & Sharma, P. (2020). Assessement of Blood Lead and Cadmium Levels in Occupationally Exposed Workers of Jodhpur, Rajasthan. *Indian Journal of Clinical Biochemistry*. Advance online publication. doi:10.100712291-020-00878-6

Grafton, R. Q., Williams, J., Perry, C. J., Molle, F., Ringler, C., Steduto, P., Udall, B., Wheeler, S. A., Wang, Y., Garrick, D., & Allen, R. G. (2018). The paradox of irrigation efficiency. *Science*, *361*(6404), 748–750. doi:10.1126cience.aat9314 PMID:30139857

Graham, A. (2018). *Why Should Schools Teach Climate Education?* UN CC: Learn. Retrieved on December 23, 2020, from: https://medium.com/uncclearn/why-should-schools-teach-climate-education-f1e101ebc56e

Gray, T., & Birrell, C. (2015). 'Touched by the Earth': A place-based outdoor learning program incorporating the Arts. *Journal of Adventure Education and Outdoor Learning*, *15*(4), 330–349. doi:10.1080/14729679.2015.1035293

Green, C. J. (2015). Toward Young children as active researchers: A critical review of the methodologies and methods in early childhood environmental education. *The Journal of Environmental Education*, *46*(4), 207–229. doi:10.1080/00958964.2015.1050345

Grosseck, G., Tiru, L., & Bran, R. (2019). Education for sustainable development evolution and perspectives: A bibliometric review of research, 1992-2018. *Sustainability*, *11*(21), 6136. doi:10.3390u11216136

Groves, H., & Pugh, F. (1999). Elementary preservice teacher perception of the greenhouse effect. *Journal of Science and Technology*, *8*(1), 75–81.

GTZ & IWA. (2003). Ecosan - closing the loop. In *Proceedings of the 2nd international symposium*. Deutsche GesellschaftfürTechnischeZusammenarbeit (GTZ) GmbH and International Water Association (IWA).

Guthman, J. (2004). *Agrarian Dreams: The Paradox of Organic Farming in California*. University of California Press. Retrieved November 19, 2020, from http://www.jstor.org/stable/10.1525/j.ctt1pnn9k.10

Haddouchi, M. (1996). Situation actuelle et perspectives de développement du palmier dattier au Maroc. *Option méditerranéenne A, 28*, 63–79.

Haertle, J., Parkes, C., Murray, A., & Hayes, R. (2017). PRME: Building a global movement on responsible management education. *International Journal of Management Education, 15*(2), 66–72. doi:10.1016/j.ijme.2017.05.002

Hägglund, S., & Samuelsson, I. P. (2009). Early childhood education and learning for sustainable development and citizenship. *International Journal of Early Childhood, 41*(2), 49–63. doi:10.1007/BF03168878

Haight, A. (2015). *Exploring teachers' perception of environmental education preparation experiences on their incorporation of environmental education in the profession* (Unpublished Doctoral dissertation). Northern Kentucky University, Highland Heights, KY.

Haines, A., Kovats, R. S., Campbell, D., & Corvalan, C. (n.d.). Climate Change and Human Health- Impacts, vulnerability, and public health. *The Journal of the Royal Institute of Public Health and Hygiene.*

Halimoon, N., Putra, U., & Halimoon, N. (2010). Removal of Heavy Metals from Textile Wastewater using Zeolite. *Environment Asia, 3*(Special issues), 124–130. doi:10.14456/ea.2010.51

Haller, L., Moakes, S., Niggli, U., Riedel, J., Stolze, M., & Thompson, M. (2020). *Entwicklungsperspektiven der ökologischen Landwirtschaft in Deutschland.* UBA Texte 32/2020. https://www.umweltbundesamt.de/publikationen/entwicklungsperspektiven-der-oekologischen

Hamid, A., Sudrajat, A., Kawangit, R. M., Don, A. G., Huda, M., Jalal, B., ... Maseleno, A. (2018). Determining basic food quality using SAW. *IACSIT International Journal of Engineering and Technology, 7*(4), 3548–3555.

Harness, H., & Drossman, H. (2011). The environmental education through filmmaking project. *Environmental Education Research, 17*(6), 829–849. doi:10.1080/13504622.2011.618626

Hart, E. (1981). Identification of key characteristics of environmental education. *The Journal of Environmental Education, 13*(1), 12–16. doi:10.1080/00958964.1981.9943018

Harvey, G. (1977). A conceptualization of environmental education. In J. Aldrich, A. Blackburn, & G. Abel (Eds.), *A report on the North American regional seminar on environmental education* (pp. 66-72). Columbus, OH: ERIC Clearinghouse for Science, Mathematics and Environmental Education.

Haryono, A., Soemarno, Djati, M. S., & Setyoleksono, A. (2014). Learning attitude and awareness against students in cultured environmental success in Probolinggo. *Journal of Environment and Earth Science, 4*(16), 72–81.

Hazra, P., Ghosh, S.K., Maity, T.K., Pandit, M.K., & Som, M.G. (2010). *Glossary of Horticulture*. Academic Press.

Heinonen-Tanski, H., & van Wijk-Sijbesma, C. (2005). Human excreta for plant production. *Bioresource Technology, 96*(4), 403–411. doi:10.1016/j.biortech.2003.10.036 PMID:15491820

Henderson, S., & Holman, S. R. (1993). Global climate change education: Technology transfer to schools. *Climate Research, 3*(1-2), 137–140. doi:10.3354/cr003137

Henry, M., Bombelli, A., Trotta, C., Alessandrini, A., Birigazzi, L., Sola, G., Vieilledent, G., Santenoise, P., Longuetaud, F., Valentini, R., Picard, N., & Saint-André, L. (2013). GlobAllomeTree: International platform for tree allometric equations to support volume, biomass and carbon assessment. *IForest (Viterbo), 6*(6), 326–330. doi:10.3832/ifor0901-006

Herrera-Cano, C. (2016). Disaster risk management in business education entrepreneurial formation for corporate sustainability. *AD-Minister*, (28), 33–48. doi:10.17230/ad-minister.28.2

Hesselbarth, C., & Schaltegger, S. (2014). Educating change agents for sustainability - Learnings from the first sustainability management master of business administration. *Journal of Cleaner Production*, *62*, 24–36. doi:10.1016/j.jclepro.2013.03.042

Higgins-Desbiolles, F., Whyte, K. P., & Tedmanson, D. J. (2013). Tourism and environmental justice. In K. Schwab & D. Dustin (Eds.), Just leisure: Things that we believe. Academic Press.

Holden, A., & Sparrowhawk, J. (2002). Understanding the Motivations of Ecotourists: The case of trekkers in Annapurna, Nepal. *International Journal of Tourism Research*, *4*(6), 435–446. doi:10.1002/jtr.402

Hollweg, K. S., Taylor, J. R., Bybee, R. W., Marcinkowski, T. J., McBeth, W. C., & Zoido, P. (2011). *Developing a framework for assessing environmental literacy*. North American Association for Environmental Education.

Hollweg, K. S., Taylor, J., Bybee, R. W., Marcinkowski, T. J., McBeth, W. C., & Zoido, P. (2011). *Developing a Framework for Assessing Environmental Literacy*. North American Association for Environmental Education. Retrieved from http://www.naaee.net/framework

Houghton, J. T., Meira Filho, L. G., Callander, B. A., Harris, N., Kattenberg, A., & Maskel, K. (1996). *Climate change 1995: The science of climate change*. Cambridge University Press.

Howlett, C., Ferreira, J. A., & Blomfield, J. (2016). Teaching sustainable development in higher education: Building critical, reflective thinkers through an interdisciplinary approach. *International Journal of Sustainability in Higher Education*, *17*(3), 305–321. doi:10.1108/IJSHE-07-2014-0102

Hsieh, H. F., & Shannon, S. E. (2005). Three approaches to qualitative content analysis. *Qualitative Health Research*, *15*(9), 1277–1288. doi:10.1177/1049732305276687 PMID:16204405

Hsu, S. (2004). The Effects of an Environmental Education Program on Responsible Environmental Behavior and Associated Environmental Literacy Variables in Taiwanese College Students. *The Journal of Environmental Education*, *35*(2), 37–48. doi:10.3200/JOEE.35.2.37-48

Huang, T. C., Lin, W., & Yueh, H. P. (2019). How to Cultivate an Environmentally Responsible Maker? A CPS Approach to a Comprehensive Maker Education Model. *International Journal of Science and Mathematics Education*, *17*(S1), 1–16. doi:10.100710763-019-09959-2

Huda, M., Ulfatmi, Luthfi, M.J., Jasmi, K.A., Basiron, B., Mustari, M.I., Safar, A., Embong, H.W.H., Mohamad, A.M., & Mohamed, A.K. (2019d). Adaptive online learning technology: Trends in big data era. In Diverse Learning Opportunities Through Technology-Based Curriculum Design. (pp.163-195), Hershey, PA: IGI Global. doi:10.4018/978-1-5225-5519-3.ch008

Huda, M. (2019). Empowering application strategy in the technology adoption: Insights from professional and ethical engagement. *Journal of Science and Technology Policy Management*, *10*(1), 172–192. doi:10.1108/JSTPM-09-2017-0044

Huda, M. (2021). (in press). Towards Professional and Ethical Balance: Insights into Application Strategy on Media Literacy Education. *Kybernetes*.

Huda, M., Almunawar, M. N., Hananto, A. L., Rismayadi, B., Jasmi, K. A., Basiron, B., & Mustari, M. I. (2018a). Strengthening Quality Initiative for Organization Stability: Insights From Trust in Cyberspace-Based Information Quality. In *Cases on Quality Initiatives for Organizational Longevity* (pp. 140–169). IGI Global. doi:10.4018/978-1-5225-5288-8.ch006

Huda, M., Anshari, M., Almunawar, M. N., Shahrill, M., Tan, A., Jaidin, J. H., ... Masri, M. (2016). Innovative Teaching in Higher Education: The Big Data Approach. *The Turkish Online Journal of Educational Technology*, *15*(Special issue), 1210–1216.

Huda, M., Haron, Z., Ripin, M. N., Hehsan, A., & Yaacob, A. B. C. (2017a). Exploring Innovative Learning Environment (ILE): Big Data Era. *International Journal of Applied Engineering Research: IJAER*, *12*(17), 6678–6685.

Huda, M., & Hashim, A. (2021). (in press). Smart Learning Environment (SLE) in Fourth Industrial Revolution (IR 4.0). *International Journal of Asian Business and Information Management.*

Huda, M., Hehsan, A., Basuki, S., Rismayadi, B., Jasmi, K. A., Basiron, B., & Mustari, M. I. (2019a). Empowering Technology Use to Promote Virtual Violence Prevention in Higher Education Context. In *Intimacy and Developing Personal Relationships in the Virtual World* (pp. 272–291). IGI Global. doi:10.4018/978-1-5225-4047-2.ch015

Huda, M., Jasmi, K. A., Alas, Y., Qodriah, S. L., Dacholfany, M. I., & Jamsari, E. A. (2017b). Empowering Civic Responsibility: Insights From Service Learning. In S. Burton (Ed.), *Engaged Scholarship and Civic Responsibility in Higher Education* (pp. 144–165). IGI Global. doi:10.4018/978-1-5225-3649-9.ch007

Huda, M., Jasmi, K. A., Basiran, B., Mustari, M. I. B., & Sabani, A. N. (2017c). Traditional Wisdom on Sustainable Learning: An Insightful View From Al-Zarnuji's Ta 'lim al-Muta 'allim. *SAGE Open*, *7*(1), 1–8. doi:10.1177/2158244017697160

Huda, M., Jasmi, K. A., Embong, W. H., Safar, J., Mohamad, A. M., Mohamed, A. K., Muhamad, N. H., Alas, Y., & Rahman, S. K. (2017d). Nurturing Compassion-Based Empathy: Innovative Approach in Higher Education. In M. Badea & M. Suditu (Eds.), *Violence Prevention and Safety Promotion in Higher Education Settings* (pp. 154–173). IGI Global. doi:10.4018/978-1-5225-2960-6.ch009

Huda, M., Jasmi, K. A., Hehsan, A., Shahrill, M., Mustari, M. I., Basiron, B., & Gassama, S. K. (2017e). Empowering Children with Adaptive Technology Skills: Careful Engagement in the Digital Information Age. *International Electronic Journal of Elementary Education*, *9*(3), 693–708.

Huda, M., Jasmi, K. A., Mustari, M. I., Basiron, B., Mohamed, A. K., Embong, W., ... Safar, J. (2017f). Innovative E-Therapy Service in Higher Education: Mobile Application Design. *International Journal of Interactive Mobile Technologies*, *11*(4), 83–94. doi:10.3991/ijim.v11i4.6734

Huda, M., Maseleno, A., Jasmi, K. A., Mustari, I., & Basiron, B. (2017g). Strengthening Interaction from Direct to Virtual Basis: Insights from Ethical and Professional Empowerment. *International Journal of Applied Engineering Research: IJAER*, *12*(17), 6901–6909.

Huda, M., Maseleno, A., Muhamad, N. H. N., Jasmi, K. A., Ahmad, A., Mustari, M. I., & Basiron, B. (2018b). Big Data Emerging Technology: Insights into Innovative Environment for Online Learning Resources. *International Journal of Emerging Technologies in Learning*, *13*(1), 23–36. doi:10.3991/ijet.v13i01.6990

Huda, M., Maseleno, A., Teh, K. S. M., Don, A. G., Basiron, B., Jasmi, K. A., Mustari, M. I., Nasir, B. M., & Ahmad, R. (2018c). Understanding Modern Learning Environment (MLE) in Big Data Era. *International Journal of Emerging Technologies in Learning*, *13*(5), 71–85. doi:10.3991/ijet.v13i05.8042

Huda, M., Muhamad, N. H. N., Isyanto, P., Kawangit, R. M., Marni, N., Mohamed, A. K., & Safar, A. J. (2020). *Building Harmony in Diverse Society: Insights from Practical Wisdom.* International Journal of Ethics and Systems. doi:10.1108/IJOES-11-2017-0208

Huda, M., Mulyadi, D., Hananto, A. L., Nor Muhamad, N. H., Mat Teh, K. S., & Don, A. G. (2018d). Empowering corporate social responsibility (CSR): Insights from service learning. *Social Responsibility Journal*, *14*(4), 875–894. doi:10.1108/SRJ-04-2017-0078

Huda, M., Qodriah, S. L., Rismayadi, B., Hananto, A., Kardiyati, E. N., Ruskam, A., & Nasir, B. M. (2019c). Towards Cooperative with Competitive Alliance: Insights into Performance Value in Social Entrepreneurship. In *Creating Business Value and Competitive Advantage with Social Entrepreneurship* (p. 294). IGI Global. doi:10.4018/978-1-5225-5687-9.ch014

Huda, M., Sabani, N., Shahrill, M., Jasmi, K. A., Basiron, B., & Mustari, M. I. (2017h). Empowering Learning Culture as Student Identity Construction in Higher Education. In A. Shahriar & G. Syed (Eds.), *Student Culture and Identity in Higher Education* (pp. 160–179). IGI Global. doi:10.4018/978-1-5225-2551-6.ch010

Huda, M., Shahrill, M., Maseleno, A., Jasmi, K. A., Mustari, I., & Basiron, B. (2017i). Exploring Adaptive Teaching Competencies in Big Data Era. *International Journal of Emerging Technologies in Learning*, *12*(3), 68–83. doi:10.3991/ijet.v12i03.6434

Huda, M., Sudrajat, S., Kawangit, R. M., Teh, K. S. M., & Jalal, B. (2019b). Strengthening divine values for self-regulation in religiosity: Insights from Tawakkul (trust in God). *International Journal of Ethics and Systems*, *35*(3), 323–344. doi:10.1108/IJOES-02-2018-0025

Huda, M., & Teh, K. S. M. (2018). Empowering Professional and Ethical Competence on Reflective Teaching Practice in Digital Era. In K. Dikilitas, E. Mede, & D. Atay (Eds.), *Mentorship Strategies in Teacher Education* (pp. 136–152). IGI Global. doi:10.4018/978-1-5225-4050-2.ch007

Huda, M., Teh, K. S. M., Nor, N. H. M., & Nor, M. B. M. (2018e). Transmitting Leadership Based Civic Responsibility: Insights from Service Learning. *International Journal of Ethics and Systems*, *34*(1), 20–31. doi:10.1108/IJOES-05-2017-0079

Hülsbergen, K.-J., & Rahmann, G. (2015). Klimawirkungen und Nachhaltigkeit ökologischer und konventioneller Betriebssysteme - Untersuchungen in einem Netzwerk von Pilotbetrieben: Forschungsergebnisse 2013 - 2014. Johann Heinrich von Thünen-Institut.

Hvenegaard, G. T. (2002). Using tourist typologies for Ecotourism research. *Journal of Ecotourism*, *1*(1), 70–18. doi:10.1080/14724040208668109

ICLEI. (2016, March). *Rajkot, India Enhancing 'Livability' through Urban Low Emission Development.* Retrieved Sept 2020, from ICLEI: https://urban-leds.org/wp-content/uploads/2019/resources/case_studies/ICLEI_cs_175-Rajkot-UrbanLEDS_2016-web.pdf

IFOAM. (2017).*Guidelines for public support to organic agriculture.* Available at: https://www.ifoam.bio/sites/default/files/2020-05/policy_toolkit_main_report.pdf

IFOAM. (2020). *Organic Academy by IFOAM Organics International.* Available at: https://www.ifoam.bio/our-work/how/training

Igbokwe, B. (2016). *Environmental literacy assessment: assessing the strength of an environmental education program (EcoSchools) in Ontario secondary schools for environmental literacy acquisition* (PhD Thesis). University of Windsor.

Independent Commission on Environmental Education. (1997). *Are we building environmental literacy?* Author.

INS. (2016). Le Niger en Chiffres 2016. Niamey.

INS. (2017). Annuaire Statistique Regional 2012 - 2016. Niamey.

Intergovernmental Panel on Climate Change (IPCC). (2006). *2006 IPCC Guidelines for National Greenhouse Inventories – A primer.* Prepared by the National Greenhouse Gas Inventories Programme, Intergovernmental Panel on Climate Change National Greenhouse Gas Inventories Programme.

Intergovernmental Panel on Climate Change (IPCC). (2014). *Climate Change 2014: Synthesis Report. Contribution of Working Groups I, II and III to the Fifth Assessment Report of the Intergovernmental Panel on Climate Change, Core Writing Team, R.K. Pachauri and L.A. Meyer. Geneva, Switzerland.* doi:10.1017/CBO9781107415324.004

Intergovernmental Panel on Climate Change (IPCC). (2014). *Glossary. Intergovernmental Panel on Climate Change.* Retrieved on May 2, 2020, from https://www.ipcc.ch/site/assets/uploads/2018/02/WGIIAR5-AnnexII_FINAL.pdf

International Alliance of Leading Education Institutes (IALEI). (2009). *Climate change and sustainable development: The response from education.* Available at: https://www.intlalliance.org/fileadmin/user_upload/documents/DPU_recommendations.pdf

International Union for Conservation of Nature (IUCN). (1970). *International Working Meeting on Environmental Education in the School Curriculum, Final Report.* Gland, Switzerland: IUCN.

Ioannides, D., & Debbage, K. G. (1998). Neo-Fordism and Flexible Specialization. In D. Ioannides & K. G. Debbage (Eds.), *The Economic Geography of the Tourist Industry*. Routledge.

IPCC. (2014). *Mitigation of Climate Change. Contribution of Working Group III to the Fifth Assessment Report of the Intergovernmental Panel on Climate Change.* Cambridge University Press.

IPCC. (2019). *Special Report on Climate Change and Land – Summary for Policy Makers.* Available at: https://www.ipcc.ch/srccl/chapter/summary-for-policymakers/

Islam, T., Rahman, M. A., & Mallick, F. (2010). Impacts of climate change induced hazards and adaptation processes in Bangladesh: An overview. In *Climate Change Adaptation and Disaster Risk Reduction: An Asian Perspective*. Emerald Group Publishing Limited. doi:10.1108/S2040-7262(2010)0000005008

Issa, T. (2017). Sustainability Perspective and Awareness Amongst Higher Education in Australia. *Sustainability, Green IT and Education Strategies in the Twenty-first Century*, 17-33.

Issaka, H. (2013). Privilège urbain:conditions de vie et santé au Niger. *Revue-Urbanites*, 42–52.

Issa, T., Isaias, P., & Issa, T. (Eds.). (2017). *Sustainability, Green IT and Education Strategies in the Twenty-first Century.* Springer. doi:10.1007/978-3-319-57070-9

Iuliana, M., Cornelia, P., Ioan, B., & Anda, M. I. (2017). Issues Concerning the Classification of Ecotourism Destinations. *Lucrari Stiintifice*, *9*(1), 167–172.

Jafari, J. (2000). *Encyclopedia of Tourism*. Routledge.

Jaishankar, M., Tseten, T., Anbalagan, N., Mathew, B. B., & Beeregowda, K. N. (2014). Toxicity, mechanism and health effects of some heavy metals. *Interdisciplinary Toxicology*, *7*(2), 60–72. doi:10.2478/intox-2014-0009 PMID:26109881

Jaishree, & Khan, T. I. (2014). Monitoring of Heavy Metal in Textile Waste Water of Sanganer, Jaipur (Rajasthan). *International Journal of Scientific and Research Publications*, *4*(3), 1–4.

Jaiwan, S., & Ajay, K. S. (2011). Effects of Heavy Metals on Soil, Plants Human Health and Aquatic Life. *International Journal of Research in Chemistry and Environment*, *1*(2), 15–21.

Jamir & Amba. (2004). *Environmental education in school system-overview and challenges.* UNESCO/JAPAN Asia pacific environmental education research seminar.

Janaranjani, M., Varunprasath, K., Priya, R., Sutharsan, L., & Lakeshmanaswamy, M. (2017). Status of Wetlands in the Coimbatore District, Tamilnadu, India. *Asian Journal of Environment & Ecology*, *4*(4), 1–12. doi:10.9734/AJEE/2017/36421

Jefferson, G. M., Ciro, P. M., & María Andrea, M. S. (2016). Environmental education and the Bogotá River: An intervention to be carried out in Cundinamarca (Colombia). *International Research in Geographical and Environmental Education*, *26*(4), 281–296. doi:10.1080/10382046.2016.1262510

Jenelapriscy, J., & Jesvinshobini, S., & ReyaIssac. (2017). Heavy Metal Pollution in Wetlands aroundCoimbatore. *International Journal of Chemtech Research*, *10*(10), 119–128.

Jennings, N., Swidler, S., & Koliba, C. (2005). Place-based education in the standards-based reform era—Conflict or complement? *American Journal of Education*, *112*(1), 44–65. doi:10.1086/444522

Jennings, S. B., Brown, N. D., & Sheil, D. (1999). Assessing forest canopies and understorey illumination : Canopy closure, canopy cover and other measures. *Forestry*, *72*(1), 72. doi:10.1093/forestry/72.1.59

Jinyang, D., & Jian, L. (2015). Self-identification of ecotourists. *Journal of Sustainable Tourism*, 255–279.

Johnson-Pynn, S., & Johnson, L. (2005). Successes and challenges in East African conservation education. *The Journal of Environmental Education*, *36*(2), 25–39. doi:10.3200/JOEE.36.2.25-39

Johnson, R. M., Henderson, S., Gardiner, L., Russell, R., Ward, D., Foster, S., Meymaris, K., Hatheway, B., Carbone, L., & Eastburn, T. (2008). Lessons learned through our climate change professional development program for middle and high school teachers. *Physical Geography*, *29*(6), 500–511. doi:10.2747/0272-3646.29.6.500

Jonassen, D. H., Peck, K. L., & Wilson, B. G. (1998). *Learning with technology: A constructivist perspective*. Prentice-Hall.

Jonassen, D. H., & Reeves, T. C. (1996). Learning with technology: Using computers as cognitive tools. In D. H. Jonassen (Ed.), *Handbook of research for educational communications and technology* (pp. 693–719). Macmillan.

Jones, C. J. (2010). *Earth reinforcement and soil structures*. T. Telford.

Jony & Fei. (2012). *Climate change, Mainstreaming gender and climate change in Nepal*. International Institute for Environment and Development. https://www.jstor.org/stable/resrep01234.6

Jose, P. (2016). Sustainability education in indian business schools: A status review. *AD-Minister*, *28*(28), 255–272. doi:10.17230/ad-minister.28.13

Joyner, L., Lackey, N. Q., & Bricker, K. (2019). Community Engagement: An Appreciative Inquiry Case Study with Theodore Roosevelt National Park Gateway Communities. *Sustainability*, *11*(24), 7147. doi:10.3390u11247147

Jung, R., & Schmidtke, K. (2019). Bodenfruchtbarkeit. In Leistungen des ökologischen Landbaus für Umwelt und Gesellschaft (Vol. 65). Braunschweig: Johann Heinrich von Thünen-Institut.

Kanene, K. (2016). The impact of environmental education on the environmental perceptions/attitudes of students in selected secondary schools of Botswana. *European Journal of Alternative Education Studies*, *1*(2), 1–19.

Kannojia, P., Sharma, P. K., & Sharma, K. (2019). Climate Change and Soil Dynamics: Effects on Soil Microbes and Fertility of Soil. In Climate Change and Agricultural Ecosystems (pp. 43-64). Woodhead Publishing. doi:10.1016/B978-0-12-816483-9.00003-7

Kanthimathi, L. (2017, July 14). Government Schools raise kitchen gardens. *The Hindu*.

Kapp, K. W. (2018). Environmental disruption: General issues and methodological problems. In *Green Accounting* (pp. 31–48). Routledge. doi:10.4324/9781315197715-2

Karmaoui, A., Messouli, M., Khebiza, Y. M., & Ifaadassan, I. (2014). Environmental Vulnerability to climate change and anthropogenic impacts in dryland (Pilot study: Middle Draa Valley, South Morocco). *J Earth Sci Clim Change*, S11. doi:10.4172/2157-7617.S11-002

Karmaoui, A. (2019). Drought and desertification in Moroccan Pre-Sahara, Draa valleys: Exploring from the perspective of young people. *Geoenvironmental Disasters*, *6*(2), 1–13. doi:10.118640677-019-0118-8

Karmaoui, A., Minucci, G., Messouli, M., Khebiza, M. Y., Ifaadassan, I., & Babqiqi, A. (2019). Climate Change Impacts on Water Supply System of the Middle Draa Valley in South Morocco. In *Climate Change, Food Security and Natural Resource Management* (pp. 163–178). Springer. doi:10.1007/978-3-319-97091-2_8

Karns, G. L. (2005). An update of marketing student perceptions of learning activities: Structure, preferences, and effectiveness. *Journal of Marketing Education, 27*(2), 163-171. doi:10.1177/0273475305276641

Karpiak, C. P., & Baril, G. L. (2008). Moral reasoning and concern for the environment. *Journal of Environmental Psychology*, *28*(3), 203–208. doi:10.1016/j.jenvp.2007.12.001

Karpudewan, M., & Roth, W. M. (2018). Changes in primary students' informal reasoning during an environment-related curriculum on socio-scientific issues. *International Journal of Science and Mathematics Education*, *16*(3), 401–419. doi:10.100710763-016-9787-x

Karthiyayini, R., Ponnammal, N. R., & Joseph, R. (2005). Air Pollution Tolerance Index of Certain Plants of Coimbatore-Ooty Highways, Near I.T.I Area, Coimbatore, Tamilnadu. *Pollution Research*, *24*(2), 363–365.

Katz, L. G. (2010). STEM in the early years. *Early Childhood Research & Practice*, *12*(2), 11–19.

Kaur, H., & Singh, S. P. (2018). Environmentally sustainable stochastic procurement model. *Management of Environmental Quality*, *29*(3), 472–498. doi:10.1108/MEQ-04-2017-0039

Kelley, S., & Nahser, R. (2014). Developing Sustainable Strategies: Foundations, Method, and Pedagogy. *Journal of Business Ethics*, *123*(4), 631–644. doi:10.100710551-013-2014-6

Keskitalo, P. (2019). Place and space in Sámi education. *Policy Futures in Education*, *17*(4), 560–574. doi:10.1177/1478210319848530

Khan, S., Cao, Q., Zheng, Y., Huang, Y., & Zhu, Y. (2008). Health risks of heavy metals in contaminated soils and food crops irrigated with wastewater in Beijing, China. *Environmental Pollution*, *152*(3), 686–692. doi:10.1016/j.envpol.2007.06.056 PMID:17720286

Kholili, H. (in press). Adaptive Learning Technology Through Social Network Site (SNS)- Strategy Enhancement. *Walailak Journal of Science and Technology*.

Kilelu, C. W., Klerkx, L., & Leeuwis, C. (2013). Unravelling the role of innovation platforms in supporting co-evolution of innovation: Contributions and tensions in a smallholder dairy development programme. *Agricultural Systems*, *118*, 65–77. doi:10.1016/j.agsy.2013.03.003

Kim, H., Lee, K., Lee, B., Kim, S., & Kim, K. (2012). Comparative study of simultaneous removal of As, Cu, and Pb using different combinations of electrokinetics with bioleaching by Acidithiobacillus ferrooxidans. *Water Research*, *46*(17), 5591–5599. doi:10.1016/j.watres.2012.07.044 PMID:22921395

Kim, S. Y., & Hamdan Alghamdi, A. K. (2019). Female secondary students' and their teachers' perceptions of science learning environments within the context of science education reform in Saudi Arabia. *International Journal of Science and Mathematics Education*, *17*(8), 1–23. doi:10.100710763-018-09946-z

Kinslow, A. T., Sadler, T. D., & Nguyen, H. T. (2019). Socio-scientific reasoning and environmental literacy in a field-based ecology class. *Environmental Education Research*, *25*(3), 388–410. doi:10.1080/13504622.2018.1442418

Kiper, T. (2013). Role of Ecotourism in Sustainable Development. In M. Özyavuz (Ed.), *Advances in Landscape Architetures*. doi:10.5772/55749

Kisi, N. (2019). A Strategic Approach to Sustainable Tourism Development using the A'WOT Hybrid Method: A case study of Zonguldak, Turkey. *Sustainability*, *11*(4), 964. doi:10.3390u11040964

Knierim, A., Garster-Bentaya, M., Mekonnen Birke, F., Bae, S., & Kelly, T. (2020). *Innovation advisors for interactive innovation process: Conceptual grounds and common understandings*. i2connect project. https://i2connect-h2020.eu/resources/library/

Kobayashi, T. (1991). A suggestion about environmental education using the five senses. *Marine Pollution Bulletin*, *23*, 623–626. doi:10.1016/0025-326X(91)90745-E

Kohlberg, L. (1986). A current statement of some theoretical issues. In S. Modgil & C. Modgil (Eds.), *Lawrence Kohlberg: Consensus and Controversy* (pp. 485–546). Falmer.

Kollmuss, A., & Agyeman, J. (2002). Mind the Gap: Why do people act environmentally and what are the barriers to pro-environmental behaviour? Mind the Gap: why do people act environmentally and what are the barriers to. *Environmental Education Research*, *8*(3), 239–260. doi:10.1080/13504620220145401

Kontogeorgopoulos, N. (2004). Ecotourism and Mass Tourism in Southern Thailand: Spatial Interdependence, Structural Connections, and Staged Authenticity. *GeoJournal*, *61*(1), 1–11. doi:10.100710708-005-8631-6

Kopnina, H., Washington, H., Taylor, B., & Piccolo, J. J. (2018). Anthropocentrism: More than just a misunderstood problem. *Journal of Agricultural & Environmental Ethics*, *31*(1), 109–127. doi:10.100710806-018-9711-1

Korashy, H. M., & El-Kadi, A. O. (2008). Modulation of TCDD-mediated induction of cytochrome P450 1A1 by mercury, lead, and copper in human HepG2 cell line. *Toxicology In Vitro*, *22*(1), 154–158. doi:10.1016/j.tiv.2007.08.003 PMID:17889500

Kortenkamp, K. V., & Moore, C. F. (2001). Ecocentrism and anthropocentrism: Moral reasoning about ecological commons dilemmas. *Journal of Environmental Psychology*, *21*(3), 261–272. doi:10.1006/jevp.2001.0205

Koslowski, B. (1996). *Theory and evidence: The development of scientific reasoning*. MIT Press.

Koutroulis, A. G. (2019). Dryland changes under different levels of global warming. *The Science of the Total Environment*, *655*, 482–511. doi:10.1016/j.scitotenv.2018.11.215

Kovbasyuk, O., & Blessinger, P. (Eds.). (2013). *Meaning-centered education: International perspectives and explorations in higher education*. Routledge. doi:10.4324/9780203115084

Kriebel, D., Tickner, J., Epstein, P., Lemons, J., Levins, R., Loechler, E. L., ... Stoto, M. (2001). The precautionary principle in environmental science. *Environmental Health Perspectives*, *109*(9), 871–876. doi:10.1289/ehp.01109871

Kroma, M. (2006). Organic Farmer Networks: Facilitating Learning and Innovation for Sustainable Agriculture. *Journal of Sustainable Agriculture*, *28*(4), 5–28. doi:10.1300/J064v28n04_03

Kucharcikova, A., Miciak, M., Malichova, E., Durisova, M., & Tokarcikova, E. (2019). The Motivation of Students at Universities as a Prerequisite of the Education's Sustainability within the Business Value Generation Context. *Sustainability*, *11*(20), 5577. doi:10.3390u11205577

Kuhn, D., Amsel, E., O'Loughlin, M., Schauble, L., Leadbeater, B., & Yotive, W. (1988). *The development of scientific thinking skills*. Academic Press.

Kuhn, D., & Ho, V. (1980). Self-directed activity and cognitive development. *Journal of Applied Developmental Psychology*, *1*(2), 119–133. doi:10.1016/0193-3973(80)90003-9

Kulkarni, M.N. (2015). Towards Sustainable Agricultural Systems. *Terra Green, 7*(11).

Kumar, V. S., Elangovan, K., & Srimathi, N. (2019). GIS-based air pollution distribution mapping of emissions from public buses for Coimbatore Corporation, Tamil Nadu, India. *Indian Journal of Geo-Marine Sciences*, *48*(6), 947–951.

Kunwar, R. R. (2018). Tourism Education, Curriculum Spaces, Knowledge Production, and Disciplinary Pluralism. *The Gaze: Journal of Tourism and Hospitality*, *9*, 83–155. doi:10.3126/gaze.v9i0.19724

Kuruneri-Chitepo, C., & Shackleton, C. M. (2011). The distribution, abundance and composition of street trees in selected towns of the Eastern Cape, South Africa. *Urban Forestry & Urban Greening*, *10*(3), 247–254. doi:10.1016/j.ufug.2011.06.001

Lagisz, M., & Laskowski, R. (2007). Evidence for between-generation effects in carabids exposed to heavy metals pollution. *Ecotoxicology (London, England)*, *17*(1), 59–66. doi:10.100710646-007-0176-7 PMID:17940866

Langergraber, G., & Muellegger, E. (2005). Ecological Sanitation—A way to solve global sanitation problems? *Environment International*, *31*(3), 433–444. doi:10.1016/j.envint.2004.08.006 PMID:15734195

Larsen, T. A., Lienert, J., Joss, A., & Siegrist, H. (2004). How to avoid pharmaceuticals in the aquatic environment. *Journal of Biotechnology*, *113*(1-3), 295–304. doi:10.1016/j.jbiotec.2004.03.033 PMID:15380662

Larwanou, M., & Saadou, M. (2005). Biodiversity of ligneous species in semi-arid to arid zones of southwestern Niger according to anthropogenic and natural factors. *Agriculture, Ecosystems & Environment*, *105*(1-2), 267–271. doi:10.1016/j.agee.2004.03.004

Laurie, R., Nonoyama-Tarumi, Y., Mckeown, R., & Hopkins, C. (2016). Contributions of education for sustainable development (ESD) to quality education: A synthesis of research. *Journal of Education for Sustainable Development*, *10*(2), 226–242. doi:10.1177/0973408216661442

Lawton, L. (2009). Birding festivals, sustainability, and Ecotourism: An ambiguous relationship. *Journal of Travel Research*, *48*(2), 259–267. doi:10.1177/0047287509332330

Leal Filho, W., & Rayman-Bacchus, L. (2019). Sustainability Governance in Traditional Crafts Communities: A Project Proposition. In *Social Responsibility and Sustainability* (pp. 277–287). Springer. doi:10.1007/978-3-030-03562-4_14

Ledley, T. S., Rooney-Varga, J., & Niepold, F. (2017). Addressing Climate Change Through Education. In *Oxford Encyclopaedia of Environmental Science*. Oxford University Press. doi:10.1093/acrefore/9780199389414.013.56

Leedy, P., & Ormrod, J. (2001). Practical research: Planning and design (7th ed.). Upper Saddle River, NJ: Academic Press.

Lee, T. H. (2007). An Ecotourism behavioural model of national forest recreation areas in Taiwan. *International Forestry Review*, *9*(3), 771–785. doi:10.1505/ifor.9.3.771

Leicht, A., Heiss, J., & Byun, W. J. (Eds.). (2018). *Issues and trends in Education for Sustainable Development.* Retrieved Sept 2020, from UNESCO: https://bangkok.unesco.org/content/unesco-publication-issues-and-trends-education-sustainable-development

Leiserowitz, A., Smith, N., & Marlon, J. R. (2011). *American Teens' Knowledge of Climate Change. Yale Project on Climate Change*. Yale University.

Levin, K., Brandhuber, R., Freibauer, A., & Wiesinger, K. (2019). Klimaanpassung. In Leistungen des ökologischen Landbaus für Umwelt und Gesellschaft (Vol. 65). Braunschweig: Johann Heinrich von Thünen-Institut.

Liang, X., & Kim, C. (2004). Classroom assessment in a web-based institutional environment: Instructors' experiences. *Practical Assessment, Research & Evaluation*, *9*(7). Retrieved June 12, 2020, from https://scholarworks.umass.edu/pare/vol9/iss1/7/

Li, D., Zhao, L., Ma, S., Shao, S., & Zhang, L. (2019). What influences an individual's pro-environmental behavior? A literature review. *Resources, Conservation and Recycling*, *146*, 28–34. doi:10.1016/j.resconrec.2019.03.024

Liefländer, A. K., Fröhlich, G., Bogner, F. X., & Schultz, P. W. (2013). Promoting connectedness with nature through environmental education. *Environmental Education Research*, *19*(3), 370–384. doi:10.1080/13504622.2012.697545

Lim, J. M., Salido, A. L., & Butcher, D. J. (2004). Phytoremediation of lead using Indian mustard (*Brassica juncea*) with EDTA and electrodics. *Microchemical Journal*, *76*(1–2), 3–9. doi:10.1016/j.microc.2003.10.002

Lindberg, K. (1991). *Policies for maximizing Nature Tourism's Ecological and Economic benefits*. World Resources Institute.

Lin, E., & Shi, Q. (2014). Exploring individual and school-related factors and environmental literacy: Comparing U.S. and Canada using PISA 2006. *International Journal of Science and Mathematics Education*, *12*(1), 73–97. doi:10.100710763-012-9396-2

Linkov, I., Trump, B., Poinsatte-Jones, K., & Florin, M. V. (2018). Governance strategies for a sustainable digital world. *Sustainability*, *10*(2), 440. doi:10.3390u10020440

Lin, Y., Chu, H., Wu, C., Chang, T., & Chen, C. (2010). Hotspot Analysis of Spatial Environmental Pollutants Using Kernel Density Estimation and Geostatistical Techniques. *International Journal of Environmental Research and Public Health*, *8*(1), 75–88. doi:10.3390/ijerph8010075 PMID:21318015

Littledyke, M. (2004). Primary children's views on science and environmental issues: Examples of environmental cognitive and moral development. *Environmental Education Research*, *10*(2), 217–235. doi:10.1080/1350462042000198186

Liu, H. L., Willems, P., Bao, A. M., Wang, L., & Chen, X. (2016). Effect of climate change on the vulnerability of a socio-ecological system in an arid area. *Global and Planetary Change*, *137*, 1–9. doi:10.1016/j.gloplacha.2015.12.014

Liu, S., & Guo, L. (2018). Based on Environmental Education to Study the Correlation between Environmental Knowledge and Environmental Value. *Eurasia Journal of Mathematics, Science and Technology Education*, *14*(7), 3311–3319. doi:10.29333/ejmste/91246

Li, X., Lee, S., Wong, S., Shi, W., & Thornton, I. (2004). The study of metal contamination in urban soils of Hong Kong using a GIS-based approach. *Environmental Pollution*, *129*(1), 113–124. doi:10.1016/j.envpol.2003.09.030 PMID:14749075

Li, Y., Chen, Y., & Li, Z. (2019). Dry/wet pattern changes in global dryland areas over the past six decades. *Global and Planetary Change*, *178*, 184–192. doi:10.1016/j.gloplacha.2019.04.017

Lizin, S., Van Dael, M., & Van Passel, S. (2017). Battery pack recycling: Behaviour change interventions derived from an integrative theory of planned behaviour study. *Resources, Conservation and Recycling*, *122*, 66–82. doi:10.1016/j.resconrec.2017.02.003

Lokeshwari, H., & Chandrappa, G. (2006). Impact of heavy metal contamination of Bellandur Lake on soil and cultivated vegetation. *Current Science*, *91*(5), 622–627.

López-Mosquera, N., Lera-López, F., & Sánchez, M. (2015). Key factors to explain recycling, car use and environmentally responsible purchase behaviors: A comparative perspective. *Resources, Conservation and Recycling*, *99*, 29–39. doi:10.1016/j.resconrec.2015.03.007

Lorenzoni, I., Nicholson-Cole, S., & Whitmarsh, L. (2007). Barriers perceived to engaging with climate change among the UK public and their policy implications. *Global Environmental Change*, *17*(3-4), 445–459. doi:10.1016/j.gloenvcha.2007.01.004

Loria, E. (2018). The Great Inequity of Climate Change. *One Natural Energy*. Retrieved on December 19, 2020, from https://www.onlynaturalenergy.com/the-great-inequity-of-climate-change/

Lotfy, S., & Mostafa, A. (2014). Phytoremediation of contaminated soil with cobalt and chromium. *Journal of Geochemical Exploration*, *144*, 367–373. doi:10.1016/j.gexplo.2013.07.003

Loughland, T., Reid, A., Walker, K., & Petocz, P. (2003). Factors influencing young people's conceptions of environment. *Environmental Education Research*, *9*(1), 3–19. doi:10.1080/13504620303471

Lueg, R., & Radlach, R. (2016). Managing sustainable development with management control systems: A literature review. *European Management Journal*, *34*(2), 158–171. doi:10.1016/j.emj.2015.11.005

Lüthi, C., Panasar, A., Schütze, T., Norström, A., McConville, J., & Saywell, D. (2010). *Sustainable Sanitation in Cities: A Framework for Action.* Available on http://publications.lib.chalmers.se/records/fulltext/192469/local_192469.pdf

Lyth, A., Nichols, S., & Tilbury, D. (2007). *Shifting towards sustainability: Education for climate change adaptation in the built environment sector.* A report prepared by the Australian Research Institute in Education for Sustainability. Retrieved on April 15, 2020, from http://aries.mq.edu.au/projects/ClimateChange/files/climateChange.pdf

Mackey, A., & Gass, S. M. (2005). *Second language research: Methodology and design.* Lawrence Erlbaum Associates Publishers.

Magurran, A.E. (2004). *Measuring biological diversity: Rejoinder, Environmental and Ecological Statistics.* Blackwell Science Ltd. doi:10.2989/16085910409503825

Mahat, H., Saleh, Y., Hashim, M., & Nayan, N. (2016). Model development on awareness of education for sustainable schools development in Malaysia. *The Indonesian Journal of Geography*, *48*(1), 37–46. doi:10.22146/ijg.12446

Makrakis, V., & Kostoulas-Makrakis, N. (2012). The challenges of ICTs to online climate change education for sustainable development: The ExConTra learning paradigm. In S. A. Anwar (Ed.), *Proceedings of the 5th Conference on eLearning Excellence in the Middle East – Sustainable Innovation in Education* (pp. 594–605). Dubai, UAE: Hamdan Bin Mohammed e-University.

Makrakis, V., Larios, N., & Kaliantzi, G. (2012). ICT-enabled climate change education for sustainable development across the school curriculum. *Journal of Teacher Education for Sustainability*, *14*(2), 54–72. doi:10.2478/v10099-012-0009-5

Malarkodi, M., Krishnasamy, R., Kumaraperumal, R., & Chitdeshwari, T. (2007). Characterization of Heavy Metal Contaminated Soils of Coimbatore District of Tamil Nadu. *Journal of Agronomy*, *6*(1), 147–151. doi:10.3923/ja.2007.147.151

Malik, S., Fatima, F., Imran, A., Chuah, L. F., Klemeš, J. J., Khaliq, I. H., ... Akbar, M. M. (2019). Improved project control for sustainable development of construction sector to reduce environment risks. *Journal of Cleaner Production*, *240*, 118214. doi:10.1016/j.jclepro.2019.118214

Manish kumar, B., Anusiya dev, K., Lekeshmanaswamy, M., & Vasuki, C.A. (2016). Studies on the bacterial contamination of Sanganur canal from Nanjundapuram origin to Noyyal end point, Coimbatore District, Tamilnadu, India. *Kongunadu Research Journal*, *3*(1), 70–72. doi:10.26524/krj132

Manju, A., Kalaiselvi, K., Dhananjayan, V., Palanivel, M., Banupriya, G. S., Vidhya, M. H., Panjakumar, K., & Ravichandran, B. (2018). Spatio-seasonal variation in ambient air pollutants and influence of meteorological factors in Coimbatore, Southern India. *Air Quality, Atmosphere & Health*, *11*(10), 1179–1189. doi:10.100711869-018-0617-x

Mao, X., Han, F. X., Shao, X., Guo, K., Mccomb, J., Arslan, Z., & Zhang, Z. (2016). Electro-kinetic remediation coupled with phytoremediation to remove lead, arsenic and cesium from contaminated paddy soil. *Ecotoxicology and Environmental Safety*, *125*, 16–24. doi:10.1016/j.ecoenv.2015.11.021 PMID:26650421

Marbun, D., Bachtiar, P. P., & Mawardi, S. (2015). *Iklim Usaha di Kabupaten Timor Tengah Utara (TTU): Kajian Kondisi Perekonomian dan Regulasi Usaha*. SMERU Research Institute.

Marcinkowski, T. (2009). Contemporary challenges and opportunities in environmental education: Where are we headed and what deserves our attention? *The Journal of Environmental Education*, *41*(1), 34–54. doi:10.1080/00958960903210015

Marjaine, S., Zsoka, A. & Széchy, A. (2011). Consumer behaviour and lifestyle patterns of Hungarian students with regard to environmental awareness. *Society and Economy, 33*(1), 89-110.

Martins, L. L., & Kellermanns, F. W. (2004). A model of business school students' acceptance of a web-based course management system. *Academy of Management Learning & Education*, *3*(1), 7–26. doi:10.5465/amle.2004.12436815

Marzuki, M., Wan Hussin, W. N., Amin, N. M., & Huda, M. (2020). The Sustainability of Passenger Boat Services in Malaysia. *Solid State Technology*, *63*(5), 4361–4372.

Maseleno, A., Pardimin, Huda, M., Ramlan, Hehsan, A., Yusof, Y.M., Haron, Z., Ripin, M.N., Nor, N.H.M., & Junaidi, J. (2018b). Mathematical Theory of Evidence to Subject Expertise Diagnostic. *ICIC Express Letters*, *12*(4), 369. doi:10.24507/icicel.12.04.369

Maseleno, A., Huda, M., Jasmi, K. A., Basiron, B., Mustari, I., Don, A. G., & Ahmad, R. (2018a). *Hau-Kashyap approach for student's level of expertise*. Egyptian Informatics Journal., doi:10.1016/j.eij.2018.04.001

Maseleno, A., Sabani, N., Huda, M., Ahmad, R., Jasmi, K. A., & Basiron, B. (2018c). Demystifying Learning Analytics in Personalised Learning. *IACSIT International Journal of Engineering and Technology*, *7*(3), 1124–1129. doi:10.14419/ijet.v7i3.9789

Maseleno, A., Shankar, K., Huda, M., Othman, M., Khoir, P., & Muslihudin, M. (2019). *Citizen Economic Level (CEL) using SAW. In Expert Systems in Finance: Smart Financial Applications in Big Data Environments, 97*. Routledge.

Masindi & Muedi. (2018). Environmental Contamination by Heavy Metals. *Heavy Metals*. Available from: https://www.intechopen.com/books/heavy-metals/environmental-contamination-by-heavy-metals doi:10.5772/intechopen.76082

Mathew, M., Kumar, M. S., Azeez, P. A., Sivakumar, R., & Pattabi, S. (2002). Sediment Quality of Wetlands in Coimbatore, Tamilnadu, India. *Bulletin of Environmental Contamination and Toxicology*, *68*(3), 389–393. doi:10.1007001280266 PMID:11993814

Matthew McConnachie, M., Shackleton, C. M., & McGregor, G. K. (2008). The extent of public green space and alien plant species in 10 small towns of the Sub-Tropical Thicket Biome, South Africa. *Urban Forestry & Urban Greening*, *7*(1), 1–13. doi:10.1016/j.ufug.2007.12.003

Matthies, S. A., Rüter, S., Schaarschmidt, F., & Prasse, R. (2017). Determinants of species richness within and across taxonomic groups in urban green spaces. *Urban Ecosystems*, *20*(4), 897–909. doi:10.100711252-017-0642-9

Maurer, M., & Bogner, F. (2019). How freshmen perceive environmental education (EE) and education for sustainable development (ESD). *PLoS One*, *14*(1), e0208910. doi:10.1371/journal.pone.0208910 PMID:30640908

Mbaiwa, J., & Stronza, A. (2010). The effects of tourism development on rural livelihoods in the Okavango Del ta, Botswana. *Journal of Sustainable Tourism*, *18*(5), 635–656. doi:10.1080/09669581003653500

Mcdonald, R. I., Kareiva, P., & Forman, R. T. T. (2008). The implications of current and future urbanization for global protected areas and biodiversity conservation. *Biological Conservation*, *141*(6), 1695–1703. doi:10.1016/j.biocon.2008.04.025

McKinney, M. L. (2002). Urbanization, Biodiversity, and Conservation. *Bioscience*, *52*(10), 883–890. doi:10.1641/0006-3568(2002)052[0883:UBAC]2.0.CO;2

Mckinney, M. L. (2006). Urbanization as a major cause of biotic homogenization. *Biological Conservation, 3*(3), 247–260. doi:10.1016/j.biocon.2005.09.005

McPherson, E.G., & Simpson, J.R. (2003). Potential energy savings in buildings by an urban tree planting programme in California. *Urban For. Urban Green, 2*, 73–86. doi:10.1078/1618-8667-00025

Mees, P. (2009). How dense are we? Another look at urban density and transport patterns in Australia, Canada and the USA. *Road & Transport Research: A Journal of Australian and New Zealand Research and Practice, 18*(4), 58.

Meinhold, J. L., & Malkus, A. J. (2005). Adolescent environmental behaviors: Can knowledge, attitudes, and self-efficacy make a difference? *Environment and Behavior, 37*(4), 511–532. doi:10.1177/0013916504269665

Mensah, I. & Ernest, A. (2013). Community Participation in Ecotourism: The Case of Bobiri Forest Reserve and Butterfly Sanctuary in Ashanti Region of Ghan. *American Journal of Tourism Management, 2*(A), 34-42.

Mensah, S., Ayarkwa, J., & Nani, G. (2018). A theoretical framework for conceptualizing contractors' adaptation to environmentally sustainable construction. *International Journal of Construction Management*, 1-11.

Merchant, C. (1992). *Radical ecology: The search for a livable world.* Routledge, Chapman & Hall.

Merriam, S. B. (1998). *Qualitative research and case study applications in education.* Jossey-Bass.

Merrington, G., Winder, L., & Green, I. (1997). The uptake of cadmium and zinc by the bird-cherry oat aphid *Rhopalosiphum padi* (Homoptera:Aphididae) feeding on wheat grown on sewage sludge amended agricultural soil. *Environmental Pollution, 96*(1), 111–114. doi:10.1016/S0269-7491(96)00149-2 PMID:15093439

Messouli, M., Messana, G., Yacoubi-Khebiza, M., El Filali, A. E. A., Boughrous, A. A., & Boulanouar, M. (2008). The evolution of groundwater exploration methods in the Moroccan Oases through history, and managing ecological risk of their present pollution. In *Climatic Changes and Water Resources in the Middle East and North Africa* (pp. 267–288). Springer. doi:10.1007/978-3-540-85047-2_19

Michalos, C., Creech, H., McDonald, C., & Hatch Kahlke, P. M. (2009). *Measuring Knowledge, Attitudes and Behaviours towards Sustainable Development: two Exploratory Studies.* International Institute for Sustainable Development.

Mikola, P. (1987). Mycorrhizae under tropical stresses. *Journal of Applied Botany and Food Quality, 61*, 15–23.

Miles, M. B., & Huberman, A. (1994). *Qualitative data analysis: An expanded sourcebook.* Sage.

Miller, L. K. (1904). *Children's Garden for School and Home, A Manual of Cooperative Gardening*. School and Home.

Milton, A., Cooke, J. A., & Johnson, M. S. (2003). Accumulation of lead, zinc, and cadmium in a wild population of *Clethrionomys glareolus* from an abandoned lead mine. *Archives of Environmental Contamination and Toxicology, 44*(3), 405–411. doi:10.100700244-002-2014-5 PMID:12712302

Ministry of Environment and Forests. (2009). *Bangladesh climate change strategy and action plan 2009.* Author.

Ministry of Urban Development. (2015, June). *Smart cities Mission statement and guidelines.* Retrieved Sept 2020, from Ministry of Urban Development, Government of India: http://smartcities.gov.in/upload/uploadfiles/files/SmartCityGuidelines(1).pdf

Ministry, H. R. D. Government of India. (2019, May). *School Nutrition(Kitchen Garden) Guidelines.* http://www.schooleducation.kar.nic.in/SSA/pdfdocs/SwasthPlus.pdf

Minucci, G., & Karmaoui, A. (2016). Exploring the water-food-energy and climate nexus: Insights from the Moroccan Draa Valley. In A. Colucci A., M. Magoni, & S. Menoni (Eds.), *Peri-Urban Areas and Food-Energy-Water Nexus* (pp. 89-97). Springer. http://doi-org-443.webvpn.fjmu.edu.cn/10.1007/978-3-319-41022-7_11

Minucci, G., & Karmaoui, A. (2017). Exploring the water-food-energy and climate nexus: insights from the Moroccan Draa Valley. In Peri-Urban Areas and Food-Energy-Water Nexus Sustainablity and Resilience Strategies in the Age of Climate Change. Springer International Publishing AG. doi:10.1007/978-3-319-41022-7_11

Mitchell, I. K., & Walinga, J. (2017). The creative imperative: The role of creativity, creative problem solving and insight as key drivers for sustainability. *Journal of Cleaner Production, 140*, 1872–1884. doi:10.1016/j.jclepro.2016.09.162

Mitchell, S., Foulger, T. S., Wetzel, K., & Rathkey, C. (2009). The negotiated project approach: Project-based learning without leaving the standards behind. *Early Childhood Education Journal, 36*(4), 339–346. doi:10.100710643-008-0295-7

Mogensen, F., & Schnack, K. (2010). The Action Competence Approach and the "New" Discourses of Education for Sustainable Development, Competence and Quality Criteria. *Environmental Education Research, 16*(1), 59–74. doi:10.1080/13504620903504032

Mohanraj, R. (2000). Pollution Status of Wetlands in Urban Coimbatore, Tamilnadu, India. *Bulletin of Environmental Contamination and Toxicology, 64*(5), 638–643. doi:10.1007001280000051 PMID:10857445

Mohanraj, R., & Azeez, P. A. (2005). Urban development and particulate air pollution in Coimbatore city, India. *The International Journal of Environmental Studies, 62*(1), 69–78. doi:10.1080/0020723042000261713

Mohanraj, R., Azeez, P. A., & Priscilla, T. (2004). Heavy metals in airborne particulate matter of urban Coimbatore. *Archives of Environmental Contamination and Toxicology, 47*(2), 162–167. doi:10.100700244-004-3054-9 PMID:15386140

Moksin, A. I., Shahrill, M., Anshari, M., Huda, M., & Tengah, K. A. (2018). The Learning of Integration in Calculus Using the Autograph Technology. *Advanced Science Letters, 24*(1), 550–552. doi:10.1166/asl.2018.12067

Molderez, I., & Fonseca, E. (2018). The efficacy of real-world experiences and service learning for fostering competences for sustainable development in higher education. *Journal of Cleaner Production, 172*, 4397–4410. doi:10.1016/j.jclepro.2017.04.062

Mondino, E., & Beery, T. (2019). Ecotourism as a learning tool for sustainable development. The case of Monviso Transboundary Biosphere Reserve, Italy. *Journal of Ecotourism, 18*(2), 107–121. doi:10.1080/14724049.2018.1462371

Morgan, K., & Murdoch, J. (2000). Organic vs. conventional agriculture: Knowledge, power and innovation in the food chain. *Geoforum, 31*(2), 159–173. doi:10.1016/S0016-7185(99)00029-9

Morin, O., Tytler, R., Barraza, L., Simonneaux, L., & Simonneaux, J. (2013). Cross cultural exchange to support reasoning about socio-scientific sustainability issues. *Teaching Science, 59*(1), 16–22.

Moschitz, H., Muller, A., Kretzschmar, U., Haller, L., de Porras, M., Pfeifer, C., Oehen, B., Willer, H. & Stolz H. (2021). *How can the EU Farm to Fork strategy deliver on its organic promises? Some critical reflections.* (unpublished)

Moschitz, H., & Frick, R. (2020). City food flow analysis. A new method to study local consumption. *Renewable Agriculture and Food Systems, 35*, 1–13. doi:10.1017/S1742170520000150

MSU. (2020). *Organic Farmer Training Program.* Available at: http://www.msuorganicfarm.org/organic-farmer-training-program.html

Mulawarman, A., Sudrajat, A., Hendri, N., Kamar, K., Mulyadi, D., Budiyanto, G., & Huda, M., and Maseleno. (2018). FMADM for determining superior commodity at agroindustry area. *IACSIT International Journal of Engineering and Technology*, *7*(4), 4667–4673.

Muller, A., Schader, Ch., Scialabba, N. E.-H., Brüggemann, J., Isensee, A., Erb, K.-H., Smith, P., Klocke, P., Leiber, F., & Stolze, M. (2017). Strategies for feeding the world more sustainably with organic agriculture. *Nature Communications*, *8*(1), 1290. doi:10.103841467-017-01410-w PMID:29138387

Muluneh, A. (2020). Impact of climate change on soil water balance, maize production, and potential adaptation measures in the Rift Valley drylands of Ethiopia. *Journal of Arid Environments*, *179*, 104195. doi:10.1016/j.jaridenv.2020.104195

Murtaza, T., & Mahmood, M. (2018). Active learning through project based learning approach in English language lessons for early age groups. *Journal of Early Childhood Care and Education. Islamabad, 2*. Retrieved from https://search.proquest.com/docview/2362277270/abstract/4F2C79042A944842PQ/1

Muslihudin, M., Ilayaraja, M., Kumar, K. S., Shankar, K., Jamilah, J., Novitasari, D., & ... & Maseleno. (2019). A. Decision Support System in Kindergarten Selection using TOPSIS Method. *International Journal of Recent Technology and Engineering*, *8*(1), 3291–3298.

National Council of Educational Research and Training (NCERT). (1975). *The curriculum for the ten-year school-A framework*. Author.

National Curriculum and Textbook Board (NCTB). (2020, May). *Mission and Vision of NCTB*. Retrieved on May 18, 2020, from http://www.nctb.gov.bd/

National Oceanic and Atmospheric Administration (NOAA). (2011). *Increasing public demand for NOAA'S climate-related data and services*. United States Department of Commerce: NOAA. Retrieved on May 28, 2020, from https://www.noaa.gov/climate.html

National Oceanic and Atmospheric Administration. (2016). *Talking the Children About Climate Change*. https://oceanservice.noaa.gov/education/climate-stewards/talking-about.html

Nawab, B., Nyborg, I. L., Esser, K. B., & Jenssen, P. D. (2006). Cultural preferences in designing ecological sanitation systems in North West Frontier Province, Pakistan. *Journal of Environmental Psychology*, *26*(3), 236–246. doi:10.1016/j.jenvp.2006.07.005

Negev, M., Sagy, G., Garb, Y., Salzberg, A., & Tal, A. (2008). Evaluating the environmental literacy of Israeli elementary and high school students. *The Journal of Environmental Education*, *39*(2), 3–20. doi:10.3200/JOEE.39.2.3-20

Nero, B. F., Callo-Concha, D., Anning, A., & Denich, M. (2017a). Urban Green Spaces Enhance Climate Change Mitigation in Cities of the Global South: The Case of Kumasi, Ghana. *Procedia Engineering*, *198*, 69–83. Advance online publication. doi:10.1016/j.proeng.2017.07.074

Nero, B. F., Campion, B. B., Agbo, N., Callo-concha, D., & Denich, M. (2017b). Tree and trait diversity, species co-existence, and diversity- functional relations of green spaces in Kumasi, Ghana. *Procedia Engineering*, *198*, 99–115. doi:10.1016/j.proeng.2017.07.164

Newton, P., & Meyer, D. (2013). Exploring the attitudes-action gap in household resource consumption: Does "environmental lifestyle" segmentation align with consumer behaviour? *Sustainability*, *5*(3), 1211–1233. doi:10.3390u5031211

Nickerson, N. P., Jorgenson, J., & Boley, B. B. (2016). Are sustainable tourists a higher spending market? *Tourism Management*, *54*, 170–177. doi:10.1016/j.tourman.2015.11.009

Nigh, R., & Bertely, M. (2018). Conocimiento y educación indígena en Chiapas, México: Un método intercultural. *Diálogos Sobre Educación, 9*(16). Retrieved from http://www.scielo.org.mx/scielo.php?script=sci_abstract&pid=S2007-21712018000100003&lng=es&nrm=iso&tlng=es

Nirmal Bhargavi, V. L., & Sudha, P. N. (2015). Removal of heavy metal ions from soil by electrokinetic assisted phytoremediation method. *International Journal of Chemtech Research*, *8*(5), 192–202.

Nonet, G., Kassel, K., & Meijs, L. (2016). Understanding Responsible Management: Emerging Themes and Variations from European Business School Programs. *Journal of Business Ethics*, *139*(4), 717–736. doi:10.100710551-016-3149-z

North American Association for Environmental Education. (2004). *Guidelines for the preparation and professional development of environmental educators*. Author.

Nowak, D.J., Green, E.J., Hoehn, R.E., & Lapoint, E. (2013b). *Carbon storage and sequestration by trees in urban and community areas of the United States*. Academic Press.

Nowak, D. J., Green, E. J., Hoehn, R. E., & Lapoint, E. (2013a). Carbon storage and sequestration by trees in urban and community areas of the United States. *Environ. Pollut. J.*, *178*, 229–236. doi:10.1016/j.envpol.2013.03.019 PMID:23583943

Nowak, D. J., Hoehn, R. E., Bodine, A. R., Greenfield, E. J., & Neil-dunne, J. O. (2016a). Urban forest structure, ecosystem services and change in Syracuse, NY. *Urban Ecosystems*, *19*(4), 1455–1477. doi:10.100711252-013-0326-z

Nwagwu, D. I. (2020). Driving sustainable banking in Nigeria through responsible management education: The case of Lagos Business School. *International Journal of Management Education*, *18*(1), 100332. doi:10.1016/j.ijme.2019.100332

O'Donnell, C., & Bascopé, M. (2019). *Using science to do social good: STEM education for sustainable development.* Position paper developed in preparation for the second "International Dialogue on STEM Education" (IDoS) in Berlin. Retrieved from www.haus-der-kleinen-forscher.de

Oakes, K. (2020, Feb 26). *How cutting your food waste can help the climate*. Retrieved Sept 2020, from BBC: https://www.bbc.com/future/article/20200224-how-cutting-your-food-waste-can-help-the-climate#:~:text=Onepercent20study percent20estimatedpercent20thatpercent20the,frompercent20puttingpercent20itpercent20inpercent20landfill

Ochieng, M. A., & Koske, J. (2013). The level of climate change awareness and perception among primary school teachers in Kisumu Municipality, Kenya. *International Journal of Humanities and Social Science*, *3*(21), 174–179.

Øian, H., Fredman, P., Sandell, K., Sæþórsdóttir, A. D., Tyrväinen, L., & Jensen, F. S. (2018). *Tourism, nature and sustainability. A review of policy instruments in the Nordic countries. Nordic Council of Ministers 2018*. Rosendahls.

Oktafianto, Sudrajat, A., Kawangit, R.M., Don, A.G., Huda, M., Saputri, A.D., & Maseleno, A. (2018). Determining housing location using weighted product. *IACSIT International Journal of Engineering and Technology*, *7*(4), 3563–3568.

Oppenheimer, M., & Anttila-Hughes, J. K. (2016). The science of climate change. *Children and Climate Change*, *26*(1), 11–30.

Organic Farm Knowledge. (2020). *Organic Farm Knowledge Platform*. Available at: https://organic-farmknowledge.org/

Orłowska, E., Godzik, B., & Turnau, K. (2012). Effect of different arbuscular mycorrhizal fungal isolates on growth and arsenic accumulation in *Plantago lanceolata* L. *Environmental Pollution*, *168*, 121–130. doi:10.1016/j.envpol.2012.04.026 PMID:22609863

Ortiz-de-Mandojana, N., Aguilera-Caracuel, J., & Morales-Raya, M. (2016). Corporate governance and environmental sustainability: The moderating role of the national institutional context. *Corporate Social Responsibility and Environmental Management*, *23*(3), 150–164. doi:10.1002/csr.1367

Orts, E. W., & Smith, N. C. (Eds.). (2017). *The moral responsibility of firms*. Oxford University Press. doi:10.1093/oso/9780198738534.001.0001

OTA. (2020). *Organic Education Webinars*. Available at: https://ota.com/programs-events/organic-education-webinars

Othman, R., Shahrill, M., Mundia, L., Tan, A., & Huda, M. (2016). Investigating the Relationship Between the Student's Ability and Learning Preferences: Evidence from Year 7 Mathematics Students. *The New Educational Review*, *44*(2), 125–138. doi:10.15804/tner.2016.44.2.10

Oyekanmi, A. A., Latiff, A. A., Daud, Z., Mohamed, R. M., Ismail, N., Aziz, A. A., ... Abiodun, A. K. (2019). Adsorption of cadmium and lead from palm oil mill effluent using bone-composite: Optimisation and isotherm studies. *International Journal of Environmental Analytical Chemistry*, *99*(8), 707–725. doi:10.1080/03067319.2019.1607318

Ozden, M. (2008). Environmental Awareness and Attitudes of Student Teachers: An Empirical Research. *International Research in Geographical and Environmental Education*, *17*(1), 40–55. doi:10.2167/irgee227.0

Pachauri, R.K. (2015 Feb 1). Climate Change Adaptation and Mitigation, The SDG Nexus. *Terra Green, 7*(11).

Palinkas, L. A., Holloway, I. W., Rice, E., Fuentes, D., Wu, Q., & Chamberlain, P. (2011). Social networks and implementation of evidence-based practices in public youth-serving systems: A mixed methods study. *Implementation Science; IS*, *6*(1), 113. doi:10.1186/1748-5908-6-113 PMID:21958674

Palmer, J. (1998). Environmental Education. In *The 21ST Century Theory, practice, progress and promise*. Routledge Publishers.

Pandey, J., & Pandey, U. (2009). Accumulation of heavy metals in dietary vegetables and cultivated soil horizon in organic farming system in relation to atmospheric deposition in a seasonally dry tropical region of India. *Environmental Monitoring and Assessment*, *148*(1–4), 61–74. doi:10.100710661-007-0139-8 PMID:18202901

Pane, M. M., & Patriana, R. (2016). The significance of environmental contents in character education for quality of life. *Procedia: Social and Behavioral Sciences*, *222*, 244–252. doi:10.1016/j.sbspro.2016.05.153

Pardimin, A., Ninsiana, W., Dacholfany, M.I., Kamar, K., Teh, K.S.M., Huda, M., Hananto, A.L., Muslihudin, M., Shankar, K., & Maseleno, A. (2018). Developing Multimedia Application Model for Basic Mathematics Learning. *Journal of Advanced Research in Dynamical and Control Systems*, *10*(14), 1347–1356.

Pavithrapriya, S., Mahimairaja, S., & Sivasubramanian, K. (2015). Pollution due to Heavy Metals in Coimbatore Wetlands, India. *Research Journal of Agriculture and Forestry Sciences*, *3*(6), 1–5.

Pearce, M., Willis, E., Mamerow, L., Jorgensen, B., & Martin, J. (2014). The prestige of sustainable living: Implications for water use in A ustralia. *The Geographical Journal*, *180*(2), 161–174. doi:10.1111/geoj.12016

Pearson, T.R.H., Brown, S.L., & Birdsey, R.A. (2007). Measurement Guidelines for the Sequestration of Forest Carbon. *North. Res. Stn.*, 42.

Peng, L., Chen, X., Zhang, Y., Du, Y., Huang, M., & Wang, J. (2015). Remediation of metal contamination by electrokinetics coupled with electrospun polyacrylonitrile nanofiber membrane. *Process Safety and Environmental Protection*, *98*, 1–10. doi:10.1016/j.psep.2015.06.003

Perdinan, Santikayasa, I. P., Sugiarto Y., Amri, A., Atmaja, T., Adi, R. F., Basit, R. A., & Arini, E. Y. (2016). *Studi Baseline Adaptasi Perubahan Iklim Fokus Anak (4CA) Wilayah Kajian Timur Tengah Utara dan Lembata Provinsi Nusa tenggara Timur*. Generasi Hijau Indonesia.

Perera, P., Vlosky, R. P., & Sampath, B. W. (2012). Motivational and Behavioral Profiling of Visitors to Forest-based Recreational Destinations in Sri Lanka. *Asia Pacific Journal of Tourism Research, 17*(4), 451–467. doi:10.1080/10941665.2011.627353

Perkins, H., & Grace, D. A. (2009). Ecotourism: Supply of nature or tourist demand? *Journal of Ecotourism, 8*(3), 223–236. doi:10.1080/14724040802534992

Peterlin, J., Dimovski, V., Tvaronavičienė, M., Grah, B., & Kaklauskas, A. (2018). The strategic process of developing social aspects of sustainability through the vision reflection in business education. *Technological and Economic Development of Economy, 24*(4), 1718–1736. doi:10.3846/tede.2018.5198

Pham-Duc, P., Nguyen-Viet, H., Hattendorf, J., Zinsstag, J., Phung-Dac, C., Zurbrügg, C., & Odermatt, P. (2013). Ascaris lumbricoides and Trichuris trichiura infections associated with wastewater and human excreta use in agriculture in Vietnam. *Parasitology International, 62*(2), 172–180. doi:10.1016/j.parint.2012.12.007 PMID:23305757

Pickett, S. T. A., Cadenasso, M. L., Grove, J. M., Nilon, C. H., Pouyat, R. V., Zipperer, W. C., & Costanza, R. (2001). Urban Ecological Systems: Linking Terrestrial Ecological, Physical,and Socioeconomic Components of Metropolitan Areas. *Annual Review of Ecology and Systematics, 32*(1), 127–157. doi:10.1146/annurev.ecolsys.32.081501.114012

Pitpitunge, A. D. (2013). Students' perceptions about climate change. *Asian Journal of Biology Education, 7*, 2–11.

Planning Commission. (2014, May 12). *Report of the Task Force on Waste to Energy (Volume9 I) (In the context of Integrated MSW Management).* Retrieved Sept 2020, from MNRE: https://mnre.gov.in/img/documents/uploads/7c1d4e00bb994cffbdd32d12cae627ce.pdf

Plutzer, E., & Hannah, A. L. (2018). Teaching Climate Change in middle schools and high schools: Investigating STEM education's deficit model. *Climatic Change, 149*(3-4), 305–317. doi:10.100710584-018-2253-8

Polk, M. (Ed.). (2015). *Co-producing knowledge for sustainable cities: Joining forces for change.* Routledge. doi:10.4324/9781315748030

Ponisio, L. C., M'Gonigle, L. K., Mace, K. C., Palomino, J., de Valpine, P., & Kremen, C. (2015). Diversification practices reduce organic to conventional yield gap. *Proceedings of the Royal Society B: Biological Sciences, 282*(1799). 10.1098/rspb.2014.1396

Pooley, J. A., & O'Connor, M. (2000). Environmental education and attitudes: Emotions and beliefs are what is needed. *Environment and Behavior, 32*(5), 711–723. doi:10.1177/0013916500325007

Poon, A. (1993). *Tourism, Technology and Competitive Strategies.* CABI.

Poornima, K., Vasudevan, K., & Venkateshwarlu, M. (2011). Evaluation of Heavy metals in Textile effluents in relation to Soil and Pond water. *Electronic Journal of Environmental Sciences, 4*, 79–84.

Portegies, A., De Haan, T., & Platenkamp, V. (2014). Embedded research: A pragmatic design for contextual learning - from fieldtrip to fieldwork to field research in Australasia. *Tourism Planning & Development, 15*(5), 600–613.

Pothitou, M., Hanna, R. F., & Chalvatzis, K. J. (2016). Environmental knowledge, pro-environmental behaviour and energy savings in households: An empirical study. *Applied Energy, 184*, 1217–1229. doi:10.1016/j.apenergy.2016.06.017

Prasad, M. N. V., & Hagemeyer, J. (Eds.). (1999). *Heavy metal stress in plants. From molecules to ecosystems.* Springer.

Prince, M., & Felder, R. (2006). Inductive teaching and learning methods: Definitions, comparisons, and research bases. *Journal of Engineering Education, 95*(2), 123–138. doi:10.1002/j.2168-9830.2006.tb00884.x

Priya, & Gabriela Jennifer, Thankam, Thankam, & Mathew. (2010). Monitoring the Pollution Intensity of Wetlands of Coimbatore, Tamil Nadu, India. *Nature Environmentand Pollution Technology*, *10*(3), 447–454.

PRME. (2013). *Principles for Responsible Management Education, an initiative supported by the United Nations*. Retrieved from https://www.unprme.org/index.php

Puspitasari, R., Riadi, D., Huda, M., & Maseleno, A. (2020). Self-Management of Youth Transmigrant in the Use of Information Technology. *European Journal of Molecular & Clinical Medicine*, *7*(11), 247–249.

Putra, D. A. D., Jasmi, K. A., Basiron, B., Huda, M., Maseleno, A., Shankar, K., & Aminudin, N. (2018). Tactical Steps for E-Government Development. *International Journal of Pure and Applied Mathematics*, *119*(15), 2251–2258.

Pysek, P. (1998). Alien and native species in Central European urban floras: A quantitative comparison. *Journal of Biogeography*, *25*(1), 155–163. doi:10.1046/j.1365-2699.1998.251177.x

Radha, R. V., Kumutha, K., & Marimuthu, P. (2014). Assessment of Cadmimum Contamination of Soils in Sewage Disposal Areas of Coimbaotre District, Tamil Nadu, India. *Current World Environment*, *9*(2), 379–386. doi:10.12944/CWE.9.2.18

Rahman, S. M. A., Tasmin, S., Uddin, M. K., Islam, M. T., & Sujauddin, M. (2014). Climate change awareness among the high school students: Case study from a climate vulnerable country. *International Journal of Built Environment and Sustainability*, *1*(1), 18–26.

Raj, C. (2009, Dec. 21). Enemies of the Earth, Copenhagen Summit. *India Today*.

Rajagopalan, R. (2015). Environmental studies: from crisis to cure (No. 3). Oxford University Press.

Rajagopalan, R. (2018). *Environmental Studies, From Crisis to Cure* (3rd ed.). Oxford University Press.

Rajbhandari, K. (2008). Ecological sanitation latrines: The experience of Nepal. *Beyond Constructionuse by all: A collection of case studies from sanitaion and hygiene promotion practitioners in South Asia*.

Rajmanickam, R., & Nagan, S. (2018). Assessment of Comprehensive Environmental Pollution Index of Kurichi Industrial Cluster, Coimbatore District, Tamil Nadu, India - a Case Study. *Journal of Ecological Engineering*, *19*(1), 191–199. doi:10.12911/22998993/78747

Rakotomamonjy, S. N., Jones, J. P. G., Razafimanahaka, J. H., Ramamonjisoa, B., & Williams, S. J. (2015). The effects of environmental education on children's and parents' knowledge and attitudes towards lemurs in rural Madagascar. *Animal Conservation*, *18*(2), 157–166. doi:10.1111/acv.12153

Rands, G. P. (2009). A principle-attribute matrix for environmentally sustainable management education and its application. *Journal of Management Education, 33*(3), 296-322. doi:10.1177/1052562908323191

Rathod, H. S., Chauhan, R., Huda, M., & Maseleno, A. (2020). A Conceptual Research Study on Identification of Biases with towards Assets Allocation and Portfolio Management of Investors. *European Journal of Molecular & Clinical Medicine*, *7*(10), 985–994.

Razzaq, A., Samiha, Y., & Anshari, M. (2018). Smartphone Habits and Behaviors in Supporting Students Self-Efficacy. *International Journal of Emerging Technologies in Learning*, *13*(2), 94–109. doi:10.3991/ijet.v13i02.7685

Rees & Anthony. (2015, Nov 6). *Division of Data, Research and Policy, Strategy and Networks Section, Unless we act now, The impact of climate change on children*. Academic Press.

Reid, A. (2019). Climate change education and research: Possibilities and potentials versus problems and perils? *Environmental Education Research*, *25*(6), 767–790. doi:10.1080/13504622.2019.1664075

Reilly, S, Petrillo, D. & Demchik, M. (2008). Environmental Education's Role in Sustainable Development: Three Case Studies from India, South Africa & the United States. *International Journal of Human Resources Management.*

Reynolds, C. C., Escobedo, F. J., Clerici, N., & Zea-Camaño, J. (2017). Does "greening" of neotropical cities considerably mitigate carbon dioxide emissions? The case of Medellin, Colombia. *Sustain., 9*(5), 785. Advance online publication. doi:10.3390u9050785

Rickinson, M. (2001). Learners and learning in environmental education: A critical review of the evidence. *Environmental Education Research, 7*(3), 207–320. doi:10.1080/13504620120065230

Ristiani, P., Teh, K.S.M., Fauzi, A., Hananto, A.L., Huda, M., Muslihudin, M., Shankar, K., & Maseleno, A. (2018). Decision Support System Model for Selection of Best Formula Milk for Toddlers Using Fuzzy Multiple Attribute Decision Making. *Journal of Advanced Research in Dynamical and Control Systems, 2*(Special issue), 2075–2088.

RMC. (2020). *Existing infrastracture.* Retrieved Sept 2020, from Rajkot Municipal Corporation: http://www.rmc.gov.in/rmcwebsite/infrastructure.aspx

Roczen, N., Kaiser, F., Bogner, F., & Wilson, M. (2014). A competence model for environmental education. *Environment and Behavior, 46*(8), 972–992. doi:10.1177/0013916513492416

Rodier, K. R. (1984). *L'analyse de l'eau, eaux naturelles, eaux résiduaires, eau de mer* (7th ed.). Dunod.

Roos, J. (2017). Practical wisdom: Making and teaching the governance case for sustainability. *Journal of Cleaner Production, 140,* 117–124. doi:10.1016/j.jclepro.2015.10.135

Rosa, A. T. R., Pustokhina, I. V., Lydia, E. L., Shankar, K., & Huda, M. (2019). Concept of Electronic Document Management System (EDMS) as an Efficient Tool for Storing Document. *Journal of Critical Reviews, 6*(5), 85–90.

Rosaleen, J. (2012). *Outdoor learning: past and present: Past and Present.* Open University Press.

Rosestolato, D., Bagatin, R., & Ferro, S. (2015). Electrokinetic remediation of soils polluted by heavy metals (mercury in particular). *Chemical Engineering Journal, 264,* 16–23. doi:10.1016/j.cej.2014.11.074

Rosli, M. R. B., Salamon, H. B., & Huda, M. (2018). Distribution Management of Zakat Fund: Recommended Proposal for Asnaf Riqab in Malaysia. *International Journal of Civil Engineering and Technology, 9*(3), 56–64.

Rosset, P. M., & Altieri, M. A. (1997). Agroecology versus input substitution: A fundamental contradiction of sustainable agriculture. *Society & Natural Resources, 10*(3), 283–295. doi:10.1080/08941929709381027

Rössler, M., Kirscht, H., Rademacher, C., Platt, S., Kemmerling, B., & Linstädter, A. (2010). *Migration and resource management in the Drâa Valley, Southern Morocco. In Impacts of global change on the hydrological cycle in West and Northwest Africa.* Springer.

Roth, C. E. (1992). *Environmental literacy: its roots, evolution and directions in the 1990s.* ERIC Clearinghouse for Science, Mathematics and Environmental Education. Retrieved from https://files.eric.ed.gov/fulltext/ED348235.pdf

Roth, S. (1999). The state of Deign Research. *Design Issues, 15*(12), 18–26. doi:10.2307/1511839

Ruiz-Mallen, I., Barraza, L., Bodenhorn, B., & Reyes-García, V. (2009). Evaluating the impact of an environmental education programme: An empirical study in Mexico. *Environmental Education Research, 15*(3), 371–387. doi:10.1080/13504620902906766

Saadou, M. (1990). *La végétation des milieux drainés nigériens à l'Est du fleuve Niger.* Université de Niamey.

Sacchetti, S., & Tortia, E. (2016). The extended governance of cooperative firms: Inter-firm coordination and consistency of values. *Annals of Public and Cooperative Economics*, *87*(1), 93–116. doi:10.1111/apce.12058

Sadler, T. D. (2004). Moral sensitivity and its contribution to the resolution of socio-scientific issues. *Journal of Moral Education*, *33*(3), 339–358. doi:10.1080/0305724042000733091 PMID:16317839

Sadler, T. D., & Zeidler, D. L. (2004). The morality of socioscientic issues: Construal and res- olution of genetic engineering dilemmas. *Science Education*, *88*(1), 4–27. doi:10.1002ce.10101

Şahin, H., & Erkal, S. (2017). An Investigation of University Students' Attitudes toward Environmental Sustainability. *European Journal of Sustainable Development*, *6*(4), 147–154. doi:10.14207/ejsd.2017.v6n4p147

Salamah, P., Jandra, M., Sentono, T., Huda, M., & Maseleno, A. (2019). The effects of Emotional Intelligence, Family Environment and Learning Styles on Social-Science Learning Outcomes: An Empirical Analysis. *TEST Engineering and Management*, *81*, 4374 - 4386.

Salazar, N. B. (2006). Building a "Culture of Peace" trough Tourism: Reflexive and analytical notes and queries. *Universitas Humanistica*, *62*(Julio-Diciembre), 319–333.

Sall, M. L., Diaw, A. K., Gningue-Sall, D., Aaron, S. E., & Aaron, J. (2020). Toxic heavy metals: Impact on the environment and human health, and treatment with conducting organic polymers, a review. *Environmental Science and Pollution Research International*, *27*(24), 29927–29942. doi:10.100711356-020-09354-3 PMID:32506411

Salmon, J. (2000). Are we building environmental literacy? *The Journal of Environmental Education*, *31*(4), 4–10. doi:10.1080/00958960009598645

Salt, D. E., Smith, R. D., & Raskin, I. (1998). Phytoremediation. *Plant Physiol. Plant Mol. Biol.*, *49*, 643–668.

Sambyal, S. S. (2018, Jan 25). *Delhi's Solid Waste Management Bye-Laws focus on individual accountability*. Retrieved Sept 2020, from Downtoearth: https://www.downtoearth.org.in/news/governance/delhi-gets-it-new-byelaws-on-solid-waste-management-59559

Saminathan, B. (2015). The Syllabus adopted from 2015-16 onwards for B. Ed(II)Year, Environmental Education.

Sammut. (2007). *Speciation du cadmium, du plomb et du zinc dans les poussieres d'emissions atmospheriques d'origine siderurgique Approche de l'impact toxicologique des poussières* (Thesis).

Sanders, J., & Heß, J. (2019). Leistungen des ökologischen Landbaus für Umwelt und Gesellschaft (Vol. 65). Braunschweig: Johann Heinrich von Thünen-Institut.

Santhosh, P., & Sridevi, A. (2013). A lab-scale study on reduction of heavy metals from electroplating effluent using conventional chemical precipitation. *Journal of Environmental Research and Development*, *8*(1), 102–108.

Santos Alves, C. (2009). *A importância do ecoturismo no património arquitectónico de Cabo Verde: das potencialidades à prática*. Instituto Superior Técnico, Universidade Técnica de Lisboa.

Santosh, K., Swamy, S., Prasad, A., & Singh, D. B. (2011). Teaching Learning Strategies or methods being adopted for teaching EVS at Primary Level. *International Journal of Transformation in Business Management*. https://www.ijtbm.com/

Saphores, J.-D. M., Ogunseitan, O. A., & Shapiro, A. A. (2012). Willingness to engage in a pro-environmental behavior: An analysis of e-waste recycling based on a national survey of US households. *Resources, Conservation and Recycling*, *60*, 49–63. doi:10.1016/j.resconrec.2011.12.003

Saraiva Santos, F. A. (2017). *Ecoturismo - desenvolvimento turístico e sustentabilidade social no parque natural da serra da estrela (Concelho de Seia - estudo de caso)*. Tese de Doutoramento, Universidade de Coimbra.

Sarathamani, T., Govindarajalu, K., & Divya, K. R. (2014). Economic Impact of Water Pollution on Agricultural and Rural Households. *IOSR Journal of Economics and Finance*, *3*(5), 54–57. doi:10.9790/5933-0355457

Saravanakumar, R., Sivalingam, S., & Elangovan, S. (2016). Assessment of air quality index of Coimbatore City in Tamil Nadu. *Indian Journal of Science and Technology*, *9*(41). Advance online publication. doi:10.17485/ijst/2016/v9i41/99185

Saribas, D., Kucuk, Z. D., & Ertepinar, H. (2017). Implementation of an environmental education course to improve pre-service elementary teachers' environmental literacy and self-efficacy beliefs. *International Research in Geographical and Environmental Education*, *26*(4), 311–326. doi:10.1080/10382046.2016.1262512

Sari, N.Y., Huda, M., & Teh, K.S.M., Ristiani, & Maseleno, A. (2018). Decision support system model for selection of best formula milk for toddlers using fuzzy multiple attribute decision making. *IACSIT International Journal of Engineering and Technology*, *7*(4), 3556–3562.

Sarinas, B. G., Gellada, L. D., Jamolangue, E. B., Teruñez, M. R., & Flores, J. R. (2014). Assessment of Heavy Metals in Sediments of Iloilo Batiano River, Philippines. *International Journal of Environmental Sciences and Development*, *5*(6), 543–546. doi:10.7763/IJESD.2014.V5.542

Sauvé, L., Brunelle, R., & Berryman, T. (2005). Influence of the globalized and globalizing sustainable development framework on national policies related to environmental education. *Policy Futures in Education*, *3*(3), 271–283. doi:10.2304/pfie.2005.3.3.5

Savadogo. (2008). *Session Focat Ecosan; Design/Dimensionnement*. CREPA Siège.

Sawitri, D. R. (2017). Education for sustainable development: How early is too early? *Advanced Science Letters*, *23*(3), 2559–2560. doi:10.1166/asl.2017.8699

Schader, C., Muller, A., Scialabba, N. E., Hecht, J., Isensee, A., Erb, K. H., Smith, P., Makkar, H. P. S., Klocke, P., Leiber, F., Schwegler, P., Stolze, M., & Niggli, U. (2015). Impacts of feeding less food-competing feedstuffs to livestock on global food system sustainability. *Journal of the Royal Society, Interface*, *12*(113), 20150891. doi:10.1098/rsif.2015.0891 PMID:26674194

Schaltegger, S., & Burritt, R. (2017). *Contemporary environmental accounting: issues, concepts and practice*. Routledge. doi:10.4324/9781351282529

Schaltegger, S., & Wagner, M. (2011). Sustainable entrepreneurship and sustainability innovation: Categories and interactions. *Business Strategy and the Environment*, *20*(4), 222–237. doi:10.1002/bse.682

Schilling, J., Freier, K. P., Hertig, E., & Scheffran, J. (2012). Climate change, vulnerability and adaptation in North Africa with focus on Morocco. *Agriculture, Ecosystems & Environment*, *156*, 12–26. doi:10.1016/j.agee.2012.04.021

Schultz, P. W., & Zelezny, L. (1999). Values as predictors of environ- mental attitudes: Evidence for consistency across 14 countries. *Journal of Environmental Psychology*, *28*(3), 255–265. doi:10.1006/jevp.1999.0129

Sedarati, P., Santos, S. P., & Pintassilgo, P. (2018). System Dynamics in Tourism Planning and Development. *Tourism Planning & Development*.

Selvam, S., Ravindran, A. A., Venkatramanan, S., & Singaraja, C. (2015). Assessment of heavy metal and bacterial pollution in coastal aquifers from SIPCOT industrial zones, Gulf of Mannar, South Coast of Tamil Nadu, India. *Applied Water Science*, *7*(2), 897–913. doi:10.100713201-015-0301-3

Selvi, A., & Aruliah, R. (2018). A statistical approach of zinc remediation using acidophilic bacterium via an integrated approach of bioleaching enhanced electrokinetic remediation (BEER) technology. *Chemosphere*, *207*, 753–763. doi:10.1016/j.chemosphere.2018.05.144 PMID:29859487

Selvi, A., Rajasekar, A., Theerthagiri, J., Ananthaselvam, A., Sathishkumar, K., Madhavan, J., & Rahman, P. K. (2019). Integrated Remediation Processes Toward Heavy Metal Removal/Recovery From Various Environments-A Review. *Frontiers in Environmental Science*, *7*, 66. Advance online publication. doi:10.3389/fenvs.2019.00066

Septiropa, Z., Osman, M.H., Abd. Rahman, A.B., Ariffin, M.A.M., Huda, M., & Maseleno, A. (2018). Profile of cold-formed steel for compression member design a basic combination performance. *International Journal of Engineering and Technology*, *7*(2.27), 284-290.

Seto, K. C., Guneralp, B., & Hutyra, L. R. (2012). Global forecasts of urban expansion to 2030 and direct impacts on biodiversity and carbon pools. *Proceedings of the National Academy of Sciences of the United States of America*, *109*(40), 16083–16088. doi:10.1073/pnas.1211658109 PMID:22988086

Seto, K. C., & Shepherd, J. (2009). Global urban land-use trends and climate impacts. *Current Opinion in Environmental Sustainability*, *1*(1), 89–95. doi:10.1016/j.cosust.2009.07.012

Shafer, M. A. (2008). Climate literacy and a national climate service. *Physical Geography*, *29*(6), 561–574. doi:10.2747/0272-3646.29.6.561

Shah, K., & Nongkynrih, J. M. (2007). Metal hyperaccumulation and bioremediation. *Biologia Plantarum*, *51*, 618–634.

Shams, S., Mahruf, M., Shohel, C., & Ahsan, A. (2014). Housing problems for middle and low income people in Bangladesh: Challenges of Dhaka megacity. *Environment & Urbanization Asia*, *5*(1), 175–184. doi:10.1177/0975425314521538

Shams, S., & Shohel, M. M. C. (2016). Food security and livelihood in coastal area under increased salinity and frequent tidal surge. *Environment & Urbanization Asia*, *7*(1), 22–37. doi:10.1177/0975425315619046

Shangwa, A., & Morgan, P. (2009). Teaching ecological sanitation in a school environment in Zimbabwe. *Waterlines*, *28*(4), 327–332. doi:10.3362/1756-3488.2009.032

Sharma, B., & Chhetri, M. K. (1970). Monitoring of heavy metals in vegetables and soil of agricultural fields of Kathmandu Valley. *Ecoprint: An International Journal of Ecology*, *12*, 1–9. doi:10.3126/eco.v12i0.3189

Sharma, M., & Sharma, R. G. (2013). Identification, purification and quantification of Andrographolide from *Andrographis paniculata* (Burm. F.) nees by HPTLC at different stages of life cycle of crop. *Journal of Current Chemical & Pharmaceutical Sciences*, *3*(1), 23–32.

Sharma, P. (2016). *Learning outcomes for environmental literacy: A guide for tectbook writers, teachers and researchers*. Centre for Environmental Education.

Sharma, S., & Malaviya, P. (2014). Bioremediation of tannery wastewater by chromium resistant fungal isolate fusarium chlamydosporium SPFS2-g. *Current World Environment*, *9*(3), 721–727. doi:10.12944/CWE.9.3.21

Sharpley, R. (2009). Shedding light on dark tourism: an introduction. In R. Sharpley & P. R. Stone (Eds.), *The darker side of travel: The theory and practice of dark tourism (3-22). Tonawanda: Channel View.*

Shenai, D., & Quadros, G. (2014). Comparison between a manmade lake and a natural lake in mumbai using select water quality parameters. *Ecology. Environmental Conservation*, *20*(2), 777–782.

Shepardson, D. P., Choi, S., Niyogi, D., & Charusombat, U. (2009). Seventh grade students' mental models of the greenhouse effect. *Environmental Education Research*, *15*(5), 549–570. doi:10.1080/13504620903114592

Shepardson, D. P., Wee, B., Priddy, M., & Harbor, J. (2007). Students' mental models of the environment. *Journal of Research in Science Teaching*, *44*(2), 327–348. doi:10.1002/tea.20161

Sherene, T. (2009). Heavy Metals of Soils in Industrial Belts of Coimbatore District, Tamil Nadu. *Nature Environment and Pollution Technology*, *8*(3), 613–618.

Shi, X., & Xu, Z. (2018). Environmental regulation and firm exports: Evidence from the eleventh Five-Year Plan in China. *Journal of Environmental Economics and Management*, *89*, 187–200. doi:10.1016/j.jeem.2018.03.003

Shohel, M. M. C. & Howes, A. J. (2008). Informality of teaching and learning in nonformal schools: Sociocultural process as meso-systems of student development. *Education 3-13*, *36*(3), 293-309.

Shohel, M. M. C., Cann, R., & Atherton, S. (2020). Enhancing students' engagement using a blended learning approach: Case studies of first-year undergraduate students. *International Journal of Mobile and Blended Learning*, *12*(4), 51–68. doi:10.4018/IJMBL.2020100104

Shohel, M. M. C., & Howes, A. (2011). Models of education for sustainable development and nonformal primary education in Bangladesh. *Journal of Education for Sustainable Development*, *5*(1), 129–139. doi:10.1177/097340821000500115

Shukla, P. R., Skea, J., Slade, R., van Diemen, R., Haughey, E., Malley, J., Pathak, M., & Portugal Pereira, J. (2019). Technical Summary, 2019. In Climate Change and Land: an IPCC special report on climate change, desertification, land degradation, sustainable land management, food security, and greenhouse gas fluxes in terrestrial ecosystems. Academic Press.

Siegner, A., & Stapert, N. (2019). Climate change education in the humanities classroom: A case study of the Lowell school curriculum pilot. *Environmental Education Research*.

Silo, N. (2015). Using Science as a Tool for infusing Environmental Education in the School Curriculum–a Barrier to Education for Sustainable Development. *European Journal of Educational Studies*, *7*(1), 11–22.

Simha, P., & Ganesapillai, M. (2017). Ecological sanitation and nutrient recovery from human urine: How far have we come? A review. *Sustainable Environment Research*, *27*(3), 107–116. doi:10.1016/j.serj.2016.12.001

Singer, F. S., & Seitz, F. (1998). *Hot talk, cold science: Global warming's unfinished debate*. Washington, DC: The Independent Institute.

Singh, C. K., Kumar, A., & Roy, S. S. (2018, Feb 13). Retrieved Sept 2020, from https://www.ncbi.nlm.nih.gov/pmc/articles/PMC5811459/

Singh, P., Kumar, A., & Borthakur, A. (2020). *Abatement of environmental pollutants*. Academic Press.

Singhal, N., Gupta, H., & Mittal, G. (2018). Importance-performance Analysis to Identify Effective Learning Approaches for Sustainability in an Indian Business School. *Vision: The Journal of Business Perspective*, *22*(3), 276–283. doi:10.1177/0972262918785963

Singh, O. V., Labana, S., Pandey, G., Budhiraja, R., & Jain, R. K. (2003). Phytoremediation: An overview of metallic ion decontamination from soil. *Applied Microbiology and Biotechnology*, *61*, 405–412.

Sivaram, M., Shanmugapriya, S., Yuvaraj, D., Porkodi, V., Akbari, A., Hashim, W., & Huda, M. (2020). Decision Support System for Determining Academic Advisor Using Simple Additive Weighting. In *Cognitive Informatics and Soft Computing* (pp. 149–156). Springer. doi:10.1007/978-981-15-1451-7_16

Skalidou, D., & Oya, C. (2018). The challenges of screening and synthesising qualitative research in a mixed-methods systematic review. The case of the impact of agricultural certification schemes. *Journal of Development Effectiveness*, *10*(1), 39–60. doi:10.1080/19439342.2018.1438495

Skamp, K., Boyes, E., & Stanisstreet, M. (2013). Beliefs and willingness to act about global warming: Where to focus science pedagogy? *Science Education*, *97*(2), 191–217. doi:10.1002ce.21050

Smith, P., Martino, D., Cai, Z., Gwary, D., Janzen, H., Kumar, P., McCarl, B., Ogle, S., O'Mara, F., Rice, C., Scholes, B., Sirotenko, O., Howden, M., McAllister, T., Pan, G., Romanenkov, V., Schneider, U., Towprayoon, S., Wattenbach, M., & Smith, J. (2008). Greenhouse gas mitigation in agriculture. *Philosophical Transactions of the Royal Society of London. Series B, Biological Sciences*, *363*(1492), 789–813. doi:10.1098/rstb.2007.2184 PMID:17827109

Sodian, B., Zaitchik, D., & Carey, S. (1991). Young children's differentiation of hypothetical beliefs from evidence. *Child Development*, *62*(4), 753–766. doi:10.2307/1131175

Sohal, A. S. (2013). Developing competencies of supply chain professionals in Australia: Collaboration between businesses, universities and industry associations. *Supply Chain Management*, *18*(4), 429–439. doi:10.1108/SCM-07-2012-0228

Srinivasan, V., Suresh Kumar, D., Chinnasamy, P., Sulagna, S., Sakthivel, D., Paramasivam, P., & Lele, S. (2014, February 1). *Water Management in the Noyyal River Basin A Situation Analysis*. Retrieved from http://eprints.atree.org/52/

Stanis, S., Supriharyono, & Bambang, A. N. (2007). Pengelolaan sumberdaya pesisir dan laut melalui pemberdayaan kearifan lokal di Kabupaten Lembata. *Jurnal Pasir Laut*, *2*(2), 67–82.

Steg, L., & Vlek, C. (2009). Encouraging pro-environmental behaviour: An integrative review and research agenda. *Journal of Environmental Psychology*, *29*(3), 309–317. doi:10.1016/j.jenvp.2008.10.004

Steinig, S. Y., & Butts, D. M. (2009). *Generations Going Green: Intergenerational Programs Connecting Young and Old to Improve Our Environment*. Retrieved Sept 2020, from American Society on Aging: https://www.asaging.org/blog/generations-going-green-intergenerational-programs-connecting-young-and-old-improve-our

Stephens, J. C., & Graham, A. C. (2008). Climate science to citizen action: Energizing nonformal climate science education. *Eos (Washington, D.C.)*, *89*(22), 204–205. doi:10.1029/2008EO220010

Sternäng, L., & Lundholm, C. (2011). Climate change and morality: Students' perspectives on the individual and society. *International Journal of Science Education*, *33*(8), 1131–1148. doi:10.1080/09500693.2010.503765

Stern, M. J., Powell, R. B., & Ardoin, N. M. (2011). Evaluating a constructivist and culturally responsive approach to environmental education for diverse audiences. *The Journal of Environmental Education*, *42*(2), 109–122. doi:10.1080/00958961003796849

Stern, M., Powell, B., & Hill, D. (2014). Environmental education program evaluation in the new millennium: What do we measure and what have we learned? *Environmental Education Research*, *20*(5), 581–611. doi:10.1080/13504622.2013.838749

Stern, P. C. (2000). New environmental theories: Toward a coherent theory of environmentally significant behavior. *The Journal of Social Issues*, *56*(3), 407–424. doi:10.1111/0022-4537.00175

Stern, P. C., & Dietz, T. (1994). The value basis of environmental concern. *The Journal of Social Issues*, *50*(3), 65–84. doi:10.1111/j.1540-4560.1994.tb02420.x

Stevenson, K. T., Carrier, S. J., & Peterson, M. N. (2014). Evaluating strategies for inclusion of environmental literacy in the elementary school classroom. *The Electronic Journal for Research in Science & Mathematics Education*, *18*(8), 1–17.

Stevenson, K. T., Peterson, M. N., Bondell, H. D., Mertig, A. G., & Moore, S. E. (2013). Environmental, institutional, and demographic predictors of environmental literacy among middle school children. *PLoS One*, *8*(3), e59519. doi:10.1371/journal.pone.0059519 PMID:23533631

Stevenson, R. B., Nicholls, J., & Whitehouse, H. (2017). What is Climate Change Education? *Curriculum Perspectives*, *37*(3), 67–71. doi:10.100741297-017-0015-9

Stevenson, R. B., Nicholls, J., & Whitehouse, H. (2017). *What is Climate Change Education?* Springer Link.

Stewart, C., Norton, D., & Fergusson, J. (1991). Historical monitoring of heavy metals in kahikatea ring wood in Christchurch, New Zealand. *The Science of the Total Environment*, *105*, 171–190. doi:10.1016/0048-9697(91)90340-K

Stobrawa, K., & Lorenc-Plucińska, G. (2008). Thresholds of heavy-metal toxicity in cuttings of European black poplar (*Populus nigra* L.) determined according to antioxidant status of fine roots and morphometrical disorders. *The Science of the Total Environment*, *390*(1), 86–96. doi:10.1016/j.scitotenv.2007.09.024 PMID:17963819

Stoffberg, G. H., Van Rooyen, M. W., Van Der Linde, M. J., & Groeneveld, H. T. (2010). Carbon sequestration estimates of indigenous street trees in the City of Tshwane, South Africa. *Urban Forestry & Urban Greening*, *9*(1), 9–14. doi:10.1016/j.ufug.2009.09.004

Strong, C. (1998). The impact of environmental education on children's knowledge and awareness of environmental concerns. *Marketing Intelligence & Planning*, *16*(6), 349–355. doi:10.1108/02634509810237523

Sugiyarti, E., Jasmi, K. A., Basiron, B., Huda, M., Shankar, K., & Maseleno, A. (2018). Decision support system of scholarship grantee selection using data mining. *International Journal of Pure and Applied Mathematics*, *119*(15), 2239–2249.

Sugiyono. (2007). *Statistika untuk Penelitian*. ALFABETA.

Sukhia, S.P. (2005-06). *Educational Administration, Organization and Health Education*. Academic Press.

Su, M. M., & Wall, G. (2018). Chinese university students' perceptions of ecotourism. *International Journal of Services Technology and Management*, *24*(4), 356. doi:10.1504/IJSTM.2018.093345

Sundar, I. (2010). Teaching & Learning Methods in Environmental Education. Sarup Book Publisher.

Sundar, M. L., & Saseetharan, M. K. (2008). Ground water quality in Coimbatore, Tamil Nadu along Noyyal River. *Journal of Environmental Science & Engineering*, *50*(3), 187–190. PMID:19552071

Sund, P., & Wickman, P. (2008). Teachers' objects of responsibility: Something to care about in education for sustainable development? *Environmental Education Research*, *14*(2), 145–163. doi:10.1080/13504620801951681

Susilowati, T., Dacholfany, M.I., Aminin, S., Ikhwan, A., Nasir, B.M., Huda, M., Prasetyo, A., Maseleno, A., Satria, F., Hartati, S., & Wulandari. (2018a). Getting parents involved in child's school: using attendance application system based on SMS gateway. *International Journal of Engineering and Technology*, *7*(2.27), 167-174.

Susilowati, T., Teh, K.S.M., Nasir, B.M., Don, A.G., Huda, M., Hensafitri, T., Maseleno, A., Oktafianto, & Irawan, D. (2018c). Learning application of Lampung language based on multimedia software. *International Journal of Engineering and Technology, 7*(2.27), 175-181.

Susilowati, T., Jasmi, K.A., Basiron, B., Huda, M., Shankar, K., Maseleno, A., & Julia, A., & Sucipto. (2018b). Determination of Scholarship Recipients Using Simple Additive Weighting Method. *International Journal of Pure and Applied Mathematics*, *119*(15), 2231–2238.

Susilowati, T., Manickam, P., Devika, G., Shankar, K., Muslihudin, M., Hashim, W., ... Maseleno, A. (2019). Decision support system for determining lecturer scholarships for doctoral study using CBR (Case-based reasoning) method. *International Journal of Recent Technology and Engineering*, *8*(1), 3281–3290.

Swaminathan, M. (2018, Apr 20). *How Can India's Waste Problem See a Systemic Change?* Retrieved Sept 2020, from EPW: https://www.epw.in/engage/article/institutional-framework-implementing-solid-waste-management-india-macro-analysis

Swarbrooke, J., & Horner, S. (2007). *Consumer Behavior in Tourism*. Butterworth Heinemann. doi:10.4324/9780080466958

Sykes, H. (2008). Editorial: Narratives in aboriginal, history and place-based education. *Curriculum Inquiry*, *38*(5), 541–544. http://www.jstor.org/stable/25475937

Tan, B. C., & Lau, T. C. (2009). Examining sustainable consumption patterns of young consumer: Is there a cause for concern? *The Journal of International Social Research*, *2*(9).

Tang, Y., Chen, A., & Zhao, S. (2016). Carbon Storage and Sequestration of Urban Street Trees in Beijing, China Carbon Storage and Sequestration of Urban Street Trees in Beijing, China. *Frontiers in Ecology and Evolution*, *4*, 1–11. doi:10.3389/fevo.2016.00053

Tashakkori, A., & Teddlie, C. (1998). *Mixed methodology: Combining qualitative and quantitative approaches*. Sage.

Tatenda, M. (2016). *Curriculum Reforms: Essential Tool for Effective Implementation of Climate Change Adaptation Strategies*. Retrieved on May 3, 2020, from https://climatestrategies.org/wp-content/uploads/2016/04/Tatenda-Curriculum-Reforms-for-Climate-Adaptation-Strategies.pdf

Taylor, S. A., Humphreys, M., Singley, R., & Hunter, G. L. (2004). Business student preferences: Exploring the relative importance of web management in course design. *Journal of Marketing Education*, *26*(1), 42–49. doi:10.1177/0273475303262352

Tchounwou, P. B., Yedjou, C. G., Patlolla, A. K., & Sutton, D. J. (2012). Heavy Metal Toxicity and the Environment. *Experientia Supplementum Molecular, Clinical and Environmental Toxicology*, 133-164. doi:10.1007/978-3-7643-8340-4_6

Teddlie, C., & Tashakkori, A. (2006). A general typology of research designs featuring mixed methods. *Research in the Schools*, *13*(1), 12–28.

Teksoz, G., Sahin, E., & Ertepinar, H. (2010). Environmental literacy, pre-service teachers, and a sustainable future. *Hacettepe Journal of Education*, *39*(39), 307–320.

The Daily Star. (2020). *Prepare yourselves as worthy citizens: Zakir Hossen to primary students*. Retrieved on May 18, 2020, from https://www.thedailystar.net/country/textbook-distribution-festival-2020-held-amid-enthusiasm-1848082

The Guardian. (2018). *I'm striking from school to protest inaction on climate change – you should too*. Retrieved on December 19, 2020, from https://www.theguardian.com/commentisfree/2018/nov/26/im-striking-from-school-for-climate-change-too-save-the-world-australians-students-should-too

The International Ecotourism Society - TIES. (2018). Available on www.ecotourism.org

The International Ecotourism Society - TIES. (2020). *Principles of Ecotourism*. The International Ecotourism Society. Available on https://Ecotourism.org/

The Mohammed VI Foundation for Environmental Protection site. (n.d.). https://www.fm6e.org/

The National Policy on Education. (1986). Government of India.

Thiombiano, A.R., Glele KaKai, P., Bayen, J.I.B.A., & Mahamane, A. (2016). Méthodes de collecte et d'analyse des données de terrain pour l'évaluation et le suivi de la végétation en Afrique Methods for sampling and analysis of field data to evaluate and monitor vegetation in Africa. *Annales des sciences agronomiques, FSA/UAC*.

Thomas, G. (2005). Facilitation in Education for the Environment. *Australian Journal of Environmental Education, 21*, 107–116. doi:10.1017/S0814062600000999

Thompson, S. C. G., & Barton, M. A. (1994). Ecocentric and anthropocentric attitudes toward the environment. *Journal of Environmental Psychology, 14*(2), 149–157. doi:10.1016/S0272-4944(05)80168-9

Thoresen, V. W. (2019). Empathy, Adaptability, Moderation and Sharing. *Teacher Education for Sustainable Development and Global Citizenship: Critical Perspectives on Values, Curriculum and Assessment*, 217.

Tilbury, D. (1995). Environmental education for sustainability: Defining the new focus of environmental education in the 1990s. *Environmental Education Research, 1*(2), 195–212. doi:10.1080/1350462950010206

Tilbury, D. (2004). Environmental education for sustainability: A force for change in higher education. In P. B. Corcoran & A. E. J. Wals (Eds.), *Higher education and the challenge of sustainability* (pp. 97–112). Kluwer Academic Publishers. doi:10.1007/0-306-48515-X_9

Tilbury, D., & Wortman, D. (2008). How is community education contributing to sustainability in practice? *Applied Environmental Education and Communication, 7*(3), 83–93. doi:10.1080/15330150802502171

Tilley, E. (2014). *Compendium of sanitation systems and technologies*. Eawag.

Too, L., & Bajracharya, B. (2015). Sustainable campus: Engaging the community in sustainability. *International Journal of Sustainability in Higher Education, 16*(1), 57–71. doi:10.1108/IJSHE-07-2013-0080

Torkshavand, Z., Gholami, M., Farzadkia, M., & Esrafili, A. (2014). Adsorption of Cu^{2+} from aqueous solution onto modified glass beads with 3aminopropyltriethoxysilane. *Iranian Journal of Health, Safety and Environment, 1*(3), 101–110.

Torres, R. (2002). Cancun's Tourism Development from a Fordist Spectrum of Analysis. *Tourist Studies, 2*(1), 87–116. doi:10.1177/1468797602002001098

Tremel-Schaub & Feix. (2005). *Delivered Contamination des sols transferts des sols vers les plantes*. Academic Press.

Trevino, L. K., & Nelson, K. A. (2016). *Managing business ethics: Straight talk about how to do it right*. John Wiley & Sons.

Trochim, W. (2020). *Research methods data base*. Retrieved from https://conjointly.com/kb/

Trotta, G. (2018). Factors affecting energy-saving behaviours and energy efficiency investments in British households. *Energy Policy, 114*, 529–539. doi:10.1016/j.enpol.2017.12.042

Trumpowe, D., & Sarwar, G. S. (2010). Concept maps: Making learning meaningful. In A. J. Sánchez, A. J. Cañas & J. D. Novak (Eds.), *Formative structural assessment: Using concept maps as assessment for learning. Proceedings of the 4th International Conference on Concept Mapping* (pp. 132–136). Lom Ediciones SA Publisher.

Tubiello, F. N., Salvatore, M., Rossi, S., Ferrara, A., Fitton, N., & Smith, P. (2013). The FAOSTAT database of greenhouse gas emissions from agriculture. *Environmental Research Letters, 8*. doi:10.1088/1748-9326/8/1/015009

Tumwebaze, I. K., Orach, C. G., Nakayaga, J. K., Karamagi, C., Luethi, C., & Niwagaba, C. (2011). Ecological sanitation coverage and factors affecting its uptake in Kabale municipality, western Uganda. *International Journal of Environmental Health Research, 21*(4), 294–305. doi:10.1080/09603123.2010.550036 PMID:21480021

Tuncay, B., Yılmaz-Tuzun, O., & Tuncer-Teksoz, G. (2012). Moral reasoning patterns and influential factors in the context of environmental problems. *Environmental Education Research, 18*(4), 485–505. doi:10.1080/13504622.2011.630576

Tuncer, G., Boone, J. W., Tuzun, O. Y., & Oztekin, C. (2014). An evaluation of the environmental literacy of pre-service teachers in Turkey through Rasch analysis. *Environmental Education Research, 20*(2), 202–227. doi:10.1080/13504622.2013.768604

Tuncer, G., Tekkaya, C., Sungur, S., Cakiroglu, J., Ertepinar, H., & Kaplowitz, M. (2009). Assessing pre-service teachers' environmental literacy in Turkey as a mean to develop teacher education programs. *International Journal of Educational Development, 29*(4), 426–436. doi:10.1016/j.ijedudev.2008.10.003

Uberoi, M. (2002). *The Penguin Book of Gardening in India.* Academic Press.

UN Conference on Environment and Development (UNCED). (1992). *Agenda 21.* New York: UNCED. Retrieved from https://sustainabledevelopment.un.org/content/documents/Agenda21.pdf

UN. (2017). *Goal 11-Sustaimnable Cities and Communities*, https://www.unenvironment.org/explore-topics/sustainable-development-goals/why-do sustainable-development-goals-matter/goal-11

UN. (2020a). *Sustainable Development Goals.* Available at: https://sustainabledevelopment.un.org/sdgs

UN. (2020b). *Communication material Sustainable Development Goals.* Available at: https://www.un.org/sustainabledevelopment/news/communications-material/

Unal, N. (2008). *Preservice teachers' perceptions toward global versus local environmental issues* (Unpublished Master thesis), Middle East Technical University, Ankara, Turkey.

UNCC Learn. (2013). *Resource Guide for Advanced Learning on Integrating Climate Change in Education at Primary and Secondary Level.* United Nations Institute for Training and Research (UNITAR).

UNEP. (1972). *Declaration of the United Nations Conference on the Human Environment.* Retrieved from https://www.unep.org/Documents.Multilingual/Default.asp

UNEP. (1983). *The state of the environment.* Nairobi: Author.

UNESCO. (2002). *Education for sustainability, From Rio to Johannesburg: Lessons learnt from a decade of commitment.* Report presented at the Johannesburg World Summit for Sustainable Development, Paris: UNESCO. Retrieved from https://unesdoc.unesco.org/ark:/48223/pf0000127100

UNESCO. (2014). *UNESCO Education strategy 2014-2021.* Retrieved Sept 2020, from UNESCO: https://unesdoc.unesco.org/ark:/48223/pf0000231288

UNESCO-UNEP. (1978). The Tbilisi Declaration: Final report of Intergovernmental Conference on Environmental Education. *Connect, 3*(1), 1–8.

UNESCO-UNEP. (1997). The oceans: Diminishing resources, degraded environment and loss of biodiversity. *Connect, 22*(3/4), 1–33.

UNESCO-United Nations Environmental Program (UNEP). (1978). *The Tbilisi Declaration: Final report intergovernmental Conference on Environmental Education.* Paris, France: UNESCO ED/MD/49.

UNFCCC. (2019). *Climate Change Threatens Progress Across Sustainable Development.* https://unfccc.int/news/climate-change-threatens-progress-across-sustainable-development-warns-new-un-report

UNFCCC. (2020). *Global Warming Potentials (IPCC Second Assessment Report).* Retrieved Sept 2020, from UNFCCC: https://unfccc.int/process/transparency-and-reporting/greenhouse-gas-data/greenhouse-gas-data-unfccc/global-warming-potentials

United Nations (UN). (1992). *Rio Declaration on Environment and Development—Preamble.* UN.

United Nations Children's Fund. (2011). *Education Kit on Climate Change and Child Rights.* United Nations Children's Fund.

United Nations Educational Scientific and Cultural Organisation (UNESCO). (1972). *The Stockholm Declaration.* UNESCO.

United Nations Educational, Scientific and Cultural Organisation (UNESCO). (2002). *UNESCO universal declaration on cultural diversity.* UNESCO.

United Nations Educational, Scientific and Cultural Organisation (UNESCO). (2010). *The UNESCO climate change initiative: Climate change education for sustainable development.* Retrieved May 14, 2020, from https://unesdoc.unesco.org/images/0019/001901/190101e.pdf

United Nations Educational, Scientific and Cultural Organisation (UNESCO). (2015). *Climate change education for sustainable development in Bangladesh. UNESCO programme and meeting document.* Retrieved on June 28, 2020, from https://unesdoc.unesco.org/ark:/48223/pf0000246779

United Nations Educational, Scientific and Cultural Organisation (UNESCO). (n.d.). *Climate change education.* Retrieved on January 2, 2020, from https://en.unesco.org/themes/education-sustainable-development/cce

United Nations Environment Programme and World Tourism Organisation – UNEP. (2002). *The world ecotourism summit final report.* UNEP and WTO.

Urry, J. (2003). *Global Complexity.* Policy.

USAID. (2018, Sept). *Greenhouse emissions in India.* Retrieved Sept 2020, from USAID: https://www.climatelinks.org/sites/default/files/asset/document/India percent20GHG percent20Emissions percent20Factsheet percent20FINAL.pdf

Usharani, K., Umarani, K., Ayyasamy, P., Shanthi, K., & Lakshmanaperumalsamy, P. (2010). Physico-Chemical and Bacteriological Characteristics of Noyyal River and Ground Water Quality of Perur, India. *Journal of Applied Science & Environmental Management, 14*(2). Advance online publication. doi:10.4314/jasem.v14i2.57830

van Liere, K. D., & Dunlap, R. E. (1980). The social bases of environmental concern: A review of hypotheses, explanations and empirical evidence. *Public Opinion Quarterly, 44*(2), 181–199. doi:10.1086/268583

Vandecasteele, B., Samyn, J., Quataert, P., Muys, B., & Tack, F. M. (2004). Earthworm biomass as additional information for risk assessment of heavy metal biomagnification: A case study for dredged sediment-derived soils and polluted floodplain soils. *Environmental Pollution, 129*(3), 363–375. doi:10.1016/j.envpol.2003.12.007 PMID:15016458

Varia, J., Martínez, S. S., Orta, S. V., Bull, S., & Roy, S. (2013). Bioelectrochemical metal remediation and recovery of Au3, Co2 and Fe3 metal ions. *Electrochimica Acta, 95*, 125–131. doi:10.1016/j.electacta.2013.02.051

Vecchi, V. (2010). *Art and creativity in Reggio Emilia: Exploring the role and potential of ateliers in early childhood education.* Routledge.

Vedamadavan & Sarithabanuraman. (2012). Assessment of Ambient Air Quality in Coimbatore City. *Civil and Environmental Research, 2*(1), 1–6.

Velempini, K., Martin, B., Smucker, T., Ward Randolph, A., & Henning, J. (2017). Environmental education in southern Africa: A case study of a secondary school in the Okavango Delta of Botswana. *Environmental Education Research, 24*(7), 1000–1016. doi:10.1080/13504622.2017.1377158

Velmpini, K. (2016). *The integration of environmental education in the secondary school curriculum: A case study of a 10th grade junior secondary school curriculum in the Okavango Delta, Botswana* (PhD Thesis). University of Ohio.

Verma, R. K., Shivay, Y. S., Prasanna, R., Choudhary, M., & Ghasal, P. C. (2020). Nutrient management options modulating soil physico-chemical and biological properties under direct-seeded rice-based cropping systems. *Archives of Agronomy and Soil Science*, *0*(0), 1–16. Advance online publication. doi:10.1080/03650340.2020.1808627

Verma, R., & Dwivedi, P. (2013). Heavy Metals Water Pollution - A case study. *Recent Research in Science and Technology*, *5*(5), 98–99.

Verma, V. (1994). *A Textbook of Plant Physiology*. Emkay Publications.

Von Münch, E., & Ingle, R. (2012). *Compilation of 25 case studies on sustainable sanitation projects from Africa. In Sustainable Sanitation Alliance (SuSanA) and Deutsche Gesellschaft für Internationale Zusammenarbeit (GIZ)*. GmbH.

Wahyuni, N., Kulik, A. A., Lydia, E. L., Shankar, K., & Huda, M. (2020). Developing Region to Reduce Economic Gap and to Support Large Environment Activities. *Journal of Environmental Treatment Techniques*, *8*(1), 540–545.

Waite, C. (2020). Spotlight on Project-Based Learning: Seeing the forest and the trees. *Childhood Education*, *96*(2), 30–41. doi:10.1080/00094056.2020.1733860

Wals, A. E. (Ed.). (2009). *Review of contexts and structures for education for sustainable development: 2009*. Unesco.

Wals, A. E. J. (2017). Sustainability by default: Co-creating care and relationality through early childhood education. *International Journal of Early Childhood*, *49*(2), 155–164. doi:10.100713158-017-0193-5

Wang, H., Marshall, C. W., Cheng, M., Xu, H., Li, H., Yang, X., & Zheng, T. (2017). Changes in land use driven by urbanization impact nitrogen cycling and the microbial community composition in soils. *Scientific Reports*, *7*(1), 1–12. doi:10.1038rep44049 PMID:28281565

Wang, Y., Li, S., Qin, S., Guo, H., Yang, D., & Lam, H. M. (2020). How can drip irrigation save water and reduce evapotranspiration compared to border irrigation in arid regions in northwest China. *Agricultural Water Management*, *239*, 106256. doi:10.1016/j.agwat.2020.106256

Wanjugi, P., Sivaganesan, M., Korajkic, A., Kelty, C. A., McMinn, B., Ulrich, R., Harwood, V. J., & Shanks, O. C. (2016). Differential decomposition of bacterial and viral fecal indicators in common human pollution types. *Water Research*, *105*, 591–601. doi:10.1016/j.watres.2016.09.041 PMID:27693971

Wan, Y., & Bi, H. (2019). What major "socio-scientific topics" should the science curriculum focused on? A Delphi study of the expert community in China. *International Journal of Science and Mathematics Education*.

Warner, M., & Sullivan, R. (Eds.). (2017). *Putting partnerships to work: Strategic alliances for development between government, the private sector and civil society*. Routledge. doi:10.4324/9781351281249

Watch my waste. (2020). *Food Waste Greenhouse Gas Calculator*. Retrieved Sept 2020, from Watch my waste: https://watchmywaste.com.au/food-waste-greenhouse-gas-calculator/

Wearing, S., & Neil, J. (2001). *Ecoturismo: impactos, potencialidades e possibilidades*. Manole.

Wearing, S., & Neil, J. (2009). *Ecotourism: Impacts, Potentials and Possibilities*. Butterworth-Heinemann. doi:10.4324/9780080940182

Weaver, D., & Lawton, L. (2007). Twenty years on: The state of contemporary ecotourism research. *Tourism Management*, *28*(5), 1168–1179. doi:10.1016/j.tourman.2007.03.004

Weckenbrock, P., Sanchez-Gellert, H. S., & Gattinger, A. (2019). Klimaschutz. In Leistungen des ökologischen Landbaus für Umwelt und Gesellschaft (Vol. 65). Braunschweig: Johann Heinrich von Thünen-Institut.

Weissert, L. F., Salmond, J. A., & Schwendenmann, L. (2014). A review of the current progress in quantifying the potential of urban forests to mitigate urban CO2emissions. *Urban Climate*, *8*, 100–125. doi:10.1016/j.uclim.2014.01.002

Welsch, H., & Kühling, J. (2009). Determinants of pro-environmental consumption: The role of reference groups and routine behavior. *Ecological Economics*, *69*(1), 166–176. doi:10.1016/j.ecolecon.2009.08.009

Wendt, C. F., Nunes, A., Verble, R., Santini, G., Boieiro, M., & Branquinho, C. (2020). Using a space-for-time approach to select the best biodiversity-based indicators to assess the effects of aridity on Mediterranean drylands. *Ecological Indicators*, *113*, 106250. doi:10.1016/j.ecolind.2020.106250

Wenger, E. (1998). *Communities of practice: Learning as a social system* (Vol. 9). Cambridge University Press. doi:10.1017/CBO9780511803932

Werner, C., Panesar, A., Rüd, S. B., & Olt, C. U. (2009). Ecological sanitation: Principles, technologies and project examples for sustainable wastewater and excreta management. *Desalination*, *248*(1-3), 392–401. doi:10.1016/j.desal.2008.05.080

Wheeler, S. (2008). The barriers to further adoption of organic farming and genetic engineering in Australia: Views of agricultural professionals and their information sources. *Renewable Agriculture and Food Systems*, *23*(2), 161–170. doi:10.1017/S1742170507002128

White, K., Habib, R., & Hardisty, D. J. (2019). How to SHIFT consumer behaviors to be more sustainable: A literature review and guiding framework. *Journal of Marketing*, *83*(3), 22–49. doi:10.1177/0022242919825649

White, T., & Wall, R. (2008). National, regional and local attitudes towards climate change: Identifying appropriate target audiences for communications. *Local Environment*, *13*(7), 589–607. doi:10.1080/13549830802260126

Whitmarsh, L. (2009). Behavioural responses to climate change: Asymmetry of intentions and impacts. *Journal of Environmental Psychology*, *29*(1), 13–23. doi:10.1016/j.jenvp.2008.05.003

Whitmarsh, L., & O'Neill, S. (2010). Green identity, green living? The role of pro-environmental self-identity in determining consistency across diverse pro-environmental behaviours. *Journal of Environmental Psychology*, *30*(3), 305–314. doi:10.1016/j.jenvp.2010.01.003

Wight, P. (2001). Ecotourists: Not a homogeneous market segment. *The encyclopedia of Ecotourism*, 37-62.

Willer, H., Schlatter, B., Travnicek, J., Kemper, L., & Lernoud, J. (2020). The World of Organic Agriculture. Statistics and Emerging Trends 2020. Research Institute of Organic Agriculture (FiBL), Frick, and IFOAM – Organics International, Bonn.

Williams, C. (2011). Research methods. *Journal of Business & Economics Research*, *5*(3). Advance online publication. doi:10.19030/jber.v5i3.2532 PMID:21543382

Winder, L., Merrington, G., & Green, I. (1999). The tri-trophic transfer of Zn from the agricultural use of sewage sludge. *The Science of the Total Environment*, *229*(1–2), 73–81. doi:10.1016/S0048-9697(99)00070-4

Wingfield, S. S., & Black, G. (2005). Active versus passive course designs: The impact on student outcomes. *Journal of Education for Business*, *81*(2), 119–123. doi:10.3200/JOEB.81.2.119-128

Wittmer, D. P. (2016). Developing a behavioral model for ethical decision making in organizations: Conceptual and empirical research. In *Ethics in public management* (pp. 57–77). Routledge.

Wohlmuther, C. & Winstersteiner, W. (2014). *International Handbook on Tourism and Peace*. Centre for peace Reasearch & UNWTO.

Wood, M. (2002). Ecotourism: principles, practices & policies for sustainability. UNEP, United Nations Environment Programme Division of Technology, Industry and Economics, Paris, France.

World Bank. (2013, Jan 22). *New World Bank report offers policymakers roadmap for building resilient sustainable cities*. Retrieved Sept 2020, from The World Bank: https://www.worldbank.org/en/news/press-release/2013/01/22/new-world-bank-report-offers-policymakers-roadmap-building-resilient-sustainable-cities

World Bank. (2014, April 11). *Urban Development: Sector Results Profile*. Retrieved Sept 2020, from The World Bank: https://www.worldbank.org/en/results/2013/04/14/urban-development-results-profile

World Commission on Environment and Development. (1987). *Our common future*. Oxford University Press. Retrieved from https://sustainabledevelopment.un.org/content/documents/5987our-common-future.pdf

World Food Programme (WFP). (2020). *Country Brief: Bangladesh 2020*. Retrieved June 26, 2020, from https://docs.wfp.org/api/documents/WFP-0000116570/download/?_ga=2.131938381.319260259.1593310272-46580275.1593310272

World Health Organization. (2016). *Sanitation Safety Planning: Manual for Safe Use and Disposal of Wastewater Greywater and Excreta*. WHO.

World Health Organization. (2017). Progress on drinking water, sanitation and hygiene: 2017 update and SDG baselines. Geneva: World Health Organization (WHO) and the United Nations Children's Fund (UNICEF).

WTTC. (2016). *Tourism as a driver of peace. Quantitative analysis on the link between peace and Tourism, May 2016*. World Travel & Tourism Council and Institute for Economics and Peace.

Wu, F. Y., Ye, Z. H., & Wong, M. H. (2009). Intraspecific differences of arbuscular mycorrhizal fungi in their impacts on arsenic accumulation by *Pteris vittata* L. *Chemosphere*, *76*(9), 1258–1264. doi:10.1016/j.chemosphere.2009.05.020 PMID:19535126

Wu, G. C. (2017). Effects of socially responsible supplier development and sustainability-oriented innovation on sustainable development: Empirical evidence from SMEs. *Corporate Social Responsibility and Environmental Management*, *24*(6), 661–675. doi:10.1002/csr.1435

Wu, J., & Si, S. (2018). Poverty reduction through entrepreneurship: Incentives, social networks, and sustainability. *Asian Business & Management*, *17*(4), 243–259. doi:10.105741291-018-0039-5

Wulandari, Aminin, S., Dacholfany, M.I., Mujib, A., Huda, M., Nasir, B.M., Maseleno, A., Sundari, E., Fauzi, & Masrur, M. (2018). Design of library application system. *International Journal of Engineering and Technology*, *7*(2.27), 199-204.

WWFIndia. (2013, Dec 20). *Paryavaran Mitra (Janvikas) and Ors. Vs Gujarat State Pollution Control Board and Ors.* Retrieved Sept 2020, from WWFIndia: https://www.wwfindia.org/?10463/Paryavaran-Mitra

Xu, F., & Fox, D. (2014). Modelling attitudes to nature, tourism and sustainable development in national parks: A survey of visitors in China and the UK. *Tourism Management*, *45*, 142–158. doi:10.1016/j.tourman.2014.03.005

Yadlapalli, A., Rahman, S., & Gunasekaran, A. (2018). Socially responsible governance mechanisms for manufacturing firms in apparel supply chains. *International Journal of Production Economics*, *196*, 135–149. doi:10.1016/j.ijpe.2017.11.016

Yan, P., & Yang, J. (2017). Species diversity of urban forests in China. *Urban Forestry & Urban Greening*, *28*, 160–166. doi:10.1016/j.ufug.2017.09.005

Yates, K., Reefer, A., Robertson, D., Hubbard-Sanchez, J., Huss, J., & Wilder, M. (2019). Educators' perceptions of environmental education and professional development in teacher preparation programs. *Applied Environmental Education and Communication*, *18*(3), 207–218. doi:10.1080/1533015X.2018.1451411

YEE. (2018). *Uniting Youth for ecotourism.Youth for Ecotourism.* Youth and Environment Europe.

Ye, W., & Xue, X. (2008). The differences in ecotourism between China and the West. *Current Issues in Tourism, 11*(6), 567–586. doi:10.1080/13683500802475927

Yin, R. K. (2003). *Case study research: Design and Methods.* Sage Publications.

Youkhana, E., Leifkes, C., & Leon-Sicard, T. (2018). Epistemic marginality, higher and environmental education in Colombia. *Gestion y Ambiente, 21*(2), 15–29. doi:10.15446/ga.v21n2supl.77752

Youth Non-Governmental Organisation (YOUNGO). (2011). *Enhanced youth participation and education. In climate change: The Article 6 implementation toolkit. A collaborative document created by international youth.* Retrieved September 14, 2012, from the YOUNGO Article 6 Working Group. Available at: https://unfccc.int/cc_inet/files/cc_inet/information_pool/application/pdf/enhanced_youth_participation.pdf

Zachariou, A., & Symeou, L. (2009). The local community as a means for promoting education for sustainable development. *Applied Environmental Education and Communication, 7*(4), 129–143. doi:10.1080/15330150902744152

Zeidler, D. L., & Schafer, L. E. (1984). Identifying mediating factors of moral reasoning in science education. *Journal of Research in Science Teaching, 21*(1), 1–15. doi:10.1002/tea.3660210102

Zhuang, P., Zou, H., & Shu, W. (2009). Biotransfer of heavy metals along a soil-plant-insect-chicken food chain: Field study. *Journal of Environmental Sciences (China), 21*(6), 849–853. doi:10.1016/S1001-0742(08)62351-7 PMID:19803093

Zhu, H. N., Robert, C., & Kamran, A. (2015). Modeling air pollutant removal, carbon storage, and CO2 sequestration potential of urban forests in Scotlandville, Louisiana, USA. *Biogeosciences For., 009*, 1–8. doi:10.3832/ifor1845-009

Zocco, D. (2009). Risk theory and student course selection. *Research in Higher Education, 3*, 1–29.

Zsoka, A., Szerenyi, Z., Szechy, A., & Kocsis, T. (2013). Greening due to environmental education? Environmental knowledge, attitudes, consumer behaviour and everyday pro-environmental activities of Hungarian high school and university students. *Journal of Cleaner Production, 48*, 126–138. doi:10.1016/j.jclepro.2012.11.030

About the Contributors

Ahmed Karmaoui started his academic career as Research Fellow at Cadi Ayyad University and awarded Ph.D. from the Faculty of Semlalia (UCA); graduated from Publons Academy, and awarded Top 1% of international Reviewers (2018) by Publons. He has published various peer-reviewed international journal articles, book chapters, and books on a range of environmental topics. He is currently the president of the Southern Center for Culture and Science, Morocco, and an assistant professor at FSTE (UMI). Broad areas of his research interests include ecosystem services assessment.

Abdelkrim Ben Salem is an Assistant Professor of Ecology and Environment in the Mohammed V University, Faculty of Sciences in Rabat, Department of Biology. He joined the University of Mohammed V after ten years as a teacher of qualifying secondary education in Draa Tafilalet Academy. Broadly, her methodological research focuses on vulnerability and adaptation to climate change. Within, Ben Salem is a PhD in environmental sciences, currently working on socioeconomic vulnerability of the agriculture and water sector to climate change. He has published interdisciplinary projects across varied outlets, including, Parasites & Vectors Journal, International Journal of Water Resources and Arid Environments and International Publisher of Information Science and Technology Research (IGI-Global).

* * *

Tassim Abu Bakar is senior lecturer at Universiti Brunei Darussalam, Brunei Darussalam.

Atm Alam is a Lecturer (Assistant Professor) in the School of Electronic Engineering and Computer Science (EECS) at the Queen Mary University of London. He received the PhD degree in Computing and Communications from the Open University, the UK in 2015. He is also an Associate Fellow of Higher Education Academy (AFHEA). Previously, he worked at the 5G Innovation Centre (5GIC), the University of Surrey as a Research Fellow as well as at the University of Bradford for a number of H2020 and Innovate UK projects.

Yabit Alas is senior assistant professor at Universiti Brunei Darussalam, Brunei Darussalam.

Martín Bascopé is a sociologist Ms in economics with a mention in public policies and currently a PhD candidate at TUM School of Education. He also serves as professor in primary and early childhood teacher training at Campus Villarrica in Pontificia Universidad Católica de Chile. His research has focused on initial and continuous teacher training, with a focus on citizen education and education for sustainable development. He is currently leading the line of education for sustainable development in the Center for Local Development in his University (CEDEL UC).

Scott Baum is Professor in the Cities Research Institute, Griffith University Australia. Trained in Economics and Sociology, he has published widely in the areas of climate change, socio-economic inequality and labour markets.

María de las Mercedes Capobianco-Uriarte is Bioengineer degree from Nacional University of Entre Ríos Business Administration degree from Miguel Hernández University. Doctor of Almeria University, Spain.

María del Pilar Casado-Belmonte is an assistant professor of Accounting and Finance in the Department of Economics and Business at Almeria University, Spain. Her research interests include accounting, family firms and corporate governance. She has published her latest research findings in international journals such Journal of Cleaner Production, Complexity, International Journal of Entrepreneurship and Small Business, and Psychology.

Umran Betul Cebesoy completed her PhD at Middle East Technical University in 2014. She explored science teachers' genetics literacy levels and their decision-making skills in genetics related socioscientific issues in her dissertation. She was a guest researcher in Stockholm University during 2019 and investigated pre-service science teachers' decision-making skills using the SEE-SEP model and developed a professional development program for enriching science teachers' pedagogical content knowledge in teaching SSI. She has authored numerous international and national peer-reviewed journal articles in various platforms. She has been working in Usak University as Associate Professor since 2018. Her research areas include socioscientific issues, environmental education and STEM.

Fauziah Che Leh is Associate Professor of Geography, Faculty of Human Sciences, Universiti Pendidikan Sultan Idris.

Mohd Aderi Che Noh is Professor at Faculty of Human Sciences, Universiti Pendidikan Sultan Idris Malaysia.

Tanjeeba Chowdhury is currently working as an Education Team Leader at DanChurchAid. Prior to this, she worked as a Research Associate at the BRAC Institute of Educational Development and BRAC Research and Evaluation Division. Mrs Chowdhury completed her Bachelor and Master's degree in Education from the Institute of Education and Research, University of Dhaka, Bangladesh. She has specialized in both qualitative and quantitative research methods. Her research interest includes marginalization in education, teaching and learning process, curriculum, and assessment.

Josefina Cortés is a researcher at the ECOS (Ecology-Complexity-Society) Laboratory; Center for Local Development (CEDEL), Villarrica Campus, Pontificia Universidad Católica de Chile. She is also member of the Chilean Society of Socioecology and Ethnoecology (SOSOET). Her research focuses on human ecology, social-ecological systems and environmental education.

Teresa Dieguez is a Doctor on Economics, MBA on Prospective and Organizational Strategy, Master on Innovation and Technological Transfer, Specialist on Strategy and Entrepreneurship, PhD Student on Social Development and Sustainability.

Jamie Fico is a graduate student in the Department of Geography at Syracuse University pursuing a Master of Arts degree. She previously worked on environmental education projects with students in the Zagora province of Morocco from 2018-2020 with the United States Peace Corps.

Pablo Gutierrez is an industrial designer and Master's degree in curricular framework and educational projects. Participate in the UC Local Development Center. Certified facilitator of the Lego © Serious Play ™ and INDEX: Design to Improve Life® methodology. Specialist in games and educational resources. Participates in the initial training of teachers in early child an primary education programs. Integrates research and development teams in serious gaming, belief mobility, learning environments, project-based learning, place-based education, and funds of knowledge.

Azmil Hashim is Associate Professor at Faculty of Human Sciences, Universiti Pendidikan Sultan Idris Malaysia.

Chinmai Hemani is a climate change and sustainability professional and founder of social enterprise – RuChiNi Environment Sustainability Planners. She has a rich experience of over six years in the field of Climate change and sustainable development. Her expertise is in Vulnerability assessment, adaptation planning for semi-arid and coastal regions, Policy analysis wrt Climate Change and Sustainable Development, sustainable lifestyle and living, Corporate Social Responsibility(CSR) initiatives focussing environment and sustainability and is engaged with many NGOs and other organisations. She is an avid nature lover, believer of frugal living, a conservationist at heart and a traveller. She sincerely believes that wisdoms from traditional knowledge, religion and ancient India can show way to the world on Sustainable living.

Saidai Hitam is senior lecturer at Universiti Brunei Darussalam, Brunei Darussalam.

Miftachul Huda is a senior lecturer at Faculty of Human Sciences, Universiti Pendidikan Sultan Idris Malaysia. His research interest includes the field of educational studies: moral, Islamic and multicultural education, learning theory, adaptive teaching and learning technology. He has been experienced in working on research for more than five years. Contributing several works, he is currently the member of the advisory editorial board in some international journals. Traveling in some countries with scholarly intent makes him extensive experience on his research expertise in international trend and issues. Thus, he commits to contributing his knowledge to the benefit of society through professional and social activities.

Mohd Hairy Ibrahim is Associate Professor at Faculty of Human Sciences, Universiti Pendidikan Sultan Idris, Malaysia.

Sari Laelatul Qodriah is senior lecturer at Universitas Muhammadiyah Cirebon, Indonesia.

Gema M. Marín-Carrillo is Lecturer at the Department of Economics and Business, and Researcher of Marketing, Sustainable Entrepreneurship at University of Almería. Currently, she is in charge of the Marketing Degree at University of Almería (Spain). She has published in impact journals such as: Industrial Marketing Services, Journal of Cleaner Production, Service Industries Journal and Managing Service Quality.

FM Mwema is a lecturer and a postdoctoral research fellow at the departments of mechanical engineering at Dedan Kimathi University of Technology, Kenya and University of Johannesburg, South Africa respectively. His research interest is the field of manufacturing and materials, specifically in surface engineering and coating technologies, severe plastic deformation, advanced materials and characterizations. He has extensively worked on thin film preparation through RF magnetron sputtering and published his works in several peer reviewed journals, book chapters and conferences. He has over 45 publications. He also has interest in additive manufacturing technologies including laser based manufacturing and fused deposition modeling. He is currently supervising 3 masters and 2 PhD students on additive manufacturing, surface engineering and advanced recycling of polymer-based composites for green building technologies.

Joan Nyika is an Environmental Science PhD student at the University of South Africa and a Tutorial Fellow at the Technical University of Kenya, Geoscience and the Environment Department. She obtained a Bsc. degree in Biochemistry and Molecular Biology from the Jomo Kenyatta University of Agriculture and Technology in 2011. She also obtained an Msc degree in Land and Water Management from the University of Nairobi and, Kenya in 2017. Her PhD work revolves around solid waste management and the effects of landfill leachate on soil and groundwater. She has also done research on the use of microbes for pollutant bioremediation and phytoremediation. She has interests in sustainable environmental management, water management, ecological and ecosystem management. Joan has published more than 15 articles and 15 book chapters in peer reviewed journals and books, respectively. She has and is mentoring both undergraduate and diploma students and has more than 5 years experience teaching and training in environmental science.

Raghavendra Rao is Professor and Head of Department of Environmental Sciences at SRM University, Sonipat, Haryana, India.

Kristina Reiss (Technical University of Munich, Germany) is dean of TUM School of Education and mathematics educator with a strong background in empirical and interdisciplinary research. Her research interests encompass in particular the teaching and learning of mathematics at the secondary school level as well as teacher training in the STEM subjects. Recent research aims in particular at the links between pre-service teacher education at the university and in-service training in the classroom.

Budi Rismayadi is senior lecturer at faculty of business and economics, Universitas Buana Perjuangan Karawang, Indonesia.

Goutam Roy is working as an Assistant Professor in Education at the Institute of Education and Research (IER), University of Rajshahi, Rajshahi, Bangladesh. Mr Roy is mainly responsible for teaching courses on research methods and evaluation at both undergraduate and post-graduate levels in his current role. Topics in the courses include research methodology, assessment and evaluation in education, action research in education, statistics in educational research etc. His particular research interests include nonformal and primary education, community participation in schools, assessment of student learning, teaching-learning activities, program evaluation and baseline-midline-endline surveys. During his career in education for around 11 years, Mr Roy has presented a few academic papers on some of the abovementioned areas at different national and international conferences. He has also contributed to several books, book chapters and journal articles. Before starting his career in academia, he worked for two international NGOs. Mr Roy was Research and Evaluation Coordinator at Plan International Bangladesh. He carried out several research projects on primary education, teaching and learning process, learning achievement of students, access and gender equity in education, and program evaluation on school improvement and program effectiveness. Before Plan International Bangladesh, he worked at BRAC Research and Evaluation Division as Research Associate. There he conducted many studies on nonformal education, need assessment study, curriculum assessment and students learning performance issues. Mr Roy was also involved with some nation-wide surveys on primary education of Bangladesh conducted by BRAC and Campaign for Popular Education (CAMPE). Furthermore, he has the experience to do collaborative works with some non-governmental organisations (NGO) and different government bodies.

Abdul Rozak is senior lecturer at Universitas Swadaya Gunung Jati, Cirebon, Indonesia.

Rajendrakumar Selvaraj is very good in research and teaching environment and sustainability issues at the graduate, postgraduate and doctoral students. He has 16 years of experience, completed 26 research projects, 22 training programs, and organized 1 international conference, and 1 national workshop. The research publications are 17 research papers in national and an international journal, 1 book with international publishers, and contributed 3 chapters for a book. Currently, he serving as an Assistant professor at the Centre for Sustainable Future of Amrita Vishwa Vidyapeetham, Coimbatore, Tamil Nadu, India. He is an expert in Environmental Planning and Impacts Assessments, Air, Soil, Water pollution control, Climate change, Sustainable Development, Natural Resource Management, Habitat Restoration, Plant, and Environment Relationships, Plant Community and Conservation of Biodiversity.

Shahriar Shams is currently working as an Assistant Professor in Civil Engineering Programme Area, Faculty of Engineering, Universiti Teknologi Brunei (UTB) since December 2012. Before joining UTB, he worked as an Assistant Professor in the Department of Civil and Environmental Engineering at Islamic University of Technology (IUT), a subsidiary organ of OIC. He obtained his PhD from The University of Manchester (UK) in 2008 and his M.Sc. in Environmental Engineering and Sustainable Infrastructure from Royal Institute of Technology (KTH) Sweden sponsored by Swedish International Development Agency (SIDA). He has published over 50 refereed journals and conference papers, including peer-reviewed Book Chapters. His research interest includes SWM, IWRM and Climate Change.

M. Mahruf C. Shohel is an academic researcher with special interests in education, childhood studies and international development. Throughout his academic career, he worked for several British universities including the Universities of Cambridge, Glasgow, Leicester, Manchester, Sunderland, Sussex, Bucks New University and The Open University. Dr Shohel has written extensively on development issues in the Global South and conducted research on schooling and disadvantaged children including socioeconomically deprived children, street children, and sex worker's children. Currently, he is engaged in the fields of education in emergencies, emerging technologies in education, and teaching and learning in higher education.

Exzayrani Sulaiman is senior lecturer at Universiti Brunei Darussalam, Brunei Darussalam.

Eduardo Terán-Yépez is Research Technician and PhD Candidate at the University of Almeria. His main research interests include international entrepreneurship, business internationalization, foreign trade, international marketing and sustainable entrepreneurship. He has published in journals such as Journal of Cleaner Production, Journal of International Entrepreneurship, and Anais da Academia Brasileira de Ciências. He has been reviewer in journals such as PlosOne and Journal of Environmental Planning and Management.

Ramachandran V. S. is an ecologist working towards the conservation of ecosystem and biodiversity. He has 17 years of research experience and has published 14 research papers. He is working as Assistant Professor in Amrita Vishwa Vidyapeetham (University managed by Mata Amritanandamayi Math) in its Coimbatore Campus. He teaches Environmental Studies for Undergraduate Students of the Science Programme of the University. He also takes care of gardens and greenery in the Campus.

Sobha Vinod is an Environmentalist, Teacher with teaching experience at the Primary and Secondary level in three prestigious schools of India which are Central School, Sacred Heart Convent, and Delhi Public School, India. She is currently pursuing research from SRM University in the field of Climate Change and Environmental Pollution due to Stubble burning, which is a serious issue responsible for the smog of N.Delhi in winters. Her main research interests are in the field of Agricultural Sciences and she is working to improve the stubble burning problem of India. She is a good educationist with a passion to improve environmental problems. She has won several medals in National Science Congress School Projects and has conducted several exhibitions, workshops, and webinars in the three prestigious Schools of India in which she has worked. Her main aim is to integrate the educational sector with the environment and to work towards mitigating environmental problems faced by India.

Index

R

S

T

U

V

W

Y

www.ingramcontent.com/pod-product-compliance
Lightning Source LLC
LaVergne TN
LVHW062111300926
849161LV00010B/2429
* 9 7 8 1 7 9 9 8 7 5 1 2 3 *